Mathematical Foundations of Computer Science

MATHEMATICAL FOUNDATIONS OF COMPUTER SCIENCE

REVISED & UPDATED

G. Shanker Rao
Dept. of Mathematics
University College of Engineering
Osmania University
(formerly H.O.D. Mathematics
Government Girraj P.G. College
Nizamabad, A.P.)

TECHSAR™
NEW DELHI

Published and Printed by

TechSar Pvt. Ltd.
4435-36/7, Ansari Road, Daryaganj,
New Delhi-110002 (India)
E-mail: tech_info@techsarworld.com
Website: www.techsarworld.com

ISBN: 978-81-88237-49-4

Reprint 2025

To

My parents and my wife Usha Rani

Preface

This book explains some of the fundamental concepts in mathematics. It can be used by the students in computer science as an introduction to the fundamental ideas of mathematics for computer science. The topics mathematical logic, predicates, relations, functions, combinatorics, algebraic structures and graph theory have been discussed in this book. Throughout, I have made an extensive use of worked examples to develop the general ideas.

Chapter 1: Deals with mathematical logic. Propositions, well formed formulas, logical equivalence, tautologies, fallacies, and duality principle were briefly discussed in this chapter.

Chapter 2: This chapter provides an introduction to predicate logic. Rules of inference and different methods were briefly discussed in this chapter.

Chapter 3: Deals with set theory. Properties of set containment, union of sets. Intersection of sets, complement of a set, difference of sets and their properties were discussed. This chapter presents the notions of relations and functions. It also deals with partially ordered sets, lattices and Hasse diagrams.

Chapter 4: Deals with algebraic system. In this chapter we give the basic theory of graphs semigraphs, normal subgraphs, cosets. Permutation groups, Homomorphism and Isomorphism were briefly discussed in this chapter.

Chapter 5: Covers elementary combinatorics. Permutations, Combinations and Binomial theorem have been discussed in this chapter.

Chapter 6: Deals with recurrence relations. Generating functions, and the methods of solving recurrence relations were discussed in this chapter.

Chapter 7: It is devoted to graph theory. This chapter presents the terminology. Connected graphs, bipartite graphs, planarity, trees, spanning trees and binary trees were discussed.

Chapter 8: Deals with applications of graph theory. Euler's, graphs and Hamiltonian graphs have been discussed. Coloring of graphs, and Chromatic polynomial have been explored in this chapter.

The book has a wide variety of exercises at all levels and it can be used for either a one or two-semester course.

I am much indebted to Sri Shiv Kumar Regional Manager A.P., I.K. International Publishing House Pvt. Ltd. whose suggestions helped me in writing the book.

I am thankful to Sonia Mamgain, Sponsoring Editor, I.K. International Publishing House Pvt. Ltd. for her encouragement, interest and cooperation during the production of the book.

Finally, a sincere expression of thanks goes to Sri Krishan Makhijani, Managing Director, I.K. International Publishing House Pvt. Ltd. for his encouragement and support.

Any suggestions for future improvements of this book will be gratefully accepted.

G. Shanker Rao

Contents

CHAPTER 1

Mathematical Logic

1.1 INTRODUCTION

In this chapter we shall study mathematical logic which is concerned with all kinds of reasoning. Mathematical logic has two aspects. On the one hand, it is analytical theory of the art of reasoning whose goal is to systamatize and codify the principles of valid reasoning. It may be used to judge the correctness of statements which make up the chain. In this aspect logic may be called classical mathematical logic. The other aspect of mathematical logic is inter-related with problems relating to the foundation of mathematics. Logic has many applications in computer science. It is used in problem analysis and programming.

1.2 MATHEMATICAL SENTENCE

Definition 1. A sentence that states a fact is called a mathematical sentence.

Example 1. $2x + 7 = 13$

Example 2. $\dfrac{x+1}{x+2} = 5$

1.3 STATEMENT

Definition 2. A statement is a declarative sentence which is either true or false but not both.

The truth or falsity of a statement is called its truth value. The truth values 'true' and 'false' of a statement are denoted by T and F respectively. They are also denoted by 1 and 0 respectively.

Example 1. Charminar is in Hyderabad

Example 2. $7 + 4 = 14$

Example 3. John likes science

1.4 NOTATIONS AND CONNECTIVES

Statements are usually denoted by the letters p, q, r.... The capital letters A, B, C, ..., P, Q, ..., with the exception of T and F are also used.

Statements can be connected by the words, like 'not', 'and' etc. These words are known as logical connectives. The statements which do not contain any of the connectives are called atomic statements or simple statements.

The common connectives used are: 'negation', 'and', 'or', 'if... then', 'if and only if' and 'equivalence'.

'negation' is denoted by the symbol $\sim$ [or $\neg$]

'and' is denoted by $\wedge$, 'or' is denoted by $\vee$, the connective 'if ... then' is denoted by $\rightarrow$. 'if and only if' is denoted by $\leftrightarrow$, and 'equivalence' is denoted by the symbol $\equiv$ (or by $\Leftrightarrow$). In this chapter we will use these connectives along with their symbols to combine various simple statements.

1.5 COMPOUND STATEMENTS

Definition 3. A statement that is formed from atomic statements through the use of sentential connectives is called a compound statement.

1.6 TRUTH TABLE

Definition 4. A table showing the truth values of a statement formula is called a 'truth table'. It is a compact way of listing symbols to show all possible Truth values for a set of sentences.

1.7 CONJUNCTION

Definition 5. Conjunction is a compound statement formed by using the word 'and' to combine two simple sentences.

If p and q represent two simple statements, the conjunction of p and q is written symbolically as $p \wedge q$. For example,

p: Rajan likes Algebra

q: Ramani likes Physics

$p \wedge q$: Rajan likes Algebra and Ramani likes Physics

The conjunction of p and q is true only when both parts are true i.e. $p \wedge q$ is true, wherever both p and q are true, otherwise $p \wedge q$ is false.

Table 1 Truth table for conjunction

p	q	$p \wedge q$
T	T	T
T	F	F
F	T	F
F	F	F

Note: The order used to list truth values for the first two columns is also used for other truth tables in this chapter.

1.8 DISJUNCTION

Definition 6. Disjunction is a compound sentence formed by using the word 'or' to combine two simple statements.

Thus when p and q represent simple statements, the disjunction p or q is written symbolically as $p \vee q$. For example,

$$p: 3 + 5 = 8$$

$$q: 10 + 12 = 21$$

$$p \vee q: 3 + 5 = 8 \text{ or } 10 + 12 = 21$$

The disjunction p or q is true when any part of compound sentence is true : p is true, or q is true, or both p and q are true.
and p or q is false, when both p and q are false.

Table 2 Truth table for disjunction

p	q	$p \vee q$
T	T	T
T	F	T
F	T	T
F	F	F

1.9 NEGATION

Definition 7. Let p be any proposition. Then we write this negation of p as $\sim p$ (or $\neg p$) and define it to be the proposition 'it is false that p'.

The negation of a statement is also obtained by placing the word 'not' within the original or given statements.

There are many ways of placing the word 'not' into a statement to form its negation.

Example. Let p be the statement

'all men are honest'

Then negation of p is the statement:

$\sim p$: 'It is false that all men are honest'

This can also be stated as follows

$\sim p$: not all men are honest

or $\sim p$: all men are not honest

Example 2. p: There are seven days in every week

$\sim p$: There are not seven days in every week

or $\sim p$: It is not the case that there are seven days in every week.

A statement and its negation have opposite truth values.

Table 3 Truth table for negation

p	$\sim p$
T	F
F	T

1.10 SOLVED EXAMPLES

Example 1. 10.1. Which of the following are propositions (statements)

(a) Triangle $P\ Q\ R$ is equilateral and its angles are not equal

(b) A cow has four legs.

(c) $3 + 5 = 8$

Solution. (a) asserts, contradictory properties.

Hence (a) is not a statement

(b) and (c) are statements

Example 1.10.2. Let p be 'it is cold' and q be 'it is raining'. Give a simple verbal sentence which describes each of the following

(i) $\sim p$ (ii) $\sim p \wedge \sim q$

Solution. (i) $\sim p$: it is not cold

(ii) $\sim p \wedge \sim q$: it is not cold and it is not raining

Example 1.10.3. Let p be 'He is tall', and q be 'He is *handsome*'. Write the following statements in symbolic form using p and q

(i) He is tall and handsome

(ii) He is neither tall nor handsome

Solution. (i) $p \wedge q$ (ii) $\sim p \wedge \sim q$

Example 1.10.5. Determine the truth value each of the following statements

(i) $4 + 3 = 6$ and $3 + 3 = 6$

(ii) $5 + 3 = 8$ or $3 + 1 = 5$

(iii) Agra is in England or $2 + 8 = 9$

Solution.

(i) Let $p: 4 + 3 = 6,$

$q: 3 + 3 = 6$

Then p is false and q is true.

Therefore $p \wedge q$ is false

(ii) Let $p: 5 + 3 = 8, q: 3 + 1 = 5$

Then p is True, q is False

P or q is True

i.e. $p \vee q$ is true

(iii) Let p: Agra is in England

$q: 2 + 8 = 9$

Here p is false and q is false

Therefore p or q is false

i.e. $p \vee q$ is false

Example 1.10.6. Construct the truth table for $p \vee \sim q$

Solution. We construct Truth table for $p \vee \sim q$ as follows

Table 4 Truth table for $p \vee \sim q$

p	q	$\sim q$	$p \vee \sim q$
T	T	F	T
T	F	T	T
F	T	F	F
F	F	T	T

Example 1.10.7. Find the truth table for $p \wedge (q \vee r)$

Solution. The truth table for $p \wedge (q \vee r)$ is given ahead

Table 5 Truth table for $p \wedge (q \vee r)$

p	q	r	$q \vee r$	$p \wedge (q \vee r)$
T	T	T	T	T
T	T	F	T	T
T	F	T	T	T
T	F	F	F	F
F	T	T	T	F
F	T	F	T	F
F	F	T	T	F
F	F	F	F	F

1.11 DOUBLE NEGATION

Definition 8. Let p be any simple statement $\sim p$ be the negation of p. The negation of $\sim p$ is called double negation of p. It is denoted by $\sim (\sim p)$.

Table 6 Truth table for double negation

p	$\sim p$	$\sim(\sim p)$
T	F	T
F	T	F

EXERCISE 1.1

1. Determine the truth value of each of the following
   (a) $3 + 5 = 8$ and $2 + 1 = 9$
   (b) $2 + 5 = 7$ and $3 + 3 = 6$
   (c) Taj Mahal is in Agra or $5 + 3 = 4$

Ans. (a) F (b) T (c) T

2. Let p: Mark is rich
   q: Mark is happy
   write each of the following in symbolic form
   (a) Mark is poor but happy
   (b) Mark is neither rich nor happy
   (c) Mark is either rich or happy

Ans. (a) $\sim p \wedge q$ (b) $\sim p \wedge \sim q$ (c) $p \vee q$

3. Let p: It is cold
   q: It is raining
   Give a simple verbal sentence which describes each of the following statements

(i) $\sim p$ (ii) $p \wedge q$ (iii) $p \vee q$

(iv) $\sim p \wedge \sim q$

Ans. (i) It is not cold

(ii) It is cold and raining

(iii) It is cold or it is raining

(iv) It is cold and it is not raining.

4. Let p be 'He is tall' and q be 'He is handsome'. Write each of the following in symbolic form using p and q

(a) He is tall and handsome

(b) He is tall but not handsome

(c) He is neither tall nor handsome

Ans. (a) $p \wedge q$ (b) $p \wedge \sim q$ (c) $\sim p \wedge \sim q$

5. Translate the following sentences into propositional forms

(a) It is raining and I will not go to a movie

(b) It is not raining

(c) I will not go to movie

Ans.

Hint: Let p: It is raining; q: I will go to a movie

(a) $p \wedge \sim q$ (b) $\sim p$ (c) $\sim q$

1.12 CONDITIONAL STATEMENTS

1.12.1 Conditional $p \rightarrow q$

Definition 9. If p and q are any two statements then the statement $p \rightarrow q$ which is read as 'If p, then q' is called a conditional statement.

A conditional statement is a compound statement formed by using the words 'if ... then' to combine two simple statements. In the above definition p and q are simple statements.

A conditional is sometimes called an implication. Thus, we may also read the symbol for conditional $p \rightarrow q$ as p implies q.

In the compound statement $p \rightarrow q$;

p is called the hypothesis or an antecedent and q is called the conclusion or consequent.

The antecedent usually follows the word 'if' and the consequent usually follows the word "then".

The condition if p then q or p implies q, is false when a true hypothesis p leads to a false conclusion q. In all other cases, the conditional if p then q will be true.

The truth table for the conditional $p \rightarrow q$ is given ahead.

Table 7 Truth table for $p \to q$

p	q	$p \to q$
T	T	T
T	F	F
F	T	T
F	F	T

The implication $p \to q$ can also be read as:

(i) q is p

(ii) q is implied by p

(iii) q is necessary for p

Example 1. If Taj Mahal is in Agra, then $5 + 7 = 12$

Example 2. Let p: I study

q: I will pass the test

Then $p \to q$: If I study, then I will pass the test

Example 3. Let p: It is raining

q: I get wet

Then $p \to q$: If it is raining, then I get wet

1.12.2 Biconditional

Definition 10. A biconditional is a compound sentence formed by combining the two conditionals $p \to q$ and $q \to p$ under a conjunction 'and'.

The biconditional 'p implies q and q implies p' is written symbolically as $(p \to q) \wedge (q \to p)$. It is also denoted by $p \leftrightarrow q$. It is read as p if and only if q.

A biconditional statement contains the connective 'if and only if' and has two conditions. It can also be read as 'p only if q' and 'p if q'.

The biconditional 'p if and only if q' is true when p and q are both true or both false.

Truth table for the biconditional is given below.

Table 8 Truth table for $p \leftrightarrow q$

p	q	$p \leftrightarrow q$
T	T	T
T	F	F
F	T	F
F	F	T

From the above truth table it is clear that, $p \leftrightarrow q$ is true, if p and q have the same truth values and false when p and q have different truth values.

Example 1. $4 + 4 = 8$ if and only if $2 + 5 = 7$

Example 2. Hyderabad in India if and only if $2 + 8 = 10$

1.12.3 Converse, Inverse and Contrapositive Propositions

Definition 11. If $p \to q$ is a conditional statement then

(i) $q \to p$ is called its converse

(ii) $\sim p \to \sim q$, is called its inverse

(iii) $\sim q \to \sim p$ is called its contrapositive

The truth values of these propositions are given below.

Truth table for the converse of $p \to q$:

Table 9 Truth table for $q \to p$ (converse of $p \to q$)

p	q	$p \to q$	$q \to p$
T	T	T	T
T	F	F	T
F	T	T	F
F	F	T	T

Truth table for the inverse of $p \to q$:

Table 10 Truth table for $\sim p \to \sim q$ (inverse of $p \to q$)

p	q	$\sim p$	$\sim q$	$\sim p \to \sim q$
T	T	F	F	T
T	F	F	T	T
F	T	T	F	F
F	F	T	T	T

Truth table for contraposition:

Table 11

p	q	$\sim p$	$\sim q$	$\sim q \to \sim p$
T	T	F	F	T
T	F	F	T	F
F	T	T	F	T
F	F	T	T	T

Example 1. Let p: It's – 6°

q: It's cold

Then the converse of $p \to q$ is

$q \to p$: If it's cold, then it's – 6°

the inverse of $p \to q$ is

$\sim p \to \sim q$: If it's not – 6°, then it is not cold

and the contrapositive of $p \to q$ is

$\sim q \to \sim p$: if it's not cold, then it is not – 6°

Example 2. Write the inverse, converse and contrapositive of the proposition. If two numbers are not equal, then their squares are not equal.

Solution. Inverse: If n is not an even integer then $2n$ is not divisible by 4
Converse: If $2n$ is divisible by 4, then n is an even integer
Contrapositive: If $2n$ is not divisible by 4, then n is not an even integer

Example 3. Write the inverse, converse and contrapositive of the following propositions.

(i) $p \to \sim q$

(ii) $p \to (q \to r)$

(iii) $[p \wedge (p \to q)] \to q$

Solution.

(i) Inverse: $\sim p \to q$

Converse: $\sim q \to p$

Contrapositive: $q \to \sim p$

(ii) Inverse: $\sim p \to (q \wedge \sim r)$

Converse: $(q \to r) \to p$

Contrapositive: $(q \wedge \sim r) \to \sim p$

(iii) Inverse: $\sim p \vee (p \wedge q) \to \sim q$

Converse: $q \to [p \wedge (p \to q)]$

Contrapositive: $\sim q \to [\sim p \vee (p \wedge \sim q)]$

Example. Write the contrapositive of the implication 'if it is raining, then I get wet'.

Solution. Let p: It is raining

q: I get wet

Then the contrapositive is

$\sim q \to \sim p$: If I do not get wet, then it is not raining

Example 5. Write the inverse, converse and contrapositive of the statement: If you don't practise, you will never learn how to play your horn

Solution. Inverse: If you practice, then you will learn how to play your horn

Converse: If you don't learn how to play your horn, then you did not practice

Contrapositive: If you learn how to play your horn, then you have practiced

1.12.4 Method of Finding the Truth Values of Compound Statements

1. Substitute the given truth values into the given compound sentence
2. Simplify the truth values within parentheses or other grouping, always working from the innermost group
3. Simplify negations
4. Simplify other connectives working from left to right

Example. Let p represent 'There is school on Independence day'

q represent 'Independence day is a holiday'

r represent 'We work on Independence day'

Given that p is false, q is true and r is false. Find the truth value of the statement

'If there is no School on independence day and indepence day is a holiday then we do not work on independence day'.

Given compound sentence can be written as

$$(\sim p \wedge q) \rightarrow \sim r$$
$$= (\sim F \wedge T) \rightarrow \sim F$$
$$= (T \wedge T) \rightarrow \sim F$$
$$= T \rightarrow \sim F$$
$$= T \rightarrow T$$
$$= T$$

Hence, the truth value of the given sentence is T.

1.13 WELL FORMED FORMULAS

Statement formulas contain one or more simple statements and some connectives.

If p and q are any two statements, then

$$p \vee q, \quad p \wedge q, \quad \sim p \wedge q$$

are some statement formulas derived from the statement variables p and q where p and q are called components of the statement formulas. A statement formula has no truth value. It is only when the statement variables in a statement formula are replaced by definite statements that we get a statement, which has a truth value that depends upon the truth values of his statements used in replacing the variables. A statement formula is a string consisting of variables, parentheses and connective symbols. A statement formula is called a well formed formula (WFF) if it can be generated by the following rules:

1. A statement variable p standing alone is a well formed formula
2. If p is a well formed formula, then $\sim p$ is a well formed formula
3. If p and q are well formed formulas; then $p \wedge q, p \vee q, p \rightarrow q$ and $p \leftrightarrow q$ are well formed formulas
4. A string of symbols is a well formed formula if and only if it is obtained by finite by many applications of the rules 1, 2 and 3.

According to the above recursion definition of a well formed formula, the formulas

$$\sim (p \vee q), \ \sim p \wedge q, \ p \rightarrow (p \wedge q)$$

are well formed formulas.

A statement formula is not a statement and has no truth values. But if we substitute definite statements in place of variables in a given formula we get a statement. The truth value of this resulting statement depends upon the truth values of the statements substituted for the variables, which appears as one of the entries in the final column of the truth table constructed. Therefore, the truth value of a well formed formula is the summary of truth values of the resulting statements for all possible assignments of truth values to the variables appearing in the formula. The final column entries of the truth table of a well formed formula gives the truth values of the formulas.

1.14 TAUTOLOGY

(universally true formula)

Definition 12. A statement formula (well formed formula) whose truth value is T for all possible assignments of truth values to the propositional variables is a called a Tautology.

Example 1. $(p \wedge q) \rightarrow (q \vee p)$ is a tautology

2. $p \vee \sim p$ is a tautology

1.15 CONTRADICTION (ABSURDITY)

Definition 13. A statement formula (i.e. well formed formula) whose truth value is F for all possible assignments of truth values to the propositional variables is called a contradiction.

Example $p \wedge \sim p$ is a tautology

Note: If α is a well formed formula; $\sim \alpha$ is a tautology then α is an absurdity i.e. α is an absurdity if and only if $\sim \alpha$ is a tautology.

1.16 CONTINGENCY

Definition 14. A statement formula which is neither a tautology nor a contradiction is called a contingency

For example, $(p \to q) \wedge (p \wedge q)$ is a contingency.

Example 1. Using truth table show that the statement $[\sim q \wedge (p \to q)] \to \sim p$ is a tautology.

Solution.

Table 12

p	q	$p \to q$	$\sim q \wedge (p \to q)$	$\sim q \wedge (p \to q) \to \sim p$
T	T	T	F	T
T	F	F	F	T
F	T	T	F	T
F	F	T	T	T

From the above truth table $\sim q \wedge (p \to q) \to \sim p$ is always.

Therefore, $\sim q \wedge (p \to q) \to \sim p$ is a tautology.

Example 2. By constructing the truth table determine whether each of the following is a tautology or a contingency or a contradiction.

(i) $p \to (q \to p)$ (V.T.U. MCA Aug. 2000)

(ii) $(q \wedge p) \vee (q \vee \sim p)$

(iii) $q \vee (\sim q \wedge p)$

Solution (i)

Table 13

p	q	$q \to p$	$p \to (q \to p)$
T	T	T	T
T	F	T	T
F	T	F	T
F	F	T	T

Therefore, $p \to (q \to p)$ is a tautology

(ii)

Table 14

p	q	$\sim p$	$q \wedge p$	$q \wedge \sim p$	$(q \wedge p) \vee (q \wedge \sim p)$
T	T	F	T	F	T
T	F	F	F	F	F
F	T	T	F	T	T
F	F	T	F	F	T

From the table; it is clear that $(q \wedge p) \vee (q \wedge \sim p)$ is a contingency.

(iii)

Table 15

p	q	$\sim p$	$q \wedge p$	$q \wedge \sim p$	$(q \wedge p) \wedge (q \wedge \sim p)$
T	T	F	T	F	T
T	F	F	F	F	F
F	T	T	F	T	T
F	F	T	F	F	T

From the table the given statement is neither true (*T*) nor false (*F*), hence a contingency.

Example 3. Show that $(p \wedge q) \rightarrow (p \vee q)$ is a tautology.

Solution. Let us construct the truth table

Table 16

p	q	$p \wedge q$	$p \vee q$	$p \wedge q \rightarrow p \vee q$
T	T	T	T	T
T	F	F	T	T
F	T	F	T	T
F	F	F	F	T

All the entries in the last column of the truth table are true (*T*). Hence, the given proposition is a Tautology.

Example 4. Show that $p \wedge \sim p$ is a contradiction

Solution. The truth table for $p \wedge \sim p$ is given below.

Table 17

p	$\sim p$	$p \wedge \sim p$
T	F	F
F	T	F

1.17 LOGICAL EQUIVALENCE

Definition 15. Two propositions *P* and *Q* are said to be logically equivalent or simply equivalent if $P \rightarrow Q$ is a tautology.

Example. $\sim (p \wedge q)$ and $\sim p \wedge \sim q$ are logically equivalent

Two formulas may be equivalent even if they do not contain the same variables. Two statement formulas *P* and *Q* are equivalent if $P \rightleftarrows Q$ is a tautology and conversely if $P \rightleftarrows Q$ is a tautology then P and Q are equivalent. We can also define logical equivalence as follows:

Definition 16. Two propositions P and Q are said to be logically equivalent if and only if they have same truth values under all circumstances.

If "P is equivalent to Q", then we can represent equivalence by writing "$P \Leftrightarrow Q$". Which can also be written as $P \Leftrightarrow Q$. The symbol "$\Leftrightarrow$" is not a connective. Usually we drop quotation marks. If P and Q are logically equivalent we also write $P \equiv Q$ and read "P is equivalent to Q". There are a number of fundamental equivalences which are useful in proofs.

1.18 LAWS OF LOGIC

1. Idempotent laws

   (a) $p \vee p \equiv p$ (b) $p \wedge p \equiv p$

2. Commutative laws

   (a) $p \vee q \equiv q \vee p$ (b) $p \wedge q \equiv q \wedge p$

3. Associative laws

   (a) $(p \vee q) \vee r \equiv p \wedge (q \vee r)$

   (b) $(p \wedge q) \wedge r \equiv p \wedge (q \wedge r)$

4. Distributive laws

   (a) $p \vee (q \wedge r) \equiv (p \vee q) \wedge (p \vee r)$

   (b) $p \wedge (q \vee r) \equiv (p \vee q) \vee (p \wedge r)$

5. Identity laws

   (a) (i) $p \vee f \equiv p$ (ii) $p \vee t \equiv t$

   (b) (i) $p \wedge f \equiv f$ (ii) $p \wedge t \equiv p$

6. Complement laws

   (a) (i) $p \wedge \sim p \equiv t$ (ii) $p \vee \sim p \equiv t$

   (b) $\sim (\sim p) \equiv p$ (ii) $\sim t \equiv f, \sim f \equiv t$

7. De Morgan's laws

   (a) $\sim (p \vee q) \equiv \sim p \wedge \sim q$

   (b) $\sim (q \wedge q) \equiv \sim p \vee \sim q$

Where t and f are used to denote the variables which are restricted to the truth values true and false respectively.

Example 1. Show that $\sim (p \rightarrow q) \equiv p \wedge \sim q$

Solution. Let us construct the truth table for the given propositions.

Table 18

p	q	$p \to q$	$\sim (p \to q)$	$\sim q$	$p \wedge \sim q$
T	T	T	F	F	F
T	F	F	T	T	T
F	T	T	F	F	F
F	F	T	F	T	F

From the table it is clear that the truth values of $\sim (p \to q)$ and $p \wedge \sim q$ are identical

$$\therefore \qquad \sim (p \to q) \equiv \wedge p \sim q$$

Example 2. Prove the logical equivalence of

$$(p \vee q) \wedge \sim p \equiv \sim p \wedge q$$

Solution. We construct the truth table for the given propositions

Table 19

p	q	$p \vee q$	$\sim p$	$(p \vee q) \wedge \sim p$	$\sim p \wedge q$
T	T	T	F	F	F
T	F	T	F	F	F
F	T	T	T	T	T
F	F	F	T	F	F

Example 3. Show that $\sim (p \to q) \equiv p \wedge \sim q$

Solution. We construct the truth table for given propositions

Table 20

p	q	$p \to q$	$\sim (p \to q)$	$\sim q$	$p \wedge \sim q$
T	T	T	F	F	F
T	F	F	T	T	T
F	T	T	F	F	F
F	F	T	F	T	F

From the truth table it is clear that the truth values of $\sim (p \to q)$ and $p \wedge \sim q$ are identical.

Example 4. Obtain the formulas having the simplest possible forms which are equivalent to the formulas given below

(i) $((P \to Q) \leftrightarrow (\sim P \to \sim Q)) \wedge R$

(ii) $(P \wedge (Q \wedge S)) \vee (\sim P \wedge (Q \wedge S))$

Solution.

(i) $(P \to Q) \leftrightarrow (\sim P \wedge \sim Q) \wedge R \Leftrightarrow (P \to Q) \leftrightarrow ((P \to Q) \wedge R)$

(by contrapositive rule)

$\Leftrightarrow (P \to Q) \wedge ((P \to Q) \wedge R) \vee (\sim (P \to Q) \wedge \sim (P \to Q) \wedge R)$

(Since, $P \leftrightarrow Q \Leftrightarrow P \wedge Q \vee (\sim P \wedge \sim Q)$)

$\Leftrightarrow ((P \to Q) \wedge R) \vee ((P \wedge \sim Q) \wedge (P \wedge \sim Q) \vee \sim R))$

$\Leftrightarrow (\sim P \vee Q \wedge R) \vee (P \wedge \sim Q) \wedge ((P \wedge \sim Q) \wedge P \wedge \sim Q)$

$\vee (P \wedge \sim Q \wedge \sim R)$

Example 4. Show that $\sim (p \vee (\sim p \wedge q))$ and $\sim p \wedge \sim q$ are logically equivalent.

Solution. Consider $\sim (p \vee (\sim p \wedge q))$

$\sim (p \vee (\sim p \wedge q)) \equiv \sim p \wedge \sim (\sim p \wedge q)$ (from De Morgan's law)

$\equiv \sim p \wedge (\sim (\sim p) \vee \sim q)$ (from De Morgan's law).

$\equiv \sim p \wedge (p \vee \sim q)$ ($\because \sim (\sim p) = p$ by double negation)

$\equiv (\sim p \wedge p) \vee (\sim p \wedge \sim q)$

$\equiv f \vee (\sim p \wedge \sim q)$ ($\because \sim p \wedge p$ is false) ($\because \sim p \wedge p \equiv f$)

$\equiv \sim p \wedge \sim q$

Therefore, $\sim (p \vee \sim p \wedge q))$ and $\sim p \wedge \sim q$ are logically equivalent.

1.19 LOGICAL IMPLICATION

We state the following theorem

Theorem 1.1 Let $P(p_1, p_2, \ldots)$ and $Q(p_1, p_2, \ldots)$ be two propositions. Then the following conditions are equivalent

1. $\sim P(p_1, p_2, \ldots) \vee Q(p_1, p_2, \ldots)$ is a tautology
2. $P(p_1, p_2, \ldots) \wedge Q(p_1, p_2, \ldots)$ is a contradiction
3. $P(p_1, p_2, \ldots) \to Q(p_1, p_2, \ldots)$ is a tautology

Definition 17. A proposition $P(p_1, p_2, \ldots)$ is said to logically imply a proposition $Q(p_1, p_2, \ldots)$ if one of the conditions in the theorem 1.1 holds.

If $P(p_1, p_2, \ldots)$ logically implies $Q(p_1, p_2, \ldots)$ then we symbolically denote it by writing $P(p_1, p_2, \ldots) \Rightarrow Q(p_1, p_2, \ldots)$,

Example 1. $(p \wedge q) \wedge \sim (p \vee q)$ is a contradiction

Hence $p \wedge q \Rightarrow p \vee q$

Example 2. $(p \wedge q) \wedge (q \to r) \to (p \to r)$ is a tautology

Hence $(p \to q) \wedge (q \to r) \Rightarrow p \to r$

Theorem. The relation in propositions defined by $P(p_1, p_2, \ldots) \Rightarrow Q(p_1, p_2, \ldots)$ is reflexive, anti-symmetric and transitive.

Note: The symbols $\rightarrow$, $\Rightarrow$ are not the same.

$\Rightarrow$ is not a connective nor $p \Rightarrow Q$ is a statement formula (proposition). $P \Rightarrow Q$ defines a relation in composite proposition $p \rightarrow Q$. The symbol $\rightarrow$ is a connective and $p \rightarrow Q$ is just a proposition.

1.20 SOLVED EXAMPLES

Example 1.20.1. Write the negation of the following statement "If x is an integer, then x is a rational number".

Solution. Let p: x is an integer

q: x is rational number

The given proposition is $p \rightarrow q$.

negation of the given proposition is $\sim (p \rightarrow q)$

$\sim (p \rightarrow q)$ and $p \wedge \sim q$ are logically equivalent

Hence, $p \wedge \sim q$: "x is an integer and x is not a rational number"

Example 1.20.2. Show that $p \wedge (p \vee q) \equiv p$

Solution. $p \wedge (p \vee q) \equiv (p \vee f) \wedge (p \vee q)$

$\equiv p \vee (f \vee q)$ (f is fallacy)

$\equiv p \vee f$

$\equiv p$

Table 21

p	q	$\sim p$	$\sim p \vee q$	$p \rightarrow q$	$(p \rightarrow q) \leftrightarrow (\sim p \vee q)$
T	T	F	T	T	T
T	F	F	F	F	T
F	T	T	T	T	T
F	F	T	T	T	T

From the table $(p \rightarrow q) \leftrightarrow (\sim p \vee q)$ is a tautology

Thus, $p \rightarrow q$ and $\sim p \vee q$ are logically equivalent

Example 1.20.3. Using the truth table show that

$$[\sim q \wedge (p \rightarrow q)] \rightarrow \sim p$$

is a tautology

Solution. We construct the truth table for the given statement

Table 22

p	q	$p \to q$	$\sim q \wedge (p \to q)$	$[\sim q \wedge (p \to q)] \to \sim p$
T	T	T	F	T
T	F	F	F	T
F	T	T	F	T
F	F	T	T	T

from the truth table it is clear that

$\sim q \wedge (p \to q) \to \sim p$ is always true. Therefore,

$\sim q \wedge (p \to q) \to \sim p$ is a tautology.

Example 1.20.4. Show that the following statements are logically equivalent "If I study, then I'll pass the test" and "I don't study" or "I'll pass the lest"

Solution. Let p: I study

q: I'll pass the test

Then $\sim p$: I don't study.

Symbolically, the given statements can be

$p \to q$: If I study then I'll pass the test

$\sim p \vee q$: I don't study or I'll pass the test

Example 1.20.5. Show that

$$(p \wedge (\sim p \vee q)) \vee (q \wedge \sim (p \wedge q)) \equiv q$$

Solution. Consider L.H.S

$$(p \wedge (\sim p \vee q)) \vee (q \wedge \sim (p \wedge q))$$

$$\equiv ((p \wedge \sim p) \vee (p \wedge q)) \vee (q \wedge \sim p \vee \sim q))$$

$$\equiv f \vee (p \wedge q) \vee (q \wedge \sim p) \vee (q \wedge \sim p) \wedge (q \wedge \sim q) \qquad (\because \; p \wedge \sim p = f)$$

$$\equiv (p \wedge q) \vee (q \wedge \sim p) \vee f$$

$$\equiv (p \wedge q) \vee (q \wedge \sim p)$$

$$\equiv (q \wedge p) \vee (q \wedge \sim p)$$

$$\equiv q \wedge (p \vee \sim p)$$

$$\equiv q \wedge t \qquad (\because p \vee \sim p \equiv t)$$

$$\equiv q$$

Hence, $(p \wedge (\sim p \vee q)) \vee (q \wedge \sim p \wedge q)) \equiv q$

Example 1.20.6. Show that $(p \vee \sim q) \wedge (\sim p \vee \sim q) \vee q$ is a tautology

Solution. Consider

$$((p \vee \sim q) \wedge (\sim p \vee \sim q)) \vee q$$
$$\equiv (p \vee \sim q) \wedge \sim p \vee ((p \vee \sim q) \wedge \sim q) \vee q$$
$$\equiv (p \wedge \sim p) \vee (\sim q \wedge \sim p) \vee (p \wedge \sim q) \vee (\sim q \wedge \sim q) \vee q$$
$$\equiv f \vee (\sim q \wedge \sim p) \vee ((p \wedge \sim q) \vee \sim q) \vee q$$
$$\equiv (\sim q \wedge \sim p) \vee (p \wedge \sim q) \vee \sim q \vee q$$
$$\equiv (\sim q \wedge \sim p) \vee (p \wedge \sim q) \vee t \qquad (\because \sim q \vee q \equiv t)$$
$$\equiv t$$

Hence $((p \vee \sim q) \wedge (\sim p \wedge \sim q)) \vee q$ is a tautology.

EXERCISE 1.2

1. There are two restaurants next to each other, one has a sign that says "Good food is not cheep" and the other sign that says "cheap food is not good" Are the signs saying the same thing? if so, verify.
2. Using truth table show that
$$[\sim q \wedge (p \rightarrow q)] \rightarrow \sim p$$
is a tautology.
3. Determine whether the following one is a tautology or contradiction
$(p \vee q) \vee (\sim p \vee \sim q)$
4. Determine whether each of the following statements is a tautology, a contingency or an absurdity
   (i) $p \rightarrow (q \rightarrow p)$
   (ii) $(p \wedge q) \rightarrow p$
   (iii) $(p \wedge (p \rightarrow q)) \rightarrow q$
   (iv) $(q \wedge (\sim p)) \wedge r$
5. Prove the logical equivalence of the following statements
   (i) $(p \vee q) \wedge \sim p \equiv \sim p \wedge q$
   (ii) $p \vee (p \wedge q) \equiv p$
   (iii) $p \wedge (\sim p \vee q) \equiv p \wedge q$
   (iv) $p \wedge (p \vee q) \equiv p$

ANSWERS 1.2

1. Both the signs are saying the same thing
2. Neither a tautology nor a contradiction.
3. (i), (ii), (iii) are tautologies
   (iv) contingency.

1.21 THE DUALITY PRINCIPLE

The principle of duality states that any established result involving statement formulas and connectives $\vee$ and $\wedge$ give a corresponding dual result by replacing $\wedge$ by $\vee$ and $\vee$ by $\wedge$. If the formula contains special variables t and f. The corresponding dual is obtained by replacing t by f and f by t. The connectives $\wedge$ and $\vee$ are called duals of each other.

Definition. Two statement formulas P and P^* are said to be duals of each other if one can be obtained from the other by replacing $\wedge$ and $\vee$ and $\vee$ by $\wedge$.

Example. Write the duals of

(i) $(p \wedge q) \vee r$ (ii) $(p \wedge q) \vee r$ (c) $\sim (p \wedge q)$

Solution. The duals are

(i) $(p \vee q) \wedge r$ (ii) $(p \vee q) \wedge r$ (c) $\sim (p \vee q)$

1.22 OTHER CONNECTIVES

We now introduce the connectives NAND, NOR which have useful applications in the design of computers.

The word NAND is a combination of "NOT" and "AND" where "NOT" stands for negation and "AND" stands for the conjunction. It is denoted by the symbol $\uparrow$.

If P and Q are two formulas then $P \uparrow q \rightarrow \sim (p \wedge q)$. The connective $\uparrow$ has the following equivalence

$$p \uparrow p \leftrightarrow \sim (p \wedge p) \leftrightarrow \sim p \vee \sim p \Leftrightarrow \sim p$$

$$(p \uparrow q) \uparrow (p \uparrow q) \leftrightarrow \sim (p \uparrow q) \leftrightarrow p \wedge q$$

$$(p \uparrow p) \uparrow (q \uparrow q) \leftrightarrow \sim p \uparrow \sim q \Leftrightarrow \sim (\sim p \wedge \sim q) \leftrightarrow p \vee q$$

The connective NAND is commutative but not associative i.e.

$$P \uparrow Q \leftrightarrow Q \uparrow P$$

NAND not associative, since $P \uparrow (Q \uparrow R) \leftrightarrow \sim P \vee (Q \wedge R)$ and $(P \uparrow Q) \uparrow R \leftrightarrow \sim (P \wedge Q) \vee \sim R$.

The connective NOR is a combination of "NOT" and "OR"; where NOT stands for negation and OR stands for the disjunction. The connective NOR is denoted by the symbol $\downarrow$, and is called joint denial $P \downarrow Q$ is read as "Neither P nor Q"

The connective NOR has following equivalence

$$P \downarrow P \leftrightarrow \sim (P \vee P) \leftrightarrow \sim P \wedge \sim P \Leftrightarrow \sim P$$

$$(P \downarrow Q) \downarrow (P \downarrow Q) \leftrightarrow \sim (P \downarrow Q) \leftrightarrow P \vee Q$$

$$(P \downarrow P) \downarrow (Q \downarrow Q) \leftrightarrow \sim P \uparrow \sim Q \leftrightarrow P \wedge Q$$

The connective $\downarrow$ is commutative but not associative.

The connectives $\wedge$, $\vee$ and ~ can be expressed in terms of the connective as follows:

(i) $\sim p \equiv p \downarrow p$

(ii) $\sim q \equiv q \downarrow q$

(iii) $p \wedge q \equiv (p \downarrow p) \downarrow (q \downarrow q)$

(iv) $p \vee q \equiv (p \downarrow q) \downarrow (p \downarrow q)$

Example 1. Write a formula which is equivalent to the formula $P \wedge (Q \rightleftarrows R)$ and contains the connective NAND ($\uparrow$) only obtain an equivalent formula which contains the connective NOR ($\downarrow$) only. **(JNTU. B. Tech 2002)**

Solution. $P(Q \rightleftarrows R) \equiv P \wedge ((Q \rightarrow R) \wedge R \rightarrow Q)$

$\equiv P \wedge ((\sim Q \vee R) \wedge (\sim R \vee Q))$ $\qquad (\because P \rightarrow Q \equiv \sim P \vee Q)$

$\equiv P \wedge (\sim Q \vee R) \wedge \sim (R \wedge \sim Q)$ (By De Morgan's Laws)

$\equiv P \wedge ((Q \uparrow \sim R) \vee \uparrow (R \uparrow \sim Q))$

$\equiv P \sim ((Q \uparrow \sim R) \uparrow (R \uparrow \sim Q))$

$\equiv \sim (P \uparrow \sim ((Q \uparrow \sim R) \uparrow (R \uparrow \sim Q)))$

$P \wedge (Q \rightleftarrows R)$ using NOR only:

$P \wedge (Q \rightleftarrows R) \equiv P \vee ((Q \rightarrow R) \wedge (R \rightarrow Q))$

$\equiv P \wedge ((\sim Q \vee R) \wedge (\sim R \vee Q))$

$\equiv P (\sim (\sim Q \downarrow R) \wedge \sim (\sim R \wedge Q)$

$\equiv P \wedge \sim (\sim (Q \downarrow R) \vee (\sim R \downarrow Q))$

$\equiv P \wedge (\sim (Q \downarrow R) \downarrow (\sim R \downarrow Q))$

$\equiv \sim (\sim P \vee \sim ((\sim Q \downarrow R) \downarrow (\sim R \downarrow Q)$

$\equiv \sim ((P \downarrow P) \vee \sim ((\sim Q \downarrow R) \downarrow (\sim R \downarrow Q))$

$\equiv (P \downarrow P) \downarrow \sim (\sim Q \downarrow R) \downarrow (\sim R \downarrow Q))$

Example 2. Construct a truth table for $p \downarrow q$.

Solution. The truth table for $p \downarrow q$ is as follows

p	q	$p \downarrow q$
T	T	F
T	F	F
F	T	F
F	F	T

from the table, $p \downarrow q$ is true if neither p is true nor q is false.

Example 3. Show that the connectives >, $\wedge$ and $\vee$ may be expressed in terms of the connective $\downarrow$ as follows

(i) $\sim p \equiv p \downarrow p$ (ii) $p \wedge q \equiv (p \downarrow q) \downarrow (q \downarrow q_0)$ (iii) $p \vee q \equiv (p \downarrow q) \downarrow (p \downarrow q)$

Solution. (i) We have

p	$\sim p$	$p \downarrow p$
T	F	F
F	T	T

from the above table it is clear that $\sim p$ and $p \downarrow p$ have same truth values. Therefore

$$\sim p \equiv p \downarrow p$$

(ii)

p	q	$p \wedge q$	$p \downarrow q$	$q \downarrow q$	$(p \downarrow p) \downarrow (q \downarrow q)$
T	T	T	F	F	T
T	F	F	F	F	F
F	T	F	T	F	F
F	F	F	T	T	F

Since, the columns shown are identical $p \wedge q$ and $(p \downarrow q) \downarrow (q \downarrow q)$ have same truth values

Hence, $p \wedge q \equiv (p \downarrow p) \downarrow (q \downarrow q)$

(iii)

p	q	$p \vee q$	$p \downarrow q$	$p \downarrow q$	$(p + q) \downarrow (p \downarrow q)$
T	T	T	F		T
T	F	T	F		T
F	T	T	F		T
F	F	F	T		F

$p \vee q$, $(p \downarrow q) \downarrow (p \downarrow q)$ have same truth values for all the combinations

Therefore, $p \vee q \equiv (p \downarrow q) \downarrow (p \downarrow q)$

1.22.1 Exclusive Disjunction

The proportional connective is $\veebar$ called exclusive disjunction:

$p \veebar q$ is read "p or q but not both".

The truth table for exclusive disjunction is as follows

p	q	$p \veebar q$
T	T	F
T	F	T
F	T	T
F	F	F

$p \veebar q$ is true if p is true or q is true but not if both p and q are true.

Example. Show that $p \veebar q \equiv (p \veebar q) \wedge \sim (p \wedge q)$

Solution. We construct the earlier table to prove the equivalence

p	q	$p \veebar q$	$p \vee q$	$p \wedge q$	$\sim (p \wedge q)$	$(p \vee q) \wedge \sim (p \wedge q)$
T	T	F	T	T	F	F
T	F	T	T	F	T	T
F	T	T	T	F	T	T
F	F	F	F	F	T	F

Since the truth tables of $p \veebar q$ and $(p \vee q) \vee \sim (q \wedge q)$ are identical

$$p \veebar q \equiv (p \vee q) \wedge \sim (p \wedge q)$$

1.23 NORMAL FORMS

If a given statement formula A $(p_1, p_2, ..., p_n)$ involves n atomic variables, we have 2^n possible combinations of truth values of statements replacing the variables.

The formula A is a tautology if A has the truth value T for all possible assignments of the truth values of the variables p_1, p_2, ..., p_n and A is called a contradiction if A has the truth value F for all possible assignments of the truth values of the n variables. A is said to be satisfiable if A has the truth value T for at least one combination of truth values assigned to p_1, p_2, ..., p_n. The problem of determining whether a given statement formula is a tautology or a contradiction is called a decision problem.

The construction of truth table involves a finite number of steps, but the construction may not be practical. We, therefore, reduce the given statement formula to normal form and final whether a given statement formula is a tautology or contradiction or at least satisfiable.

Definition 18. A formula which is a product (conjunction) of the variables and their negations is called an elementary product.

If p and q are atomic variables then p, $\sim p$, $\sim p \wedge q$ and $p \wedge \sim p$ are same examples of elementary products.

The sum (disjunction) of variables and their negations in a formula is called an elementary sum.

If p and q are any two atomic variables p, $\sim p \vee q$, $p \vee \sim p$ and $\sim q \vee p$ are some examples of elementary sums.

1.23.1 Disjunctive Normal Form

Definition 19. Let A denote a given formula. Another formula B which is equivalent to A is called a disjunctive normal form of A if B is a sum of elementary products.

A disjunctive normal form of a given formula is constructed as follows:

(i) replace '$\rightarrow$', '$\leftrightarrow$' by using the logical connectives $\wedge$, $\vee$ and $\sim$

(ii) use Demorgan's laws to eliminate $\sim$ before sum or products

(iii) Apply distributive laws repeatedly and eliminate product of variables to obtain the required normal form

Example 1. Obtain disjunctive normal form of $p \wedge (p \rightarrow q)$

Solution. $p \wedge (p \rightarrow q) \equiv p \wedge (\sim p \vee q)$

$$\equiv (p \wedge \sim p) \vee (p \wedge q)$$

Example 2. Obtain disjunctive normal form of

$$p \vee (\sim p \rightarrow (q \vee (q \rightarrow \sim r)))$$

Solution. $p \vee (\sim p \rightarrow q \vee (q \rightarrow \sim r))$

$$\equiv p \vee (\sim p \rightarrow q \vee (\sim q \vee \sim r)))$$

$$\equiv p \vee (p \vee (q \vee (\sim q \vee \sim r)))$$

$$\equiv p \vee p \vee q \vee \sim q \vee \sim r$$

$$\equiv p \vee q \vee \sim q \vee \sim r$$

Hence, the disjunctive normal form of

$p \vee (\sim p \rightarrow (q \vee (\sim q \vee \sim r)))$ is $p \vee q \vee \sim q \vee \sim r$

1.23.2 Conjunctive Normal Form

Definition 20. Let A denotet a given formula. Another formula B which is equivalent to A is called a conjunctive normal formula if B is a product of an elementary sum.

Example. The conjunctive normal form of $p \wedge (p \rightarrow q)$ is $p \wedge (\sim p \vee q)$

1.23.3 Principal Disjunctive Normal Form

Let p and q be two statement variables. Then $p \wedge q, p \wedge \sim q, \sim p \wedge q$ and $\sim p \wedge \sim q$ are called minterms of p and q. They are called boolean conjunctives of p and q. Each minterm has the truth value T for exactly one combination of truth values of the variables p and q. There are 2^n possible minterms for the two variables p and q. Note that none of the minterms should contain both a variable and its negation. The number of minterms in n variables is 2^n.

We now introduce one more normal form called the principal disjunctive normal form of formula A.

Definition 21. If A is a given formula, then an equivalent formula B of A consisting of disjunctives of minterms only is called the principal disjunctive normal form of the formula A.

The principal disjunctive normal formula of A is also called the sum-of-products canonical form of A.

Example. Obtain the principal disjunctive normal form of

$$(\sim p \vee \sim q) \rightarrow (\sim p \vee r)$$

Solution. $(\sim p \vee \sim q) \rightarrow (\sim p \wedge r) \Leftrightarrow \sim(\sim p \vee \sim q) \vee (\sim p \wedge r)$

$$\Leftrightarrow \sim(\sim(p \wedge q)) \vee (\sim p \wedge r)$$

$$\Leftrightarrow (p \wedge q) \vee (-p \wedge r)$$

$$\Leftrightarrow ((p \wedge q) \wedge (r \vee \sim r)) \vee (\sim p \wedge r \wedge (q \vee \sim q))$$

$$\Leftrightarrow (p \wedge q \wedge r) \vee (p \wedge q \wedge \sim r) \vee (\sim p \wedge r \wedge q) \vee (\sim p \wedge r \wedge \sim q)$$

The principal disjunctive normal form of the given formula is

$$(p \wedge q \wedge r) \vee (p \wedge q \wedge \sim r) \vee (\sim p \wedge q \wedge r) \vee (\sim p \wedge \sim q \wedge r)$$

1.23.4 Principal Conjunctive Normal Form

The dual of a minterm is called a Maxterm. For a given number of variables the maxterm consists of disjunctions in which each variable or its negation, but not both, appears only once. Each of the maxterm has the truth value F for exactly one combination of the truth values of the variables. Now we define the principal conjunctive normal form.

Definition 22. If A is a given formula, then an equivalent formula B is called principal conjunctive normal form of A if B is a product of maxterms. To obtain the principal conjunctive normal form of A, we first find the principal disjunctive normal form of $\sim A$ and apply $\sim$. The principal conjunctive normal form of A is also called the product-of-sums canonical form.

Example. Obtain the principal conjunctive normal form of

$$(p \wedge q) \vee (\sim p \wedge r)$$

Solution. $(p \wedge q) \vee (\sim p \wedge r) \Leftrightarrow ((p \wedge q) \vee \sim p) \vee ((p \wedge q) \vee r)$

$$\Leftrightarrow (p \vee \sim p) \wedge (q \vee \sim p) \wedge ((p \vee r) \wedge (q \vee r)$$

$$\Leftrightarrow ((q \vee \sim p) \vee (r \wedge \sim r)) \wedge (p \vee r \vee (q \wedge \sim q)) \wedge (q \vee r \vee (p \wedge \sim p))$$

$$\Leftrightarrow (q \vee \sim p \vee r) \wedge (q \vee \sim p \wedge \sim r) \wedge (p \vee r \vee q)$$

$$\wedge (p \vee r \vee \sim q) \wedge (q \vee r \vee p) \wedge (q \vee r \vee \sim p)$$

$$\Leftrightarrow (\sim p \vee q \vee r) \wedge (\sim p \vee q \vee \sim r) \wedge (p \vee q \vee r) \wedge (p \vee \sim q \vee r)$$

1.24 SOLVED EXAMPLES

Example 1.24.1. Obtain the disjunctive normal form of $p \vee (\sim p \to q \to \sim r))$

Solution. $p \vee (\sim p \to (q \vee (q \to \sim r)))$

$$\equiv p \vee (p \vee q \vee \sim q \vee \sim r)$$

$$\equiv p \vee p \vee q \vee \sim q \vee \sim r$$

$$\equiv p \vee q \vee \sim q \vee \sim r$$

Example 1.24.2. Obtain the principal disjunctive normal form of $\sim p \vee q$

Solution. $\sim p \vee q \equiv (\sim p \wedge (q \vee \sim q)) \vee (q \wedge (p \vee \sim p))$

$$\equiv (\sim p \wedge q) \vee (\sim p \wedge \sim q) \vee (q \wedge p) \vee (q \wedge \sim p)$$

$$\equiv (\sim p \wedge q) \vee (\sim p \wedge \sim q) \vee (p \wedge q) \vee (\sim p \wedge q)$$

$$\equiv (\sim p \wedge q) \vee (\sim p \wedge \sim q) \vee (p \wedge q)$$

Hence $(\sim p \wedge q) \wedge (\sim p \wedge \sim q) \wedge (p \wedge q)$ in the required principal disjunctive normal form

Example 1.24.3. Obtain the disjunctive normal form of

$$(p \wedge \sim (q \wedge r)) \vee (p \to q)$$

Solution. $p \wedge \sim (q \wedge r) \vee (p \to q) \equiv (p \wedge \sim (q \wedge r)) \vee (\sim p \vee q)$

$$\equiv (p \wedge (\sim q \vee \sim r)) \vee (\sim p \vee q)$$

$$\equiv (p \wedge \sim q) \vee (p \wedge \sim r) \vee (\sim p \vee q)$$

The disjunctive normal form of $(p \wedge (\sim q \wedge r)) \vee (p \to q)$ is

$(p \wedge \sim q) \vee (p \wedge \sim r) \vee (\sim p \vee q)$

Example 1.24.4. Find the principal conjunctive normal form of $p \vee (q \to r)$

Solution. Let $A \equiv p \vee (q \to r)$

Now $\qquad \sim A \equiv \sim (p \vee (q \to r))$

$$\equiv \sim (p \vee (\sim q \vee r))$$

$$\equiv \sim p \wedge \sim (\sim q \vee r)$$

$$\equiv \sim p \wedge (\sim (\sim q)) \wedge \sim r)$$

$$\equiv \sim p \wedge q \wedge \sim r$$

∴ The principal conjunctive normal form of A is

$$\sim(\sim A) \equiv \sim(\sim p \wedge q \wedge \sim r) \equiv p \vee \sim q \vee r$$

Example 1.24.5. Obtain principal disjunctive normal form of

$$P \vee (\sim P \rightarrow (Q \vee (\sim Q \rightarrow R)))$$

Solution. $P \vee (\sim P \rightarrow (Q \vee (\sim Q \rightarrow R)))$

$$\equiv P \vee (\sim P \rightarrow Q \vee (\sim\sim Q \vee R))$$

$$\equiv P \vee (\sim P \rightarrow Q \vee (Q \vee R))$$

$$\equiv P \vee (\sim(\sim P)) \vee Q \vee R$$

$$\equiv (P \vee P) \vee (Q \vee R)$$

$$\equiv P \vee (Q \vee R)$$

$$\equiv P \wedge (Q \vee \sim Q) \wedge (R \vee \sim R) \vee (Q \vee R) \wedge (P \wedge \sim P)$$

$$\equiv (P \wedge Q \wedge R) \vee (P \wedge \sim Q \wedge R) \vee (P \wedge \sim Q \wedge \sim R) \wedge (P \wedge Q \vee R) \vee (\sim P \vee Q \vee R)$$

Example 1.24.6. Obtain the principal disjunctive normal form of $P \rightarrow Q$, by constructing truth table.

Solution. Truth table for $P \rightarrow Q$

P	Q	P→Q	min term
T	T	T	P ∧ Q
T	F	F	P ∧ ~ Q
F	T	T	~ P ∧ Q
F	F	T	~ P ∧ ~ Q

From the truth table, the principal normal disjunctive normal of $P \rightarrow Q$ is

$$(P \wedge Q) \vee (\sim P \wedge Q) \vee (\sim P \wedge \sim Q) \quad \therefore$$

$$P \rightarrow Q \Leftrightarrow (P \wedge Q) \vee (\sim P \vee Q) \vee (\sim P \wedge \sim Q)$$

Example 1.24.7 Obtain the principal disjunctive normal form of $p \vee (q \wedge q)$

Solution: $p \vee (p \wedge q) \equiv (p \vee (q \vee \sim q) \vee (p \wedge q)$

$$\equiv (p \wedge q) \vee (p \wedge \sim q) \vee (p \wedge q)$$

$$\equiv (p \wedge q) \vee (p \wedge \sim q) \vee (p \wedge q)$$

$$\equiv (p \wedge q) \vee (p \wedge \sim q)$$

Example 1.24.8 Obtain the principal conjunctive normal form of $(p \wedge q) \vee (\sim p \wedge q)$

Solution: We have

$$\begin{aligned}(p \wedge q) \vee (\sim p \wedge q) &\equiv (p \wedge q) \vee \sim p) \wedge (p \wedge q) \vee q)\\ &\equiv (p \vee \sim p) \wedge (q \vee \sim p) \wedge ((p \wedge q) \vee (q \vee q))\\ &\equiv k \wedge (q \vee \sim p) \wedge [(p \vee q) \wedge t]\\ &\equiv (q \vee \sim p) \wedge [p \vee q]\end{aligned}$$

Hence, the principal conjunctive normal form of $(p \wedge q) \vee (\sim p \wedge q)$ is

$$(p \vee q) \wedge (q \vee \sim p)$$

EXERCISE 1.3

1. By means of a truth table, prove that
$$p \wedge q \equiv (p \downarrow q) \downarrow (q \downarrow p) \equiv \sim p \downarrow \sim q$$
2. Obtain disjunctive normal form of
(i) $p \wedge (p \to q)$ (ii) $\sim (p \vee q) \leftrightarrow (p \wedge q)$
3. Obtain the principal disjunctive normal forms of
(i) $\sim p \vee q$ (ii) $(p \wedge q) \vee (\sim p \wedge r) \vee (q \wedge r)$
(iii) $p \to ((p \to q) \wedge \sim (\sim q \vee \sim p))$
4. Obtain the principal conjunctive normal forms of
(i) $(\sim p \to r) \wedge (q \leftrightarrow p)$
(ii) $(q \to p) \wedge (\sim p \wedge q)$
5. Obtain principal conjunctive normal form of
$$p \vee (\sim P \to (Q \vee (\sim Q \to R)))$$
6. Write a formula which is equivalent to the formula $P \uparrow (Q \rightleftarrows R)$ and contains the connective NAND ($\uparrow$) only. Obtain equivalent formula which contains NOR ($\downarrow$) only.

ANSWER 1.3

2. (i) $(p \wedge \sim p) \vee (p \wedge q)$
(ii) $(\sim p \wedge \sim q \wedge p \wedge q) \vee (p \wedge \sim p) \vee (q \wedge \sim p)$
3. (i) $(p \wedge q) \vee (\sim p \wedge q) \vee (\sim p \wedge \sim p)$
(ii) $(p \wedge q \wedge r) \vee (p \wedge q \wedge \sim r) \vee (\sim p \wedge q \wedge r) \vee (\sim p \wedge \sim q \wedge r)$
(iii) $(p \wedge q) \vee (\sim p \wedge q) \vee (\sim p \wedge \sim q)$
4. (i) $(p \vee q \vee r) \wedge (p \vee \sim q \vee r) \wedge (p \vee \sim q \vee \sim r) \wedge (\sim p \vee q \vee r)$
(ii) $(\sim q \vee p) \wedge \sim (p \vee \sim q)$
5. $(P \vee Q \vee R) \wedge (P \vee \sim Q \vee \sim R) \wedge (Q \vee \sim P \vee \sim R) \wedge (R \wedge \sim P \wedge \sim Q)$
6. $P \vee (Q \rightleftarrows R) \equiv \sim (P \uparrow \sim ((Q \uparrow \sim R) \uparrow (R \uparrow \sim Q)))$
$\equiv (P \downarrow P) \downarrow \sim ((\sim Q \downarrow R) \downarrow (\sim R \downarrow Q))$

CHAPTER 2

Predicate Logic

2.1 INTRODUCTION

Logic is the study of general conditions of valid inference, and an inference is a judgement derived from another judgement or from other judgements. In this chapter we discuss, the rules of inference. We also discuss two logical notions, called Quantifiers. In the precious chapter, we have discussed propositions in which each statement has been about a particular object. In this chapter we shall see how to write propositions that are about whole classes of objects.

2.2 PREDICATES

In grammar, a predicate is the word in a sentence which expresses what is said of the object. It is a part of a declarative sentence describing the properties of an object or relation among objects. For example, 'is a tennis player', 'is an Architect' are predicates.

In logic, the word predicate has a broader role than in grammar. The basis for this is the observation that a predicate is supplemented by including a variable x as place holder, for the intended subject and the result behaves as a statement function in the sense that for each value of x, a statement result consider the statement

$$p : x \text{ is an odd number}$$

The truth value of p depends on the value of x. The statement p is 'true' when $x = 5$ and 'false' when $x = 8$. In this section, we extend the system of logic to include such statements.

Definition 2.1. A statement which would be a proposition except for the fact that it includes variables whose values are not specified is called a Predicate (or an open sentence)

In grammar 'John loves' is a not a predicate. If 'x' is introduced as place-holder for the object, then we get the result as 'John loves x', Which is a statement function. Thus we can define a predicate $p(x)$ as an expression having the quality that on an assignment of values to the variables x, from an appropriate domain a statement results.

Examples (i) $x + y = 8$ (ii) The sum of the first natural numbers is $\frac{n(n+1)}{2}$ (iii) If $x < y$ then $x^3 < y^3$; are all predicates.

Definition 2.2. Let $p(x)$ be a statement involving a variables x and a set D. We call $P(x)$; a propositional function if for each x in D; $P(x)$ is a proposition.

The set D is called the domain of discourse (or the universe of discourse) of P. It is the set of all possible values which can be assigned to variables in statements involving predicates.

For example:

(i) The domain of discourse for the propositional function

$P(x)$: x is cricket player

Can be taken as the set of all human beings.

(ii) The domain of discourse for the propositional function

$Q(x)$: $x^2 - 2x - 8 = 0$

is the set of all real numbers.

2.2.1 *n*-place Predicates

Definition 2.3. A predicate which involves n variables is called an n-place predicate.

A predicate which involves one variable is called a one-place predicate and a predicate which involves two variables is called a two-place Predicate.

Predicates are generally denoted by capital letters such as P, Q, R, etc. and following the name of a predicate, we list in parentheses the variables which are used by the predicate.

2.2.2 Satisfiable Predicate

Definition 2.4. An n-place predicate is said to be satisfiable if there exists an n-tuple which satisfies it.

2.2.3 Valid Predicate

Definition 2.5. An n-place predicate is said to be valid if all n-tuples satisfy it.

2.2.4. Truth Set of a Predicate

Let $P(x)$ be a predicate on a set S. Then the set of all those values of $x \in S$, which make $P(x)$ a true proposition is called the truth set of $P(x)$. It is denoted by $T[(Px)]$.

2.2.4 Equivalent Predicates

Definition 2.6. Two predicates are said to be equivalent, if they have same truth value for possible values of their variables.

Thus, two predicates $P(x)$ and $Q(x)$ are equivalent if $P(x) \leftrightarrow Q(x)$.

2.3 QUANTIFIERS

The process of creating a proposition from a propositional function is known as quantification. In this section we will discuss two of types quantification, namely, universal quantification and existential quantification.

2.3.1 Universal Quantifier

Consider the proposition: "All odd prime numbers greater than 2." The word "all" in this proposition is a logical quantifier. This proposition can be translated as follows:

"For every x, if x is an odd prime then x is greater than 2". Similarly; the proposition.

"Every natural number is a real number" may be translated as "For every x, if x is a natural number. Then x is a real number". The phrase 'for every x' is called a universal quantifier. In symbols it is denoted by $\forall x$.

The phrases 'for every x', 'for all x', and 'for each x', have the same meaning and we can symbolize each by $\forall x$.

If $P(x)$ denotes a predicate, (propositional function) then the universal quantification for $P(x)$ is the statement:

"For all values of x; $P(x)$ is true"

Note: A universal quantifier is generally followed by $\rightarrow$.

Example 1. Let $A = \{x: x \text{ is a natural number less than } 12\}$

Here $P(x)$ is the sentence "x is a natural number less than 12".

"is a natural number less than 12" in the sentence is the common property.

$P(1)$ is true, Therefore, $1 \in A$

$P(2)$ is true, Therefore, $2 \in A$

But $P(14)$ is not true, Therefore, $14 \notin A$.

Example 2. Let $P(x)$: $x + 7 < 13$, then for all $x \geq 0$, $P(x)$ is a false statement; because $P(8)$ is not true.

Example 3. For every real number x, $x^2 \geq 0$.

This can be written as $\forall x\, [x^2 \geq 0]$

2.3.2 Existential Quantifier

In some situations we only require that there be at least one value for which the predicate is true. This can be done by prefixing $P(x)$ with the phrase "There exists an x".

The phrase 'there exists an x' is called existential quantifier. The existential quantification for a predicate is the statement "there exists a value x" for which $P(x)$

The symbol $\exists$ is used to denote the logical quantifier 'there exists'. The phrases 'there exists an x'; 'for same x' and 'for at least one x', have the same meaning.

$\exists x\, P(x)$ means that there is at least one value in the universe of the variable which satisfies $P(x)$:

Note: An existential quantifier is usually followed by $\wedge$.

Example 1. Consider $\exists x\, P(x)$ where $P(x)$ is $x^2 = 6$.

If the universe of discourse is the set of all real numbers, then $P(x)$ is true

Example 2. The proposition:

There is a dog without a tail can be written as ($\exists$ a dog) (the dog without tail).

Example 3. The proposition:

There is an integer between 3 and 12 inclusive may be written as

($\exists$ an integer) (the integer between 3 and 12)

2.4 FREE AND BOUND VARIABLES

Definition 2.7. A variable which has been either quantified or substituted for in a predicate is said to be a bound variable.

The truth value of a predicate is may be determined with the binding of the variables.

Definition 2.8. Variables which have not been bound by quantifies are called free variables.

In the statement $\exists\, x\, P(x, y)$, the variable x is bound by the existential quantification $\exists x$, but the variable y is free because it is not bound by a quantifier.

Example. Consider the statement $\exists x\, P(x, y)$.

The variable x is bound by the existential quantification $\exists x$. The given statement, but the variable y is not bound by a quantifier. Therefore, y is a free variable.

2.5 NEGATIONS OF QUANTIFIED PREDICATES

Consider the universally quantified predicate $\forall\, x\, P(x)$.

The negation of $\forall\, x\, P(x)$ is $\sim \forall\, x\, P(x)$

Clearly $\sim \forall\, x\, P(x)$ is equivalent to $\exists\, x \sim P(x)$

i.e. $\sim \forall\, x\, P(x) \equiv \exists\, x \sim P(x)$

Example. The negation of "No rectangle is a Square"
is "Some rectangles are Squares"

Similarly the negation of the existential quantified predicate $\exists x P(x)$ is
$\exists x P(x)$

$\sim \exists x P(x)$ is equivalent to $\forall x \sim P(x)$.

i.e. $\sim \exists x P(x) \equiv \forall x \sim P(x)$

The negation of $\exists x (x^2 = 5)$ is equivalent to $\forall x \sim (x^2 = 5)$

This can be written as $\forall x (x^2 \neq 5)$

2.6 NESTED QUANTIFIERS

Quantifiers that occur with the scope of other quantifiers are called nested quantifiers.

Example. $\forall x \exists y (x + y = 1)$: where the universe of discourse is the set of real numbers.

Table 2.1 Equivalences involving quantifiers

I_1'	Distributivity of $\exists$ over $\vee$
	$\exists x (P(x) \vee Q(x)) \equiv \exists x P(x) \vee \exists x Q(x)$
	$\exists x (P \vee Q(x)) \equiv P \exists x Q(x)$
I_2'	Distributivity of $\forall$ over $\wedge$
	$\forall x (P(x) \wedge Q(x)) \equiv \forall x P(x) \wedge \forall x Q(x)$
	$\forall x (P \wedge Q(x)) \equiv P \wedge \forall (x Q(x))$
I_3'	$\sim (\exists x P(x)) \equiv \forall x \sim P(x)$
I_4'	$\sim (\forall x P(x)) \equiv \exists x \sim (P(x))$
I_5'	$\exists x (P \wedge Q(x)) \equiv P \wedge (\exists x Q(x))$
I_6'	$\forall x (P \vee Q(x)) \equiv P \vee \forall (x Q(x))$
I_7'	$\forall x P(x) \Rightarrow \exists x P(x)$
I_8'	$\forall x P(x) \forall x Q(x) \Rightarrow \forall (P(x) \vee Q(n))$
I_9'	$\exists (P(x) \wedge Q(x)) \Rightarrow \exists x P(x) \wedge \exists x Q(x)$

2.7 SOLVED EXAMPLES

Example 2.7.1. Let $P(x)$ denote the statement $x \leq 8$. What are the truth values of (i) $P(0)$ (ii) (5) (iii) $P(10)$

Solution. (i) $P(0)$ is true (T)

(ii) $P(5)$ is true (T)

(iii) $P(10)$ is false (F)

Example 2.7.2. Express each of the following statements in symbolic form: It is given that the universe discourse is the set of all students.

(i) All students are clever

(ii) Some students are not successful

(iii) Every clever student is successful
(iv) There are some successful students who are not clever
(v) Some students are clever and successful

Solution. Let $P(x)$ denote "x is clever"
$Q(x)$ denote "x is successful"

The given sentences can be written as follows:

(i) $\forall\, x\, P(x)$
(ii) $\exists\, x(\sim Q(x))$
(iii) $\forall\, x\, (P(x) \Rightarrow Q(x))$
(iv) $\exists\, x\, (Q(x) \wedge \sim P(x))$
(v) $\exists\, x\, (P(x) \wedge Q(x))$

Example 2.7.3. Let $P(x)$ be the statement "$x = x^2$". If the universe of discourse consists of the integers, what are the truth values?

(i) $P(0)$ (ii) $P(1)$ (iii) $P(4)$ (iv) $P(-2)$ (v) $\exists\, x\, P(x)$ (vi) $\forall\, x\, P(x)$

Solution. (i) T (ii) T (iii) F (iv) F (v) T (vi) F

Example 2.7.4. It is given the universe discourse of the propositional function $P(x)$ consists of the integers 0, 1, 2, 3 and 4. Write each of the following propositions using disjunctions, conjunctions, and negations.

(a) $\exists\, P(x)$ (b) $\forall\, x\, P(x)$ (c) $\exists\, x \sim P(x)$
(d) $\forall\, x \sim P(x)$ (e) $\sim \exists\, x\, P(x)$ (f) $\sim \forall\, x\, P(x)$.

Solution.

(a) $P(0) \vee P(1) \vee P(2) \vee P(3) \vee P(4)$
(b) $P(0) \wedge P(1) \wedge P(2) \wedge P(3) \wedge P(4)$
(c) $\sim P(0) \vee \sim P(1) \vee \sim P(2) \vee \sim P(3) \vee \sim P(4)$
(d) $\sim P(0) \wedge \sim P(1) \wedge \sim P(2) \wedge \sim P(3) \wedge \sim P(4)$
(e) $\sim (P(0) \vee (P(1) \vee (P(2) \vee (P(3) \vee (P(4))$
(f) $\sim (P(0) \wedge (P(1) \wedge (P(2) \wedge (P(3) \wedge (P(4))$

Example 2.7.5. Negate the proposition. "All integers greater than 8".

Solution. The given proposition can be written as

$$(\forall \text{ integers } x)\ (x > 8)$$

The negation of the gives statement is ($\exists$ an integer x) $(x \leq 8)$

i.e. There is an integer less than or equal to 8.

Example 2.7.6. Translate the statement "The sum of two positive integers is positive" into a logical expression.

Solution. We rewrite the given statement as follows:

"For every two positive integers, the sum of these integers is positive"

Introducing variables x and y; the above statement can be written as: "For all positive integers x and y, $x + y$ is positive".

The logical expression is

$$\forall\, x\, \forall\, y\ (x > 0) \wedge (y > 0) \rightarrow (x + y > 0)$$

Example 2.7.7. Find the truth set of the proposition $P(x)$: $x + 7 = 0$

Solution. Solving $x + 7 = 0$, we get $x = -7$

$\therefore \qquad T[P(x)] = \{-7\}$

Example 2.7.8. Find the truth values of each of the following statements R is the universe of discourse.

(i) $\forall x, |x| = x$

(ii) $\exists x, x^2 = x$

(iii) $\forall x, x + 2 > x$

(iv) $\forall x, x + 5 = x$

Solution. (i) False, since $|-5| = 5$ i.e. when $x = -5$, $|x| = x$ does not hold

(ii) True

(iii) True

(iv) False, since the equation $x + 5 = x$ has no solution

Example 2.7.9. Translate the following into symbolic form.

All men are mortal

Socrates is a man

Using these prove that Socrates is mortal. (VTU, MCA. Aug. 2000)

Solution. Let $P(x) = x$ is a man

$Q(x) = x$ is mortal

$P(x) \wedge Q(x)$ = Man is mortal

$\forall x\, [P(x) \wedge Q(x)]$: All men are mortal

Let x: Socrates, then

$P(x)$: Socrates is a man

$Q(x)$: Socrates is mortal

then $P(x) \wedge Q(x)$: Socrates is mortal

Example 2.7.10. Symbolize: "All cats are animals". (VTU, MCA. M. 2001)

Solution. Let $P(x)$: x is a cat

$Q(x)$: x is an animal

$P(x) \rightarrow Q(x)$: Cat is an animal

$\therefore \qquad \forall x\, [P(x) \rightarrow Q(x)]$: All cats are animals

Example 2.7.11. If $P(x)$: x is a person (VTU, MCA. M. 2001)

$Q(x, y)$: x is a father of y

$R(x, y)$: x is a mother of y

Symbolize the statement "x is a father of the mother of y"

Solution. Consider the variable z such that "z is mother of y"

Then, $R(z, y)$: z is the mother of y

$Q(x, z)$: x is the father of z

$Q(x, z) \wedge R(z, y)$: x is the father of z and z is the mother of y. If such a person exists

$\therefore \qquad P(x) \wedge Q(x, z) \wedge R(z, y)$

$(\exists x)\, (P(x) \wedge Q(x, z) \wedge R(z, y))$: x is the father of mother of y.

Example 2.7.12. Negate $(\forall x \, \exists y) \, (P(x) \vee Q(y))$

Solution. $\sim (\forall x \, \exists y) \, (P(x) \vee Q(y))$

$$\equiv \exists x \, \forall y \sim (P(x) \vee Q(y))$$

$$\equiv \exists x \, \forall y \, (\sim P(x) \wedge \sim Q(y))$$

Example 2.7.13. Express the following proposition in symbolic form, and find its negation.

"If all triangles are right-angled, than no triangle is equiangular."

Solution. Let $p(x)$: x is right-angled

$q(x)$: x is equiangular

and T be the set of all triangles.

The required symbolic form of the given proportion is

$$[\forall x \in T, \; p(x)] \rightarrow [\forall x \in T, \sim q(x)]$$

Example 2.7.14. If $p(x) : x^2 - 7x + 10 = 0$, $q(x) : x^n - 2x - 3 = 0$, $r(n) : x < 0$ determine the truth value of the following statements given that the universe contains only the integers 2 and 5.

(i) $\forall x, p(x) \rightarrow \sim r(x)$ (ii) $\exists x, q(x) \rightarrow r(x)$

Solution. (i) Let the universe be denoted by $\cup$ $p(x)$ is true when $x = 2$ and $x = 5$

Since the universe contains only the elements 2 and 5 $p(x)$ is true for all

$$x \in \cup = \{2, 5\}$$

$r(x) = x^2 - 2x - 3 = (x - 3)(x + 1)$ is true for $x = 3$ and $x = -1$, which are not in $\cup$, $q(x)$ is false for all $x \in \cup$

Thus $\sim q(x)$ is true for all $x \in \cup$

i.e., $p(x)$ is true for all $x \in \cup$ and $\sim q(x)$ in true for all $x \in \cup$.

Hence, $\forall x, p(x) \rightarrow \sim r(x)$ is true

(ii) $q(x)$ is true for $x = 2$ and $r(x)$ is true for $x = 2$

Hence, $\exists x, q(x) \rightarrow r(n)$ is true.

EXERCISE 2.1

1. If the universe of discourse for the following statements, then find the truth values of the each.
   (a) $\forall x, x^2 = x$
   (b) $\exists x; 5x = x$
   (c) $\forall x_i; x - 8 < x$
   (d) $\exists x; x^2 - x + 9 = 0$
2. Determine the truth value of each of these statements. If the universe of discourse for each variable consists of all real numbers
   (a) $\forall x, x - 3 < x$ (b) $\exists x, x^2 - 5x + 6 = 0$
   (c) $\exists x, x^2 - 2x + 7 = 0$
3. Determine the truth values of each of the following statements where $x = \{1, 2, 3\}$ is the universal set.

(i) $\exists x\ \forall y, x^2 < y + 1$

(ii) $\exists x\ \forall y, x^2 + y^2 < 12$

(iii) $\forall x, \forall y, x^2 + y^2 < 12$

4. Symbolize using quantifiers

(i) Every even number is divisible by 2

(ii) There is the no prime number between 23 and 29

5. Negate $\forall x\ \exists y\ (P(x) \vee Q(y))$

6. Negate:

(a) $\forall x : |x| = x$ (b) $\exists x, x^2 = x$

(c) $\exists, x + 3 > x$ (d) $\exists x, x + 4 = x$

7. Negate: $\exists x\ \forall y\ (P(x, y) \rightarrow Q(x, y))$

8. Negate the following

(i) $(\exists x)\ (\forall y)\ [P(x) \vee Q(y)]$

(ii) $\forall x\ [P(x) \rightarrow Q(x)]$

9. Demonstrate the implication $\sim ((\exists x))\ P(x) \wedge Q(a) \Rightarrow (\exists x)\ P(x) \rightarrow \sim Q(a)$

10. If $P(x)$; odd numbers are Prime numbers

$Q(x)$: Prime numbers are even

Then negate the following statement

"All odd numbers are not prime numbers and some prime numbers are even".

11. Symbolize using a quantifier: Everyone in the room spoke, French or German".

12. $\forall x\ (R(x) \rightarrow (F(x) \vee G(x))$

where $R(x) = x$ was person in the room

$F(x) = x$ spoke French

$G(x) = x$ spoke German

13. Suppose that P and Q are formulae and x, y are variables such that x does not bound in P and not at all in Q, and y does not occur is either P or Q then show that

(i) $\sim \forall x\ P(x) \equiv \exists x \sim Q(x)$, $\quad \sim \exists\ x\ P(x) \equiv \forall x \sim P(x)$

(ii) $(\forall x\ P) \wedge Q \equiv \forall x\ (P \wedge Q)$, $\quad (\exists x\ P) \wedge Q \equiv \exists\ x(P \wedge Q)$

ANSWERS 2.1

1. (a) F (b) T (c) T
   (d) F
2. (a) T (b) T (c) F
3. (a) T (b) T (c) F
4. (i) $(\forall x)\ (\exists(x)/2$: where $\exists(x) = x$ is an even number

(ii) $P(x)$: x is not a prime number between 23 and 29

or

$P(x)$: There is no prime number between 23 and 29

5. $\exists x\ \forall y\ (-P(x) \wedge \sim Q(y))$

6. (a) $\exists x, |x| \neq x$ (b) $\forall x, x^2 \neq x$ (c) $\exists x, x + 3 < x$

(d) $\forall x, x + 4 \neq x$

7. $\forall x\ \exists y\ (P(x, y) \wedge \sim Q(x, y))$

8. (i) $(\forall x)\ (\exists y) \sim [P(x) \vee Q(y)]$

(ii) $\exists x\ [P(x) \wedge \sim Q(x)]$

10. $\exists x\ P(x) \vee (\forall x) \sim Q(x)$

2.8 RULES OF INFERENCE FOR STATEMENT CALCULUS

Logic is the study of inferences. An inference is a judgement derived from another judgement or from other judgements. The terms inference and reasoning (critical inference) are used as synonyms. Reasoning is of two types. Reasoning, which starts from some assertions or statements and seeks to unfold their implications is formal, and the reasoning which starts from observed facts to discover their character is inductive. The term reasoning is applied for both inference and proof. The term inference and proof appear almost synonymous in the definition of logic.

Logic is the branch of mathematics that tells whether an argument is valid or invalid. In every argument there are two things.

(i) The premises (i.e. series of statements)

(ii) The conclusion (i.e. the inference)

If we start from the premises and proceed to the conclusion the process is called **inference.** So prove that a statement is true we need a sequence of sentences. When we are given a sequence of premises that are true and we are given a sequence of sentences. When we are given a sequence of premises that are true and we apply hours of reasoning to reach conclusion that is true; we say that we are proving an argument by means of formal proof. The statements used in proof include axioms or postulates. We draw conclusions by using the rules of inference. We use the term inference to designate a set of premises accompanied by a suggested conclusion regardless of whether not the conclusion is a consequence of the premises. It was stated that logic is concerned with valid inferences. A valid inference is defined as a 'correct' or "accurate" inference. It may or may not be true. An inference is valid when it is reasonably justified, by judegements from which it is derived. An inference is true if it is in accordance with relevant facts. In this section, we discuss rules of inference.

2.8.1 Law of Detachment

The law of Detachment states that when two given premises are true; one a conditional and the other the hypothesis of that conditional, it then follows the conclusion of the conditional is true.

The law of Detachment is also known as the Modus Ponens symbolically, we can write the law of Detachment as follows:

$$\frac{p \to q}{\therefore q}$$

The symbol $\therefore$ is used to mean the word "Therefore". We use the symbol $\therefore$ or the word therefore only when we reach a conclusion.

In symbols, we can also write the law of Detachment as

$$(p \wedge (p \to q)) \to q$$

where q is called valid conclusion. $(P \wedge (P \to q) \to q)$ is a tautology.

"If n is greater than 5, then n^2 is greater than 25" is a true statement. Then by Modus Ponens, it follows that n^2 is greater than 25.

Example 2. Assume that the following propositions are true

p: two triangles are similar

$p \to q$: If two triangles are similar then their corresponding sides are proportional

by Modus Ponens : the conclusion is q: The corresponding sides of the triangles of the triangles are proportional is true.

2.8.2 The Law of Modus Tollens

The law of Modus Tollens States that when two given premises are true, one a conditional and the other the negation of the conclusion of that conditional, it then follows that the negation of the hypothesis of the conditional is true.

Symbolically, the Law many be written as:

$$P \to q$$

$$\frac{\sim q}{\therefore \sim P}$$

The tautology $[(p \to q) \wedge \sim q] \to \sim p$ is the basis of the law of Modus Tollens.

2.8.3 Rule of Trasitivity

(Law of the Syllogism or Chain rule)

If $p \to q$ and $q \to r$ are accepted as true than $p \to r$ must be accepted as true. i.e.

$$\frac{\begin{array}{c} p \to q \\ q \to r \end{array}}{\therefore p \to r}$$

we can extend the transitive rule as follows:

$$\begin{array}{c} p \to q \\ q \to r \\ r \to s \\ \therefore p \to s \end{array} \qquad \frac{\begin{array}{c} p_1 \to p_2 \\ p_2 \to p_3 \\ p_3 \to p_4 \\ p_4 \to p_5 \end{array}}{\therefore p_1 \to p_5}$$

Example. We use the Chain rule to solve equations; in Algebra

Consider If $3x + 7 = 16$, Then $3x = 9$

If $3x = 9$, Then $x = 3$

We get the conclusion as follows:

$$\begin{array}{llll} & p \to q: & \text{If } 3x + 7 = 16, & \text{Then } 3x = 9 \\ & q \to r: & \text{If } 3x = 9, & \text{Then } x = 3 \\ \hline \therefore & p \to r: & \text{If } 3x + 7 = 16, & \text{Then } x = 3 \end{array}$$

Symbolically, the Chain rule may be written as follows:

$$[(p \to q) \wedge (q \to r)] \to (p \to r)$$

2.8.4 The Law of Contrapositive

The law of contrapositive states that when a conditional premise is true, it follows that the contrapositive of the premise is also true.

Symbolically, the law can be written as:

$$\frac{p \to q}{\therefore \sim q \to \sim p}$$

The tautology $(p \to q) \leftrightarrow (\sim q \to \sim p)$ is the basis of the rule of inference.

Example Use a contrapositive argument to verify the following valid inference

$$\frac{w \to (r \to s)}{\therefore (w \wedge r) \to s}$$

Suppose $\sim S \to \sim (w \wedge r)$

Now $\sim S \to \sim (w \wedge r) \equiv \sim (\sim s) \vee (w \wedge r)$

$(\because p \to q \equiv \sim p \vee q)$

$\equiv S \vee \sim (w \wedge r)$

$\equiv \sim (w \wedge r) \vee s$ $(\because \vee$ is commutative$)$

$\equiv w \wedge r \to s$

Table 2.3 Rules of inference

Implications:		
I_1	$P \wedge Q \Rightarrow P$	(Simplification)
I_2	$P \wedge Q \Rightarrow Q$	(Simplification)
I_3	$P \Rightarrow P \vee Q$	(Addition)
I_4	$Q \Rightarrow P \vee Q$	(Addition)
I_5	$\sim P \Rightarrow P \rightarrow Q$	
I_6	$Q \Rightarrow P \rightarrow Q$	
I_7	$\sim (P \rightarrow Q) \Rightarrow P$	
I_8	$\sim (P \rightarrow Q) \rightarrow \sim Q$	
I_9	$P: Q \Rightarrow P \wedge Q$	
I_{10}	$\sim P, P \vee Q \Rightarrow Q$	(Disjunctive syllogism)
I_{11}	$P, P \rightarrow Q \Rightarrow Q$	
I_{12}	$\sim Q, P \rightarrow Q \Rightarrow \sim P$	(Modus Tollens)
I_{13}	$P \rightarrow R, Q \rightarrow R \Rightarrow P \rightarrow R$	(Hypothetical syllogism)
I_{14}	$P \vee Q, P \rightarrow R, Q \rightarrow R \Rightarrow R$	(Dilemma)
Equivalences:		
E_1	$\sim \sim P \Leftrightarrow P$	Double negation
E_2	$P \wedge Q \Leftrightarrow Q \Leftrightarrow P$	(Commutative law)
E_3	$P \vee Q \Leftrightarrow Q \vee P$	(Commutative law)
E_4	$(P \wedge Q) \wedge R \Leftrightarrow P \wedge (Q \wedge R)$	(Associative law)
E_5	$(P \vee Q) \vee R \Leftrightarrow P \vee (Q \vee R)$	(Associative law)
E_6	$P \wedge (Q \vee R) \Leftrightarrow (P \wedge Q) \vee (P \wedge R)$	(Distributive law)
E_7	$P \vee (Q \wedge R) \Leftrightarrow (P \vee Q) \wedge (P \vee R)$	(Distributive law)
E_8	$\sim (P \wedge Q) \Leftrightarrow \sim P \vee \sim Q$	(De Morgan's law)
E_9	$\sim (P \vee Q) \Leftrightarrow \sim P \wedge \sim Q$	(De Morgan's law)
E_{10}	$P \vee P \Leftrightarrow P$	
E_{11}	$P \wedge P \Leftrightarrow P$	
E_{12}	$R \vee (P \wedge \sim P) \Leftrightarrow R$	
E_{13}	$R \vee (P \vee P) \Leftrightarrow R$	
E_{14}	$R \vee (P \vee \sim P) \Leftrightarrow T$	
E_{15}	$R \wedge (P \wedge \sim P) \Leftrightarrow F$	
E_{16}	$P \rightarrow Q \Leftrightarrow \sim P \wedge Q$	
E_{17}	$\sim (P \rightarrow Q) \Leftrightarrow \sim P \vee Q$	
E_{18}	$(P \rightarrow Q) \Leftrightarrow \sim Q \rightarrow \sim P$	
E_{19}	$P \rightarrow (Q \rightarrow R) \Leftrightarrow (P \wedge Q) \rightarrow R$	
E_{20}	$\sim (P \rightleftarrows Q) \Leftrightarrow P \rightleftarrows \sim Q$	
E_{21}	$P \rightleftarrows Q \Leftrightarrow (P \rightarrow Q) \wedge (Q \wedge P)$	
E_{22}	$P \rightleftarrows Q \Leftrightarrow (P \wedge Q) \vee (\sim P \wedge \sim Q)$	

2.9 ARGUMENT

An argument is an assertion that, given a set of propositions $p_1, p_2, \ldots, p_n$ called premises, has as a consequence another proposition q. It is denoted by $p_1, p_2, \ldots, p_n \vdash q$ (or by $p_1 \wedge p_2, \wedge \ldots \wedge p_n) \rightarrow q$. If $p_1, p_2, \ldots p_n$ yields q, then q is called the conclusion (or consequence).

In this case we say that q logically follows from $p_1, p_2, \ldots p_n$. Arguments may be categorical, hypothetical, disjunctive or relational. In a text book devoted to logic, names for many types of arguments will be

2.9.1. Valid Argument

Definition 2.9. An argument $p_1, p_2, \ldots, p_n \vdash q$ is true if q is true, whenever all the premises $p_1, p_2, \ldots p_n$ are true. An argument which is true is called a valid argument.

2.9.2 False Argument

Definition 2.10 If an argument is not true, then it is said to be a false argument.

2.9.3 Fallacy

Definition 2.11 If an argument is false, then it is called fallacy.

Fallacies are invalid arguments. They Violate some logical principle or principles. There are fifteen types of fallacies. Aristotle arranged them in two groups. Namely,

(i) Verbal fallacies and (ii) non-verbal fallacies

In this text, we arrange the fallacies in three groups.

1. The fallacy of affirming the consequent or affirming the converse. It can be expressed symbolically as follows:

$$\begin{array}{c} p \rightarrow q \\ \underline{q \text{ fallacy}} \\ \therefore \quad p \end{array}$$

2. The fallacy denying anticedent. This can be expressed as

$$\begin{array}{c} p \rightarrow q \\ \underline{\sim p \text{ fallacy}} \\ \therefore \quad \sim q \end{array}$$

3. Non-sequitur fallacy: It takes the form

$$\begin{array}{c} \underline{\quad p \quad} \\ \therefore \quad q \end{array}$$

2.9.4 Rules *P* and *T*

Rule **P**: A premise may be introduced at any point in the derivation

Rule **T**: A formula *S* may be introduced in a derivation, if *S* is a Tautology implied by anyone or more of the preceding formulas in the derivation.

The above rules are used in the process of derivation by which one demonstrates that particular formula is a valid consequence of a given set of Premises.

The most commonly used proof in logic involves the columns. The first column cosists of statements and the second column consists of reasons that allow us to make the statements. Let us study same examples.

Example 1. Prove $\sim Q, P \rightarrow Q \Rightarrow \sim P$

Solution.

Statements	Reasons
1. $P \rightarrow Q$	rule **P**
2. $\sim Q \rightarrow \sim \mathbf{p}$	rule **T**, (1) and $E18$
3. $\sim Q$	rule **P**
4. $\sim$ P	rule **T**, (2), (3) and I_{11}

Example 2. Given the premises

$$p \rightarrow r$$
$$\sim p \rightarrow q$$
$$\sim r$$

Prove the conclusion q

Solution.

Statements	Reasons
1. $p \rightarrow r$	given
2. $\sim p \rightarrow q$	given
3. $\sim r$	given
4. $\sim p$	The law of Modus Tollens, (1), (3)
5. q	The law of Detachment, (2), (4)

Examples 3. Prove that R is a valid inference from the premises $P \rightarrow Q$, $Q \rightarrow R$, and P

Solution.

1. $P \rightarrow Q$	Rule **P**
2. P	Rule **P**
3. Q	Rule **T**, (1), (2) and I_{11}
4. $Q \rightarrow R$	Rule **P**
5. R	Rule **T**, (3), (4) and I_{11}

$\therefore$ R follows logically from the given premises.

2.9.5 Deduction Theorem: (Rule of Conditional Proof)

Rule **CP**: If we can drive S from R and a set of premises, then we can derive $R \rightarrow S$ from the set of premises alone.

Example. Using deduction theorem show that $R \rightarrow S$ can be derived from the premises

$$P \rightarrow (Q \rightarrow S), \sim R \cup P, \text{ and } Q.$$

Solution. We include R as an additional premise and show S, so that $R \rightarrow S$ can be derived

1. $\sim R \vee P$	rule **P**
2. R	rule **P** (assumed premise)
3. P	rule **T**, (1), (2) and I_{10}
4. $P \rightarrow (Q \rightarrow S)$	rule **P**
5. $Q \rightarrow S$	rule **T**, (3), (4) and I_{11}
6. Q	rule **P**
7. S	rule **T**, (5), (6), and I_{11}
8. $R \rightarrow S$	**CP**

2.10 CONSISTENCY

Consistency is an extremely important notion in mathematical logic.

A set of formulas $H_1, H_2, \ldots, H_n$ is said to be consistent if their conjunction $H_1 \wedge H_2 \wedge \ldots \wedge H_n$ has the truth value T for same assignment of the truth values to the atomic variables appearing in $H_1, H_2, \ldots, H_n$. And a set of formulae $H_1, H_2, \ldots, H_n$ is inconsistent if their conjunction $H_1 \wedge H_2 \wedge \ldots \wedge H_n$ implies a contradiction i.e. $H_1 \wedge H_2 \wedge \ldots \wedge H_n \Rightarrow S \wedge \sim S$ (a contradiction) where S in any formula.

Definition 2.12. A collection of statements is consistent if the statements can all be true simultaneously by

We use the notion inconsistency of in a method of proof called proof by contradiction (or indirect proof).

Example. Show that $(R \rightarrow \sim Q), R \vee S, S \rightarrow \sim Q, P \rightarrow Q \Leftrightarrow \sim P$ are in consistent.

Solution.

1. P	(assumed)
2. $P \rightarrow Q$	Rule **P**
3. Q	(1) and (2)
4. $S \rightarrow \sim Q$	P
5. $Q \rightarrow \sim S$	$P \rightarrow Q \rightleftarrows \sim Q \rightarrow \sim P$
6. $\sim S$	(3), (5)

7. $R \vee S$	P
8. $\sim R \to S$	$P \to Q \rightleftarrows \sim P \vee \sim P$
9. $\sim S \to R$	$P \to Q \rightleftarrows \sim Q \to \sim P$
10. R	(6), (9)
11. $R \to \sim Q$	P
12. $\sim Q$	(10), (11)
13. $Q \wedge \sim Q$	(3), (12)

inconsistent

2.11 AUTOMATIC THEOREM PROVING

In this section we describe a procedure of derivation which can be conducted mechanically. The formulation of proof is based on the work of Hao Wang.

Hao Wang Rules

1. Variables:

   The capital letters $A, B, C, \ldots, P, Q, R \ldots$ are used as statement variables.

2. Connectives:

   The connectives $\sim$, $\wedge$, $\vee$, $\to$ and $\rightleftarrows$ appear in the formulas with the order of precedence as given

3. A string of formulas:

   A string of formulas is defined as follows.

   (i) Any formula is a string of formulas

   (ii) If α and β are strings of formulas. Then α, β, and β, α are strings of formulas

   (iii) Only those strings which are obtained by steps.

   (i) and (ii) are strings of formulas, with the exception of the empty string which is also a string of formulas.

4. Sequent:

   If α and β are strings of formulas, then $\alpha \xrightarrow{s} \beta$ is called a sequent in which α is denoted the antecedent and β the consequent of sequent.

   A sequent $\alpha \xrightarrow{s} \beta$ is true if only if either at least one of the formulas of the antecedent is false or at least one of the formulas of the consequent is true.

5. Axiom Schema:

   If α and β are strings of formulas such that every formula in both α and β is a variable only, the sequent $\alpha \xrightarrow{s} \beta$ is an axiom if and only if α and β have at least one variable in common.

If $\alpha \xrightarrow{s} \beta$ is an axiom, then $\alpha \overset{s}{\Rightarrow} \beta$.

6. Theorem:

   The following sequents are theorems of the system

   (i) Every axiom is a theorem

   (ii) If a sequent α is a theorem. If a sequent β results from α through the use of one of the rules of the system which are given in O, then β is also a theorem

   (iii) The sequents obtained by obtained by (i) and (ii) are the only theorems.

7. Rules:

   If $\alpha, \beta, \gamma, \ldots$ denote strings of formulas, and x, y denote formulas to which the connectives are applied, then the following rules are used to combine formulas within strings by introducing connectives.

Antecedent rules:

Rule $\sim \Rightarrow$: If $\alpha, \beta \overset{s}{\Rightarrow} x, \gamma$, Then $\alpha, \sim x, \beta \overset{s}{\Rightarrow} \gamma$

Rule $\wedge \Rightarrow$: If $x, y, \alpha, \beta \overset{s}{\Rightarrow} \gamma$, Then, $\alpha, x \wedge y, \beta \overset{s}{\Rightarrow} \gamma$

Rule $\vee \Rightarrow$: If $x, \alpha, \beta \overset{s}{\Rightarrow} \gamma$ and $y, \alpha, \beta \overset{s}{\Rightarrow} \gamma$, then $\alpha, x \vee y, \beta \overset{s}{\Rightarrow} \gamma$

Rule $\rightarrow \Rightarrow$: If $y, \alpha, \beta \overset{s}{\Rightarrow} \gamma$, and $\alpha, \beta \overset{s}{\Rightarrow} x, \gamma$, then $\alpha, x \rightarrow y, \beta \overset{s}{\Rightarrow} \gamma$

Rule $\rightleftarrows \Rightarrow$: If $x, y, \alpha, \beta \overset{s}{\Rightarrow} \gamma$, and $\alpha, \beta \overset{s}{\Rightarrow} x, y, \gamma$, then $\alpha, x \rightleftarrows y, \beta \overset{s}{\Rightarrow} \gamma$

Consequent rules.

Rule $\Rightarrow \sim$: If $x, \alpha \overset{s}{\Rightarrow} \beta, \gamma$, then $\alpha \overset{s}{\Rightarrow} \beta, \sim x, \gamma$

Rule $\Rightarrow \wedge$: If $\alpha \overset{s}{\Rightarrow} x, \beta, \gamma$ and $\alpha \overset{s}{\Rightarrow} y, \beta, r$, then $\alpha \overset{s}{\Rightarrow} \beta, x \wedge y, \gamma$

Rule $\Rightarrow \vee$: If $\alpha \overset{s}{\Rightarrow} x, y, \beta, \gamma$, then $\alpha \overset{s}{\Rightarrow} \beta, x \vee y, \gamma$

Rule $\Rightarrow \rightarrow$: If $x, \alpha \overset{s}{\Rightarrow} y, \beta, \gamma$, then $\alpha \overset{s}{\Rightarrow} \beta, x \rightarrow y, \gamma$

Rule $\Rightarrow \rightleftarrows$: If $x, \alpha \overset{s}{\Rightarrow} y, \beta, \gamma$, and $y, \alpha \overset{s}{\Rightarrow} x, \beta, \gamma$, then $\alpha \overset{s}{\Rightarrow} \beta, x \rightleftarrows y, \gamma$

In the above rules the symbol "$\xrightarrow{s}$" is a generalization of the connective $\rightarrow$ to strings of formulas. And the symbol "$\overset{s}{\Rightarrow}$" is applied to strings of formulas as a generalization of the symbol $\Rightarrow$.

$\therefore$ $\alpha \rightarrow \beta$ means "α implies β or" $\alpha \rightarrow \beta$ is a Tautology and $\alpha \overset{s}{\Rightarrow} \beta$ means that $\alpha \xrightarrow{s} \beta$ is true.

Also, $A, B, C \xrightarrow{s} D, E, F$ is true if $A \wedge B \wedge C \rightarrow D \vee E \vee F$ is true.

In the method of derivation explained earlier, we showed a conclusion **C** follows from the premises $H_1, H_2, \ldots H_n$ by constructing a derivation whose last step was **C**, and the Premises $H_1, H_2, \ldots, H_n$ we introduced at various stages by using rule **P**. In the new formulation: to show that **C** follows $H_1, H_2, \ldots, H_n$. We establish that

$$\xrightarrow{s} H_1 \rightarrow (H_2 \rightarrow H_3 \ldots (H_n \rightarrow C) \ldots)) \qquad \ldots(1)$$

is a theorem.

That

$$\overset{s}{\Rightarrow} H_1 \rightarrow (H_2 \rightarrow (H_3 \ldots (H_n \rightarrow C) \ldots)) \qquad \ldots(2)$$

and show that the assumption (2) is justified. If the assumption is not justified then (1) is not a theorem. This task is accomplished by working backward from (2), using the rules and showing that (2) holds if some simples sequent is a theorem. We continue to work backward we arrive at the simplest possible sequents. If these sequents are axioms, then we have justified our assumption of (2). If at least one of the simplest sequents is not justified, then assumption of (2) is not justified and the conclusion C does not follow from the premises $H_1, H_2, \ldots, H_n$.

Example 1. Show that P does not follow from $P \vee Q$.

Solution.

Assume

(1) $\overset{s}{\Rightarrow} (P \vee Q) \rightarrow P$

(1) If (2) $P \vee Q \overset{s}{\Rightarrow} P$ $(\Rightarrow \rightarrow)$

(2) If (3) $P \overset{s}{\Rightarrow} P$ and (4) $Q \overset{s}{\Rightarrow} P$ $(\vee \Rightarrow)$

(3) Is an axiom

but (4) is not an axiom

Hence P does not follow from $P \vee Q$.

i.e. $\xrightarrow{s} (P \vee Q) \rightarrow P$ is not a theorem.

Example 2. Show that $P \vee Q$ follows from P.

Solution. To prove that $P \vee Q$ follows form P, we need to show that

(1) $\overset{s}{\Rightarrow} P \rightarrow (P \vee Q)$

(1) if (2) $P \overset{s}{\Rightarrow} P \overset{\vee}{\Rightarrow} Q$ $(\Rightarrow \rightarrow)$ (The connective $\rightarrow$ in (1) is eliminated)

(2) if (3) $P \overset{s}{\Rightarrow} P, Q$ $(\Rightarrow \vee)$ ($p \overset{s}{\Rightarrow} P \vee Q$ is named by (2))

(3) is an axiom, hence (3) is a theorem

(3) is an axiom. Starting from (3);

The reversal of the above derivation leads to

$$\overset{s}{\Rightarrow} P \rightarrow (P \vee Q)$$

(i)	$P \overset{s}{\Rightarrow} P, Q$	Axiom
(ii)	$P \overset{s}{\Rightarrow} P \vee Q$	Rule $(\Rightarrow \vee)$, (i)
(iii)	$\overset{s}{\Rightarrow} P \rightarrow (P \vee Q)$,	Rule $(\Rightarrow \rightarrow)$, (ii)

Table 2.3 Rules of inference for addition and deletion of Quantifies

R_1 **US**:	Universal Specification (instantiation): $\dfrac{\forall x\, P(x)}{\therefore\ P(k)}$ k is same element of the universe
R_2 **ES**:	Existential Specification (instantiation): $\dfrac{\exists x\, P(x)}{\therefore\ P(k)}$ k is same element for which $P(k)$ is true
R_3: **UG**:	Universal Generalization: $\dfrac{P(x)}{\forall P(x)}$
R_4 **EG**:	Existential Generalization: $\dfrac{P(k)}{\therefore \exists x\, P(x)}$

Example 1. Show that $\exists x\, M(x)$ follows logically from the premises

$$(x)\ (H(x) \rightarrow M(x)) \text{ and } \exists x\, H(x)$$

Solution.

{1}	(1) $\exists x\, H(x)$	**P**
{1}	(2) $H(y)$	**E**s, (1)
{3}	(3) $x\ (H(x) \Rightarrow M(x))$	**P**
{3}	(4) $H(y) \rightarrow M(y)$	**US**, (3)
{1, 3}	(5) $M(y)$	**T**, (2), (4), I_{11}
{1, 3}	(6) $\exists x\, M(x)$	**EG**, (5)

Example 2. Consider the 'Socrates argument' given by

All men are mortal
Socrates is a man
Therefore Socrates is a mortal

In symbols

If $H(x)$: x is a man
$M(x)$: x is a mortal

and S: Socrates

Show that $x(H(x) \rightarrow M(x)) \wedge H(s) \Rightarrow M(s)$

Solution.

{1}	(1)	$x(H(x) \rightarrow M(x))$	**P**
{1}	(2)	$H(s) \rightarrow M(s)$	**US**, (1)
{3}	(3)	$H(s)$	**P**
{1, 3}	(4)	$M(s)$	**T**, (2), (3), I_{11}

2.12 METHODS OF PROOF

There are nine methods of proof. In this section we briefly discuss: direct proof, indirect proof, proof by counter example and proof by cases.

2.12.1 Direct Proof

We assume that P is true; and from the available information the conclusion Q is shown to be true by valid reference. In this method of proof we construct a chain of statements $P, P_1, P_2, \ldots, P_n, Q$ where P is either a hypothesis of the theorem or an axiom and to each of the implication $P \Rightarrow P_1, P_1 \Rightarrow P_2, \ldots, P_n \Rightarrow Q$ is either an axiom or is implied by the implication preceding it.

Example 1. If x and y are odd integers then $x + y$ is an even integer.

Solution. (Direct Proof):

An odd integer is of the form $2m + 1$, where m is an integer

x and y are odd integers; therefore

$$x = 2m_1 + 1, y = 2m_2 + 1 \text{ for some integers } m_1 \text{ and } m_2$$

Hence, $x + y = (2m_1 + 1) + (2m_2 + 1)$

$$= 2m_1 + 1 + 2m_2 + 1$$

$$= 2m_1 + 2m_2 + 1$$

$$= 2\,(m_1 + m_2 + 1)$$

$\Rightarrow$ $x + y$ is even.

Example 2. Prove that if $|a| > |b|$ then $a^2 > b^2$, by direct method

Solution. Since $|a| > |b|$,

we have $|a|^2 > |b|^2$ ...(1)

but $|a|^2 = a^2$, and $|b|^2 = b^2$

Hence $a^2 > b^2$ (from (1))

2.12.2 Method of Contraposition

(Indirect Proof):

This method of proof is very useful and is powerful at all levels of the subject of mathematics. Indirect method follows from the tautology $(p \to q) \leftrightarrow ((\sim q) \to (\sim p))$. This states that the implication $p \Rightarrow q$ is equivalent to $\sim q \Rightarrow \sim p$.

To prove $p \Rightarrow q$ indirectly, we assume that q is false and then show that p is false.

Example Prove: If α^2 is an even integer, then α is an even integer

Solution. Let p: α^2 is an even integer

q: α is an even integer

Let $\sim q$ be true.

α is not an even integer. Therefore a must be odd and α is of the form $\alpha = 2m + 1$ for some integer m now $\alpha = 2m + 1$

Therefore, $\alpha^2 = (2m + 1)^2 = 4m^2 + 4m + 1 = 2(2m^2 + 2m) + 1$

i.e., α^2 is of the form $2n + 1$ where $n = 2m^2 + 2m$

Hence, α^2 is odd

Thus we have $\sim q \Rightarrow \sim p$

Hence by contraposition α is even.

2.12.3 Proof by Contradiction

This method is based on the tautology $(p \to q) \wedge (\sim q \to \sim p)$. In this method, the proof can be constructed as follows

(i) Assume $p \wedge (\sim q)$ is true

(ii) On the basis of assumption find some conclusion that is false

(iii) The contradiction is step (ii) leads to the conclusion that $P \wedge \sim q$ is false which proves that $p \to q$ is true.

The method is known as reduction or absurdum or proof by contradiction.

Example. Show that $\sqrt{2}$ is not a rational number

Solution. Let us assume that $\sqrt{2}$ is rational. Then we can find integers such that

$$\sqrt{2} = \frac{p}{q}$$

where p and q have no common factor. (If there are any common factors, we cancel them)

Squaring on both sides, we get

$$2 = \frac{p^2}{q^2} \quad \Rightarrow p^2 = q^2$$

$$\Rightarrow p^2 \text{ is even}$$

$$\Rightarrow p \text{ is even}$$

$$\Rightarrow p = 2m \text{ for some integer } m$$

$$\Rightarrow (2m)^2 = 2q^2$$

$$\Rightarrow 4m^2 = 2q^2$$

$$\Rightarrow q^2 = 2m^2$$

$$\Rightarrow q \text{ is even}$$

Now, p is even, and q is even

$\Rightarrow p$ and q have 2 as the common factor which is a contradiction to the statement that p and q have no common factor.

Hence our assumption $\sqrt{2}$ is rational leads to a contradiction, thus $\sqrt{2}$ is irrational.

Example 2. Suppose that the integers 1, 2, 3, ..., 10 are radomly positioned around a circular wheel. Show that the sum of some set of three consecutively positioned numbers is at least 15.

Solution. (Proof by contradiction)

Let a_r represents the integer at position r on the wheel. Then we are to prove that

$$\left.\begin{array}{l} a_1 + a_2 + a_3 \geq 15 \\ a_2 + a_3 + a_4 \geq 15 \\ \vdots \\ a_{10} + a_1 + a_2 \geq 15 \end{array}\right\} \qquad \ldots(1)$$

where $a_1 + a_2 + \ldots + a_{10} = 1 + 2 + 3 + \ldots + 10$ ...(2)

Let us assume that, the above conclusion (1) is false. Then we must have

$$a_1 + a_2 + a_3 < 15$$

$$a_2 + a_3 + a_4 < 15$$

$$\vdots$$

$$a_{10} + a_1 + a_2 < 15$$

We can write the above inequalities as

$$a_1 + a_2 + a_3 \leq 14$$

$$a_2 + a_3 + a_4 \leq 14$$

$$\vdots$$

$$a_{10} + a_1 + a_2 \leq 14$$

Taking the sum we set

$$3(a_1 + a_2 + \dots + a_{10}) \le 140$$

$$\Rightarrow \quad 3(1 + 2 + \dots + 10) \le 140$$

$$\Rightarrow \quad \frac{3.10.(10+1)}{2} \le 140$$

$$\Rightarrow \quad 3.5.11 \le 140$$

$$\Rightarrow \quad 165 \le 140$$

a contradiction.

Hence the proposition (1) is true.

2.12.4 Proof by Counter Example

To show that $\forall x, p(x)$, it is sufficient to give specific example k, in the universe such that $p(k)$ is false. Where the object k is called a counter example to the assertion $\forall x, p(x)$.

Example. Prove or disprove the statement

If x and y are real numbers then

$$(x^2 = y^2) \Leftrightarrow (x = y)$$

Solution. – 5, 5 are real numbers such that

$$(-5)^2 = 5^2 \quad \text{but} -5 \neq 5$$

$\therefore$ The result is false, hence the implication is false.

2.12.5 Proof by Cases

To prove $p \to q$ by cases, we take p to be in the form $p_1 \vee p_2 \vee \dots \vee p_n$ by proving separately, each of the following $p_1 \to p_2, p_2 \to q, \dots, p_n \to q$, we can establish $(p_1 \vee p_2 \vee \dots \vee p_n) \to q$.

SOLVED EXAMPLES

Example 1. Prove that the argument given below is valid

$$\begin{array}{l} p \to r \\ \sim p \to q \\ \underline{q \to s} \\ \therefore \quad \sim r \to s \end{array}$$

Solution. Consider $(p \to r) \wedge (\sim p \to q) \wedge (q \to s)$

now $\quad (p \to r) \wedge (\sim p \to q) \wedge (q \to s)$

$\Rightarrow \quad (p \to r) \wedge (\sim p \to s)$ (by Transitive rule)

$\Leftrightarrow \quad (\sim r \to \sim p) \wedge (\sim p \to s)$ (Using law of Contrapositive)

$\Rightarrow \ \sim r \to s$ (by transitive rule)

$\therefore$ The argument is valid

Example 2. Show that the argument is valid.

$$\begin{array}{ll} & p \to r \\ & q \to r \\ \hline \therefore & (p \vee q) \to r \end{array}$$

Solution. We have

$$\begin{array}{lll} & (p \to r) \wedge (q \to r) & \\ \Leftrightarrow & (\sim p \vee r) \wedge (\sim q \vee r) & \\ \Leftrightarrow & (r \vee \sim p) \wedge (r \vee \sim q) & \text{(by commutative law)} \\ \Leftrightarrow & r \vee (\sim p \wedge q) & \text{(by distributive law)} \\ \Leftrightarrow & (\sim p \wedge \sim q) \vee r & \text{(by commutative law)} \\ \Leftrightarrow & \sim (p \vee q) \vee r & \text{(Demorgan's law)} \\ \Leftrightarrow & (p \vee q) \to r & \end{array}$$

$\therefore$ The argument is valid.

Example 3. Prove the validity of the argument

All men are fallible

All kings are men

Therefore, all kings are fallible

Solution. Let $M(x)$: x is a man

$K(x)$: x is a king

$F(x)$: x is a fallible

The proof of validity is as follows.

Statements	Reasons
1. $\forall\ x, [M(x) \to F(x)]$	Premise 1
2. $M(\alpha) \to F(\alpha)$	Step 1 and US (universal specification)
3. $\forall\ x, [K(x) \to M(x)]$	Premise 2.
4. $K(\alpha) \to M(\alpha)$	Step 3 and US
5. $K(\alpha) \to F(\alpha)$	Steps 2 and 4 Transitive rule
6. $\forall\ x, [K(x) \to F(x)]$	Step 5 and *ES* (existential specification)

Example 4. Prove the validity of the argument every living thing is a planet or an animal Dawood's dog is alive and it is not a plant all animals have hearts.

Hence, Dawood's dog has a heart

(It is given universe consists all living things)

Solution. $P(x)$: x is a plant

$A(x)$: x is an animal

$H(x)$: x has a heart

α : Dawood's dog

The proof of validity is given below

Statements	Reasons
1. $\forall x, [P(x) \vee A(x)]$	Premise
2. $\sim P(\alpha)$	Premise
3. $P(\alpha) \vee A(\alpha)$	from (1) and US
4. $A(\alpha)$	from (2) and (3) and disjunctive syllogism
5. $\forall x, [A(x) \to H(x)]$	Premise
6. $A(\alpha) \to H(\alpha)$	(5) and specification
7. $H(\alpha)$	(4), (6) and Modus Ponens

Example 5. Prove the conclusion

If Ali does not study, then he will fail

Ali did not did not fail

$\therefore$ Ali studied

Solution. Let p : Ali studied

q : Ali failed

Statement	Reasons
1. $\sim p \to q$	Premise
2. $\sim q$.	Premise
3. $\sim q \to p$	(1), law of contra positive
4. p	(2), (3) law of detachment

$\therefore$ Ali studied

It is of the form

$$\begin{array}{c} \sim p \to q \\ \sim q \\ \hline p \end{array}$$

$\therefore$ a valid conclusion

Example 6. Prove that the argument is valid. If I drive to work, then I will arrive tired.

I am not tired.

$\therefore$ I do not drive to work

Solution. Let p : *I* drive to work

q : *I* arrive late

Then the argument is of the form

$$\begin{array}{r} p \to q \\ \sim p \\ \hline \therefore \quad \sim q \end{array}$$

by Modus Tollers rule, the argument is valid. A formal proof is given below

Statement	Reasons
1. $p \to q$	Premise
2. $\sim q \to \sim p$	law of contra positive
3. $\sim q$	premise
4. $\sim p$	by(2), (3) and Modus ponens (i.e. law of detachment)

Example 7. Show that the argument is valid

$$\begin{array}{l} \sim p \to (q \to \sim r) \\ \sim p \vee s \\ \sim t \to q \\ \sim s \\ \hline \therefore \quad r \to t \end{array}$$

Solution.

Statement	Reasons
1. $\sim p \vee s$	Premise
2. $\sim s$	Premise
3. $\sim p$	by (1), (2) and disjunction syllogism
4. $\sim p \to (q \to \sim r)$	Premise
5. $q \to \sim r$	by (3), (4) and law of detachment
6. $\sim t \to q$	premise
7. $\sim t \to \sim r$	by (5), (6) and transitive rule (Hypothetical syllogism)
8. $r \to t$	by (7) and law of contra positive

Example 8. Test whether the following argument is valid

$$\begin{array}{c} p \to q \\ r \to s \\ p \vee r \\ \hline \therefore \quad q \vee s \end{array}$$

Solution. Consider $(p \to q) \wedge (r \to s) \wedge (p \vee r)$

now $\quad (p \to q) \wedge (r \to s) \wedge (p \vee r) \Leftrightarrow (p \to q) \wedge (r \to s) \wedge (\sim p \to r)$

$\Leftrightarrow \quad (p \to q) \wedge (\sim p \to r) \wedge (r \to s)$ (by commutative law)

$\Rightarrow \quad (p \to q) \wedge (\sim p \to s)$ (by transitive rule i.e. rule of syllogism)

$\Leftrightarrow \quad (\sim q \to \sim p) \wedge (\sim p \to s)$ (by law of contra positive)

$\Rightarrow \quad \sim q \to s$ (by transitive rule)

$\Leftrightarrow \quad q \vee s$

$\therefore$ The argument is valid

Example 9. If the product of two integers a and b is even then, show that a is even or b is even

Solution. Let p : ab is even

q : a is even or b is even

we are to prove $p \to q$.

we assume $\sim q$ and prove the contra positive $\sim q \to \sim t$. Assume $\sim q$

we have $\sim q$: a is odd and b is odd (De Morgan's law)

$\Rightarrow \quad a = 2m + 1$, and $b = 2n + 1$, where m and n are integers

$$\therefore \quad ab = (2m + 1)(2n + 1)$$

$$= 4mn + 2m + 2n + 1$$

$$= 2(2mn + m + n) + 1$$

$\Rightarrow ab$ is odd

$\therefore \quad \Rightarrow \sim p$

$\therefore \quad \sim q \to \sim p$

hence $\quad p \to q$

Example 10. Given the following sentences. Either Prashant joined the football team, or he signs in the chorus. If he sings in the chorus, he will need a white shirt. If he needs a white shirt, then he can not save his money, Prashant saves his money. Show that Prashant joined the football team

Let p : "Prashant joined the foot ball team"

q : "He sings in the chorus"

r : "He needs a white shirt"

s : "He saves his money"

The proof is given below

1. $p \vee q$	Premise
2. $q \to r$	Premise
3. $r \to \sim s$	Premise
4. s	Premise
5. $q \to \sim s$	(2), (1) chain rule

6. $\sim q$ (5), (4) Moduls Tollery
7. p (1), (6), disjunctive syllogism

$\therefore$ Prashant joined the football team

Hence proved

EXERCISE 2.2

1. Show that $\sim Q, P \rightarrow Q \Rightarrow \sim P$
2. Prove that R is a valid inference from the premises, $P \rightarrow Q$, $Q \rightarrow R$, and P
3. Show that $R \vee S$ follows logically from the premises $(\vee D, (C \vee D) \rightarrow \sim H$, $\sim H \rightarrow (A \wedge \sim B)$ and $(A \wedge \sim B) \rightarrow (R \wedge S)$
4. Show that $R \wedge (P \vee Q)$ is a valid conclusion from the premises $P \vee Q$, $Q \rightarrow R$, $P \rightarrow M$ and $\sim M$
5. Show that the following premises are inconsistent
   i. If Jack misses many classes through illness, then he fails in high school
   ii. If Jack fails high school, then he is uneducated
   iii. If Jack reads a lot of books, then he is not uneducated
   iv. Jack misses many classes through illness and reads a lot of books
6. Prove that $\overset{s}{\Rightarrow} \sim Q \wedge (P \rightarrow Q) \rightarrow \sim P$
7. Show that $S \vee R$ is tautologically implied by $(P \vee Q) \wedge (P \rightarrow R) \wedge (Q \rightarrow S)$
8. Show that the following premises are inconsistent $P \rightarrow Q$, $P \rightarrow R$, $Q \rightarrow \sim R$, P
9. Show that $(x)\ (P(x) \vee Q(x)) \Rightarrow x\ P(x) \vee (\exists x)\ P(x)$
10. Prove that $(\exists x)\ (P(x) \wedge Q(x)) \Rightarrow (\exists x\ P(x) \wedge (\exists x)\ Q(x)$
11. Show that from
    (i) $(\exists x)\ (F(x) \wedge S(x)) \rightarrow (y)\ (M(y) \rightarrow W(y))$
    (ii) $(\exists y)\ (M(y) \wedge \sim W(y))$
    the conclusion $(x)\ (F(x) \rightarrow \sim S(x))$ follows.
12. Show that $\sim P(a, b)$ follows logically from
    $(x)\ (y)\ (P(x, y) \rightarrow W(x, y))$ and $\sim W(a, b)$
13. Show that $P(x) \wedge Q(x) \Rightarrow (\exists x)\ (P(x) \wedge Q(x))$
14. Show that at least 4 of 22 days must fall on the same day of the week (Prove by contradiction)
15. Use a proof by cases to show that
    $|xy| = |x|\ |y|$ where x and y are real numbers
16. Give a proof by contradiction of the theorem "If $3n + 2$ is odd, then n is odd"
17. Show that the statements given below are equivalent

p: n is an even integer
q: $n - 1$ is an odd integer
r: n^2 is an even integer

18. Let $P(x)$: $x^2 > 0$. If the universe of discourse consists of all integers, then show by counter example the statement $\forall x\, P(x)$ is false

19. Give a direct proof of the theorem
"If n is an odd integer, then n^2 is an odd integer"

20. Prove that the sum of two rational numbers is rational

21. Let $P(n)$ be "If x and y are positive integers with $x \geq y$, then $x^n \geq y^n$. Show that $P(0)$ is true.

22. Prove that the following arguments are valid 3

(a) $p \to q$
$q \to r$
p
$\therefore\ r$

(b) $p \to q$
$r \to \sim q$
r
$\therefore\ \sim p$

(c) $p \wedge q$
$p \to (q \to r)$
$\therefore\ r$

(d) $p \to q$
$\sim p$
$\therefore\ \sim p$

(e) p
$p \to \sim q$
$r \to q$
$\therefore\ \sim r$

(f) $p \to q$
$r \to s$
$\sim q \vee \sim s$
$\therefore\ \sim (p \wedge r)$

(g) $p \to q$
$r \to s$
$\sim q \vee \sim s$
$\therefore\ \sim p \vee \sim r$

(h) $p \to q$
$\sim q \to p$
$\sim r$
$\therefore\ q$

(i) $p \to (q \to r)$
$\sim q \to \sim p$
p
$\therefore\ r$

23. Test the validity of the following arguments

(a) If a man is bachelor, he is unhappy
If a man is unhappy, he dies young
∴ Bachelors die young

(b) If a person is poor, he is unhappy
If a person is unhappy, he dies young
∴ Poor person die young

(c) Babies are illogical
Nobody is despised who can manage a crocodile
illogical people are despised
∴ Babies can not manage crocodiles

(d) If it rains, John will be sick
John was not sick

It did not rain

(e) If there is a strike by students, the examination will be postponed
There was no strike by the students

∴ The examination was not postponed

(**Ans.** (e) is not valid, (a), (b), (c), (d) are valid)

24. Show that $R \vee S$ follows logically from the premises $C \vee D$, $(C \vee D) \rightarrow \sim H$, $\sim H \rightarrow (A \wedge \sim B)$ and $(A \wedge \sim B) \rightarrow R \vee S$

25. Show that the argument given below is valid

$$\begin{array}{c} p \rightarrow q \\ q \rightarrow s \\ r \rightarrow \sim s \\ \sim p \veebar r \\ \hline \therefore \sim p \end{array}$$

26. Prove the validity of the following arguments

(a)
$$\begin{array}{c} p \rightarrow r \\ r \rightarrow s \\ t \vee \sim s \\ \sim t \vee u \\ \sim u \\ \hline \therefore \sim p \end{array}$$

(b)
$$\begin{array}{c} u \rightarrow r \\ (r \rightarrow s) \rightarrow (p \vee t) \\ q \rightarrow (u \wedge s) \\ \sim t \\ \hline \therefore q \rightarrow p \end{array}$$

27. Given the premises: If John saves enough money, then he will buy a computer
John cannot buy a computer prove the conclusion: John did not save enough money.

2.13 NATURAL NUMBERS

2.13.1 In this section, we examine the set of natural numbers and introduce; the methods of generating such set.

The natural numbers (including zero) are those objects which can be generated by starting with an initial object 0 (zero) and from any object n already generated passing to another uniquely determined object n^+; the successor of n. The objects differently generated are always distinct. Thus the natural numbers appear as a set of objects.

$$0^+, 0^{++}, 0^{+++}, \ldots$$

The transition to the usual notation is made upon introducing 1, 2, 3, ... 9, to stand for 0^+, 0^{++}, 0^{+++}, ... $0^{+++++++++}$ and employing the notation. The set of Natural is denoted by N.

The set of natural numbers can also be generated by starting with usual null set ϕ and notion of a successor set.

If A is a given set; then the successor set of A is the set $AU\{A\}$. It is denoted by A^+.

$$\therefore \qquad A^+ = AU\{A\}$$

Let ϕ be a null set, then find the successor sets of ϕ, these sets are:

$$\phi, \phi^+ = \phi\ U\ \{\phi\};\ \phi^{++} = \phi\ U\ \{\phi\}\ U\ \{\phi\ U\ \{\phi\}\}$$

They can also be written as ϕ, $\phi^+ = \{\phi\}$, $\phi^{++} = \{\phi, \{\phi\}\}$, renaming the set ϕ as 0 (zero),

$$\phi^+ = 0^+ = \{\phi\} = 1$$

$\phi^{++} = 1^+ = \{\phi, \{\phi\}\} = \{0, 1\} = 2, \ldots$ we get the set $\{0, 1, 2, 3, \ldots\}$ each element in the above set is a successor of previous element; except for the element 0 (0 is not the successor of any element).

2.13.2 Peano's axioms: The set of natural number has the following properties; which are called Peano's axioms for the natural number system.

1. $0 \in N$ (where $0 = \phi$)
2. If $n \in N$; then $n^+ \in N$; where $n^+ = nU\ \{n\}$
3. If a subset $S \subseteq N$, possesses the properties
   (a) $0 \in S$; and
   (b) if $n \in S$; and $n^+ \in S$ then $S = N$

Property (3) is known as the minimality property; which is the basis for the principle of Mathematical induction.

2.13.3 Mathematical Induction: In this section we describe the principle of induction which is used as a method of proof. We know that, Induction is a very valuable mathematical and Scientific tool. It can lead us to probable results which may be worth the effect of trying to prove deductively.

There are two forms of Mathematical Induction:

1. The principle of weak Mathematical Induction.
2. The principle of strong Mathematical Induction.

Many problems of Mathematical Induction may first be visualized by examining a number of special cases and seeing a general pattern emerge. From this is a general result may be conjuctured; we now state principles.

2.13.4 The Principle of Mathematical Induction (Weak Induction)

Let $P(n)$ stand for "the natural number n has the property P". To prove $P(n)$ is true for all integers $n \geq 0$; it suffices to prove:

1. $P(0)$ is true
2. For all $m \geq 0$; $P(m)$ implies $P(m + 1)$

We have 3 steps to a proof using the principle of Mathematical Induction.

(i) Basis of Induction

(ii) Induction Hypothesis

(iii) Induction step

In the basis step we show that $P(n_0)$ is true. In the Induction Hypothesis; we assume that $P(m)$ is true for $m \geq n_0$; and finally we show that $P(m + 1)$ is true on the basis of Induction Hypothesis.

2.14 SOLVED EXAMPLES

Example 1: Show that $2^{3n} - 1$, is an Integral Multiple of 7.

Solution: Let $P(n)$ be the statement "$2^{3n} - 1$, is an integral multiple of 7". Let us verify this by Mathematical Induction.

1. Basis of Induction: for $n = 0$; $P(0) = 2^0 - 1 = 1 - 1 = 0 = 0.7$
   $\therefore$ $P(0)$ is an integral multiple of 7 i.e. $P(0)$ is true.
2. Inductive Hypothesis: For some arbitrary choice $m \in N$; assume that $P(m)$ holds; i.e. $P(m)$ is an integral multiple of 7 and $P(m) = 2^{3m} - 1 = 7x$ for some integer x.
3. Inductive step: Now we show on the basis of Induction Hypothesis $P(m + 1)$ holds.

$$\begin{aligned} P(m+1) &= 2^{3(m+1)} - 1 \\ &= 2^{3m+3} - 1 \\ &= 2^{3m}.2^3 - 1 \\ &= 2^{3m}.8 - 8 + 7 \\ &= 8\,.\,7x + 7 \\ &= 7(8x + 1) \end{aligned}$$

$\therefore$ $P(m + 1)$ is an integral multiple of 7.

$\therefore$ $P(m) \Rightarrow P(m + 1)$

Hence, from the principle of Mathematical Induction; $P(n)$ is true for all $n \in N$.

Example 2: Show $P(n) : n < 2^n$

Solution: Let $P(n)$: $n < 2^n$

1. Basis of Induction: for $n = 0$; $P(0)$: $0 < 2^0 = 1$

   $\therefore$ P(0) is true.

2. Inductive Hypothesis: For some arbitrary choice $m \in N$; assume that $P(m)$ holds; that is $P(m)$: $m < 2^m$.

3. Inductive step: Consider $P(m)$: $m < 2^m$. Adding 1 to both sides; we get $m + 1 < 2^m + 1 < 2^m + 2^m = 2^{m+1}$.

   $\Rightarrow$ $P(m + 1)$; $(m + 1) < 2^{m+1}$

   $\therefore$ $P(m) \Rightarrow P(m + 1)$ that is $P(m + 1)$ is true.

Hence, from the principle of Mathematical Induction $P(n)$ is true for all $n \in N$.

Example 3: Show that $1 + 2 + 3 + \ldots + n = \dfrac{n(n+1)}{2}$ for all positive integers $n \geq 1$.

Solution: Let $P(n)$; $1 + 2 + \ldots + n = \dfrac{n(n+1)}{2}$

1. Basis Induction: for $n = 1$; $P(1) = 1 = \dfrac{1(1+1)}{2}$. The formula is true for $n = 1$; i.e. $P(1)$ is true.

2. Inductive Hypothesis: Assume that the statement $P(n)$ is true for $n = m$; i.e.

$$P(m): 1 + 2 + \ldots + m = \frac{m(m+1)}{2}$$

3. Inductive step: We have $P(m + 1)$; $1 + 2 + \ldots + m + (m + 1) = P(m) + (m + 1)$

Since $P(m) = \dfrac{m(m+1)}{2}$ by Inductive Hypothesis.

$$P(m+1):\quad P(m) + (m+1) = \frac{m(m+1)}{2} + (m+1)$$

$$= \frac{m(n+1)}{2}\left(\frac{m}{2} + 1\right)$$

$$= \frac{(m+1)(m+2)}{2}$$

$$\Rightarrow \qquad P(m+1) = \frac{(m+1)(m+2)}{2}$$

$\Rightarrow$ The formula holds for $n = m + 1$

$\therefore \qquad P(m) = P(m + 1)$

Hence, by principle of Mathematical Induction $P(n)$ is true for all positive integers $n \geq 1$.

Example 4: By the principle of Mathematical Induction prove that if $3^{2n} - 1$ is a multiple of 8 where $n \geq 1$; $n \in N$ ($\mu(8)$ denotes "a multiple of 8").

Solution: Let $P(n)$ be the statement $P(n)$: $3^{2n} - 1 = \mu(8)$

Step 1: When $n = 1$

$P(1)$. $3^{2 \cdot 1} - 1 = 3^2 - 1 = 9 - 1 = 8 = \mu(8)$

$\therefore$ P(1) is true

Step 2: Assume that the statement is true for $n = m$; that is $p(m)$; $3^{2m} - 1 = \mu(8)$.

Step 3: Consider $P(m + 1)$; $3^{2(m+1)} - 1$

Now $\quad 3^{2(m+1)} - 1 = 3^{2m+2} - 1 \quad = 3^{2m} . 3^2 - 1 = 3^{2m} . 9 - 1$

$$= 3^{2m} - 9 - 9 + 8$$

$$= 9 (3^{2m} - 1) + 8$$

$$= 9 (\mu(8)) + \mu (8)$$

(from step (2): $2^{2m} - 1 = \mu(8)$ and $8 = \mu(8)$)

Hence, $P(m + 1)$ is true.

From the principle of Mathematical Induction $P(n)$ is true for all positive integers $n \geq 1$.

Example 5: Show that $n^3 + 2n$ is divisible by 3.

Solution: Let $P(n) : n^3 + 2n$ is divisible by 3.

Step 1: $P(0) = 0$ is divisible by 3.

$\therefore P(0)$ is true.

Step 2: Let us assume that for any m, $p(m)$ is true i.e. $m^3 + 2m$ is divisible by 3.

Step 3:

$p(m + 1)$; $(m + 1)^3 + 2(m + 1) \qquad = m^3 + 3m^2 + 3m + 1 + 2m + 2$

$$= (m^3 + 2m) + 3 (m^2 + m + 1)$$

$m^3 + 2m$; is divisible by 3 and

$3(m^2 + m + 1)$ is divisible by 3

$\therefore$ $(m^3 + 2m) + 3\,(m^2 + m + 1)$ is divisible by 3

$\Rightarrow$ $(m + 1)^3 + 2(m + 1)$ is divisible by 3

$\Rightarrow$ $P(m + 1)$ is true.

$\therefore$ Hence, $p(n)$ is true for all $n \in N$.

Example 6: If x and y are two distinct integers then show that $x^n - y^n$ is divisible by $(x - y)$.

Solution: Let $P(n)$ be the statement.

$P(n)$: $x^n - y^n$ is divisible by $(x - y)$

Step 1: $P(0)$: 0 is divisible by $(x - y)$

$\Rightarrow$ $P(0)$ is true.

Step 2: Let us suppose that $p(m)$ is true that is; for any $m \in N$;

$p(m)$; $x^m - y^m$ is divisible by $(x - y)$

Step 3: Now we are to show that $p(m + 1)$ is true.

We have $p(m + 1)$: $x^{m+1} - y^{m+1}$

consider $x^{m+1} - y^{m+1} = x^{m+1} - x^m y + x^m y - y^{m+1}$

$= x^m\,(x - y) + y(x^m - y^m)$

$x^m(x - y)$ is divisible by $(x - y)$ and (by step 2) $x^m - y^m$ is divisible by $(x - y)$

$\therefore$ $x^m(x - y) + y(x^m - y^m)$ is divisible by $(x - y)$

$\Rightarrow$ $x^{m+1} - y^{m+1}$ is divisible by $(x - y)$

$\Rightarrow$ $P(m+1)$ is true.

Hence, by the principle of Mathematical Induction $p(n)$ is true for all natural numbers n.

2.15 STRONG MATHEMATICAL INDUCTION

Let $p(n)$ be a statement which, for each integer n; may be either true or false. Then $p(n)$ is true for all natural numbers; if there exists a natural number $r \geq 0$ such that:

(1) $P(0), P(1), \ldots P(r)$ are all true.

(2) where $m \geq r$; the assumption that $p(i)$ is true for all natural number $0 \leq i \leq m$ implies that $P(m + 1)$ is true.

There are 3 steps to proof by strong Mathematical Induction:

1. Basis of Induction: Show $P(0)$, $P(1)$ ... $P(r)$ are all true.
2. Strong Inductive Hypothesis: Assume $P(i)$ is true for i; such that $0 \le i \le m$; where $m \ge r$.
3. Inductive step: Show that $P(m + 1)$ is true on the basis of the strong inductive hypothesis.

Example: If F_n is the nth Fibonacci number; then show that

$$F_n = \frac{1}{\sqrt{5}}\left[\left(\frac{1+\sqrt{5}}{2}\right)^{n+1} - \left(\frac{1-\sqrt{5}}{2}\right)^{n+1}\right]$$

for all itegers $n \ge 0$

Proof: 1. Basis of Induction: when $n = 0$

$$F_0 = 1 = \frac{1}{\sqrt{5}}\left[\frac{1+\sqrt{5}}{2} - \frac{1-\sqrt{5}}{2}\right]$$

when $n = 1$

$$F_1 = 1 = \frac{1}{\sqrt{5}}\left[\left(\frac{1+\sqrt{5}}{2}\right)^{2} - \left(\frac{1-\sqrt{5}}{2}\right)^{2}\right]$$

1. The formula F_n holds for $n = 0$; and $n = 1$
2. Strong Inductive Hypothesis: for $n \ge 1$; assume

$$F_m = \frac{1}{\sqrt{5}}\left[\left(\frac{1+\sqrt{5}}{2}\right)^{m+1} - \left(\frac{1-\sqrt{5}}{2}\right)^{m+1}\right]$$

for each integer m; where $= 0 \le m \le n$.

3. Inductive step:

Let $\alpha = \frac{1+\sqrt{5}}{2}$ and $\beta = \frac{1-\sqrt{5}}{2}$

$$\Rightarrow \alpha^2 = \left(\frac{1+\sqrt{5}}{2}\right)^2 = \frac{1+5+2\sqrt{5}}{2} = \frac{3+\sqrt{5}}{2}$$

similarly;

$$\beta^2 = \frac{3-\sqrt{5}}{2}$$

$$\alpha + 1 = \frac{1+\sqrt{5}}{2} + 1 = \frac{3+\sqrt{5}}{2}$$

$\Rightarrow \alpha + 1 = \alpha^2$

Similarly $\beta + 1 = \beta^2$

Now; $F_n = \frac{1}{\sqrt{5}}\left[\alpha^{n+1} - \beta^{n+1}\right]$

$$= F_{n-1} = \frac{1}{\sqrt{5}}\left[\alpha^n = \beta^n\right]$$

We know that

$$F_{n+1} = F_n + F_{n-1}$$

$$= \frac{1}{\sqrt{5}}\left[\alpha^{n+1} - \beta^{n+1}\right] + \frac{1}{\sqrt{5}}[\alpha^n - \beta^n]$$

$$= \frac{1}{\sqrt{5}}\left[\alpha^{n+1} + \alpha^n - \beta^{n+1} + \beta^n\right]$$

$$= \frac{1}{\sqrt{5}}\left[\alpha^n (\alpha + 1) - \beta^n (\beta + 1)\right]$$

$$= \frac{1}{\sqrt{5}}\left[\alpha^n . \alpha^2 - \beta^n . \beta^2\right]$$

$$= \frac{1}{\sqrt{5}}\left[\alpha^{n+2} - \beta^{n+2}\right]$$

$\Rightarrow$ F_{n+1} is true.

$\therefore$ Hence, proved.

EXERCISE 2.3

1. Prove that $n(n + 1)(n + 2)$ is divisible by 6 wherever n is a natural number.
2. Show that $2n < m' \; \forall n \geq 4$.
3. For each integer $n \geq 5$; show that $2^n > n^{-2}$.
4. By the principle of Mathematical Induction prove that:

   (a) $1^2 + 2^2 + \ldots + n^2 = \frac{n(n+1)(2n+1)}{6}$

   (b) $1 + 4 + 7 + \ldots + (3n - 2) = \frac{n(3n-1)}{2}$

   (c) $\frac{1}{1.3} + \frac{1}{3.5} + \frac{1}{5.7} + \ldots + \frac{1}{(2n+1)(2n+1)} = \frac{n}{2n+1}$
5. Prove that $1 + 3 + 5 + \ldots + (2n - 1) = n^2 \; \forall n \in N$.

6. Prove that $\sum_{r=1}^{n}(2n-1) = n^2$

7. By the principle of Mathematical Induction show that $1^3 + 2^3 + \ldots + n^3$

$= \dfrac{n^2(n+1)^2}{4} \quad \forall n \in N.$

CHAPTER 3

Set Theory

3.1 INTRODUCTION

In this chapter we study sets, relations, functions, partially ordered sets, and studies The notion of a set is elementary to all of mathematics and every branch of a mathematics can be considered a study of objects of one kind or another. Cantor was the founder of the theory of sets. The word set is a primitive term and is reqarded as one of the basic undefined terms of mathematics. The set notion is important in programming. We begin our study with sets and their operations.

3.2 SETS

3.2.1 Set

Definition 3.1. A set is a well defined collection of objects.

In the above definition the words set and collection for all practical purposes are synonymous. We have already used the word set to define itself.

3.2.2 Notation

Each of the object in the set is called a member or an element of line set. The objects themselves can be almost anything. Elements of a set are usually denoted by lower-case letters, while sets are denoted by capital letters of English Alphabet.

The symbol $\in$ indicates the membership in a set. If x is an element of a set A, then we write $x \in A$. The symbol $\in$ is read "is a member of", or "is an elements of". The symbol $\notin$ is used to indicate that an object is not in the given set. The symbol $\notin$ is read as "is not a member of" or "is not an element of".

If x is not an element of a set A, then we write $x \notin A$.

3.2.3 Specifying Sets

One way of describing a set is to list all the members of the set be between braces.

Thus $\{a, b, c\}$ represents a set.

This method is called 'the listing method'. It is also called tabulation method. In this method, the order in which the elements are listed is immaterial and the method is used for smaller sets.

Another method of defining sets is by a description of some attribute or characteristic of the element of the set. This method is more general and involves a description of the set property.

$$A = \{x | x \text{ has the property } P\}$$

designates "the set A of all objects x such that n has the property P. The vertical bar '|' is read as "such that" the set can also be written as

$$A = \{x{:}x \text{ has the property } P\}$$

in which the symbol ':' is read as 'such that.'

This notation is called set-builder notation.

Example. (i) $A = \{x | x \text{ is complex member}\}$

(ii) $B = \{x | x \text{ is a positive integer greater than } 5\}$

We can describe a set by its characteristic function.

Example. $\mu A(x) = \begin{cases} 1 \text{ if } x \in A \\ 0 \text{ if } x \notin A \end{cases}$

In another method we describe the set by a recursive formula.

Example. Let $x_0 = 2, \quad x_1 = 1, \quad x_{i+1} = x_i + x_{i-1}, \quad i \geq 1$

and $A = \{x_i : i \geq 0\}$

We can also describe a set by an operation on some other set.

3.3 SUBSETS

Definition 3.2. If every element of a set A is also an element of a set B, then A is called a subset of the set B. If A is a subset of the set B, then we say that A is contained in B and write

$$A \subseteq B$$

Symbolically. If $x \in A \Rightarrow x \in B$, then $A \subseteq B$

Example. $A = \{1, 5, 7\}$, is subset of $B = \{1, 2, 3, 4, 5, 6, 7, 8\}$

If A is not a subset of the set B, then we write

$$A \not\subseteq B$$

Example. If $A = \{2, 4, 8\}$ and $B = \{1, 2, 3, 4, 5, 6, 7\}$

then $\quad A \subseteq B.$

$8 \in A$ but $8 \notin B$

Note:

(i) If A is a subset of B, it is possible that $A = B$, to emphasize this fact we write $A \subseteq B$.

(ii) If A is contained in B, then we may also state that B contains and write $B \supseteq A$.

(iii) Every set is a subset to itself, but no set is a member of itself

(iv) Empty set ϕ is a subset of every set

3.3.1 Proper Subset

Definition 3.3. A set A is called a proper subset of the set B, if

(i) A is a subset of B, and

(ii) B is not a subset of A

that is, A is said to be a proper subset of B, if every element of A belongs to the set B: but there is at least one element of B, which is not in A. If A is a proper subset of B, then we denote it by $A \subset B$.

Note: Every set A is a subset to itself and ϕ is a subset of A. The subsets ϕ and A are called improper subsets of A.

3.3.2 Equal Sets

Definition 3.4. If A and B are sets such that every element of A is an element of B, and every element B is also an element of A, then A and B are said to be equal sets.

If A and B are equal then we write '$A = B$'

$A = B$, is read as A and B are identical

Example. $A = \{2, 4, 6\}$. $B = \{x \in Z \mid 1 \leq x \leq 7, x \text{ is even}\}$

3.3.3 Super Set

Definition 3.5. If A is a subset of B, then B is called a super set of A.

Example. If $A = \{0, 1, 3\}$ $B = \{0, 1, 2, 3, 4\}$

B is a super set of A

3.4 NULL SET

Definition 3.6. The set with no elements in it is called an empty set or null set.

A null set is denoted by the symbol ϕ.

The null set is a subset of every set i.e. if A is any set then $\phi \subseteq A$.

Example. The set of real roots of $x^2 + 7 = 0$

3.5 SINGLETON

Definition 3.7. A set having only one element is called a singleton.

Example. $A = \{5\}$.

Theorem 3.1. Two sets A and B are equal only if and if $A \subseteq B$ and $B \subseteq A$

Proof. If $A = B$, then every elements of A is a member of B and every elements of B is a member of A

$$\therefore \qquad A \subseteq B \text{ and } B \subseteq A$$

Conversely; let us suppose that $A \subseteq B$, $B \subseteq A$ and $A \neq B$.

Since, $A \neq B$, there is either an element of A that is not in B or there is an element of B. That is not in A.

But $A \subseteq B$, so every element in A is in B since $B \subseteq A$, every element of B is in A. Therefore, our assumption that $A \neq B$, leads to a contradiction.

Hence, $\qquad A = B$.

Theorem 3.2. The empty set is unique.

3.6 FINITE SET

Definition 3.8. A set is said to be finite if it has a finite number of elements.

Example $\qquad \{2, 4, 6, 8\}$

If A is a finite set with n elements in it, then n is called the cardinality of A. If n is the cardinality of A, we write $|A| = n$.

3.7 INFINITE SET

Definition 3.9. A set infinite, if it not finite.

Example. The set of natural numbers.

3.8 UNIVERSAL SET

Definition 3.10. In many discussions all the sets under consideration to be the subsets of one particular set. This set is called the universal set for that discussion and is denoted by U.

The universal set also denoted by μ(or X)

Example. If $A = \{0, 1, 3\}$, $B = \{2, 4, 5\}$, $C = \{6, 8, 9\}$, then the universal set is

$$U = \{0,1, 2, 3, 4, 5, 6, 7, 8, 9\}$$

3.9 POWER SET

Definition 3.11. The set of all subsets of A is called the power set of A.

The power set of A is denoted by $P(A)$ or by 2^A.

If A has n elements in it, then $P(A)$ has 2^n elements.

Example. If $A = \{1, 2\}$, then

$$P(A) = \{\phi, \{1\}, \{2\}, A\}$$

Note. A set is never equal to its power set.

3.10 DISJOINT SETS

Definition 3.12. Two sets A and B are said to be disjoint, if they have no elements in common.

Example. The sets $A = \{1, 3, 5\}$, and $B = \{2, 4, 6\}$ are disjoint.

3.11 PROPERTIES OF SET CONTAINMENT

Theorem 3.3. If A is any set then A contains A

Proof. If $x \in A$, then $x \in A$, by the repetition of the statement,

$$\therefore \quad A \text{ contains } A$$

Theorem 3.4. If $A \subseteq B$ and $B \subseteq C$, then $A \subseteq C$, where A, B and C are sets.

Proof. Let $x \in A$

$$\therefore \quad x \in A \Rightarrow x \in B \quad (\because \ A \subseteq B)$$

$$\Rightarrow x \in C \quad (\because B \subseteq C)$$

$$\therefore \quad x \in A \Rightarrow x \in C$$

$$\therefore \quad A \subseteq C$$

3.12 OPERATIONS ON SETS

3.12.1 Union of Sets

Definition 3.13. Let A and B be sets. The union of sets A and B is the set that contains those elements that are either in A or B or in both.

The union of the sets A and B is denoted by $A \cup B$ and is read as "A union B".

Symbolically: $A \cup B = \{x : x \in A \text{ or } x \in B\}$

or

$$A \cup B = \{x : x \in A \vee x \in B\}$$

Example. If $A = \{2, 5, 7\}, B = \{1, 3, 5, 8\}$

Then $A \cup B = \{1, 2, 3, 5, 7, 8\}$

3.12.2 Union of More Than Two Sets

Definition 3.14. If $A_1, A_2, ..., A_n$ denote sets, then the union of these sets denoted by

$\bigcup_{i=1}^{n} A_i$ is defined as the set $\bigcup_{i=1}^{n} A_i = \{x \mid x \in A_i \text{ for at least one } i\}$

3.13 PROPERTIES OF UNION OPERATION

Theorem 3.5. The following properties hold for the union of sets A and B

(a) $A \cup A = A$ (Idempotent law)

(b) $A \cup B = B \cup A$ (Commutative law)

(c) $(A \cup B) \cup C = A \cup (B \cup C)$ (Associative law)

(d) $A \cup \phi = A$ (identity law)

(e) $A \cup U = U$ (identity law)

U represents the universal set.

Proof.

(a) $A \cup A = \{x : x \in A \text{ or } x \in A\}$

$= \{x : x \in A\} = A$

$\therefore A \cup A = A$

(b) $A \cup B = \{x : x \in A \text{ or } x \in B\}$

$= \{x : x \in B \text{ or } x \in A\}$

$= B \cup A$

$\therefore A \cup B = B \cup A$

i.e. union of sets is commutative.

(c) $A \cup (B \cup C) = \{x : x \in A \text{ or } x \in B \cup C\}$

$= \{x : x \in A \text{ or } (x \in B \text{ or } x \in C\}$

$= \{x : x \in A \text{ or } x \in B \text{ or } x \in C\}$

$= \{x : (x \in A \text{ or } x \in B) \text{ or } x \in C\}$

$= \{x : x \in A \cup B \text{ or } x \in C\}$

$= (A \cup B) \cup C$

$\therefore$ union of sets is associative

(d) $A \cup \phi = \{x : x \in A \text{ or } x \in \phi\}$

$= \{x : x \in A\}$ ($\because \phi$ has no element in it)

$= A$

$\therefore A \cup \phi = A$

(e) clearly

$A \cup U \subseteq U$: (since every set is a subset of universal set) ...(1)

Let $x \in U$ than $x \in U \Rightarrow x \in A \text{ or } x \in U$

$$\Rightarrow x \in A \cup U$$

$\therefore$ $$U \subseteq A \cup U \quad ...(2)$$

Combining (1), and (2), we have

$$A \cup U = U$$

3.14 INTERSECTION OF SETS

Definition 3.15. The intersection of sets A and B is the set whose elements are all of the elements common to both A and B.

The intersection of the sets A and B is denoted by $A \cap B$.

Symbolically: $A \cap B = \{x : x \in A \text{ and } x \in B\}$

or

$$A \cap B = \{x : A \wedge x \in B\}$$

$A \cap B$ is read as A intersection B

If $A \cap B = \phi$, then A and B are called disjoint sets.

3.14.1 Intersection of More Than Two Sets

Definition 3.16 If $A_1, A_2, ... A_n$ denote sets, then the intersection of these sets denoted by $\bigcap_{i=1}^{n} A_i$ is defined as follows

$$\bigcap_{i=1}^{n} A_i = \{x : x \in A_i : \text{for every } i (i = 1, 2, ..., n)\} = \{x : x \in A_i : \forall, i = 1, 2, ..., n\}$$

Example. If $A = \{1, 2, 4, 6\}$, $B = \{2, 4, 8, 10\}$

then $A \cap B = \{2, 4\}$

3.15 PROPERTIES OF SET INTERSECTION

Theorem 3.6 The following properties hold for the intersection of the sets.

(a) $A \cap A = A$ (Idempotent law)

(b) $A \cap B = B \cap A$ (Commutative law)

(c) $(A \cap B) \cap C = A \cap (B \cap C)$ (Associative law)

(d) $A \cap \phi = \phi$ (Identity law)

(e) $A \cap U = A$ (Identity law)

where A and B are sets.

U is the universal set.

Proof.

(a) $A \cap A = \{x: x \in A \text{ and } x \in A\}$
$= \{x: x \in A\}$
$= A$
$\therefore A \cap A = A$

(b) $A \cap B = \{x: x \in A \text{ and } x \in B\}$
$= \{x: x \in B \text{ and } x \in A\}$
$\therefore \quad = B \cap A$
$\therefore A \cap B = B \cap A$
i.e. Intersection of sets is commutative.

(c) $A \cap (B \cap C) = \{x: x \in A \text{ and } x \in (B \cap C)\}$
$= \{x: x \in A \text{ and } (x \in B \text{ and } x \in C)\}$
$= \{x: (x \in A \text{ and } x \in B) \text{ and } x \in C\}$
$= \{x: x \in A \cap B \text{ and } x \in C\}$
$= (A \cap B) \cap C$
i.e. Intersection of sets is associative.

(d) ϕ is the subset of every set
$\therefore \quad \phi \subseteq A \cap \phi$ ...(1)
and $\quad A \cap \phi = \phi$ ...(2)
form (1) and (2), we have
$A \cap \phi = \phi$

(e) clearly $\quad A \cap U \subseteq A$ ...(1)
Let x be any number of A, then
$x \in A \Rightarrow x \in U \quad \because (A \subseteq U)$
$\Rightarrow x \in A \text{ and } x \in U$
$\Rightarrow x \in A \cap U$
$A \subseteq A \cap U$ ...(2)
Combining (1) and (2), we have
$A \cap U = A$

3.16 DISTRIBUTIVE LAWS

Theorem 3.7. The following properties hold for operations of set union and intersection.

(a) $A \cup (B \cap C) = (A \cap B) \cap (A \cup C)$
(b) $A \cap (B \cup C) = (A \cap B) \cup (A \cap C)$
for all sets A, B and C

These properties are called distributive properties.

Proof. (a) Let $x \in A \cup (B \cap C)$

Then $x \in A \cup (B \cap C) \Rightarrow x \in A$ or $x \in (B \cap C)$

$\Rightarrow x \in A$ or ($x \in B$ and $x \in C$)

$\Rightarrow$ ($x \in A$ or $x \in B$) and ($x \in A$ or $x \in C$)

$\Rightarrow x \in A \cup B$ and $x \in A \cup C$

$\Rightarrow x \in (A \cup B) \cap (A \cup C)$

$\therefore \quad A \cup (B \cap C) \subseteq (A \cup B) \cap (A \cup C)$ ...(i)

Let $x \in (A \cup B) \cap (A \cup C)$

Then $x \in (A \cup B) \cap (A \cup C) \Rightarrow$ ($x \in A$ or $x \in B$) and ($x \in A$ or $x \in C$)

$\Rightarrow x \in A$ or ($x \in B$ and $x \in C$)

$\Rightarrow x \in A$ or $x \in (B \cap C)$

$\Rightarrow x \in A \cup (B \cap C)$

$\therefore (A \cup B) \cap (A \cup C) \subseteq A \cup (B \cap C)$ ...(ii)

Combining (i) and (ii), we have

$A \cup (B \cap C) = (A \cup B) \cap (A \cup C)$

(b) Let $x \in A \cap (B \cup C)$,

Then $x \in A \cap (B \cup C) \Rightarrow x \in A$ and $x \in (B \cup C)$

$\Rightarrow x \in A$ and ($x \in B$ or $x \in C$)

$\Rightarrow$ ($x \in A$ and $x \in B$) or ($x \in A$ and $x \in C$)

$\Rightarrow x \in A \cap B$ or $x \in A \cap C$

$\Rightarrow x \in (A \cap B) \cup (A \cap C)$

$\therefore \quad A \cap (B \cup C) \subseteq (A \cap B) \cup (A \cap C)$ ...(i)

Now let $x \in (A \cap B) \cup (A \cap C)$;

Then $x \in (A \cap B) \cup (A \cap C) \Rightarrow x \in (A \cap B)$ or $x \in (A \cap C)$

$\Rightarrow x \in A$ and $x \in B$ or $x \in A$ and $x \in C$

$\Rightarrow x \in A$ and ($x \in B$ or $x \in C$)

$\Rightarrow x \in A$ and $x \in B \cup C$)

$\Rightarrow x \in A \cap (B \cup C)$

$\therefore (A \cap B) \cup (A \cap C) \subseteq A \cap (B \cup C)$ ...(ii)

from (i) and (ii), we have

$A \cap (B \cup C) = (A \cap B) \cup (A \cap C)$

3.19 DIFFERENCE OF SETS: (RELATIVE COMPLEMENT)

Definition 3.16. If A and B are subsets of the universal set U, then the relative complement of B in A is the set of all elements in A which are not in A. It is denoted by $A–B$.

$$\therefore \quad A-B = \{x: x \in A \text{ and } x \notin B\}$$

Example. Let $A = \{1, 2, 3\}, B = \{2, 3, 4, 5, 6\}$;

Then $A–B = \{1\}$

3.20 COMPLEMENT OF A SET

Definition 3.17. Let U be the universal set. The complement of the set A, denoted by A^C, is the complement of A with respect to U. In other words, the complement of the set A is $U - A$

$$A^C = \{x: x \notin A\}$$

The complement of A is also denoted by $\overline{A}$. (or by A')

3.21 PROPERTIES OF COMPLEMENTATION

Theorem 3.8. If A and B are subsets of a universal set U, then the following properties hold:

(a) $\overline{U} = \phi$

(b) $\overline{\phi} = U$

(c) $A \cup \overline{A} = U$

(d) $A \cap \overline{A} = \phi$

(e) $(\overline{\overline{A}}) = A$

(f) $A \subseteq B \Rightarrow \overline{B} \subseteq \overline{A}$

(g) $(\overline{A \cap B}) = \overline{A} \cup \overline{B}$ (De Morgan's Law)

(h) $(\overline{A \cup B}) = \overline{A} \cap \overline{B}$

Proof.

(a) Clearly $\phi \subseteq \overline{U}$

conversely, $x \in \overline{U} \Rightarrow x \notin U$

$\overline{U} \subseteq \phi \Rightarrow x \in \phi$

$\therefore \phi \subseteq \overline{U}, \quad \overline{U} \subseteq \phi \Rightarrow \overline{U} = \phi$

(b) Let $x \in \overline{\phi}$, then

$x \in \overline{\phi} \Rightarrow x \notin \phi$

$\Rightarrow x \in U$

$\therefore\ \overline{\phi} \subseteq U$

Conversely; let $x \in U$, then

$x \in U \Rightarrow x \notin \phi$

$\Rightarrow x \in \overline{\phi}$

$\therefore\ U \subseteq \overline{\phi}$

Now, $\overline{\phi} \subseteq U, U \subseteq \overline{\phi} \Rightarrow \overline{\phi} = U$

The properties (c) and (d) are very simple and follow from the definition of complement. We prove the remaining.

Proof of (e).

$$x \in (\overline{\overline{A}}) \Rightarrow x \notin \overline{A}$$

$$\Rightarrow x \in A$$

$\therefore$ $$(\overline{\overline{A}}) \subseteq A \qquad \text{...(i)}$$

Conversely let $x \in A$, then

$$x \in A \Rightarrow x \notin \overline{A}$$

$$\Rightarrow x \in (\overline{\overline{A}})$$

$\therefore$ $$A \subseteq (\overline{\overline{A}}) \qquad \text{...(ii)}$$

from (i) and (ii)

$$(\overline{\overline{A}}) = A$$

Proof of (f): We have $A \subseteq B$

Let $x \in \overline{B}$, then

$$x \in \overline{B} \Rightarrow x \notin B$$

$$\Rightarrow x \notin A\ (\because A \subseteq B)$$

$$\Rightarrow x \in \overline{A}$$

Hence, $\overline{B} \subseteq \overline{A}$

Proof of (g): Let $x \in (\overline{A \cap B})$; then

$$x \in (\overline{A \cap B}) \Rightarrow x \notin (A \cap B)$$
$$\Rightarrow x \notin A \text{ or } x \notin B$$
$$\Rightarrow x \notin \overline{A} \text{ or } x \in \overline{B}$$
$$\Rightarrow x \in \overline{A} \cup \overline{B}$$

$\therefore$ $$(\overline{A \cap B}) \subseteq \overline{A} \cup \overline{B} \qquad \text{...(i)}$$

Conversely Let $x \in \overline{A} \cup \overline{B}$, then

$$x \in (\overline{A} \cup \overline{B}) \Rightarrow x \in \overline{A} \text{ or } x \in \overline{B}$$
$$\Rightarrow x \notin A \text{ or } x \notin B$$
$$\Rightarrow x \notin (A \cap B)$$
$$\Rightarrow x \in (\overline{A \cap B})$$

$$\overline{A} \cup \overline{B} \subseteq (\overline{A \cap B}) \qquad \text{...(2)}$$

Combining (1) and (2), we have

$$(\overline{A \cap B}) = \overline{A} \cup \overline{B}$$

Proof of (h): Let $x \in (\overline{A \cup B})$, then

$$x \in (\overline{A \cup B}) \Rightarrow x \notin A \cup B$$
$$\Rightarrow x \notin A \text{ and } x \notin B$$
$$\Rightarrow x \in \overline{A} \text{ and } x \in \overline{B}$$
$$\Rightarrow x \in \overline{A} \cap \overline{B}$$

$$\overline{A} \cap \overline{B} \subseteq (\overline{A \cup B}) \qquad \text{...(3)}$$

Conversly, let $x \in \overline{A} \cap \overline{B}$, then

$$x \in \overline{A} \cap \overline{B} \Rightarrow x \in \overline{A} \text{ and } x \in \overline{B}$$
$$\Rightarrow x \notin A \text{ and } x \notin B$$
$$\Rightarrow x \notin (A \cup B)$$
$$\Rightarrow x \in (\overline{A \cup B})$$

$\therefore$ $$(\overline{A} \cap \overline{B}) \subseteq (\overline{A \cup B}) \qquad \text{...(4)}$$

Combining (3) and (4), we have

$$(\overline{A \cup B}) = \overline{A} \cap \overline{B}$$

3.22 PROPERTIES OF DIFFERENCE

Theorem 3.9. If A and B are subsets of a universal set U, then

(a) $A - B = A \cap \overline{B}$

(b) $\overline{A} = U - A$

(c) $A - A = \phi$

(d) $A - \phi = A$

(e) $A - B = B - A$, if and only if $A = B$

(f) $A - B = A$, if only if $A \cap B = \phi$

(g) $A - B = \phi$, if only if $A \subseteq B$

Proof. We prove theorem 3.9(a) and leave the other as exercises.

Let $x \in A - B$, then

$$\begin{aligned} x \in A - B &\Rightarrow x \in A \text{ and } x \notin B \\ &\Rightarrow x \in A \text{ and } x \in \overline{B} \\ &\Rightarrow x \in A \cap B \end{aligned}$$

$$\therefore \qquad A - B \subseteq A \cap \overline{B} \qquad (1)$$

Conversely: Let $x \in A \cap B$, then

$$\begin{aligned} x \in A \cap \overline{B} &\Rightarrow x \in A \text{ and } x \in \overline{B} \\ &\Rightarrow x \in A \text{ and } x \notin B \\ &\Rightarrow x \in A - B \end{aligned}$$

$$\therefore \qquad A \cap \overline{B} \subseteq A - B \qquad (2)$$

Hence, from (1) and (2)

$$A - B = A \cap \overline{B}$$

3.23 SYMMETRIC DIFFERENCE

Definition 3.18. The symmetric difference of the sets A and B is the set containing those elements in either A or B, but not in both A and B.

In other words the symmetric difference of the sets A and B is the relative complement of $A \cup B$ in $A \cap B$.

The symmetric difference of A and B is denoted by $A \Delta B$ (or by $A \oplus B$)

Symbolically: $A \Delta B = \{x : x \in A \cup B, \text{ and } x \notin A \cap B\}$

from the definition, we have

$$A \Delta B = (A - B) \cup (B - A)$$

Example. Let $A = \{1, 2, 3, 4, 5, 6, 7\}$, $B = \{3, 4, a, b, c, d\}$

then $A \cup B = \{1, 2, 3, 4, 5, 6, 7, a, b, c, d\}$, $A \cap B = \{3, 4\}$

$\therefore$ $A \Delta B = \{1, 2, 5, 6, 7, a, b, c, d\}$

3.24 PROPERTIES OF SYMMETRIC DIFFERENCE

Theorem 3.10. If A and B are sets, then

(i) $A \Delta A = \phi$

(ii) $A \Delta B = B \Delta A$

(iii) $A \Delta \phi = A$

(iv) $(A \Delta B) \Delta C = A \Delta (B \Delta C)$

(v) $A \Delta B = (A \cup B) - (A \cap B) = (A - B) \cup (B - A)$

3.25 VENN DIAGRAMS

Set operations can be illustrated by Venn diagrams. A Venn diagram is a useful device to our consideration of relationships that may exist between the subsets of a given universe. The universal set U is represented by the set of points inside and on the boundary of a single closed curve (usually a rectangle). If A is a subset of U, then it is represented by points inside on the boundary of another simply closed curve (usually a circle) inside the rectangle. Several Venn diagrams together with their interpretations are shown below. (usually the label U is omitted).

Note. Using a Venn diagram, we can produce a geometrical interpretation for any expression involving sets and set operations. However, it should be noted that Venn diagrams do not prove the truth of a relationship between sets, they only illustrate Plausibility.

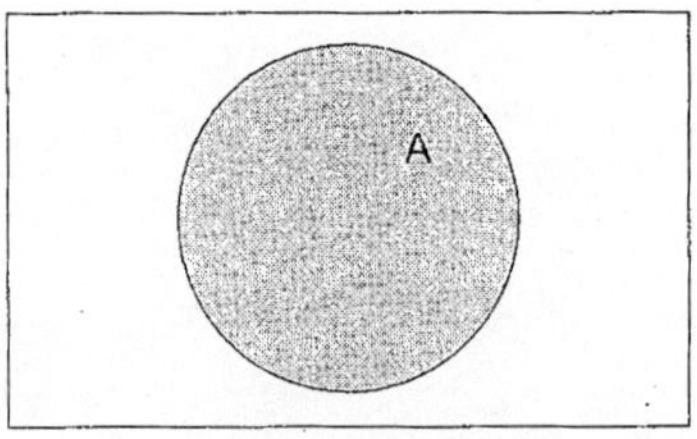

(a) The set A (shaded)

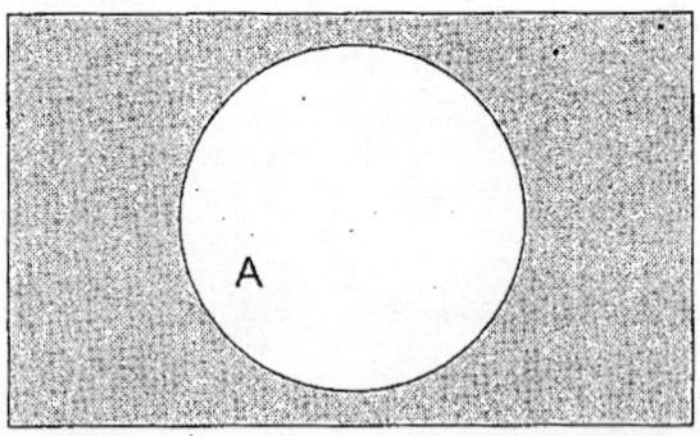

(b) The set $\overline{A}$ (shaded portion)

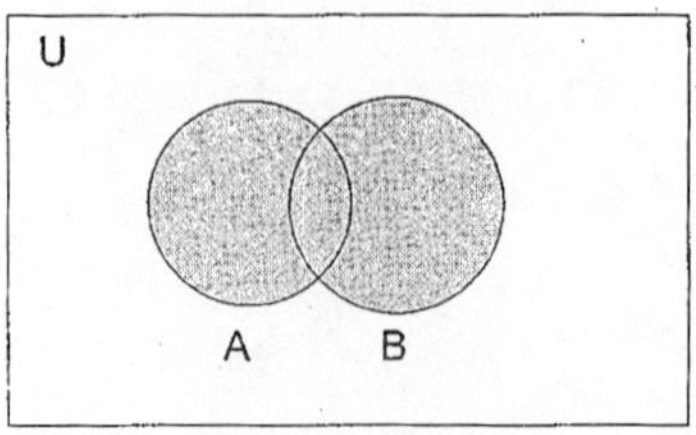

(c) $A \cup B$ (shaded)

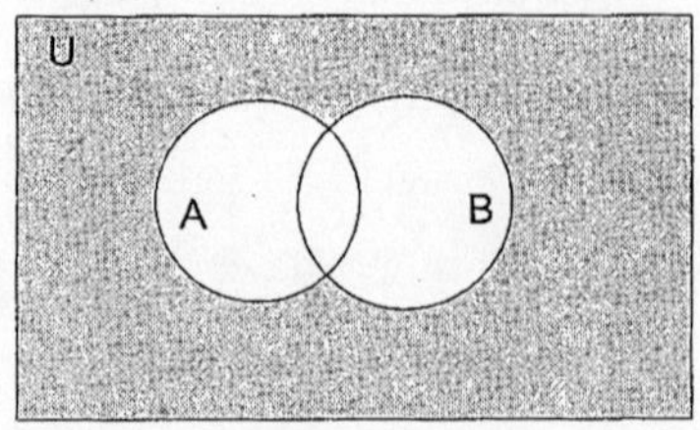

(d) $(\overline{A \cup B})$ (shaded)

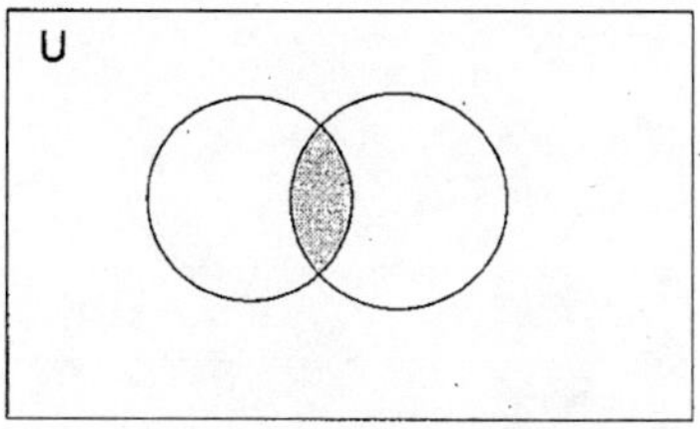

(e) $A \cap B$ (shaded portion)

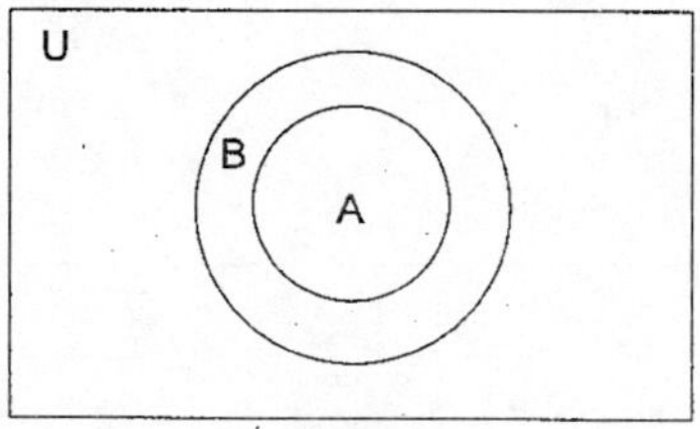

(f) $A \subset B$

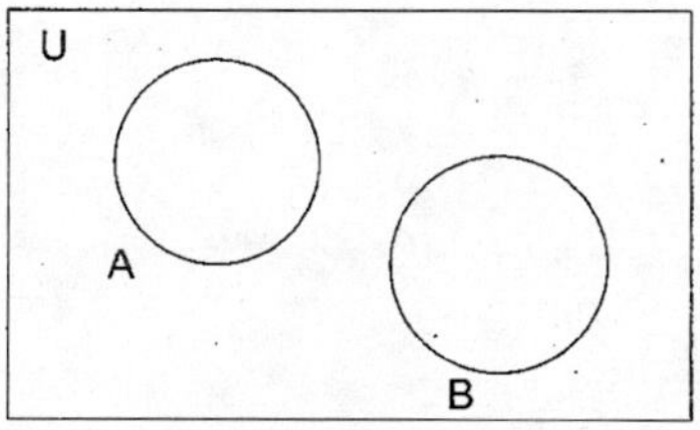

(g) A, B disjoint

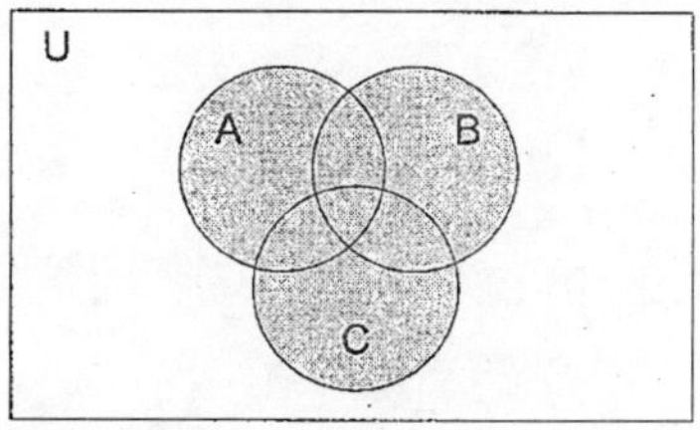

(h) $A \cup B \cup C$ (shaded)

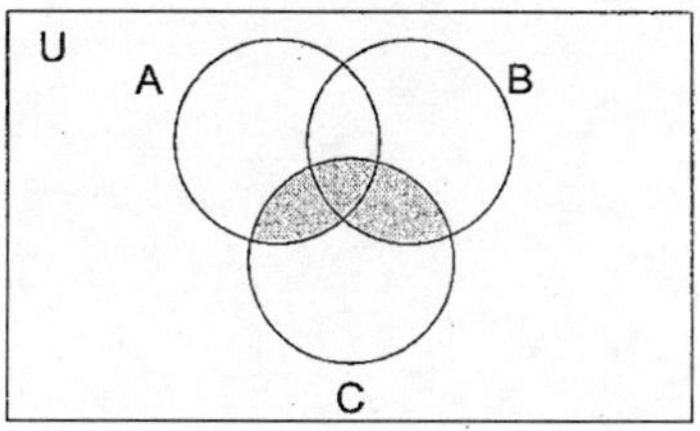

(i) $(A \cup B) \cap C$ (shaded)

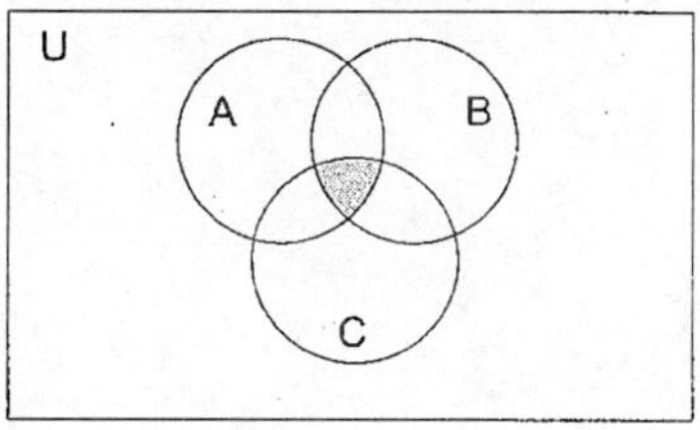

(j) $(A \cap B) \cap C$ (shaded)

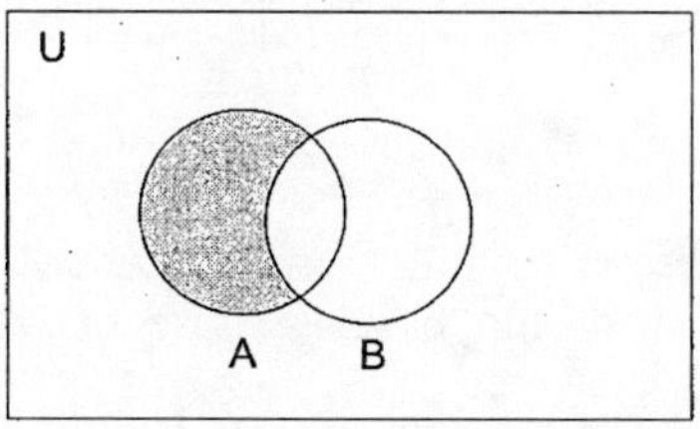

(k) $A - B$ (shaded portion)

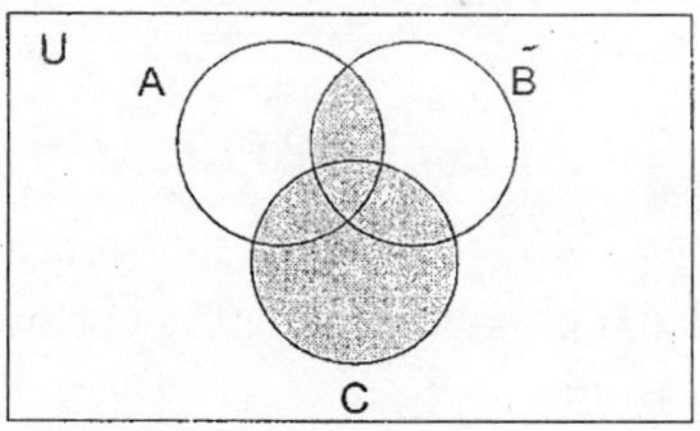

(l) $(A \cap B) \cup C$ (shaded)

Fig. 3.1 Venn diagrams

3.26 PRINCIPLE OF DUALITY

The principle of duality states that any established result involving sets and complements and operation of union and intersection give a corresponding dual result by replacing $\cup$ by ϕ , and $\cup$ by $\cap$, and vice versa.

Example. Consider $A \cup \overline{A} = U$, then applying the principle of duality, we can write

$$A \cap \overline{A} = \phi$$

Table 3.1 Set Identities

Identity	Name
1. (a) $A \cup \phi = A$	Identity laws
(b) $A \cap U = A$	
2. (a) $A \cup U = U$	Domination laws
(b) $A \cap \phi = \phi$	
3. (a) $A \cup A = A$	Idempotent law
(b) $A \cap A = A$	
4. $(\overline{\overline{A}}) = A$	Complementation law
5. (a) $A \cup B = B \cup A$	Commutative law
(b) $A \cap B = B \cap A$	
6. (a) $A \cup (B \cup C) = (A \cup B) \cup C$	Associative laws
(b) $A \cap (B \cap C) = (A \cap B) \cap C$	
7. (a) $A \cap (B \cup C) = (A \cap B) \cup (A \cap C)$	Distributive laws
(b) $A \cup (B \cap C) = (A \cup B) \cap (A \cup C)$	
8. (a) $\overline{A \cup B} = \overline{A} \cap \overline{B}$	
(b) $\overline{A \cap B} = \overline{A} \cup \overline{B}$	De Morgan's laws
9. (a) $A \cup (A \cap B) = A$	Absorption laws
(b) $A \cap (A \cup B) = A$	
10. (a) $A \cup \overline{A} = U$	Complement laws
(b) $A \cap \overline{A} = \phi$	

3.27 SOLVED EXAMPLES

Example 3.27.1. Show that the empty set is unique.

Solution. If possible let ϕ_1 and ϕ_2 be two empty sets. Empty set is a subset of every set, therefore

$$\phi_1 \subseteq \phi_2 \text{ and } \phi_2 \subseteq \phi_1$$

Hence, $\phi_1 = \phi_2$

Example 3.27.2. Prove that $A - (A - B) = A \cap B$

Solution. Consider $A - (A - B)$

We have, $A - (A - B) = A - (A \cap \overline{B}), \quad (\because A - B = A \cap \overline{B})$

$$= A \cap (\overline{A} \cup \overline{\overline{B}})$$

$$= A \cap (\overline{A} \cup B)$$

$$= (A \cap \overline{A}) \cup (A \cap B)$$

$$= \phi \cup (A \cap B)$$

$$= A \cap B$$

$$\therefore \quad A - (A - B) = A \cap B$$

Example 3.27.3. If A and B are sets, prove that $A - B = A - (A \cap B)$

Solution. Let $x \in A - B$

Then $\quad x \in A - B \Rightarrow x \in A \text{ and } x \notin B$

$$\Rightarrow x \in A \text{ and } x \notin (A \cap B)$$

$$\Rightarrow x \in A - (A \cap B)$$

$$\therefore \quad A - B \subseteq A - (A \cap B) \qquad ...(1)$$

Again, let $x \in A - (A \cap B)$, then

$$x \in A - (A \cap B) \Rightarrow x \in A \text{ and } x \notin (A \cap B)$$

$$\Rightarrow x \in A \text{ and } (x \in A \text{ and } x \notin B)$$

$$\Rightarrow x \in A \text{ and } x \notin B$$

$$\Rightarrow x \in (A - B)$$

$$\therefore \quad A - (A \cap B) \subseteq (A - B) \qquad ...(2)$$

From (1) and (2)

$$A - B = A - (A \cap B)$$

Example 3.27.4. Let $U = \{a, b, c, d, e, f, g, h, k\}$

$A = \{a, b, c, g\}$, $\quad B = \{a, e, f, g\}$

$C = \{a, c, f\}$, and $D = \{f, h, k\}$

Verify $\quad (\overline{A \cup B}) = \overline{A} \cap \overline{B}$

and $(\overline{A \cap B}) = \overline{A} \cup \overline{B}$, find $\overline{C}$, and $\overline{D}$ also find $\overline{C} \cap \overline{D}$

Solution. We have $\overline{A} = U - A = \{a, b, c, d, e, f, g, h, k\} - \{a, b, c, g\}$

$$= \{d, e, f, h, k\}$$

$$\overline{B} = U - B = \{a, b, c, d, e, f, g, h, k\} - \{d, e, f, g\}$$

$$= \{a, b, c, h, k\}$$

$$A \cup B = \{a, b, c, d\} \cup \{d, e, f, g\}$$

$$= \{a, b, c, d, e, f, g\}$$

$$(\overline{A \cup B}) = \cup - (A \cup B) = \{a, b, c, d, e, f, g, h, k\} - \{a, b, c, d, e, f, g\}$$
$$= \{h, k\} \quad ...(1)$$

$$\overline{A} \cap \overline{B} = d, e, f, h, k\} \cap \{a, b, c, h, k\} = \{h, k\} \quad ...(2)$$

From (1) and (2)

$$(\overline{A \cup B}) = \overline{A} \cap \overline{B} \text{ (Verified)}$$

Now, $A \cap B = \{a, b, c, g\} \cap \{d, e, f, g\} = \{g\}$

$\therefore$ $(\overline{A \cap B}) = U - \{g\} = \{a, b, c, d, e, f, h, k\}$ ...(3)

$$\overline{A} \cup \overline{B} = \{d, e, f, h, k\} \cup \{a, b, c, h, k\} = \{a, b, c, d, e, f, h, k\} \quad ...(4)$$

from (3) and (4)

$$(\overline{A \cap B}) = \overline{A} \cup \overline{B} \text{ (Verified)}$$

$$\overline{C} = U - C$$
$$= \{a, b, c, d, e, f, g, h, k\} - \{a, c, f\}$$
$$= \{b, d, e, g, h, k\}$$

$$\overline{D} = \cup - D = \{a, b, c, d, e, f, g, h, k - f, h, k\} = \{a, b, c, d, e, g\}$$

$\therefore$ $\overline{C} \cap \overline{D} = \{b, d, e, g, h, k\} \cap \{a, b, c, d, e, g\} = \{b, d, e, g\}$

Example 3.27.5. Given that ϕ is an empty set. find $P(\phi)$, $P(P(\phi))$, and $P(P(P(\phi)))$

Solution. $P(\phi) = \{\phi\}$

$$P(P(\phi)) = \{\phi, \{\phi\}\}$$
$$P(P(P(\phi))) = \{\phi, \{\phi\}, \{(\phi)\}, \{\phi, \{\phi, \{\phi\}\}\}$$

Example 3.27.6. Show that $[A \cap (B \cup \overline{A})] \cup B = B$

Solution. $[A \cap (B \cup \overline{A})] \cup B$

$$= [(A \cap B) \cup (A \cap \overline{A})] \cup B$$
$$= [(A \cap B) \cup \phi] \cup B$$
$$= (A \cap B) \cup B$$
$$= B$$

Example 3.27.7. Let A_k denote the sets of real number defined by

$$A_o = \{a : a \leq 1\}$$

and $$A_k = \left\{ a : a < 1 + \frac{1}{k} \right\}, k = 1, 2, 3...$$

show that $\bigcap_{k=1}^{\infty} = A_k = A_o$

Solution. $A_k = \left\{a : a < 1 + \frac{1}{k}\right\}$

$\therefore$ $A_1 = \left\{a : a < 1 + \frac{1}{1}\right\} = \{a : a < 2\}$

$A_2 = \left\{a : a < 1 + \frac{1}{2}\right\} = \left\{a : a < \frac{3}{2}\right\}$

$A_3 = \left\{a : a < 1 + \frac{1}{3}\right\} = \left\{a : a < \frac{4}{3}\right\}$

$\vdots$

$A_\infty = \left\{a : a < 1 + \frac{1}{\infty}\right\} = \{a : a \le 1\}$

Hence $A_1 \cap A_2 \cap \ldots \cap A_\infty = \{a : a \le 1\} = A_o$

$\therefore$ $\bigcap_{K=1}^{\infty} A_k = A_o$

Example 3.27.8. If A and B are subsets of the universal set μ. Show that

(i) $(\overline{A \cup B}) = \overline{A} \cap \overline{B}$

(ii) $(\overline{A \cap B}) = \overline{A} \cup \overline{B}$

Proof: (i) $(\overline{A \cup B}) = \mu - (A \cup B)$

$= (\mu - A) \cap (\mu - B) = \overline{A} \cap \overline{B}$

(ii) $(\overline{A \cap B}) = \mu - (A \cap B) = (\mu - A) \cup (\mu - B) = \overline{A} \cup \overline{B}$

Example 3.27.9. If A and B are sets such that $A \cup B = A$ and $A \cap B = A$, then prove that $A = B$

Solution. $A \cup B = A \Rightarrow B \subseteq A$ ...(1)

and $A \cap B = A \Rightarrow A \subseteq B$ ...(2)

from (1) and (2) we have

$$A = B$$

EXERCISE 3.1

1. Use Venn diagram to show that

   $A \cap (B \cup C) = (A \cap B) \cup (A \cap C)$ (O.U. MCA, 1998)

2. Prove that $A - (A \cap B) = A - B$
3. If A and B are subsets of universal set $\cup$, show that

   $A \subseteq B \Leftrightarrow A \cup B = B$ (O.U. MCA, 1999)

4. If $P(X)$ denotes the power set of X, then show that

   $P(A) \cup P(B) \subseteq P(A \cup B)$

5. Prove that for any sets A and B

   $(A \cup B) - (A \cap B) = (A - B) \cup (B - A)$ (V.T.U. Feb. 2002)

6. If A, B and C are sets, then show that

   $A \cap (B - C) = (A \cap B) - (A \cap C)$ (V.T.U. Sep. 1999)

7. If $A \subset B$ and $B \subset C$, then show that $A \subset C$
8. If $A = \{1, 2\}$, $B = \{1, 2, 4, 5\}$, $C = \{5, 7, 9, 10\}$, find

   (i) $A \cup B$ (ii) $A \cap B$ (iii) $(A \cup B) \cap C$ (iv) $(A \cap B) \cap C$

9. If $A = \{a, b, c\}$, find the power set of A
10. If $A = \{a, b, c, d\}$, list all the subsets of A
11. If $U = \{a, b, c, d, e, f, g, h\}$, $A = \{a, c, f, g\}$, $B = \{a, e\}$, $C = \{b, h\}$ compute

    (i) $A - B$ (ii) $\overline{A}$ (iii) $\overline{B}$ (iv) $\overline{A \cup B}$

12. If $A = \{1, 3, 5, 7, 8, 9\}$, $B = \{3, 5, 8\}$, then verify

    $(A - B) \cup (B - A) = (A \cup B) - (A \cap B)$

13. If $A = \{1, 3, 5, 7, 9\}$, $B = \{3, 5, 8\}$, then find $A \Delta B$
14. If $A \oplus B$ denote the symmetric difference of the sets A and B, then find $A \oplus B$ for the following

    (i) $A = \{a, b\}$, $B = \{a, c\}$

    (ii) $A = \{a, b\}$, $B = \{b, c\}$

15. Define the following, give examples

    (a) Subject (b) Union of sets (c) Intersection of sets (d) Complement of a set (e) Power set

16. Prove $A - B = A \cap \overline{B} = \overline{B} - \overline{A}$
17. If A and B are sets, then prove

(i) $A \subseteq B$ if $A \cup B = B$

(ii) $A \subseteq B$ if $A \cap B = A$

18. Show that (*a*) $A \subseteq B \Leftrightarrow A \cup B = B$

(b) $A \cap \overline{A} = \phi$; $A \cup \phi = A$ (O.U. MCA 1999)

19. Use Venn diagram to show that

$$A \cap (B \cup C) = (A \cap B) \cup (A \cap C)$$

20. Show that $A - B \subseteq C$, if and only if $A - C \subseteq B$

21. Prove the following

(i) $A \cup (B \cap C) = (A \cup B) \cap (A \cup C)$

(ii) $A - (B \cup C) = (A - B) \cap (A - C)$

Use Venn diagrams

22. If $S_1 = S_2 \cup S_3$ and $S_2 \cap S_3 = \phi$

Prove that $S_2 = S_1 - S_3$

$S_3 = S_1 - S_2$

3.28 CARDINALITY

Definition 3.19. If A is a finite set with n distinct elements, then n is called the cardinality of A. The cardinality of A is denoted by $|A|$ or by $n(A)$

Example. If $A = \{a, b, c, d, e\}$ then $|A| = 5$

3.28.1 Cardinality of Union (The Principle of Inclusion-Exclusion)

If A and B are finite sets, then the number of elements in $A \cup B$, denoted by $|A \cup B|$ is given by

$$|A \cup B| = |A| + |B| - (A \cap B)$$

$$(\text{or } n(A \cup B) = n(A) + n(B) - n(A \cap B))$$

where $n(A \cup B) = |A \cup B|$

If $A \cap B = \phi$, (i.e. A and B are disjoint), then

$$|A \cup B| = |A| + |B| \qquad \text{(Counting Principle 1)}$$

The following properties hold:

(i) $|A \cup B| = |A - B| + |B - A| + |A \cap B|$

(ii) $|A| = |A - B| + |A \cap B|$

(iii) $|B| = |B - A| + |A \cap B|$

Example. If $|A| = 2$, $|B| = 3$, $|A \cap B| = 1$, then find $n(A \cup B)$

Solution. Using the formula $|A \cup B| = |A| + |B| - |A \cap B|$

We get $|A \cup B| = 2 + 3 - 1 = 4$

$\therefore$ $n(A \cup B) = 4$

The counting principle can be extended to three sets and stated as follows:

Theorem 3.11. Let A, B and C be sets. The cardinal number of union of the these sets is given by

$$|A \cup B \cup C| = |A| + |B| + |C| - |A \cap B| - |B \cap C| - |C \cap A| + |A \cap B \cap C|$$

Example. Verify

$$|A \cup B \cup C| = |A| + |B| + |C| - |A \cap B| - |B \cap C| - |C \cap A| + |A \cap B \cap C|$$

where $A = \{1, 2, 3, 4, 5\}$, $B = \{2, 3, 4, 6, 9\}$, $C = \{3, 4, 6, 8\}$

Solution. We have $A \cup B \cup C = \{1, 2, 3, 4, 5, 6, 8, 9\} \Rightarrow |A \cup B \cup C| = 7$

$$A \cap B = \{2, 3, 4\}, B \cap C = \{3, 4, 6\}, C \cap A = \{3, 4\}$$

and $\quad A \cap B \cap C = \{3, 4\}$

$\therefore \quad |A| + |B| + |C| - |A \cap B| - |B \cap C| - |C \cap A| = 5 + 4 + 4 - 3 - 3 - 2 + 2 = 7$

$$= |A \cup B \cup C|$$

Verified

Example 2. If there are 200 faculty members that speak French, 50 that speak Russian, 100 that speak Spanish, 20 that speak French and Russian, 60 that speak French and Spanish, 35 that speak Russian and Spanish, while only 10 that speak French, Russian, and Spanish. Determine how many speak either French, Russian or Spanish. [V.T.U. Sep. 1999]

Solution. Let A be the set of faculty members who speak French, B be the set of faculty members who speak Russian and C be the set of faculty members who speak Spanish.

Then we have $\quad |A| = 200$, $|B| = 50$, $|C| = 100$, $|A \cap B| = 20$

$|A \cap C| = 60$, $|B \cap C| = 35$ and $|A \cap B \cap C| = 10$

Substituting in

$$|A \cup B \cup C| = |A| + |B| + |C| - |A \cap B| - |B \cap C| - |A \cap C| + |A \cap B \cap C|$$

We get $|A \cup B \cup C| = 200 + 50 + 100 - 20 - 35 - 60 + 10 = 245$

There are 245 faculty members who speak French, Russian or Spanish.

3.29 COMPARABLE SETS

Definition 3.20. Let A and B be any sets. A and B are said to be comparable if $A \subset B$ or $B \subset A$

Example. Let $A = \{1, 2, 3\}$, and $B = \{1, 2, 3, 4, 6, 8\}$ then A and B are comparable since $A \subset B$

3.30 NOT COMPARABLE SETS

Definition 3.21. Sets A and B are said to be not comparable if $A \not\subset B$ or $B \not\subset A$

Example. If $A = \{a, b\}$, $B = \{b, d, e, f, g\}$ then $A \not\subset B$ and $B \not\subset A$. Therefore A and B are not comparable

3.31 MULTISET

Definition 3.22. A collection of objects that are not necessarily distinct is called a multiset.

Example. $\{a, a, b, b, c, c, c\}$

3.32 MULTIPLICITY

Definition 3.23. Let S be a multiset and $x \in s$. The multiplicity of x is defined to be the number of times the element x appears in the multiset S.

Example. Let $S = \{a, b, b, c, c, c, c, d, d, e, f, f\}$

The multiplicity of a is 1
The multiplicity of b is 2
The multiplicity of c is 4
The multiplicity of d is 2
The multiplicity of e is 1
The multiplicity of f is 2

If A and B are multisets, then $A \cup B$ and $A \cap B$ are also multisets. The multiplicity of an element $x \in A \cup B$, is equal to the maximum of the multiplicity of x is A and B. The multiplicity of $x \in A \cap B$ is equal to the maximum of the multiplicities of x in A and in B.

Example. Let $A = \{1, 1, 1, 2, 2, 3, 3, 4, 4\}$

and $B = \{1, 1, 2, 2, 3, 4\}$

then $A \cap B = \{1,1, 2, 2, 3, 4\}$

and $A \cup B = \{1, 1, 1, 2, 2, 3, 3, 4, 4\}$

3.33 CARTESIAN PRODUCT OF SETS

3.33.1 Ordered Pair

Definition 3.24. An order pair (a, b) is the set $\{a, \{a, b\}\}$

Example.

$$(2, 5) = \{2, \{2, 5\}\}$$

$$(4, 4) = \{4, \{4\}\}$$

In the ordered pair (a, b) the element a is called the first element and the element b is called the second element. If (a, b) and (a', b') are ordered pair such that $(a, b) = (a', b')$ then $a = a'$ and $b = b'$

Example. If $P(x, y)$ is a point is the plane, the these coordinates of P for in a pair. The first member x, is called, the x-coordinate of P and y is called the y-coordinate of P.

3.33.2 Cartesian Product

Definition 3.25. If A and B are non-empty sets, then the Cartesian product of A and B is the set of all ordered pairs (a, b) where $a \in A$ and $b \in B$

The product $A \times B$ of set A and B is the set $\{(a, b): a \in A \text{ and } b \in B\}$

Example 1. If $A = \{a, b\}, B = \{p, q, r\}$ then

$$A \times B = \{(a, p), (a, q), (q, r), (b, p), (b, q), (b, r)\}$$

Example 2. If $A = \{0, 1, 3, 5\}, B = \{2, 3, 4\}$

then $A \times B = \{(0, 2), (0, 3), (0, 4), (1, 2), (1, 3), (1, 4), (3, 2), (3, 3), (3, 4), (5, 2), (5, 3), (5, 4)\}$

Example 3. If A and B are both sets of real numbers, then $A \times B$ is the cartisian plane.

Theorem 3.12. If A, B and C are non-empty sets then

(i) $A \times (B \cup C) = (A \times B) \cup (A \times C)$

(ii) $A \times (B \cap C) = (A \times B) \cap (A \times C)$

Proof. (i) $A \times (B \cup C) = \{(a, b): a \in A, \text{ and } b \in (B \cup C)\}$

$$= \{(a, b): a \in A, \text{ and } (b \in B \text{ or } b \in C)\}$$

$$= \{(a, b): (a \in A \text{ and } b \in B) \text{ or } (a \in A \text{ and } b \in C)\}$$

$$= \{(a, b): (a, b) \in A \times B \text{ or } (a, b) \in (A \times C)\}$$

$$= \{(a, b): (A \times B) \cup (A \times C)\}$$

$$= (A \times B) \cup (A \times C)$$

(ii) $A \times (B \cap C) = \{(a, b): a \in A, b \in (B \cap C)\}$

$$= \{(a, b): a \in A, b \in B \text{ and } b \in C\}$$

$$= \{(a, b): a \in A, b \in B \text{ and } a \in A, b \in C\}$$

$$= \{(a, b): (a, b) \in A \times B \text{ and } (a, b) \in A \times C\}$$

$$= \{(a, b): (a, b) \in (A \times B) \cap (A \times C)\}$$

$$= (A \times B) \cap (A \times C)$$

Theorem 3.13. If A, B and C are non-empty sets, then $A \subseteq B \Rightarrow A \times C \subseteq B \times C$

Proof. Let (a, b) be any element of $A \times C$, then

$$(a, b) \in A \times C \Rightarrow a \in A, b \in C$$

$$\Rightarrow a \in B, b \in C \ (\because A \subseteq B)$$

$$\Rightarrow (a, b) \in B \times C$$

$\therefore \ A \times C \subseteq B \times C$

3.33.3 Cartesian Product of ∩ Sets

Definition 3.26. Let $A_1, A_2, ..., A_n$ denote n sets where $n \geq 2$, then the Cartesian product $A_1 \times A_2 \times ... \times A_n$ is the set of all n–tuples of the form $(a_1, a_2, ..., a_n)$ where $a_1 \in A_1, a_2 \in A_2, ..., a_n \in A_n$

From the definition we have $A_1 \times A_2 \times ... \times A_n = \{(a_1, a_2, ..., a_n):$
$a_1 \in A_1, a_2 \in A_2, ..., a_n \in A_n\}$

Theorem 3.14. If A and B are sets, then $|A \times B| = |A||B|$ [Counting Principle 2]

Example. If $A = \{2, 5, 7, 9\}$, and $B = \{1, 3, 4, 8\}$

then $|A \times B| = 16, |A| = 4, |B| = 4$

$\therefore$ $|A \times B| = 16 = 4 \times 4 = |A| \cdot |B|$

The product of set $A_1, A_2, ..., A_n$ is denoted by $\prod_{i=1}^{n} A_i$

$$\therefore \quad \prod_{i=1}^{n} A_i = A_1 \times A_2 \times ... \times A_n$$

we use

$$A^2 = A \times A$$
$$A^3 = A \times A \times A$$
$$\vdots$$
$$A^n = A \times A \times ... \times A \ (n \text{ factors})$$

If R denotes the set of all real numbers, then

$R \times R = R^2$ is the Cartesian plane

$R \times R \times R = R^3$ is the Euclidean space

EXERCISE 3.2

1. If A, B, C are finite sets, prove that

$$|A - B - C| = |A| - |A \cap B| - |A \cap C| + |A \cap B \cap C|$$

2. Out of 30 students is a dormitory, 15 take an art course, 8 take biology course and 6 take a chemistry course. It is known that 3 students take all three courses, show that 7 or more students take name of the courses.

(V.T.U Feb 2002)

3. Among the integers 1 to 200 find the number of integers that are

(V.T.U. March 2001)

   (i) divisible by 2 or 5 or 9
   (ii) not divisible by 5

(iii) not divisible by 2 or 5 or 9

(iv) divisible by 5 or not by 2 or 9

Ans. (i) 129 (ii) 160 (iii) 71 (iv) 18

4. Let $A = \{1, 2, 5\}$, $B = \{0, 3\}$, find $A \times B$

Ans. $A \times B = \{(1, 0), (1, 3), (2, 0), (2, 3), (5, 0), (5, 3)\}$

5. Prove that $A \times B = \phi$ if $A = \phi$; or $B = \phi$

6. Prove that $A \times B = B \times A$ if and only if $A = B$

7. If A = $\{a, b\}$, $B = \{2, 3\}$, $C = \{3, 4\}$

Find (i) $A \times (B \cup C)$, $(A \times B) \cup (A \times C)$ and show that

$$A \times (B \cup C) = (A \times B) \cup (A \times C)$$

(ii) Find $A \times (B \cap C)$, $(A \times B) \cap (A \times C)$

and show that $A \times (B \cap C) = (A \times B) \cap (A \times C)$

8. Show by an example that

(i) $A \times B \neq B \times A$

(ii) $(A \times B) \times C = A \times (B \times C)$ (O.U. MCA 1994)

9. For the sets S, T, and U prove that

$$(S \cap T) \times U = (S \times U) \cap (T \times U)$$ (O.U. MCA 1995, 96)

10. A researcher studying snakes finds the following population of snakes in a swamp:

32 were green, 20 were water snakes, 45 were poisonous snakes, 15 were green and poisonous, 7 were green water snakes, 10 were poisonous water snakes and 5 were poisonous green water snakes.

(a) How many snakes were there all together?

(b) How many snakes in only one category?

(c) How many snakes were green water snakes but were not poisonous?

Ans. (a) 70 (b) 48 (c) 2

3.34 BINARY RELATIONS

The chapter deals primarily with the concept of a binary relation.

A relation (i.e. binary relation) may involve equality or inequatity. The mathematical concept of a binary relation deals with the way the variables are related or paired. A binary relation may signify a family relation ship such as "is the son of", "is the brother of". In mathematics the expressions like "is less than", "is greater than", "is parallel to" are relations. In this chapter we shall consider relation called binary relations.

Definition 3.27. Let A and B be sets. Then any subset R of $A \times B$ is called a binary relation from A to B

Example 1. Let $A = \{2, 4, 6\}$, $B = \{1, 3, 5\}$

then $R = \{(2, 1), (4, 3), (6, 5)\}$, is a relation from A to B

Example 2. Let $A = \{1, 2, 3, 4\}, B = \{a, b, c\}$

then $R = \{(1, a), (2, b), (4, c)\}$ is a relation

from A to B

If R is a relation from A to B, then the set of all first elements of the ordered pairs (a, b), which belong to R is called the domain of R. The range of R is the set of all seconds coordinates of the ordered pairs (a, b) which belong to R.

If R_1 and R_2 are two relations with the same domain D and same range, then we can define the relations $R_1 \cup R_2$ and $R_1 \cap R_2$ with the same domain D and same range.

Null set ϕ is a subset of every set. Therefore for any specified non-empty domain and range; ϕ is a relation called the null relation or empty relation.

If R is a relation from A to B and $(a, b) \in R$, then we use the relation $a\ R\ b$ to denote that $(a, b) \in R$.

If the ordered pair (a, b) is not an element of R, then we write $a \not R\ b$.

If $(a, \text{b}) \in R$, then we say that 'a is related to b'.

3.35 RELATION ON A SET

Definition 3.28. A relation on a set A is the relation from A to A.

If R is a relation on A, then R is a subset of $A \times A$.

Example. Let $A = \{1, 2, 3, 4, 5\}$

and $R = \{(2, 4), (4, 4)\}$

since, $R \subseteq A \times A$, R is a relation on A.

3.35.1 Number of Relations on a Finite Set

Let A be a finite set having n elements. A relation on the set A is a subset of $A \times A$.

Since $A \times A$ has n^2 elements there are 2^{n^2} subsets of $A \times A$. Thus there are 2^{n^2} relations on A, which has n elements.

Example 1. If $A = \{p, q\}$ then this number of relation on A is $2^{2^2} = 2^4 = 16$

Example 2. If $A = \{a, b, c\}$, then the number of relations on the set A is $2^{3^2} = 2^9$ $= 512$

3.36 PROPERTIES OF (BINARY) RELATIONS

3.36.1 Reflexive Relation

Definition 3.29. A relation on a set A is said to be reflexive if $(a, a) \in R\ \forall\ a \in A$, i.e. $a\ R\ a\ \forall a \in A$

Example 1. Let $A = \{a, b, c\}$

and $R = \{(a, a), (a, b), (b, a), (b, b), (b, c), (c, b), (c, c)\}$ R is a relation on A.

Since $(a, a), (b, b), (c, c), \in R$, the relation R is reflexive.

Example 2. 'Equality' is a reflexive relation, since an element is equal to itself.

Note. If $(a, a) \notin R$ for at least one element $a \in A$, then the relation R is not reflexive.

Note. If A is a non-empty set then $R = \{(a, a\}: a \in A\}$ is a called diagonal relation on A.

Example. Let $A = \{1, 2, 3\}$

and $R = \{(1, 1), (1, 3), (2, 2), (3, 1)\}$ be a relation on A

$3 \in A$ but $(3, 3) \notin R$, therefore R is not reflexive.

3.36.2 Symmetric Relation

Definition 3.30. A relation R on a set A is called symmetric if $(b, a) \in R$; wherever $(a, b) \in R$ for all $a, b \in A$.

i.e. R is symmetric on A, if

$$(a, b) \in R \Rightarrow (b, a) \in R. \qquad \forall (a, b) \in R$$

Example 1. Let R be the relation 'is perpendicular to' the set of all straight lines. then R is a symmetric relation.

Example 2. Let $A = \{1, 2, 3\}$

and $R = \{(1, 2), (2, 1), (1, 3), (3, 1),(2, 3), (3, 2)\}$

be a relation on A.

Since $(a, b) \in R \Rightarrow (b, a) \in R \quad \forall (a, b) \in R$

the relation R is symmetric.

3.36.3 Anti-symmetric Relation

Definition 3.31. A relation R on a set A is said to be anti-symmetric, if $(a, b) \in R$ and $(b, a) \in R \Rightarrow a = b$

Example 1. Let $A = \{a, b, c\}$

and $R = \{(a, a), (b, b), (c, c)\}$ be A relation on A.

R is an anti-symmetric relation on A.

Example 2. Let N denote the set of natural numbers and R be a relation in N, defined by 'a is divisor of b'

i.e. $a R b$ if a divides b.

Since, 'a divides b' and 'b divides a' implies that $a = b$,

the relation R is an anti-symmetric relation on A.

3.36.4 Transitive Relation

Definition 3.32. A relation R on a set A is a said to be transitive, if $(a, b) \in R$, $(b, c) \in R \Rightarrow (a, c) \in R$.

i.e. $\quad a R b, b R c \Rightarrow a R c \ \forall \ a, b, c \in A$

Example 1. Let $A = \{1, 2, 3)$ and $R = \{(1, 1), (2, 2), (2, 3), (3, 2), (3, 3)\}$ be a relation on A, then R is a transitive relation.

Example 2. Let A denote the set of all straight lines in a plane and R be a relation on A defined by 'is parallel to'. Then R is a transitive relation on A.

3.36.5 Equivalence Relation

Definition 3.33. A relation defined on a set A, is said to be an equivalence relation on A, if R is reflexive, symmetric and transitive. An equivalence relation on a set is usually denoted by '~'.

Example 1. Let $A = \{a, b, c\}$

and $R = \{(a, a), (a, b), (b, a), (b, b), (b, c), (c, b), (c, a), (a, c), (c,c)\}$

then R is reflexive, symmetric and transitive.

Hence R is an equivalence relation on A.

Example 2. Let A be the set of all triangles in a plane and let R be a relation in A defined by 'is congruent to', then R is a reflexive, symmetric and transitive.

$\therefore$ R is an equivalence relation on A.

Example 3. Let $A = \{1, 2, 3, 4, 5\}$ and $P = \{A \times A\}$.

R is a relational set such that $((x, y), (x', y')) \in R$ if and only if $xy' = x'y$. Show that R is an equivalence relation compute A/R. **(BE, VTU, 2001)**

Solution. We have $A = \{1, 2, 3, 4, 5\}$

$$R = \{(x, y), (x', y') : xy' = x'y\} \text{ and } P = \{A \times A\}$$

R is reflexive: Let $((x, y), (x, y)) \in R$, then $xy = yx$

Hence, R is reflexive

R is symmetric: We have $((x, y), (x', y')) \in R$

$$\Rightarrow xy' = x'y$$
$$\Rightarrow y'x = yx'$$
$$\Rightarrow ((x', y'), (x, y) \in R$$

$\therefore$ R is symmetric

R is transitive: Let $((x, y), (x', y')) \in R$ and $((x', y'), (x'', y'')) \in R$, then $xy' = x'y$ and $x'y'' = x''y'$

we have $$xy'' = x\left(\frac{x''y'}{x'}\right) = xy'\left(\frac{x''}{x'}\right) \qquad \text{(from } x'y'' = x''y')$$

$$= x'y\left(\frac{x''}{x'}\right) \qquad \text{(since } xy' = x'y)$$

$$= yx'' = x''y$$

Thus $((x, y), (x'', y'')) \in R$

$\therefore$ R is transitive.

R is reflexive, symmetric and transitive. Therefore, R is an equivalence relation.

A/R = {(1, 1), (2, 2), (3, 3), (4, 4), (5, 5)}, {(1, 2), (2, 4)}, {(2, 1), (4, 2)}, {(1, 3)}, {(1, 4)}, {(2, 3)}, {(2, 5)}, {(3, 1)}, {(3,2)}, {(3, 4), {(1, 5)}, {(3, 5)} {(4, 1)}, {(4, 3)}, {(4, 5)}, {(5,1)}, {(5, 2)}, {(5, 3)}, {(5,4)}

3.36.6 Irreflexive Relation

Definition 3.34. A relation R on a set A is irreflexive if $a \not R\ a\ \forall\ a \in A$.

Example 1. Let $A = \{1, 2, 3\}$ and $R = \{(1, 2), (2, 3), (3, 1)\}$.
The relation R is irreflexive on A.

3.36.7 Compatible Relation

Definition 3.35. A relation R defined on a set A is said to be a compatible relation; if it is reflexive and symmetric.

Example 1. If R is an equivalence relation on a set A, then R is a compatible relation on A.

Example 2. Let $A = \{a, b, c, d\}$
and $R = \{(a, a), (a, b), (a, c), (b, a), (b, b), (b, c), (b, d), (c, a), (c, b), (c, c), (c, d), (d, b), (d, d)\}$ be a relation on A.

Clearly R is reflexive and symmetric.
Since (a, b), (b, d) does not imply $(a, b) \in R$,
i.e. $(a, b) \in, R, (b, d) \in R \not\Rightarrow (a, d)$
R is not transitive.
Hence, R is a compatible relation on A.

3.36.8 Universal Relation

Definition 3.36. A relation R on a set A is said to be a universal relation if $R = A \times A$

Example 1. Let $A = \{a, b\}$, then
$$R = \{(a, a), (a, b), (b, a), (b, b)\}$$
is a universal relation on A.

Example 2. Let $A = \{1, 2, 3\}$, then
$$R = \{(1, 1), (1, 2), (1, 3), (2, 1), (2, 2), (2, 3), (3, 1), (3, 2), (3, 3)\}$$
is a universal relation on the set A.

3.36.9 Complementary Relation

Definition 3.37. Let R be a relation from A to B, then the complementary relation $\overline{R}$, is the set of ordered pairs $\{(a, b) : (a, b) \notin R\}$

Example 1. Let R be the relation
$$R = \{(a, b): \text{a divides } b\}, \text{ on the set of positive integers}$$
Then $$\overline{R} = \{(a, b): a \text{ does not divide } b\}$$

Example 2. Let $A = \{1, 2, 3\}$ and

$$R = \{(1, 1), (1, 2), (1, 3), (2, 2), (3, 3)\}$$

Then $\overline{R} = \{(2, 1), (2, 3), (3, 1), (3, 2)\}$

3.36.10 The Inverse of a Relation

Definition 3.38. Let R be a relation from a set A to B. The inverse relation from B to A is the set of ordered pairs $\{(b, a) : (a, b) \in R\}$. The inverse of R is denoted by R^{-1}

$\therefore$ $$R^{-1} = \{(b, a): (a, b) \in R\}$$

Example 1. Let R be the relation

$R = \{(a, b)$: a divides $b\}$, on the set of positive integers of R is

$R^{-1} = \{(a, b)$: b divides $a\}$

Example 2. Let $A = \{a, b, c, d\}$ and $R = \{(a, b), (b, c), (a, d), (a, c), (c, b)\}$

Then $R^{-1} = \{(b, a), (c, b), (d, a), (c, a), (b, c)\}$

3.36.11 Irreflexive Relation

Let A be a non-empty set and R be a relation on A. If no element of A is related to itself by R then R is said to be an irreflexive relation.

3.37 SOLVED EXAMPLES

Example 3.37.1 If $A = \{2, 3\}$, $B = \{3, 4, 5, 6\}$ and R is a relation from A to B defined as follows:

$(a, b) \in R$, if "a divides b". write the solution set of R

Solution. 2 divide 4, and 2 divides 6

$\therefore$ $(2, 4) \in R$, and $(2, 6) \in R$

3 divides 3, and 3 divides 6

$\therefore$ $(3, 3) \in R$ and $(3, 6) \in R$

4 divides 4

$\therefore$ $(4, 4) \in R$

$\therefore$ $R = \{(2, 4), (2, 6), (3, 3), (3, 6), (4, 4)\}$

Example 3.37.2. If R is a transitive relation on a set A, then show that R^{-1} is also transitive.

Solution. Let $(a, b) \in R^{-1}, (b, c) \in R^{-1}$

then $(b, a) \in R$, and $(c, b) \in R$

Now $(c, b) \in R, (b, a) \in R, \Rightarrow (c, a) \in R$

$\Rightarrow (a, c) \in R^{-1}$

$\therefore$ $(a, b) \in R^{-1}, (b, c) \in R^{-1} \Rightarrow (a, c) \in R^{-1}$

$\Rightarrow R^{-1}$ is transitive.

Example 3.37.3 Show that the relation 'equality' defined in any set A, is an equivalence relation.

Solution. Let R denote the relation 'equality' on A

$$a = a \ \forall \ a \in A \quad \Rightarrow \quad a\,R\,a \ \forall \ a \in R$$

R is reflexive

$$a = b,\ b = a \text{ for all } a, b \in A$$

$\therefore$ $a\,R\,b \Rightarrow b\,R\,a$

$\Rightarrow R$ is symmetric

$$a = b,\ b = c \Rightarrow a = b = c \quad \Rightarrow a = c$$

$\therefore$ $a\,R\,b,\ b\,R\,c \Rightarrow a\,R\,c$

$\therefore$ R is transitive

Then R is an equivalence relation on A.

Example 3.37.4. Let Z denote the set of integers and the relation R on Z be defined by "$a\,R\,b$", if $a - b$ is an even integer. Then show that R is an equivalence relation.

Solution. R is reflexive:

$$o = a - a \text{ is even, hence } a\,R\,a \qquad \forall \ a \in Z$$

R is symmetric:

if $a - b$ is even, then $b - a$ is also even since $b - a = -(a - b) \ \forall \ a, b \in Z$

$a\,R\,b \Rightarrow b\,R\,a \qquad \forall \ a, b \in Z$

$\therefore$ R is symmetric.

R is transitive: If $a\,R\,b,\ b\,R\,c$ then

$$a - b \text{ is even, } b - c \text{ is even}$$

Consequently; $a - c = (a - b) + (b - c)$ is also even

$\therefore$ $a\,R\,b,\ b\,R\,c \Rightarrow a\,R\,c$

$\therefore$ Thus R is reflexive, symmetric and transitive

$\therefore$ R is equivalence relation on Z.

Example 3.37.5. If A is a set with n elements, then find the number of reflexive relations on A.

Solution. A is a set with n elements

A relation R on A is a subset of $A \times A$

Consequently, a relation in A is determined by specifying whether each of the n^2 ordered pairs in $A \times A$ is in R.

If R is reflexive, then each of the n ordered pairs (a, a) for $a \in A$ must be in R. Each of the other $n(n-1)$ ordered pairs of the from (a, b) where $a \neq b$, may or may not be in R. Hence, by the product rule of counting, there are $2^{n(n-1)}$ reflexive relations on A.

Example 3.37.6. Let A be the set of triangles in the plane, and R is the relation on A defined by "a is similar to b" then show that R is an equivalence relation on A.

Solution. Every triangle is similar to itself, therefore the relation R is reflexive.

If "a is similar to b" then "b is similar to a".

i.e. $\quad (a, b) \in R \Rightarrow (b, a) \in R$

$\therefore$ R is symmetric

Clearly, if 'a is similar to b' and 'b is similar to a'

Then a is similar to c'

$\therefore$ R is transitive

Therefore, R is reflexive, symmetric and transitive, thus R is an equivalence relation.

Example 3.37.7. Let $A = \{1, 2, 3, 4\}$ and

$$R = \{(1, 1), (1, 2), (2, 1), (2, 2), (3, 1), (3, 3), (1, 3), (4, 1), (4, 4)\}$$

is R an equivalence relation on A. (V.T.U. March 2000)

Solution. $a \in A \Rightarrow (a) \in R$

$\therefore$ $\Rightarrow R$ is reflexive

$(4, 1) \in R, (1, 4) \notin R$

i.e. $(a, b) \in R$ does not imply $(b, a) \in R$

Therefore R is not symmetric.

Hence R is not an equivalence relation on A.

Example 3.37.8. X is a family of sets and R is a relation on X, defined by "x is a subset of y". Show that R is a anti-symmetric and transitive.

Solution. Let $(A, B) \in R$, and $(B, A) \in R$

then $A \subseteq B$ and $B \subseteq A$

$\therefore$ $A = B$

Thus R is anti-symmetric

Also, $A \subseteq B, B \subseteq C \Rightarrow A \subseteq C$

i.e. $(A, B) \in R, (B, C) \in R \Rightarrow (A, C) \in R$

$\therefore$ R is transitive

Example 3.37.9. Let $A = \{1, 2, 3\}, B = \{1, 2, 3, 4\}$.

If $R_1 = \{(1, 1), (2, 2), (3, 3)\}$ and $R_2 = \{(1, 1), (1, 2), (1, 3), (1, 4)\}$ are relations from A to B, then find

$$R_1 \cup R_2, R_1 \cap R_2, R_1 - R_2, R_2 - R_1$$

Solution. $R_1 \cup R_2 = \{(1, 1), (2, 2), (3, 3), (1, 2), (1, 3), (1, 4)\}$

$R_1 \cap R_2 = \{(1, 1)\}$

$R_1 - R_2 = \{(2, 2), (3, 3)\}$

$R_2 - R_1 = \{(1, 2), (1, 3), (1, 4)\}$

3.38 PARTITIONS

Definition 3.39. If A is a non-equity set; then it can be decomposed into different subsets which have no common elements and whose union is the set A itself such a collection of sets is called a partition of A.

Definition 3.40. A partition of a non–empty set A is a disjoint collection P of non-empty subsets of A whose union is A.

The subsets of partition are called the cells or blocks of partition.

A given set may have more than one partition.

If $\tau = \{X_1, X_2, ..., X_n\}$ where $X_1, X_2, ..., X_n$ are non-empty subsets of x such that (i) $X_1 \cap X_i = \phi$, if $i \not\supset j$

(ii) $X_1 \cup X_2 \cup ... \cup X_n = A$

then τ is a partition of A.

Example 1. Let $A = \{a, b, c, d\}$

Then, $\tau_1 = \{\{a, b\} \{c, d\}, \tau_2 = \{\{a\}, \{b, c\}, \{d\}\}$

and $\tau_3 = \{\{a, c\}, \{b, d\}\}$

are partitions of A

Example 2. List all the partitions of A = {1, 2, 3}

Solution. {1}, {2}, {3}, {1, 2}, {2, 3}, {1, 3} are the non-empty subsets of A. The partition are

(i) $\tau_1 = \{\{1\}, \{2\}, \{3\}\}$

(ii) $\tau_2 = \{\{1\}, \{2, 3\}\}$
$= \{(1, 1), (2, 2), (2,3), (3, 2) (3, 3)\}$

(iii) $\tau_3 = \{\{2\}, \{1, 3\}\}$
$= \{(2, 2), (1, 1), (1, 3), (3, 1), (3, 3)\}$

(iv) $\tau_4 = \{\{3\}, \{1, 2\}\}$
$= \{(3, 3), (1, 1), (1, 2), (2, 1), (2, 2)\}$

3.39 RELATION RELATED SETS

Defination 3.41. Set A be a non-empty set and R be a relation on A. Let $x \in A$, then the R-relative set of x is defined to be the set of all elements y in B, with the property that x is related to y. It is denoted by R or by $[x]_R$.

$$\therefore \quad R(x) = [x]_R = \{y \in B: x\,R\,y\}$$

Example. Let $A = \{1, 2, 3, 4\}$, $B = \{a, b, c, d\}$ and $R = \{(1, a), (2, b), (3, c), (3, d)\}$ then $R(3) = \{c, d\}$

3.40 EQUIVALENCE CLASSES

Let $A = \{1, 2, 3, 4\}$ and

$R = \{(1, 1), (2, 2), (3, 3), (4, 4), (1, 2), (2, 1)\}$

be a relation defined on A. Clearly R is reflexive symmetric and transitive. Therefore R is an equal valence relation on A.

Consider $[1]_R$, $[2]_R$, $[3]_R$ and $[4]_R$

We have $[1]_R = \{1, 2\}$

$[2]_R = \{1, 2\}$

$[3]_R = \{3\}$

and $[4]_R = \{4\}$

observe that $[1]_R = [2]_R = \{1, 2\}$

$$[1]_R \cap [3]_R = [1]_R \cap [4]_R = [2]_R \cap [3]_R = [2]_R \cap [4]_R = \phi$$

and $[1]_R \cup [2]_R \cup [3]_R \cup [4]_R = A$

The relation R is such that it gives size to subsets $[x]_R$ of A for which either $[x]_R = [y]_R$ or $[x]_R \cap [y]_R = \phi$, $\forall x, y \in A$

Thus the relation R on A, induces a partition of A.

Theorem 3.14. Every equivalence relation on a set generates a unique partition of the set. The blocks of this partition form R-equivalence classes.

Proof. Let R be an equivalence defined on a set A

For any $x \in A$, $[x]_R \subseteq A$, be given by

$$[x]_R = \{y: y \in A \text{ and } x \, R \, y]$$

The relation R is an equivalence relation

$$\Rightarrow R \text{ is reflexive}$$

$$\Rightarrow x \, R \, x \text{ is true}$$

$$\Rightarrow x \in [x]_R$$

Let $y \in A$ such that $x \, R \, y$, then $y \in [y]_R$

R is symmetric, therefore $x \, R \, y \Rightarrow y \, R \, x$

$\therefore \quad x \in [y]_R$

Let $\quad z \in [y]_R$, then

$$x \, R \, y, x \, R \, z \Rightarrow y \, R \, z$$

$$\Rightarrow [y]_R \subseteq [x]_R$$

by symmetric, $\quad [x]_R \subseteq [y]_R$

$$[x]_R \subseteq [y]_R \text{ and } [y]_R \subseteq [x]_R \Rightarrow [x]_R = [y]_R$$

If $x \, R \, y$, then it is shown that $[x]_R = [y]_R$

If $x \;\; R \;\; y$, then we must shown that $[x]_R$ and $[y]_R$ are disjoint. To prove this, Let us assume that, these is at least one element, say $z \in [x]_R \cap [y]_R$

Now $x \, R \, z$ and $y \, R \, z \Rightarrow x \, R \, y$ which is *a* contradiction

Hence, $\quad x \;\; R \;\; y \Rightarrow [x]_R \cap [y]_R = \phi$

From the above it is clear that each element of A generates an R-equivalence class which is non-empty. The R-equivalence classes generated by any two elements are either equal of disjoint and the union of R-equivalence classes generated by the elements of A is the set A. Hence, the R-equivalence classes generated by the elements of A defines a partition of A.

Note. The family of the equivalence classes generated by the elements of A is denoted by A/R and is called the quotient set of A and R. Each element of A/R is a set.

Example. Let Z be the set of integers and let R be the relation "congruent modulo 5" defined by

$$R = (x, y): x \in Z, y \in Z \text{ and } (x - y) \text{ is divisible by } 5\}$$

Determine the equivalence classes generated by the elements of Z.

Solution. The equivalence classes are:

$$[0]_R = \{..., -10, -5, 0, 5, 10,...\},$$

$$[1]_R = \{..., -9, -4, 1, 6, 11,...\},$$

$$[2]_R = \{..., -8, -3, 2, 7, 12,...\},$$

$$[3]_R = \{..., -7, -2, 3, 8, 13,...\},$$

$$[4]_R = \{..., -6, -1, 4, 9, 14, 19,...\}$$

$$\therefore \quad Z/R = \{[0]_R, (1)_R, (2)_R, (3)_R, (4)_R\}$$

3.41 RELATIONS ON COORDINATE DIAGRAMS

If A and B are the subsets of real numbers, and R is a relation from A to B; then the relation R can be displayed on a coordinate diagram of $A \times B$, in which the ordered pairs of R are represented by points in a Cartesian plane.

Example. Let $A = \{1, 2, 3, 4\}$ and $B = \{1, 3, 5\}$

$$R = \{(1, 3), (2, 3), (3, 5), (4, 5)\}$$

Then sketch R on the coordinate diagram of $A \times B$.

Solution. R can be sketched on the coordinate diagram of $A \times B$ as follows

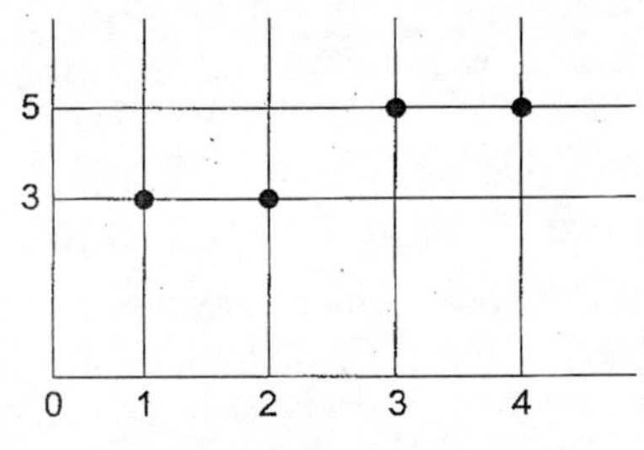

Fig. 3.2

3.42 COMPOSITION OF RELATIONS

Let R be a relation from A to B and S be a relation from B to C. Then we can define a relation, the composition of R and S written as $S \circ R$. The relation $S \circ R$ is a relation from the set A to the set C and is defined as follows

Definition 3.41. If $a \in A$, $b \in B$, $c \in C$, then $(a, c) \in S \circ R$ if and only if for some $b \in B$, we have $(a, b) \in R$ and $(b, c) \in S$.

Example. Let $A = \{1, 2, 3, 4\}$ and R, S be relations on A defined by

$$R = \{(1, 2), (1, 3), (2, 4), (3, 2)\}$$

$$S = \{(1, 4), (4, 3), (2, 3), (3, 1)\}$$

Then $(1, 2) \in R, (2, 3) \in S \Rightarrow (1, 3) \in S \, o R$

$(1, 3) \in R, (3, 1) \in S \Rightarrow (1, 3) \in S \, o R$

$(3, 2) \in R, (2, 3) \in S \Rightarrow (3, 3) \in S \, o R$

$(2, 4) \in R, (4, 3) \in S \Rightarrow (2, 3) \in S \, o R$

$\therefore$ $S \, o R = \{(1, 3), (1, 1), (3, 3), (2, 3)$

Theorem 3.15. Let A, B, C and D be sets; R be a relation from A to B. S be a relation from B to C and T be a relation from C to D. Then

$$T \, o (S \, o R) = (T \, o S) \, o R$$

Proof. Let M_R, M_S and M_T denote the matrices related to the relations R, S and T respectively. Then

$$M_{T \, o (S \, o R)} = M_{S \, o R} \cdot M_T = (M_Q \cdot M_S) \cdot M_T$$

Similarly $M_{(T \, o S) \, o R} = M_R \cdot M_{(T \, o S)}$

$M_R.(M_S.M_T)$

Since matrix multiplication is associative, we have

$$(M_R \cdot M_S) \cdot T = M_R \cdot (M_S \cdot M_T)$$

$\therefore$ $M_{T \, o (S \, o R)} = M_{(T \, o S) \, o R}$

i.e. The relations $T \, o (S \, o R)$ and $(T \, o S) \, o R$ are the same matrices

Hence, $T \, o (S \, o R) = (T \, o S) = R$

Theorem 3.16. Let R be a relation from the set A to B and C be a relation from the set B to the set C, then

$$(S \, o R)^{-1} = R^{-1} \, o \, S^{-1}$$

Proof. Let $(c, a) \in (S \, o R)^{-1}$, for some $c \in C$ and $a \in A$

Then $(c, a) \in (S \, o R)^{-1}$ if $(a, c) \in S \, o R$.

Therefore, there is an an element $b \in B$, with $(a, b) \in R$, and $(b, c) \in S$

Now, $(a, b) \in R$ and $(b, c) \in S \Rightarrow (b, a) \in R^{-1}$ and $(c, b) \in S^{-1}$

$\Rightarrow$ $(c, b) \in S^{-1}$ and $(b, a) \in R^{-1}$

$\Rightarrow$ $(c, a) \in R^{-1} o S^{-1}$

Thus $(S \, o R)^{-1} = R^{-1} o S^{-1}$

Example Let $A = \{a, b\}$

$$R = \{(a, a) \, (b, a) \, (b, b)\} \text{ and } S = \{(a, b), (b, a), (b, b)\}$$

be relations on A. Then verify

$$(S \, o R)^{-1} = R^{-1} \, o S^{-1}$$

Solution. We have

$$S \, o \, R = \{(a, b), (b, a), (b, b)$$

$\Rightarrow$ $$(S \, o \, R)^{-1} = \{(b, a), (a, b), (b, b)\}$$

and $$R^{-1} = [\, (a, a), (a, b), (b, b)\}, S^{-1} = \{(b, a)\}(a, b)(b, b)\}$$

$\Rightarrow$ $$R^{-1} \, o \, S^{-1} = \{b, a\}, (a, b), (b, b)\} \quad = (S \, o \, R)^{-1}$$

3.43 RELATION AND DIGRAPHS

A relation can be represented pictorially by drawing its graph. Let R be a relation on the set $A = \{a_1, a_2, ... a_n\}$.

The element a_i of A are represented by points (or circles) called nodes (or vertices). If $(a_i, a_j) \in R$, then we connect the vertices a_i and a_j by means of an arc and put an arrow in the direction a_i to a_j. If $(a_i, a_j) \in R$, and $a_j, a_i \in R$, then we draw two arc (edges) between a_i and a_j. If $(a_i, a_i) \in R$, then we draw one arc which starts from node a_i and ends or a_j. Such an arc is called a loop. When all the nodes corresponding to the ordered pairs in R connected by arcs (with proper arrows), we get a graph of the relation R. If R is reflexive, then we have a loop at each node in the graph of R. If R is symmetric then $(a_i, a_j) \in R$ implies $(a_j, a_i) \in R$, and the nodes a_i, a_j will be connected by two arcs (edges), one from a_j to a_i and the other from a_i to a_j.

In-degree: The number of edges terminating at a vertex in the digraph of a relation is called the in-degree of the vertex.

Out-degree: The number of edges leaving a vertex is called the out-degree of that vertex.

Example 1. Let $A = \{a, b, c\}$and R be a relation on A given by $R = \{(a, b), (a, c), (c, b), (c, c)\}$

Construct the digraph of R.

Solution. The digraph of R is shown in Fig. 3.3.

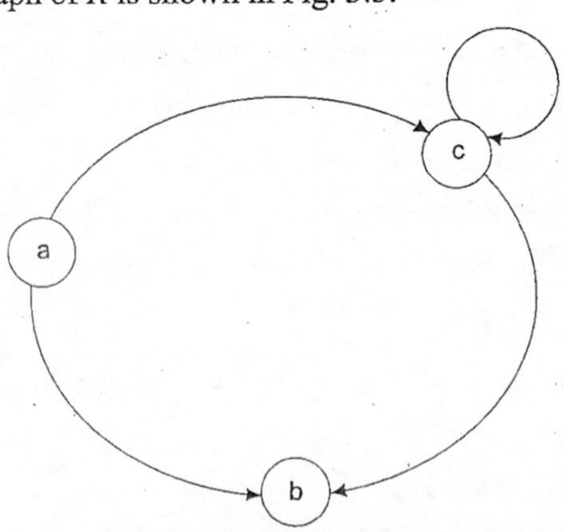

Fig. 3.3 Diagraph of Relation R

Example 2. Let $A = \{1, 2, 3, 4\}$ and

$$B = \{(1, 1), (2, 2), (3, 3), (1, 2), (2, 1), (2, 3), (2, 4), (3, 4), (4, 1)\}$$

Construct the digraph of R

Solution. The digraph of R is shown in Fig 3.4.

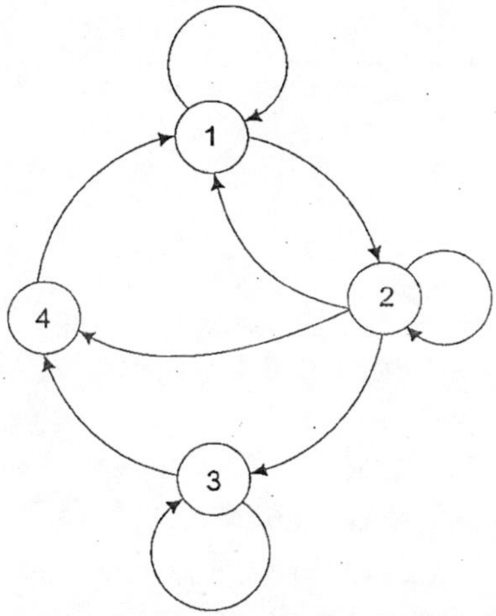

Fig. 3.4 Digraph of relation S

3.44 PATHS IN DIGRAPH OF A RELATION

If R is a relation on a set A, then a path of length n in R from a_i to a_j is a finite sequence P: $a_i, a_1 \ldots a_{n-i}\, a_j$ beginning with a_i and ending with a_j such that

$$a_i\, Ra_1,\, a_1\, Ra_2, \ldots .\, a_{n-1}\, Ra_i$$

Thus, *a* path in a digraph of the relation R is a succession of edges, where the indicated directions of the edges are followed. The length of *a* path P in the digraph of R is the number of edges occurring in it. The relation R^n: If n is a positive integer, the relation R^n on the set A can defined as following:

$(a_i, a_j) \in R^n$, if there is a path of length n from a_i to a_j in R. The relation R^∞ can be defined on A, by letting $(a_i, a_j) \in R^\infty$ mean, that there is a some path in R from a_i to a_j.

A cycle in a digraph is a path of length $n \geq 1$, from a vertex to itself.

Example 1. Let $A = \{1, 2, 3, 4, 5\}$, and R be a relation on A such that,

$R = \{(1, 1), (1, 2), (2, 3), (3, 5), (3, 4), (4, 5)\}$ compute (i) R^2 (ii) R^∞

Solution. The digraph of the relation R is shown in Fig. 3.5.

(i) $(1, 1) \in R$ and $(1, 1) \in R \Rightarrow (1, 1) \in R^2$

$(1, 1) \in R$ and $(1, 2) \in R \Rightarrow (1, 2) \in R^2$

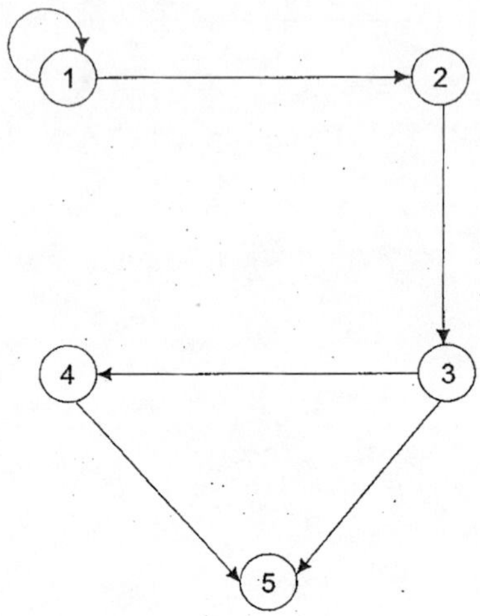

Fig. 3.5

$(1, 2) \in R$ and $(2, 3) \in R \Rightarrow (1, 3) \in R^2$

$(2, 3) \in R$ and $(3, 5) \in R \Rightarrow (2, 5) \in R^2$

$(2, 3) \in R$ and $(3, 4) \in R \Rightarrow (2, 4) \in R^2$

$(3, 4) \in R$ and $(4, 5) \in R \Rightarrow (3, 5) \in R^2$

$R^2 = \{(1,1), (1, 2), (1, 3), (2, 5), (2, 4), (3, 5)\}$

(ii) There is a path from 1 to 4 in $R \Rightarrow (1, 4) \in R^\infty$ length of the path from 1 to 4 is 3. There is a path from 1 to 5 $\Rightarrow (1, 5) \in R^\infty$ the length of the path is 3.

Similarly, $(1, 3) \in R^\infty$, $(1, 4) \in R^\infty$, $(2, 5) \in R^\infty$,

$\therefore \quad R^\infty = \{(1, 1), (1, 2), (1, 3), (1, 4), (1, 5), (2, 3), (2, 4) (2, 5), (3, 4), (3, 5), (4, 4)\}$

Example 2. Let $A = \{1, 2, 3, 4\}$ and $R = \{(1, 2), (2, 1), (2, 3), (3, 4), (4, 2)\}$ be a relation on A.

Then $R^\infty = \{(1, 1), (1, 2), (1, 3), (1, 4), (2, 1), (2, 3), (2, 4), (3, 1), (3, 2), (3, 3), (3, 4), (4, 1), (4, 2), (4, 3), (4, 4)\}$

3.45 TABULAR FORM OF A RELATION

A binary relation on a set can also be represented in the tabular form.

Example. Let $A = \{1, 2, 3, 4\}$ and

$R = \{(1, 1), (1, 2), (2, 2), (1, 4), (2, 4)\}$

be a relation on A. The tabular form of the relation R is shown in Fig 3.6.

The check marks in the cells represent the elements (ordered pairs) of R. If the cells in the main diagonal of the table contain the check marks then R is reflexive. For example, consider the relation $R = \{(1, 1), (2, 2), (3, 3), (4, 4), (1, 3), (2, 4), (3, 4)\}$.

	1	2	3	4
1	✓	✓		✓
2		✓		✓
3				
4				

Fig. 3.6

defined on $A = \{1, 2, 3, 4\}$

The tabular form of R is:

	1	2	3	4
1	✓		✓	
2		✓		✓
3			✓	✓
4				✓

Fig. 3.7

The cells in the main diagram of the table contain check marks. Thus, R is a reflexive relation on A.

If the relation R is a symmetrical relation on a set A, then the check marks, will be symmetrical with respect to the main diagonal in the table. For example, consider the relation $R = \{(1, 1), (1, 2), (2, 1), (2, 2), (2, 3), (3,2), (3, 3)$ defined on $A = \{1, 2, 3\}$

The tabular form of R is

	1	2	3
1	✓	✓	
2	✓	✓	✓
3		✓	✓

Fig. 3.8

The check marks are cells, symmetric with respect to the main diagonal. Therefore, R is symmetric.

3.4.5 Transitive Extension

Definition 3.42. Let R be a relation on the set A. Another relation R_1 defined on A is called the transitive extension of R, if R_1 contains R and

$$(a, b) \in R, (b, c) \in R \Rightarrow (a, c) \in R_1$$

Example. Let $A = \{1, 2, 3, 4\}$

$$R = \{(1, 2), (2, 3), (3, 2), (2, 4)\}$$

and $\quad R_1 = \{(1, 2), (1, 3), (2, 2), (2, 3), (2, 4), (3, 2), (3, 3), (3, 4)\}$

Clearly R_1 contains R and $(a, b) \in R, (b, c) \in R \Rightarrow (a, c) \in R_1$

$\therefore$ R_1 is a transitive extension of R

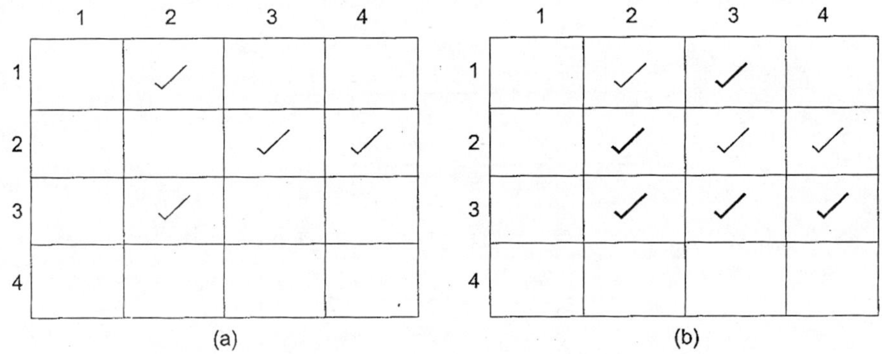

Fig. 3.9 Transitive extension

R is shown in Fig 3.9(a) and R_1 is shown in Fig. 3.9(b). Note that the ordered pairs in R_1 which are not in R are marked with heavy check lines.

3.46 MATRIX REPRESENTATION OF RELATIONS

Suppose A and B are both finite sets and R is a relation from A to B. Then R may be represented by a matrix called the relation matrix of R. It is denoted by M_R. The relation matrix M_R of R is also called the Zero-one matrix of R (or boolean matrix of R).

If $A = \{a_1, a_2, ..., a_m\}$ and $B = \{b_1 b_2, ..., b_n\}$ are finite sets containing m and n elements respectively and R is a relation from A to B, then the relation matrix of R is the $m \times n$, matrix $M_R = (m_{ij})_{m \times n}$, defined by

$$m_{i_j} = \begin{cases} 1 & \text{if } (a_i, b_j) \in R \\ 0 & \text{if } (a_i, b_j) \notin R \end{cases}$$

Where m_{i_j} is the element in the ith now and j^{th} column. M_R can be first obtained by constituting a table, whose columns are preceded by a column consisting of successive elements of A and whose rows are headed by a row consisting of successive elements of B.

If $(a_i, b_j) \in R$, then we enter i in the ith row and j^{th} column and if $(a_k, a_l) \notin R$, then we enter zero in the kth row and l^{th}, column.

Example 1. Let $A = \{1, 2, 3\}$ and $R = \{(x, y): x < y\}$. Find M_R

Solution. We have $R = \{(1, 2), (1, 3), (2, 3)\}$

The table and the corresponding relation matrix (zero-one matrix) are given below.

	1	2	3
1	0	1	1
2	0	0	1
3	0	0	0

(a)

$$M_R = \begin{bmatrix} 0 & 1 & 1 \\ 0 & 0 & 1 \\ 0 & 0 & 0 \end{bmatrix}$$

(b)

Fig. 3.10 Relation Matrix

The matrix M_R of a relation R on a set A, which is a square matrix, can be used to determine whether the relation R has certain properties. The relation R is reflexive if and only if $m_{ii} = 1$ for $i = 1, 2, ..., n$. In other words, R is reflexive if all the elements in the main diagonal of M_R are equal to 1.

Symmetric Relation: The relation R is symmetric if and only if $m_{ij} = m_{ji}$ for all pairs of integers i and j with $i = 1, 2,..., n$ and $j = 1, 2, ..., n$. In this case, we have $M_R = M_R^T$.

Anti-symmetric Relation: The relation R is anti-symmetric if $m_{ij} = 1$ with $i \neq j$, then $m_{ij} = 0$. For example, consider the relation matrix, M_R given below.

$$M_R = \begin{bmatrix} 1 & 1 & 0 \\ 0 & 1 & 1 \\ 0 & 1 & 1 \end{bmatrix}$$

The relation R is reflexive and symmetric, but R is not anti-symmetric.

3.46.1 Join of Zero-one Matrices

Definition 3.43. Let $A = [a_{i_j}]_{m \times n}$ and $B = [b_{i_j}]_{m \times n}$ be $m \times n$ zero-one matrices.

Then the join of A and B is the zero-one matrix with $(i, j)^{th}$ entry $a_{i_j} \vee b_{i_j}$. The join of A and B is denoted by $A \vee B$

Example. If $A = \begin{bmatrix} 1 & 1 & 0 \\ 0 & 1 & 1 \end{bmatrix}$, and $B = \begin{bmatrix} 0 & 0 & 1 \\ 0 & 1 & 0 \end{bmatrix}$

Then $A \vee B = \begin{bmatrix} 1 & 1 & 1 \\ 0 & 1 & 1 \end{bmatrix}$

3.46.2 The Meet of Zero-one Matrices

The meet of the matrices A and B is zero-one matrix with $a_{i_j} \wedge b_{i_j}$ as the $(i, j)^{th}$ entry. The meet of A and B is denoted by $A \wedge B$.

If $A = \begin{bmatrix} 1 & 1 & 0 \\ 0 & 1 & 1 \end{bmatrix}$ and $B = \begin{bmatrix} 0 & 1 & 0 \\ 0 & 1 & 1 \end{bmatrix}$

Then $A \wedge B = \begin{bmatrix} 0 & 1 & 0 \\ 0 & 1 & 1 \end{bmatrix}$

If M_{R_1} and M_{R_2} are matrices representing the relations R_1 and R_2 respectively. Then the matrices representing the union and intersection of these relations are

$$M_{R_1 \cup R_2} = M_{R_1} \vee M_{R_2} \text{ and } M_{R_1 \cap R_2} = M_{R_1} \wedge M_{R_2}$$

Example. Suppose $M_{R_1} = \begin{bmatrix} 1 & 1 & 0 \\ 0 & 1 & 0 \\ 0 & 0 & 1 \end{bmatrix}$ and $M_{R_2} = \begin{bmatrix} 1 & 0 & 1 \\ 0 & 1 & 1 \\ 0 & 1 & 0 \end{bmatrix}$

then, the matrices representing $R_1 \cup R_2$ and $R_1 \cap R_2$ are

$$M_{R_1 \cup R_2} = \begin{bmatrix} 1 & 1 & 1 \\ 0 & 1 & 1 \\ 0 & 1 & 1 \end{bmatrix} \text{ and } M_{R_1 \cap R_2} = \begin{bmatrix} 1 & 0 & 0 \\ 0 & 1 & 0 \\ 0 & 0 & 0 \end{bmatrix}$$

Let A, B and C be sets having m, n and p elements respectively. Suppose R be a relation from A to B, and S be a relation from B to C. Let $M_R = [a_{i_j}]_{m \times n}$, $M_s = [b_{i_j}]_{n \times p}$, denote the zero-one matrices representing the relations R and S respectively. Then the matrix representing the composite relation $S \circ R$ is denoted by $M_{S \circ R} = [c_{i_j}]_{m \times p}$.

Where $C_{ij} = 1$, if and only if $a_{l_k} = b_{k_j}$ for some k.

and we have

$$M_{S \circ R} = M_R \cdot M_S$$

We also have $M_{R_2} = M^{[2]}, \ldots, M_{R^n} = M_R^{[n]}$

Example. Let $M_R = \begin{bmatrix} 1 & 0 & 1 \\ 1 & 1 & 0 \\ 0 & 0 & 0 \end{bmatrix}$ and $M_S = \begin{bmatrix} 0 & 1 & 0 \\ 0 & 0 & 1 \\ 1 & 0 & 1 \end{bmatrix}$

denote the matrices representing the relation R and S respectively. Then the matrix for $S \circ R$ is

$$M_{S \circ R} = M_R \cdot M_S = \begin{bmatrix} 1 & 1 & 1 \\ 0 & 1 & 1 \\ 0 & 0 & 0 \end{bmatrix}$$

Note. An ordered pair (a_i, c_j) belongs to $S \circ R$, if and only if there is an element b_k such that $(a_i, b_k) \in R$, and $(b_k, c_j) \in S$.

Example. Let $A = \{1, 2, 3\}$

Let R and S be relations on A defined as follows

$$R = \{(1, 1), (1, 3), (2, 1), (2, 2), (2, 3), (3, 2)\}$$

and $$S = \{(1, 1), (2, 2), (2, 3), (3, 1), (3, 3)\}$$

Then $$S \circ R = \{(1, 1), (1, 3), (2, 1), (2, 2), (2, 3), (3, 2), (3, 3)\}$$

The matrix representation of R is

$$M_R = \begin{bmatrix} 1 & 0 & 1 \\ 1 & 1 & 1 \\ 0 & 1 & 0 \end{bmatrix}$$

and the matrix representation of S is

$$M_S = \begin{bmatrix} 1 & 0 & 0 \\ 0 & 1 & 1 \\ 1 & 0 & 1 \end{bmatrix}$$

Clearly $$M_{S \circ R} = \begin{bmatrix} 1 & 0 & 1 \\ 1 & 1 & 1 \\ 0 & 1 & 1 \end{bmatrix} = M_R \cdot M_S$$

can be easily verified.

3.47 CLOSURES OF RELATIONS

3.47.1 Reflexive Closures of a Relation

Definition 3.44. Let R be a relation on a set A. If S is reflexive relation on A, such that S contains R and S itself is contained within every reflexive relation that contains R, then S is called the reflexive closure of R.

If R is relation on a set A, the reflexive closuers of R can be obtained by adding all pairs of the form (a, a) with $a \in A$, provided (a, a) is not already in R.

The reflexive closure of R is the set $R \cup D$

where $D = \{(a, a): a \in A\}$ is the diagonal relation on A

Example. Let $A = \{1, 2, 3\}$

and $R = \{(1, 1)\}, (2, 2), (1, 2), (2, 1), (3, 2)\}$

be a relation on A.

The reflexive closure of R is obtained by adding (3,3) to R.

$\therefore$ The reflexive closure of R is $\{(1, 1)\}, (2, 2), (3, 3), (1, 2), (2, 1), (3, 2)\}$

3.47.2 Symmetric Closure

Definition 3.45. Let R be a relation on a set A. If S is any symmetric relation confaining, and S itself is contained in within every symmetric relation that contains R, then S is called the symmetric closure of R.

The symmetric closure of a relation R can be constructed by adding all ordered pairs of the form (b, a) where $(a, b) \in R$ that are not already present in R

Example. Let $R = \{(1, 1)\}, (1, 2), (2, 2), (2,3), (3, 1), (3, 2)\}$ be a relation on $A = \{1, 2, 3\}$. The symmetric closure of R is obtained by adding the ordered pairs (2, 1) and (1, 3). Since these are the only pairs of the form (b, a) with $(a, b) \in R$ that are not in R.

$\therefore$ The symmetric closure of R is the relation

$$\{(1, 1)\}, (1, 2), (2, 2), (2, 3), (3, 1), (3, 2), (2, 1), (1, 3)\}$$

Note. If R is a relation on A, and R^{-1} is its inverse then

$R \cup R^{-1}$ is the symmetric closures of R.

3.47.3 Connectivity Relation

Definition 3.46. Let R be a relation on a set A. The connectivity relation R^* consists of the ordered pairs (a_i, a_j), such that there is a path of length at least one from a_i to a_j.

R^* is the union of all the sets R^n.

i.e.
$$R^* = \bigcup_{n=1}^{\infty} R^n$$

Where R^n is the set of all pairs (a_i, a_j), such that there is a path of length n from a_i to a_j.

3.47.4 Transitive Closure

Definition 3.47. Let R be a relation on the set A.

R_1 denote the transitive extension of R, R_2 denote the transitive extension of R_1, and in general R_{i+1} denote the transitive extension of R_1 then the transitive closure of R is defined as the set union of $R, R_1, R_2, ..., R_{i+1}, ...$
It is denoted by R^+. Thus

$$R^+ = R \cup R_1 \cup R_2 \cup ... \cup R_{i+1} \cup ... = R_1 \cup R_2 \cup ... = R^*$$

i.e. $$R^+ = R^+ \cup R \cup R^2 \cup R^3 \cup ... \cup R^n \cup ...$$

$\therefore$ R^+ is the smallest transitive relation containing R and R^*.

Theorem. The transitive closure of a relation R equals the connectivity relation R^*

Example. Let $A = \{1, 2, 3, 4\}$, and $R = \{(1, 2)\}, (2, 3), (3, 4), (2, 1)\}$
then $R^+ = \{(1, 1), (1, 2), (1, 3), (1, 4), (2, 1), (2, 2), (2, 3), (2, 4), (3, 4)\}$ is the transitive closure of R.

If M_R denotes the zero-one matrix of the relation R on a set A with n elements, then the zero-one matrix of the transitive closure R^+ is the matrix.

$$M_{R^+} = M_R \vee M_R^{[2]} \vee M_R^{[3]} \vee ... \vee M_R^{[n]}$$

Example. Find the zero-one matrix of the transitive closure of the relation

$$= \begin{bmatrix} 1 & 0 & 1 \\ 0 & 1 & 0 \\ 1 & 1 & 0 \end{bmatrix}$$

Solution. Let A be the set on which R is defined. Then A is set consisting of 3 elements

$$M_{R^+} = M_R \vee M_R^{[2]} \vee M_R^{[3]}$$

$$M_R = \begin{bmatrix} 1 & 0 & 1 \\ 0 & 1 & 0 \\ 1 & 1 & 0 \end{bmatrix} \Rightarrow M_R^{[2]} = \begin{bmatrix} 1 & 1 & 1 \\ 0 & 1 & 0 \\ 1 & 1 & 1 \end{bmatrix} \text{ and } M_R^{[3]} = \begin{bmatrix} 1 & 1 & 1 \\ 0 & 1 & 0 \\ 1 & 1 & 1 \end{bmatrix}$$

$$\therefore \quad M_{R^+} = \begin{bmatrix} 1 & 0 & 1 \\ 0 & 1 & 0 \\ 1 & 1 & 0 \end{bmatrix} \vee \begin{bmatrix} 1 & 1 & 1 \\ 0 & 1 & 0 \\ 1 & 1 & 1 \end{bmatrix} \vee \begin{bmatrix} 1 & 1 & 1 \\ 0 & 1 & 0 \\ 1 & 1 & 1 \end{bmatrix}$$

$$= \begin{bmatrix} 1 & 1 & 1 \\ 0 & 1 & 0 \\ 1 & 1 & 1 \end{bmatrix}$$

EXERCISE 3.3

1. Let $A = \{1, 2, 3\}$, $B = \{1, 2, 3, 4\}$, $C = \{0, 1, 2\}$
   R be a relation from A to B, S be a relation from B to C, defined as follows
   $$R = \{(1, 1)\}, (1, 4), (2, 3), (3, 1), (3, 4)$$
   $$S = \{(1, 0)\}, (2, 0), (3, 1), (3, 2), (4, 1)$$
   Find SoR.
2. Let $R = \{(1, 1)\}, (2, 1), (3, 2), (4, 3)\}$ be a relation defined on $A = \{1, 2, 3, 4\}$. Compute R^2 and R^3.
3. How many different relations are there from a set with m elements to a set with n elements.
4. Let R_1 be the 'less than' relation on the set of real numbers and let R_2 be the 'greater than' relation on the set of real numbers i.e. $R_1 = \{(x, y): x < y\}$ and $R_2 = \{x, y: x > y)$, find $R_1 \cup R_2$, $R_1 \cap R_2$, $R_1 - R_2$, $R_2 - R_1$ and $R_1 \oplus R_2$.
5. If $M_R = \begin{bmatrix} 1 & 1 & 1 & 0 \\ 0 & 0 & 0 & 1 \\ 0 & 1 & 0 & 0 \\ 0 & 0 & 0 & 0 \end{bmatrix}$ and $M_S = \begin{bmatrix} 0 & 0 & 1 & 1 \\ 0 & 0 & 1 & 0 \\ 1 & 0 & 0 & 0 \\ 1 & 0 & 0 & 0 \end{bmatrix}$

   represent the relations R and S defined on $A = \{1, 2, 3, 4\}$, find $M_R \cdot M_S$ and SoR.
6. Let R and S be relations from $A = \{1, 2, 3, 4\}$ to $B = \{1, 2, 3\}$ where
   $$R = \{(1, 1)\}, (1, 2), (2, 1), (2, 3), (3, 2), (4, 3)$$
   $$S = \{(1, 3), (2, 3), (3, 2), (3, 3)\}$$
   Find (i) $\overline{R}$ (ii) $R \cup S$ (iii) R (iv) $R \cap S$ (v) S^{-1}
7. The graph of a relation R on $A = \{1, 2, 3, 4\}$ is given below Find R

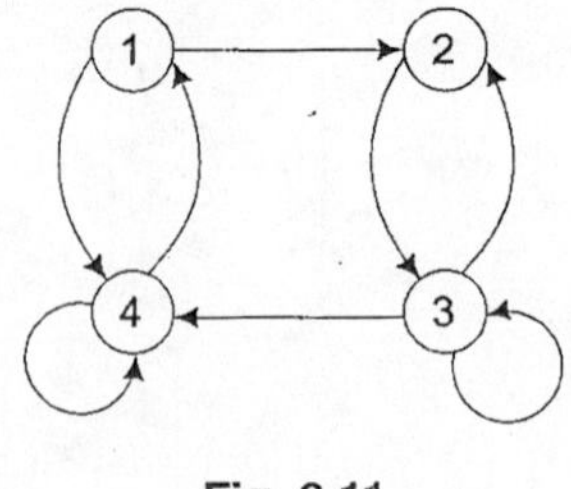

Fig. 3.11

8. Let $A = \{1, 2, 3, 4\}$, and $R = \{(1, 2), (2, 3), (3, 4), (2, 1)\}$ be a relation on A. Find the transitive closure of R.
9. Let $A = \{1, 2, 3, 4\}$. Determine whether the relations are reflexive, symmetric, anti-symmetric or transitive.

(a) $R = \{(1, 2), (2, 2), (3, 3)\}$

(b) $R = A \times A$

(c) $R = \{(1, 3), (4, 2), (2, 4), (3, 1), (2, 2)\}$

10. Let $R = \{(1, 2), (3, 4), (2, 2)\}$ and $S = \{(4, 2), (2, 5), (3, 1), (1, 3)\}$ be relations from A to B where $A = \{1, 2, 3, 4\}$ and $B = \{1, 2, 3, 4, 5\}$. Find RoS, SoR, RoR, SoS, and $RoRoR$

11. The following are relations in N, the set of natural numbers. Give their domains and range sets.

(i) $\{(1, 1), (16, 2), (81, 3), (256, 4)\}$

(ii) $\{(2, 1), (4, 2), (10, 5), (18, 9), (20, 10)\}$

12. Let $A = \{1, 2, 3\}$ and $R = \{(1, 1), (1, 4), (4, 1), (4, 4), (2, 2), (2, 3), (3, 2), (3,3)\}$

Write the matrix of R and sketch its graph.

13. Let $A = \{a, b, c, d, e, f, g, h\}$ and let $A_1 = \{a, b, c, d\}$, $A_2 = \{a, c, e, f, g, h\}$, $A_3 = \{a, c, e, g\}$, $A_4 = \{f, h\}$, $A_5 = \{b, d\}$. Find a partition of A from the sets of A_1, A_2, A_3, A_4, and A_5

14. Let $A = \{a, b, c, d\}$ and $R = \{(a, a), (a, b), (b, c), (c, a), (d, c), (c, b)\}$ be a relation on A. Find $R(a)$, $R(b)$. And if $A = \{c, d\}$. Find $R(A)$.

15. Prove that if R is a transitive relation on a set, the $R^2 \subseteq R$

(JNTU, June 2003)

ANSWER

1. $SoR = \{(1, 0), (1, 1), (2, 1), (2, 2), (3, 0), (3, 1)\}$
2. $R_2 = \{(1, 1), (2, 1), (3, 1), (4, 2)\}$

   $R_3 = \{(1, 1), (2, 1), (3, 1), (4, 1)\}$
3. 2^{mn}
4. $R_1 \cup R_2 = \{(x, y)\} : x \neq y\}$

   $R_1 \cap R_2 = \phi$, $R_1 - R_2 = R_1$, $R_2 - R_1 = R_2$, $R_1 \oplus R_2 = \{(x, y) > x \neq y\}$
5. $$M_R \cdot M_S = \begin{bmatrix} 1 & 0 & 1 & 1 \\ 1 & 0 & 0 & 0 \\ 0 & 0 & 1 & 0 \\ 0 & 0 & 0 & 0 \end{bmatrix}$$

   and $SoR = \{(1, 1), (1, 3), (1, 4), (2, 1), (3, 3)\}$
6. (i) $\{(1, 3), (2, 2), (3, 1), (3, 3), (4, 1), (4, 2)\}$

   (ii) $\{(1, 1), (1, 2), (1, 3), (2, 1), (2, 3), (3, 2), (3, 3), (4, 3)\}$

   (iii) $\{(2, 3), (3, 2)\}$

   (iv) $\{(3, 1), (3, 2), (2, 3), (3, 3)\}$
7. $R = \{(1, 1), (1, 2), (2, 3), (3, 2), (3, 3), (3, 4), (1, 4), (4, 1), (4, 4)\}$

8. $\{(1, 1), (1, 2), (1, 3), (1, 4), (2, 1), (2, 2), (2, 3), (2, 4), (3, 4)\}$

9. (a) Symmetric and transitive
   (b) Equivalence relation
   (c) Symmetric

10. $RoS = \{(4, 2), (3, 2), (1, 4)\}$

    $SoR = \{(1, 5), (3, 2), (2, 5)\}$

    $RoR = \{(1, 2)\}$, $SoS = \{(4, 5), (3, 3)\}$

    $RoRoR = \phi$

11. (i) Domain = $\{1, 16, 81, 256\}$, Range = $\{1, 2, 3, 4\}$
    (ii) Domain = $\{2, 4, 10, 18, 20\}$, Range = $\{1, 2, 5, 9, 10\}$

12. $M_R = \begin{bmatrix} 1 & 0 & 0 & 1 \\ 0 & 1 & 1 & 0 \\ 0 & 1 & 1 & 0 \\ 1 & 0 & 0 & 1 \end{bmatrix}$

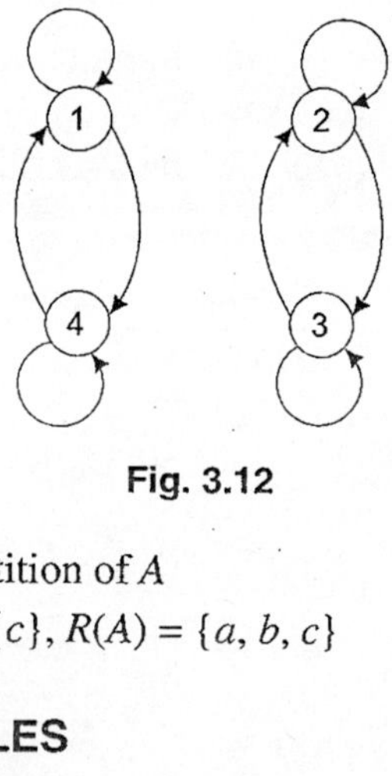

Fig. 3.12

13. $P = \{A_3, A_4, A_5)$ is partition of A

14. $R(a) = \{a, b\}$, $R(b) = \{c\}$, $R(A) = \{a, b, c\}$

3.48 SOLVED EXAMPLES

Example 3.48.1. Let $A = \{1, 2, 3, 4\}$, and R be a relation on A that has the matrix M_R where

$$M_R = \begin{bmatrix} 1 & 0 & 0 & 0 \\ 0 & 1 & 0 & 1 \\ 1 & 1 & 1 & 0 \\ 0 & 0 & 0 & 1 \end{bmatrix}$$

Construct the digraph of R, and find the in-degrees and out-degrees of all the vertices.

Solution.

	1	2	3	4
In-degree:	2	3	1	1
Out-degree:	1	1	3	2

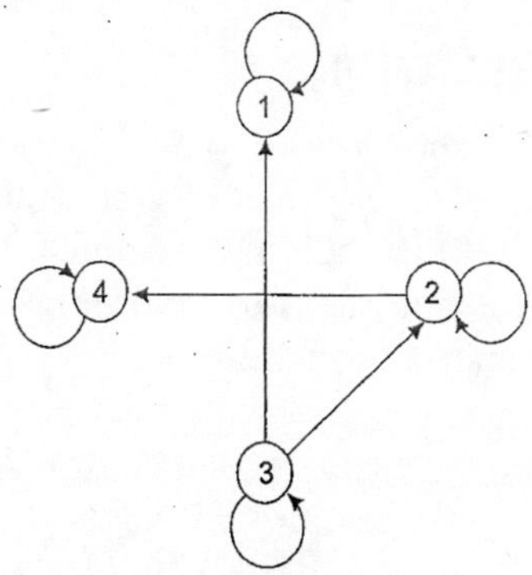

Fig. 3.13

Example 3.48.2. Find the relation determined by the Figure 3.14.

Solution. The relation R of the digraph is

$$R = \{(a, a), (a, c), (b, c), (c, b), (c, c), (d, c)\}$$

Example 3.48.3. Give an example of a non-empty set and a relation on the set that satisfies each of the following combinations of properties. Also a digraph of the relation.

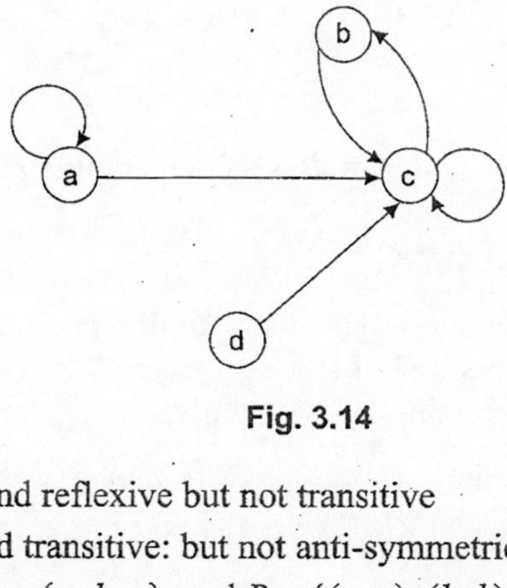

Fig. 3.14

(i) Symmetric and reflexive but not transitive

(ii) Reflexive and transitive: but not anti-symmetric

Solution. (i) Let $A = \{a, b, c\}$, and $R = \{(a, a), (b, b), (c, c), (a, b), (a, c), (b, a), (c, b)\}$

Clearly R is reflexive and symmetric, but not transitive.

(ii) Let $A = \{a, b, c\}$ and $R = \{(a, a), (b, b), (c, c), (a, b), (a, c), (b, c), (b, a), (c, b), (c, a)\}$

The relation R is reflexive and transitive but not anti-symmetric.

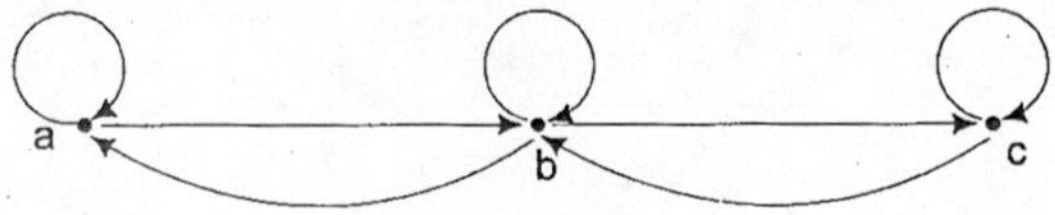

Fig. 3.15

3.49 COMPATIBILITY RELATIONS

Definition 3.48. A relation in a set A is said to be a compatibility relation if it is reflexive and symmetric. A compatibility relation is some times denoted by $\approx$. The relation matrix of a compatibility relation is symmetric and has its diagonal elements unity. Therefore it is sufficient to give only the elements of lower triangular part of the relation matrix in such a case.

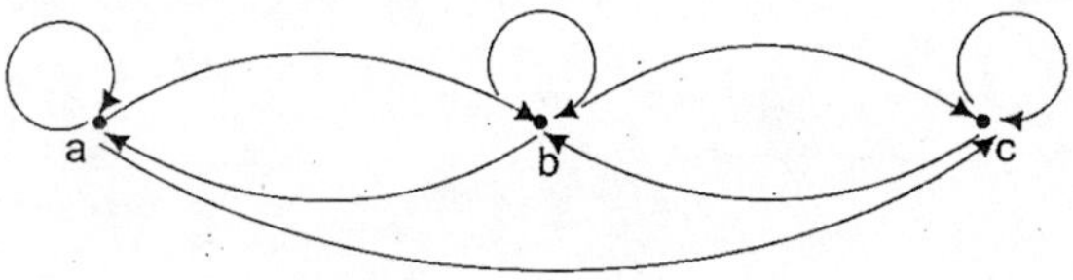

Fig. 3.16

Example 1. Let $A = \{1, 2, 3, 4\}$ and $R = \{(1, 1), (1, 2), (2, 1), (2, 2), (3, 3), (3, 4), (4, 3), (4, 4)\}$, be a relation on A. R is reflexive and symmetric, therefore R is a compability relation in A.

Example 2. Let $A = \{1, 2, 3, 4, 5\}$ and

$$R = \{(1, 1), (1, 2), (2, 1), (2, 2), (3, 3), (3, 4), (4, 3), (4, 4), (4, 5), (3, 5), (5, 3), (5, 4), (5, 5)\}$$

be a relation in A.

R is reflexive and symmetric. Hence, R is a compatibility relation in A.

3.50 COMPATIBILITY BLOCK

Definition 3.49. Let A be a set and $\approx$ be a compatibility relation. A subset $B \leq$ is called a maximal compatibility block if any element of B is compatible to every other element of B and no element of $A - B$ is compatible to all the elements of B.

Example 1. Consider the graph of the relation R, as shown in Figure 3.17

The maximal compatibility blocks of the relation R are $\{a, b, c, d\}$ $\{b, e\}$, $\{c, f\}$, $\{e, f\}$

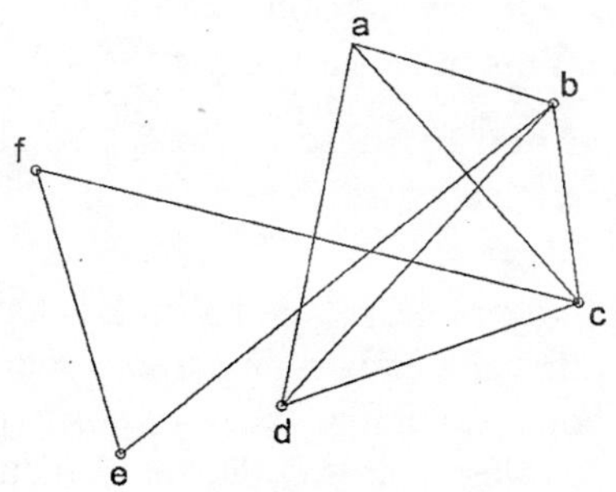

Fig. 3.17

Note. A compatibility relation defines a covering of the set on which it is defined.

Example 2. Suppose that the relation R on a set is represented by the matrix.

$$M_R = \begin{bmatrix} 1 & 1 & 0 \\ 1 & 1 & 1 \\ 0 & 1 & 1 \end{bmatrix}$$

All the diagonal elements of M_R are equal to 1, therefore the relation R is reflexive. M_R symmetric therefore R is symmetric.

Since R is reflexive and symmetric, R is a compatibility relation.

In this section we study partial order relation, and partially ordered sets. The idea of a partial order was inspired by the relation, which is a partial order on a given set.

3.51 PARTIAL ORDER RELATION

Definition 3.50. A relation R on a set A is called a partial order relation in A, if R is reflexive, anti-symmetric and transitive.

If R is a partial order on a set A, then A is said to be partially ordered by R. The partial order R on A is simply called an order relation on the set A.

Example 1. Let $A = \{1, 2, 3, 6, 12, 24, 36\}$ and R be a relation in A which is defined by "a divides b". Then R is a partial order in A.

Example 2. Let A be a non-empty set and $P(A)$ denote the power set of A. Then the relation "set inclusion" denoted by $\leq$ in $P(A)$ is a partial order relation in A.

3.52 PARTIALITY ORDERED SET (POSET)

Definition 3.51. Let P be a non-empty set and R be a partial order relation. Then P with the partial order R on it is called a partially ordered set or a Poset.

A partially ordered set is denoted by (P, R). When we want to specify the partial order relation R, usually, we denote the partial order relation by the symbol $\leq$ (The symbol $\leq$ does not necessarily mean "less than or equal to)"

Example. Let A be a collection of subsets of a set S. The relation $\subseteq$ of set inclusion is a partial order relation on A. The set $(A, \subseteq)$ is a partially ordered set (i.e. a Poset).

If Z denotes the set of integers the relation $\leq$ on Z is a partial order relation. The relation $\geq$ on the set of real numbers is partial order relation. The usual relation $\geq$ (greater than or equal to) on the set Z^+ of positive integers is a partial order relation. The relation $<$ (less than) on the set of natural numbers is not a partial order relation since it is not reflexive.

3.53 COMPARABILITY

Let R be a partial order on P and $a, b \in P$. If $a\,R\,b$ or $b\,R\,a$ then a, b are said to be comparable. For example, consider the partial order "divisibility" on Z^+ the set of all positive integers. The elements 2 and 4 are comparable since 2/4.

Definition 3.52. Let $(P, \leq)$ be a Poset. The elements $a, b \in P$ are said to be comparable if $a \leq b$ or $b \leq a$. Linearly ordered set:

If P is a partially ordered set, then some of the elements of P may be non-comparable. There are partially ordered sets in which every pair of elements is comparable and in which case the sets are called linearly ordered sets.

Definition 3.53. Let $(P, \leq)$ be a partially ordered set. If for every $a, b \in P$, we have $a \leq b$ or $b \leq a$ then $\leq$ is called linear ordering or simple ordering on P and $(P, \leq)$ is called a totally ordered set or simply ordered set or a chain.

Example. Let Z^+ be the set of all positive integers. The usual relation $\leq$ (less than or equal to) is a linear order. The $(Z^+, \leq)$ is a linearly ordered set.

3.54 DUAL ORDER

If $\leq$ is a partial order on a set P, with the converse of R is also a partial order on P, i.e. if $\leq$ is a partial order on P then $\geq$ is also a partial order on P. The partially order set $(P, \geq)$ is called the dual of $(P, \leq)$.

Example 1. Let $P = \{0, 1, 2\}$, then $(P, \leq)$ is a partial order on P. The dual of $(P, \geq)$ is $(P, \geq)$, that is the dual the set $\{(0, 0), (1,1), (2, 2), (1, 0), (2, 0), (2, 1)\}$.

Example 2. The dual of the Poset $(P, \geq)$ is the Poset $(P, \leq)$.

Example 3. The dual of the Poset (Z^+ "is a divisor of") is the Poset (Z^+, "is a multiple of").

3.55 HASSE DIAGRAM

A Hasse diagram is a pictorial representation of a finite partial order on a set. In this representation, the objects (i.e. the elements) are shown as vertices. Two related vertices in the Hasse diagram of a partial order are connected by line if and only if they are related.

Example 1. Let $A = \{a\}$ and $\leq$ be the inclusion relation on the elements *of* $P(A)$. The Hasse diagram of $(P(A), \leq)$ can be drawn as shown in figure given below.

Fig. 3.18

Example 2. Let $P = \{3, 4, 12, 24, 48, 72\}$ and the relation $\leq$ be defined on P such that $a \leq b$, if a divides b. The Hasse diagram of $(P, \leq)$ is shown in the Figure 3.19.

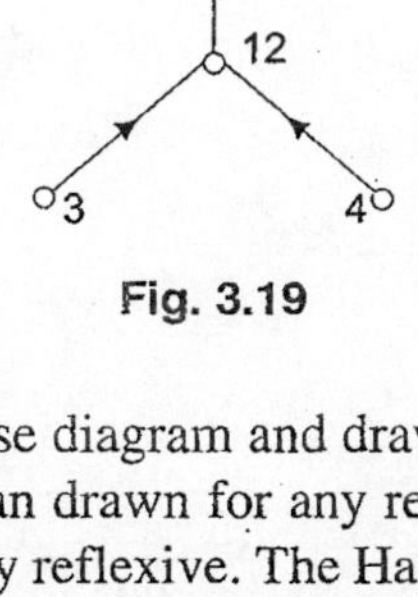

Fig. 3.19

We avoid arrows in a Hasse diagram and draw lines to show at the elements are related. Hasse diagrams can drawn for any relation which is anti-symmetric and transitive but not necessary reflexive. The Hasse diagram of a Poset need not be unique. For example, consider the set $A = \{a, b\}$. The relation set inclusion is a partial ordering on $P(A)$.

The Hasse diagrams of $(P(A), \leq)$ are shown in Fig. 3.20.

If $(P, \leq)$ is a Poset the Hasse diagram of $(P, \geq)$ can be obtained by rotating the Hasse diagram of $(P, \leq)$ through 180°; so that the points at the top become the points at the bottom.

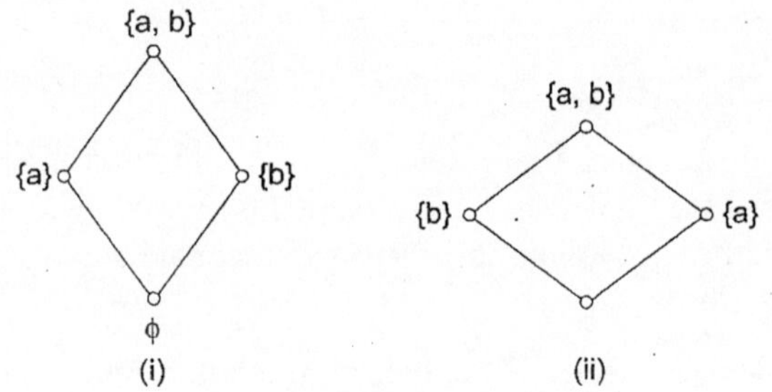

Fig. 3.20

Some Hasse diagrams may have a unique point which is above all the other points in the diagram and in some cases, the Hasse diagram may have a unique point which is below all the other points.

Definition 3.53 A relation on set A is said to be connected if for every pair of distinct elements $a, b \in A$, either $a\,R\,b$ or $b\,R\,a$. A partial ordering on a set A is said be linear ordering if it is connected.

3.56 LEXICOGRAPHIC ORDERING

Let $(A, \leq)$, and $(B, \leq)$ be two partially ordered sets. We define another partial order on $A \times B$ denoted by $\prec$ and is defined as follows:

$$(a, b) \prec (a', b') \text{ if } a < a' \text{ or if } a = a' \text{ and } b \leq b'.$$

The order $\prec$ is called Lexicographic ordering (or dictionary ordering). In this ordering, elements in the first coordinate dominates except in the case of ties. In the case of equality we consider second coordinate. The Lexicographic ordering defined above can be extended as follows:

Let $(A_1, \leq), (A_2, \leq), \ldots, (A_n, \leq)$ denote partially ordered sets. We define a partial order $\prec$ on $A_1 \times A_2 \times \ldots \times A_n$ as follows:

$(a_1, a_2, \ldots, a_n) \prec (a_1'a_2', \ldots, a_n')$ if and only if $a_1 < a_1'$ or

$a_1 = a_1'$ and $a_2 < a_2'$ or

$a_1 = a_1'$, $a_2 = a_2'$ and $a_3 < a_3'$ or

....

$a_1 = a_1'$, $a_2 = a_2', \ldots, a_{n-1} = a_{n-1}'$ and $a_n \leq a_n'$

In the above ordering, the first coordinate dominates. In the case of equality we consider the second coordinate. If the equality holds again, we pass on to the next coordinate and so on.

The order in which the words in an English dictionary appear is an example of Lexicographic ordering.

Example. Let $A = \{a, b, c, ..., z\}$ and let A be linearly ordered in the used way, $a \leq b, b \leq c, ..., y \leq z$. The set $A^n = A \times A \times ... \times A$ (n factors, can be identified with the set S^n of all words having length n. The Lexicographic ordering on A^n has the property that if $w_1 < w_2$ (where w_1 and w_2 are two words in S^n), then w_1 would precede w_2 in the dictionary listing.

Thus : bold $\prec$ bolt

Card $\prec$ cart

Park $\prec$ part

Mark $\prec$ mast

...

We can extend Lexicographic order to posets. If P is a partially ordered set and P^* denotes the set of all finite sequences of elements of P, we can extend Lexicographic order to P^* as follows.

Let $\alpha = a_1 a_2 ... a_m$, and $\beta = b_1 b_2 ... b_n$ belong to P^* with $m \leq n$. We can say that $\alpha < \beta$ if $(a_1, a_2, ..., a_m) < (b_1, b_2, ..., b_m)$ in A^n under the Lexicographic ordering P^n.

Example. Let $P = \{a, b, c, ..., z\}$ and let R be a simple ordering on P denoted by $\leq$, where $a \leq b \leq c \leq ... \leq z$.

Let $S = A \cup A^2 \cup A^3$

Then S consists of all words (strings) of three of fewer than 3 letters from P. Let $<$ denotes the lexicographic ordering on S.

We have

Be $\prec$ Bet

Leg $\prec$ Let

Peg $\prec$ Pet

Sea $\prec$ Sec

Rat $\prec$ Row

...

We can use names "lexically less than" or "lexically equal to" or "lexically greater than", to denote the lexicographic ordering.

3.56.1 Cover of an Element

Let $(P, \leq)$ be a partially ordered set. An element $b \in P$ is said to cover an element $a \in A$, if $a < b$ and if there does not exist any element $c \in P$ such that

$$a \leq c \text{ and } c \leq b$$

If b covers an element a, then a line is drawn between the elements a and b. The set pairs (a, b) such that b cover a is called the covering relation o

Example. Let $S = \{1, 2, 3\}$, then the covering relation $\{(A, B)\}$: $A \leq$. power set of S is

{ϕ, 1, ϕ, 2, ϕ, 3, 1, 1, 2, 1, 1, 3}

{2, 1, 2, 2, 2, 3, 3, 1, 3, 3, 2, 3}

{1, 2, 1, 2, 3, 1, 3, 1, 2, 3}

{2, 3, 1, 2, 3}

3.57 LEAST AND GREATEST ELEMENTS

Let $(P, \leq)$ denote a partially ordered set. If there exists an element $a \in P$, such that $a \leq x \ \forall \ x \in P$, then a is called the least member in P, relative to the partial ordering $\leq$. Similarly, if case exists an element $b \in P$. Such that $x \leq b \ \forall \ x \in P$, then b is called the greatest member in P relative to $\leq$.

The least member of P is usually denoted by 0, and the greatest member of P is denoted by 1. For a given Poset, the greatest or least member may or may not exist. The least member in a Poset P if it exist is unique and the greatest number in a Poset P if it exist is unique. In every chain the least and greatest member always exist.

Example 1. Let $A = \{a, b\}$ and $P(A)$ denote the power set of A. Then $P(A) = \{\phi, \{a\}, \{b|, \{a, b\}$. Let $\leq$ be the set inclusion on the elements of $P(A)$ already in the least member and $\{a, b\}$ is the greatest member in $(P(A), \leq)$

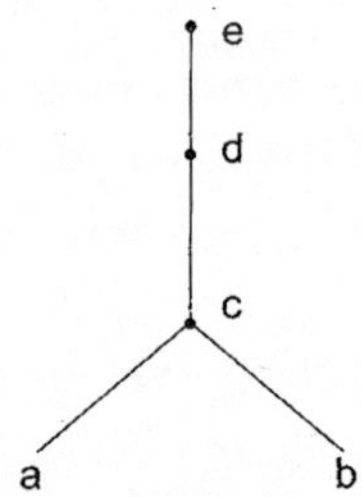

Fig. 3.21

Example 2. Consider Poset P where Hasse diagram is given in Fig. 3.21.
e is the greatest element in P. There is no least member in P.
Example. 3 Let $P\{1, 2, 3, 4, 5\}$, and $\leq$ the relation usual "less than or equal to". Then the Hasse diagram of $(P, \leq)$ is shown in the Figure 3.22.

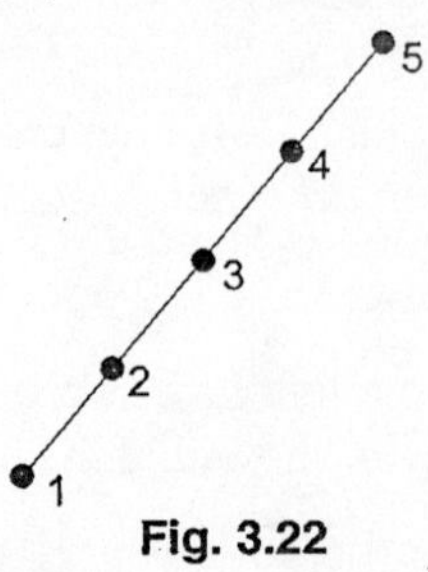

Fig. 3.22

3.58 MAXIMAL AND MINIMAL ELEMENTS

Let $(P, \leq)$ denote a partially ordered set. An element $a \in P$ is called a minimal element of a relative to $\leq$, if no $x \in P$, such that $x < a$. Similarly, an element $b \in P$ is called a maximal member of P relative to the partial ordering $\leq$ if is for no $x \in P$ such that $b < x$.

The minimal and maximal members of a partially ordered set need not be unique. Distinct minimal elements of a Poset are incomparable and distinct maximal elements of a Poset are also incomparable.

Example 1. Consider the Poset shown in the Figure 3.23.

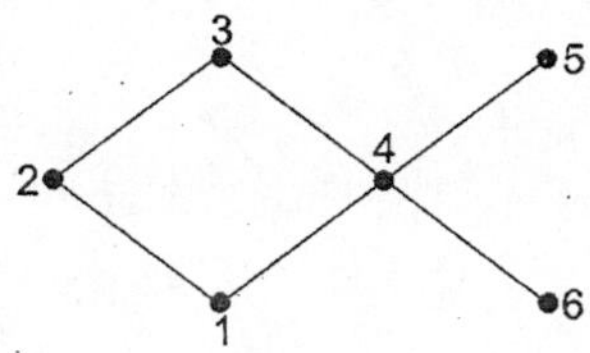

Fig. 3.23

There are two maximal elements and two minimal elements.

Example 2. For the Poset shown in the following Hasse diagram, f, g are the maximal elements and the elements a, b, c are minimal elements.

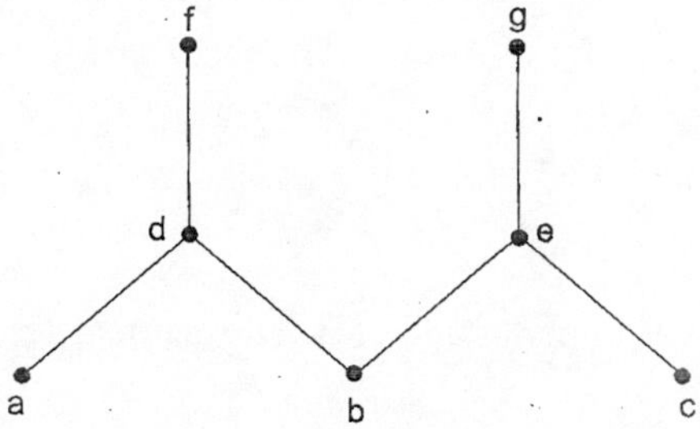

Fig. 3.24

3.59 UPPER AND LOWER BOUNDS

Some times it is possible to find an element that is greater than all the elements in a subset A of a Poset $(P, \leq)$. Any element $u \in P$ is called an upper bound for A if for all $x \in P, x \leq u$. Similarly an element $l \in P$ is called a lower bound of A, if for all $x \in A, l \leq x$

Example 1. Let $P = \{1, 2, 3, 4, 5, 6\}$ be ordered as pictured in the Figure 3.25.

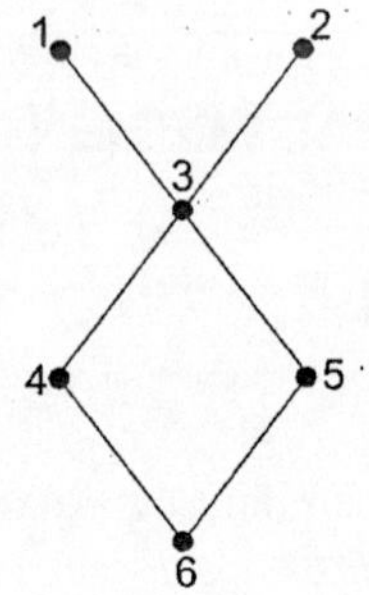

Fig. 3.25

If $A = \{4, 5\}$, then the upper bounds of A are 1, 2 and 3. The lower bound of A is 6.

Example 2. Let $A = \{a, b, c\}$ and $(P(A), \leq)$ be the partially ordered set where the partial order $\leq$ is the set inclusion. The Hasse diagram of the Poset is pictured in the figure given below.

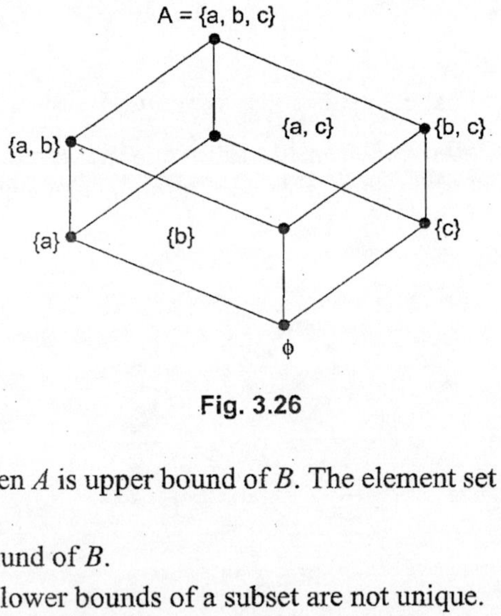

Fig. 3.26

If $B = \{a, c\}$ then A is upper bound of B. The element set is a lower bound of B.

ϕ is also a lower bound of B.

The upper and lower bounds of a subset are not unique.

3.60 LEAST UPPER BOUND (SUPREMUM)

Definition 3.54. Let P be a partially ordered set, and A be a subset of P. An element $M \in P$ is called the least upper bound of A if M is an upper bound of A and $M \leq M'$ where M' is an upper bound of $M \leq M'$. The least upper bound of a subset A is denoted by sup(A) (or Sup A).

Example. Let $P = \{a, b, c, d, e, f, g, h)$ denote a partially ordered set. Whose Hasse diagram is shown in the Figure 3.27.

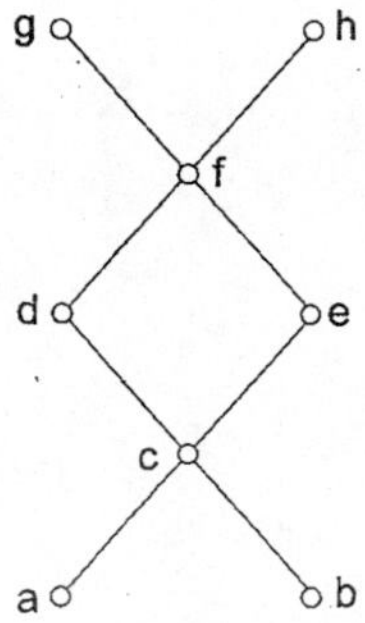

Fig. 3.27

If $A = \{c, d, e\}$, then, f, g, h are upper bounds of A. The element f is the least upper bound of A.

If $(P, \leq)$ is a poset, then its dual $(P, \geq)$ is is also a Poset. The least number of $(P, \leq)$ in the greatest member in $(P, \geq)$ relative to $\geq$ and vice versa.

Theorem 3.17. Let $(P, \leq)$ be a partially ordered set and A be a subset of P. Then

(i) The least upper bound of A, if it exists is unique

(ii) The greatest lower bound of A, if it exists is unique

Proof. (i) If possible let be b_1 and b_2 be the least upper bounds of A and let $b_1 \neq b_2$. Now b_2 is supremum and b_1 is an upper bound of A

$$\Rightarrow b_2 \leq b_1$$

Similarly b_1 is supermum and b_2 is an upper bond of A

$$\Rightarrow b_1 \leq b_2$$

Therefore by symmetric property

$$b_2 \leq b_1;\ b_1 \leq b_1 \leq b_2 \Rightarrow b_1 = b_2$$

(ii) Left as an exercise.

3.61 THE GREATEST LOWER BOUND (INFINUM)

Definition 3.55. Let P is partially ordered set and A denote a subset of P. An element m is called the greatest lower bound of A if m is a lower bound of A, and $m' \leq m$ wherever m' is a lower bound of A. It is denoted by $g\ l\ b(A)$ or $Inf\,(A)$.

The greatest lower bound is Poset if it exists is unique.

Example Consider the Poset $P = \{1, 2, 3, 4, 5, 6, 7, 8\}$. Whose Hasse diagram is shown in the Figure 3.28.

If $B = \{3, 4, 5\}$, then the elements 1, 2, 3 are lower bounds of B.

3 is the greatest lower bound of A.

The *glb* of a and b is the called the meet or product of a and b, and the $l\ u\ b$ of a and b is called the join or sum of a and b where a and b are the elements of z^+. The symbols such as * and $\oplus$ are used to denote meet and join respectively. If P is a chain, then every subset A of P has a supremum and infinum.

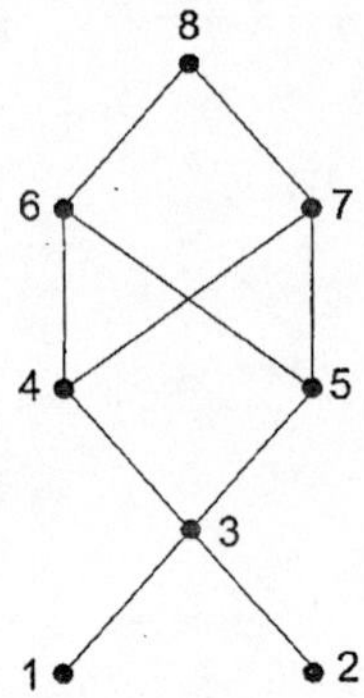

Fig. 3.28

If Z^+ denotes the set of partial integers, ordered by divisibility. If a and b are elements of N, then

$$\text{inf} \quad (a, b) = gcd\,(a, b)$$

$$\text{and Sup} \quad (a, b) = lcm\,(a, b)$$

where $gcd\,(a, b)$ is the greatest common divisor of a and b, and $lcm\,(a, b)$ is the least common multiple of a and b.

3.62 WELL ORDERED SET

A set P with an ordering relation in well-ordered, if every non-empty subset of the set P has a least element.

Example. The set of natural numbers is well-ordered and well-ordering theorem.

Theorem. Every set P can be well ordered.

3.63 N-ARY OPERATIONS

Let A be a non-empty set and f be a mapping

$$f: A \times A \to A$$

Then f is called a binary operation on A.

If $f : A^n \to A$ is a mapping, then f is called an n-ary operation. n is called the order of the operation f. If $n = 1$, then f is called a unary operation.

i.e. $f : A \to A$ is a unary operation.

Example. Addition is a binary operation the set of integers.

EXERCISE 3.4

1. Let $A = \{a, b, c, d\}$ and R = {(a, b), (b, c), (a, c), (a, d), (c, d)}, find the symmetric closure of R.

2. Let $A = \{1, 2, 3, 4, 5\}$ and $R = \{(1, 1), (1, 2), (2, 3), (3, 5), (3, 4), (4, 5)\}$ be a relation on A. compute (i) R^r (ii) (R^∞).
3. If R is a relation on a set A. Then prove that R^∞ is the transitive closure of R.
4. Let $A = \{a, b, c, d\}$ and R is a relation that has the matrix

$$M_R = \begin{bmatrix} 1 & 0 & 0 & 0 \\ 0 & 1 & 0 & 0 \\ 1 & 1 & 1 & 0 \\ 0 & 1 & 0 & 1 \end{bmatrix}$$

Construct the diagram of R and list the in-degrees and out-degrees of all vertices.

5. Let R and S be two relations on the set of positive integer I

$$R = \{(x, 2x) : x \in I\},\ S = \{(x, 7x) : x \in I\}$$

compute (i) RoR (ii) SoR (iii) $RoRoR$ (iv) $RoSoR$

6. Let $A = B = \{1, 2, 3, 4\}$ and $R = \{(1, 1), (1, 3), (2, 3), (3, 1), (4, 2), (4, 4)\}$ $S = \{(1, 2), (2, 3), (3, 1), (3, 2), (4, 3)\}$ be relations on A.

Compute (i) $M_{R \cup S}$ (ii) $M_{R \cap S}$ (iii) $M_R{-1}$ (iv) $M_{\bar{S}}$

7. If $A = \{1, 2, 3, 4\}$ and R be a relation whose zero-one matrix is

$$M_R = \begin{bmatrix} 0 & 1 & 0 & 0 \\ 1 & 0 & 1 & 0 \\ 0 & 0 & 0 & 1 \\ 0 & 0 & 0 & 1 \end{bmatrix}$$

Find R and matrix of transitive closure of R.

8. Let $A = \{1, 2, 3, 4, 5\}$ and R be a relations such that

$$M_R = \begin{bmatrix} 1 & 0 & 0 & 1 & 0 \\ 0 & 1 & 1 & 0 & 0 \\ 0 & 0 & 0 & 1 & 0 \\ 1 & 0 & 1 & 0 & 1 \\ 0 & 1 & 0 & 0 & 1 \end{bmatrix}$$

Compute the transitive closure of R.

9. Let $A = \{1, 2, 3, 6\}$. Define R on A by aRb if a divides b. Show that R is not an equivalence relations. (V.T.U. July 2002)

10. If $S = \{a, b, c\}$ and $A = P(S)$, draw the Hasse diagram of $(A, \subseteq)$.

[V.T.U. Mach 2000]

11. Find the matrix of the partial order whose Hasse diagram is

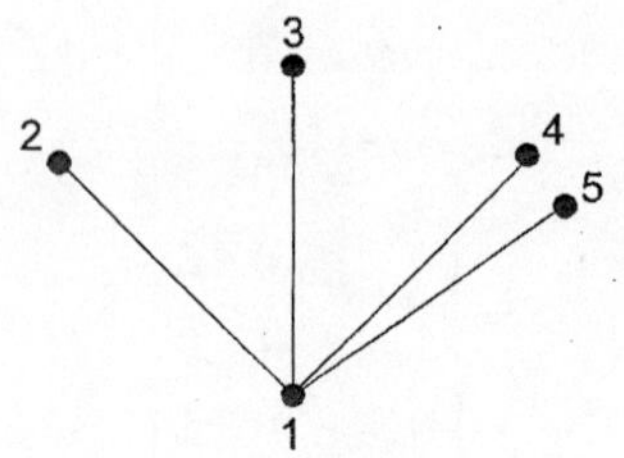

Fig. 3.29

12. Find the maximal and minimal element of the poset A whose Hasse diagram is given below

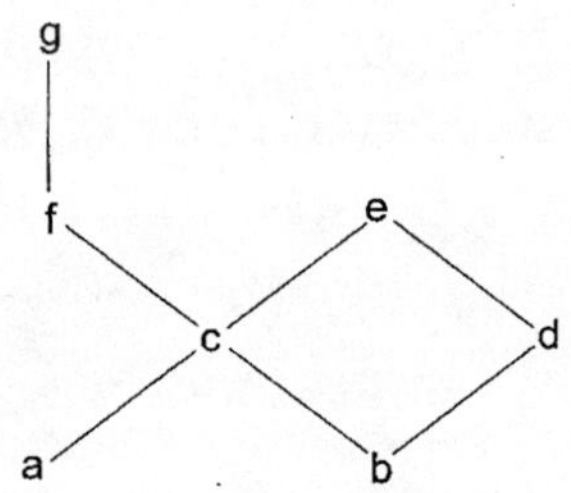

Fig. 3.30

13. Let $(A, \leq)$ be a Poset. Show that A has at least one maximal element and one minimal element. (V.T.U. Mach 2001)

14. Let $A = \{1, 2, 3, 4\}$ which of the following relations are reflexive A and Transitive on A

$$R_1 = \{(1, 1), (1, 3), (2, 1)\}$$

$$R_2 = \{(1, 1), (1, 2), (1, 4), (2, 1), (2, 2), (3, 3), (4, 1), (4, 4)\}$$

$$R_3 = \{(1, 1), (1, 2), (1, 3), (1, 4), (2, 2), (2, 3), (2, 4), (3, 3), (3, 4), (4, 4)\}$$

15. What is the reflexive closure of the relation $R = \{(a, b): a < b\}$

16. Which elements of the Poset $(\{2, 4, 5, 10, 12, 20, 25\}, |)$ are maximal and which are minimal.

17. Determine whether the Posets represented by each of the Hasse diagrams in the figures given ahead have a greatest element and a least element.

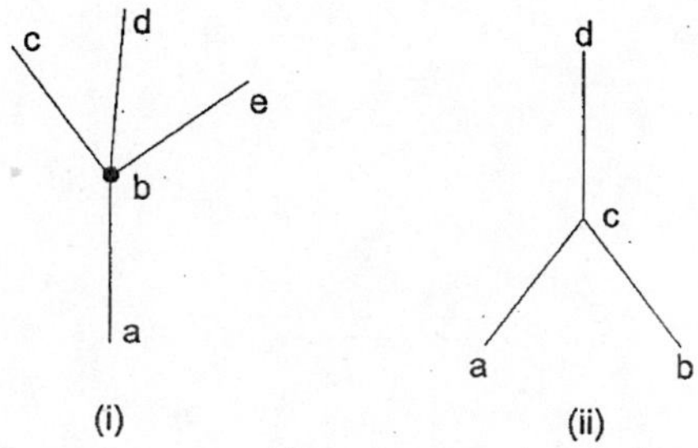

Fig. 3.31

18. Find the lower bounds of the subsets (i) {1, 2, 3} and (ii) {1, 3, 4, 6} in the Poset whose Hasse diagram is given below

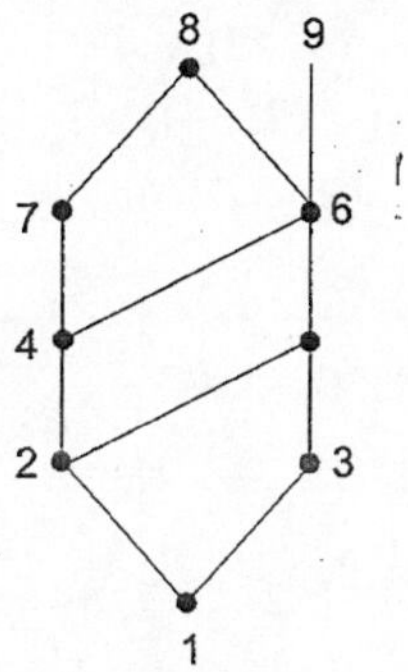

Fig. 3.32

19. Draw the Hasse diagram of the Poset ({2, 4, 5, 10, 12, 20, 25), |)

ANSWERS 3.4

1. $R \cup R^{-1} = \{(a, b), (a, c), (a, d), (b, a), (b, c), (c, a), (c, b), (c, d), (d, a), (d, c)\}$

2. $R^{\infty} = \{(1, 1), (1, 2), (1, 3), (2, 5), (2, 4), (3, 5)\}$

3. $R^{\infty} = \{(1, 1), (1, 2), (1, 3), (1, 4), (1, 5), (2, 3), (2, 4), (2, 5), (3, 4), (3, 5)$ $(4, 5)\}$

4. $R = \{(a, a), (b, b), (c, a), (c, b), (c, c), (d, b), (d, d)\}$

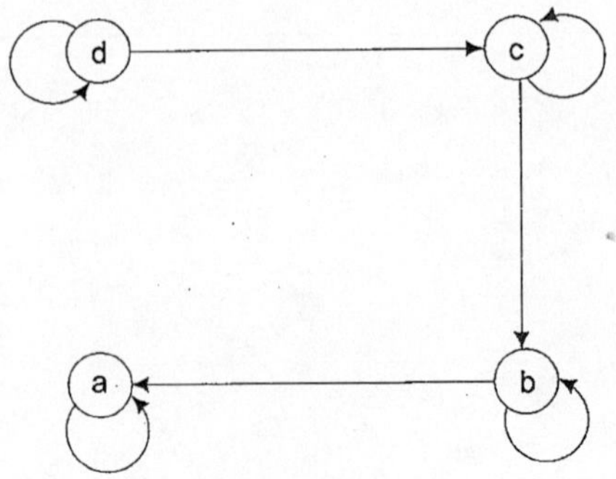

Fig. 3.33

Vertex:	a	b	c	d
In-degree:	2	3	1	1
Out-degree	1	1	3	2

5. (i) $RoS = \{(2, 28), (4, 56), (6, 84), ...\}$

(ii) $SoR = \{(1, 14), (2, 28), (3, 42), ...\}$

(iii) $RoRoR = \{(1, 8), (2, 16), (3, 34), ...\}$

(iv) $RoSoR = \{(1, 28), (1, 56), ...\}$

6. (i) $\begin{bmatrix} 1 & 1 & 1 & 0 \\ 0 & 0 & 1 & 0 \\ 1 & 1 & 0 & 0 \\ 0 & 1 & 1 & 1 \end{bmatrix}$ (ii) $\begin{bmatrix} 0 & 0 & 0 & 0 \\ 0 & 0 & 1 & 0 \\ 1 & 0 & 0 & 0 \\ 0 & 0 & 0 & 0 \end{bmatrix}$ (iii) $\begin{bmatrix} 1 & 0 & 1 & 0 \\ 0 & 0 & 0 & 1 \\ 1 & 1 & 0 & 0 \\ 0 & 0 & 0 & 1 \end{bmatrix}$

(iv) $\begin{bmatrix} 1 & 0 & 1 & 0 \\ 1 & 1 & 0 & 1 \\ 0 & 0 & 1 & 1 \\ 1 & 1 & 0 & 1 \end{bmatrix}$

7. $R = \{(1, 2), (2, 1), (2, 3), (3, 4), (4, 4)\}$

$$M_{R^\infty} = \begin{bmatrix} 1 & 1 & 1 & 1 \\ 1 & 1 & 1 & 1 \\ 0 & 0 & 0 & 1 \\ 0 & 0 & 0 & 1 \end{bmatrix}$$

8. $\begin{bmatrix} 1 & 1 & 1 & 1 & 1 \\ 1 & 1 & 1 & 1 & 1 \\ 1 & 1 & 1 & 1 & 1 \\ 1 & 1 & 1 & 1 & 1 \\ 1 & 1 & 1 & 1 & 1 \end{bmatrix}$

9.

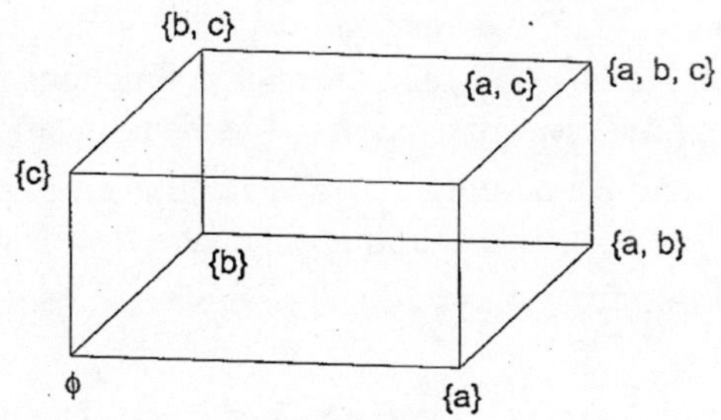

Fig. 3.34

10. $M_R = \begin{bmatrix} 1 & 1 & 1 & 1 & 1 \\ 0 & 1 & 0 & 0 & 0 \\ 0 & 0 & 1 & 0 & 0 \\ 0 & 0 & 0 & 1 & 0 \\ 0 & 0 & 0 & 0 & 1 \end{bmatrix}$

11. Minimal elements: a, b

Maximal elements: e and g

12. R_2 and R_3 are reflexive R_3 is transitive

13. $R \cup \Delta = \{(a, b), a \leq b\}$

14. The minimal elements are 2, 5

The maximal elements are 12, 20, 25

15. (i) The least element is a

The Poset has no greatest element.

(ii) d is the greatest element .

There is no least element.

16. (i) The upper bounds of $\{1, 2, 3\}$ are 5, 6, 8, 9

The only lower bound of $\{1, 2, 3\}$ is 1.

(ii) The upper bounds of $\{1, 3, 4, 6\}$ are 6, 8, 9. The lower bound is 1.

17.

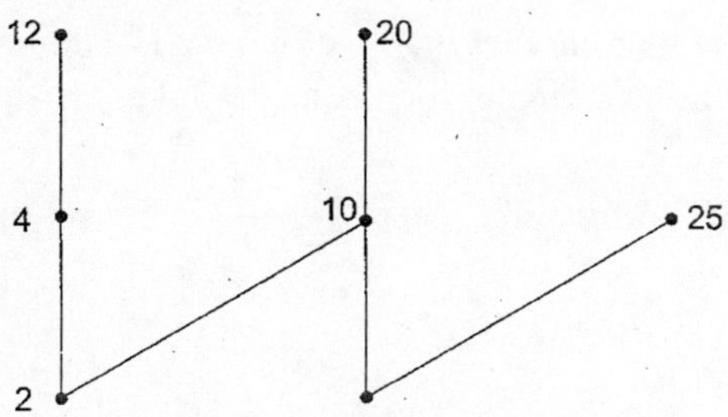

Fig. 3.35

3.64 FUNCTIONS

The concept of relation was defined very generally in the preceding section. We shall now discuss particular class of relations called functions. They are widely used in mathematics and the concept is basic to the idea of computation.

Definition 3.56. Let A and B non-empty sets. A relation from A to B is called a function if for every $a \in A$. There is a unique element $b \in B$, such that $(a, b) \in f$.

Thus, if it is a function from A to B, then f is a relation from A to B, such that

(i) Domain of $f = A$

(ii) Whenever $(a, b) \in f$, and $(a, c) \in f$, we have $b = c$. The notation $f: A \to B$ means f is a function from A to B.

Funetions are also called mappings or transformations. The terms such as 'correspondence' and 'operation' are used as synonyms for "function".

Given any function $f: A \to B$, the notation $f = b$ means $(d, b) \in f$. It is customary any to write $b = f(d)$. The element d is called an argument and $f(d)$ is called the value of the function for the argument d or the image of d under f.

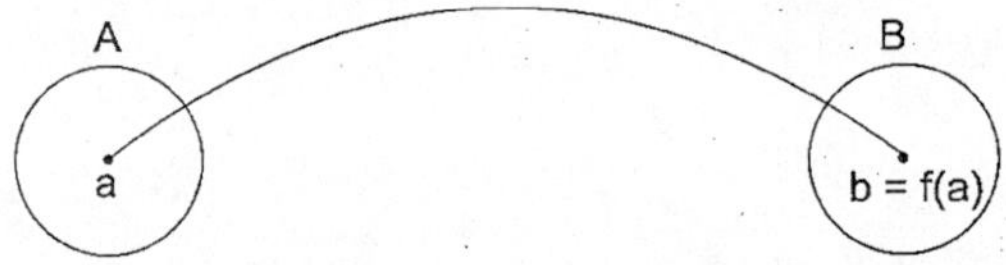

Fig. 3.36 Representation of a function

Example 1. Let $A = \{a, b, c\}$ $B = \{1, 2, 3\}$ and $f = \{(a, 1), (b, 2), (c, 3)\}$. Then, we have $f(a) = 1, f(b) = 2$, $f(c) = 3$, clearly f is a function from A to B.

Example 2. Let $A = \{a_1, a_2, a_3, a_4\}$, $B = \{b_1, b_2, b_3, b_4\}$ and $f = \{(a_1, b_1), (a_2, b_1), (a_3, b_3), (a_4, b_3)\}$. Here, every element of A is related to exactly one element of B, i.e. the association is unique. Hence, f in a function.

If f is a function from A to B, then A is called the domain of f and the set B is called the codomain of f. The set of all images under f is called the range of the function f. It is denoted by $f(A)$. The range of f is also denoted by R_f.

If D_f denotes the domain of a function $f: A \to B$ and R_f denotes the range of f, then $D_f = A$, and $R_f \subseteq B$.

Example 3 (i) Consider the function $f(x) = \sqrt{2x - 6}$. The domain of the function f is the set of all real numbers greater than equal to 3. i.e. $[3, \infty]$

(ii) The domain of the function $f(x) = \dfrac{x^2 - 4}{x + 2}$ is the set of all real numbers except -2. i.e. $D_f = R - \{-2\}$

(iii) The domain of $f(x) = \dfrac{x^4+7}{x^2+6x+5}$ is $R - \{-5, -1\}$

(iv) Consider the function $f: A \to N$ defined by

$$f(a) = 1, f(b) = 5, f(c) = 7, f(d) = 11$$

where $A = \{a, b, c, d\}$ and N is the set of the natural numbers. The range of f is the set $\{1, 5, 7, 11\}$.

A function need not be defined by a formula. While defining the property, it is customary to identify the function by formula, for example, $f(x) = x^2 + x + 1$ for $x \in R$, represents the function $f = \{x^2 + x + 1 : x \in R\}$, where R is the set of real numbers.

3.64.1 Restriction and Extension

Definition. 3.57. If $f: A \to B$ and $P \subseteq A$, then $f \cap (P \times B)$ is a function from $P \to B$, called the restriction of the function f to P. The restriction of f to P is written as f/p: $P \to B$.

Where the domain of f/p is P and $f/p(x) = f(x)\ \forall\ x \in P$.

If g is a restriction a function f, then f is called extension of g.

Example. Let $f: R \to R$ be defined by $f(x) = x\}$. If N is the set of natural numbers and N is $\{0, 1, 2, 3,...)$, then $N \subseteq R$ and $f/N = \{(0, 0), (1, 1), (2, 2)...\}$

3.64.2 Injective Functions

Definition 3.58. Let $f: A \to B$ be a function from A to B. If distinct elements of A are mapped to distinct elements of B, then f is called injective (q one-one) function. In other words, the function $f: A \to B$ is said to be injective, if whenever $a_1 \neq a_2$, then $f(a_1) \neq f(a_2)$.

If f is an injective function, we also say that f is an injection.

In order to prove that a function f is injective, we must establish that for all a_1, a_2 in A if

$$f(a_1) = f(a_2), \text{ then } a_1 = a_2$$

We assume that $f(a_1) = f(a_2)$ and obtain $a_1 = a_2$

Example 1. If $f: R \to R$ is defined by $f(x) = 5x\ \forall\ x \in R$, then f is an injective function since:

$$f(x_1) = f(x_2) \Rightarrow 5x_1 = 5x_2 \Rightarrow x_1 = x_2\ \forall\ x_1, x_2 \in R$$

Example 2. Let $A = \{x \in R : x \neq 2\}$ and define $f(x) = \dfrac{x}{x^2-2}$ for all $x \in A$. Show that f is one-one.

Solution. Let $x_1, x_2 \in A$ such that $f(x_1) = f(x_2)$, then

We have $$\frac{x_1}{x_1 - 2} = \frac{x_2}{x_2 - 2} \Rightarrow x_1 (x_2 - 2) = x_2 (x_1 - 2)$$

$$\Rightarrow x_1 = x_2$$

Therefore, f is one-one i.e. injective.

3.64.3 Onto Function: (Surjective Function)

Definition 3.59. Let $f: A \Rightarrow B$ be a function from A to B. If $f(A) = B$, that is if the range $R_f = B$. Then f is said to be a subjective function or onto mapping.

If f is a surjective function, we also say that f is a surjection.

If f is not onto, then it is said to be into mapping.

Example 1. The function $f: R \to R$, given by $f(x) = 2x \ \forall \ x \in R$ is onto.

Example 2. The function $f: Z \to Z$, defined by $f(x) = x^2$ is not onto, since there is no integer such that $x^2 = -1$.

3.64.4 Bijective Function

Definition 3.60. If a function $f: A \to B$ is both injective and surjective, then f is said to be bijective.

If f is a bijective function, then f is called a bijection or a one-to-one correspondence.

Example 1. $f: R \to R$, defined by $f(x) = 2x + 5$ for all $x \in R$, is a bijection.

3.64.5 Identity Mapping

Definition 3.61. Let A be a non-empty set and $f : A \to A$ be a mapping. If every element of A is mapped onto itself then f is called an identity mapping on A. It is denoted by I_A.

$$\therefore \qquad I_A = \{(a, a) : a \in A\}$$

3.64.6 Constant Function

Definition 3.62. If $f : A \to A$ is a function such that every element of A is mapped onto the some element of A then f is called a constant function.

If f is a constant function, then the range of f has only one element in it.

Example. If $f: R \to R$ is defined by $f(x) = 4 \ \forall \ x \in R$, then f is a constant function. We have $R_f = \{4\}$

3.64.7 Inverse of Function

Definition 3.63. If $f : A \to B$ is a bijection of A onto B, then

$$\{(b, a) \in B \times A : (a, b) \in f\}$$

is a function on B into A. This function is called the inverse function of f, and is denoted by f^{-1}. The function f^{-1} is also called the inverse of f.

If $f: A \to B$ is a bijection and $f(a) = b$.

Then $a = f^{-1}(b)$ where $a \in A$ and $b \in B$.

Example 1. Let $A = \{a, b, c\}$, $B = \{1, 2, 3\}$, and $f = \{(a, 1), (b, 3), (c, 2)\}$. Clearly f is both one-one and onto.

$\therefore$ $$f^{-1} = \{(1, a), (2, c), (3, b)\} \text{ is the inverse of } f.$$

Example 2. Let $A = \{x \in R: x \neq 1\}$, and define $f(x) = \dfrac{2x}{x-1}$

f is a bijective function with Range of f as the set $B = \{y \in R : y \neq 2\}$.

The inverse of f is given by

$$f^{-1} = \frac{y}{y-2} \qquad \text{for } y \in B$$

Example 3. $f : Q \to Q$ defined by $f(x) = 3x + 7$ is invertible and $f^{-1}(y) = \dfrac{y-7}{3}$.

Theorem 3.18. If $f: A \to B$ is both one-one and onto, then $f^{-1}: B \to A$ is both one-one and onto.

Proof. Let $f : A \to B$ be both one-one and onto. Then there exist elements $a_1, a_2 \in A$ and elements $b_1, b_2 \in B$ such that

$$f(a_1) = b_1, \text{ and } f(a_2) = b_2$$

$\Rightarrow$ $$(a_1) = f^{-1}\ (b_1), a_2 = f^{-1}(b_2)$$

now let $$f^{-1}(b_1) = f^{-1}(b_2)$$

then $$f^{-1}(b_1) = f^{-1}(b_2)$$

$\Rightarrow$ $$a_1 = a_2$$

$\Rightarrow$ $$f(a_1) = f(a_2)$$

$\Rightarrow$ $$b_1 = b_2$$

$\therefore$ f^{-1} is one-one

Again, since f is onto, for $b \in B$, there is exists an element $a \in A$, such that $b = f(a)$

$\therefore$ $$b = f(a)$$

$\Rightarrow$ $$f^{-1}(b) = a \qquad (\text{i.e. } a = f^{-1}(b))$$

$\Rightarrow$ f^{-1} is onto

Hence, f^{-1} is both one-one and onto.

3.64.8 Composition of Functions

Definition 3.64 Let $f: A \to B$ and $g: B \to C$ be two mappings. Then the composition of two mappings f and g denoted by $g \, o \, f$ is the mapping from A to C defined by

$$g \, o \, f = \{(a, c) : (a, b) \in f, (b, c) \in g \text{ for some } b \in B\}$$

i.e. $\quad g \, o \, f : A \to C$ is a mapping defined by

$$(g \, o \, f)(a) = g(f(a)) \text{ where } a \in A$$

$g \, o \, f$ cannot be defined unless the range of f is a subset of the domain of g.

In general, the computing of functions is not commutative

i.e. $\quad g \, o \, f \neq f \, o \, g$, where f and g are functions.

Example Let $f : R \to R$, $g : R \to R$ be defined by $f(x) = x + 1$, $g(x) = 2x^2 + 3$. Then

$$(g \, o \, f)(x) = g[f(x)] = g(x + 1) = 2(x + 1)^2 + 3 = 2x^2 + 4x + 5$$

and $\quad (f \, o \, g)(x) = f[g(x)] = f(2x^2 + 3) = 2x^2 + 3 + 1 = 2x^2 + 4$

$\therefore$ $g \, o \, f$ and fog are both defined but $g \, o \, f \neq f \, o \, g$

Theorem 3.19. Let $f : A \to B$, $g : B \to C$ be both are-one and onto functions. Then $g \, o \, f : A \to C$ is also one-one and onto.

Proof. Let $a_1, a_2 \in A$ Then

$$(g \, o \, f)(a_1) = (g \, o \, f)(a_2)$$

$\Rightarrow \quad g(f(a_1) = g(f(a_2)$

$\Rightarrow \quad f(a_1) = f(a_2) \quad (\because g \text{ is one-one})$

$\Rightarrow \quad a_1 = a_2 \quad (\because f \text{ is one-one})$

Hence $g \, o \, f$ is one-one

Now, from the definition $g \, o \, f : A \to C$ is a function $g : B \to C$ is onto, therefore

$c \in C \Rightarrow$ there is some element $b \in B$, such that $c = g(b)$

$f : A \to B$ is onto, therefore by definition there exists an element $a \in A$, such that $f(a) = b$

$\therefore \quad c = g(b) = g(f(a)) = (g \, o \, f)(a)$

$\Rightarrow \quad g \, o \, f : A \to C$ is onto

$\therefore$ $g \, o \, f$ is both one-one and onto (i.e. bijective)

Note. The converse of two theorem is not there i.e. if $g \, o \, f$ is one-one and onto, then it is not necessary that f and g are also one-one and onto.

3.65 ASSOCIATIVITY *OF* MAPPINGS:

If $f: A \to B$, $g : B \to C$, $h : C \to D$ are three functions, then $g \, o \, f : A \to C$, $f \, o \, g : C \to D$ are also functions. We can form $(h \, o \, g) \, o \, f: A \to D$ and $h \, o \, (g \, o \, f): A \to D$.

assuming that $a \in A$, we have

$$[(h\,o\,g)\,o\,f]\,(a) = (h\,o\,g)\,o\,f(a)$$
$$= (h\,o\,g)\,f(a)$$
$$= h(g\,(f(a)\,)$$
$$= h(gof\,(a)$$
$$= h\;o\,(g\,o\,f)\,(a)$$

$\therefore$ $$(h\,o\,g)o\,f = h\,o(g\,o\,f)$$

Thus the composition of functions is associative.

Example. Let $f(x) = x + 2$, $g(x) = x - 2$, and $h(x) = 3x$ for $x \in R$, where R is the set of real numbers. Find

(i) $g\,o\,f$ (ii) $f\,o\,g$ (iii) $h\;o\,g$ (iv) $f\,o\,g\,o\,h$ (v) $(h\,o\,g)\,o\,f$ (vi) $h\;o(g\,o\,f)$

Solution.

(i) $(g\,o\,f)\,(x) = g[\,f(x)] = g[x + 2] = [x + 2] - 2 = x$

(ii) $(f\,o\,g)\,(x) = f[g\,(x)] = f(x - 2) = (x - 2) + 2 = x$

(iii) $(h\,o\,g)\,(x) = h(g\,(x)) = h(x - 2) = 3(x - 2)$

(iv) $f\,o\,g\,o\,h = f(h(g(x)) = f(h(x - 2)) = f(3(x - 2)) = f(3x - 6) = 3x - 6 + 2$
$= 3x + 4$

(v) $(h\;o\;g)of(x) = (h\;o\;g)\,.(f(x)) = (h\;o\;g)\;(x + 2) = h\;(g(x + 2))$
$= h(x + 2 - x) = h(x) = 3x$

(vi) $h\;o(g\,o\,f)\,(x) = h[(g\,o\,f)(x)] = h[(g\,o\,f)(x)] = h[g(f(x)] = h\;(g\;(x + 2)$
$= h(x + 2 - 2) = h(x) = 3x$

Theorem 3.20. If $f: A \rightarrow B$ is an invertible mapping then

$$f\,o\,f^{-1} = I_B \text{ and } f^{-1}\;o\,f = I_A$$

where I_A is an identity mapping on A and I_B is an identity mapping on B.

Proof. f is invertible, then f^{-1} is defined by

$$f(a) = b \Leftrightarrow f^{-1}(b) = a \text{ where } a \in A \text{ and } b \in B$$

To Prove $f\,o\,f^{-1} = I_B$; let $b \in B$ and $f^{-1}(b) = a$, for some $a \in A$

then $\quad f\,o\,f^{-1}\,(b) = f(f^{-1}(b) = f(a) = b$

$\therefore \quad f\,o\,f^{-1}(b) = b \;\forall\; a \in B$

$\Rightarrow \quad f\,o\,f^{-1} = I_B$

now $\quad (f^{-1}\;o\,f)\,(a) = f^{-1}\,(f(a)) = f^{-1}(b) = a$

$\therefore \quad f^{-1}\;o\,f(a) = a \;\forall\; a \in A$

$\Rightarrow \quad f^{-1}\;o\,f = I_A$

Theorem 3.21. If $f: A \to B$, $g: B \to C$ are both one-one and onto, then $(g \, o \, f)^{-1} = f^{-1} o \, g^{-1}$.

Proof. $f: A \to B$ is one-one and onto

$g: B \to C$ is one-one and onto

$\therefore$ $gof: A \to C$ is also one-one and onto

$\Rightarrow (g \, o \, f)^{-1}: C \to A$ is one-one and onto

Let $a \in A$, then there exists an element $b \in B$ such that

$$f(a) = b \Rightarrow g = f^{-1}(b)$$

Now, $b \in B \Rightarrow$ there exists an element $c \in C$ such that

$$g(b) = c \Rightarrow b = g^{-1}(c)$$

Then $(g \, o \, f)\,(a) = g[f(a)] = g(b) = c \Rightarrow a = (g \, o \, f)^{-1}(c)$ ...(1)

and $(f^{-1} o \, g^{-1})\,(c) = f^{-1}\,(g^{-1}(c)) = f^{-1}(b) = a \Rightarrow a = f^{-1} o \, g^{-1}\,(c)$ ...(2)

combining (1) and (2) we have

$$(g \, o \, f)^{-1} = f^{-1} \, o \, g^{-1}$$

SOLVED EXAMPLES

Example 1. On which sets A will the identity function $I_A: A \to A$ be (i) one-one (ii) an onto function.

Solution. A can be any set and the identity function is always one-one and onto.

Example 2. Can a constant function be one-one and onto solution. If f is a constant function, the co-domain g f consists a single element. (i) Therefore, the constant function is one-one if the domain of f consists of a single element (ii) a constant function is always onto.

Example 3. Let f be a function from $A = \{a, b, c, d\}$ to $B = \{1, 2, 3\}$ defined by

$$f(a) = 1, f(b) = 3, f(c) = 1, f(d) = 2$$

is f onto.

Solution. $f(a) = 1, f(b) = 3, f(c) = 1, f(d) = 2$

$\therefore$ $R_f = \{1, 2, 3\} = B$ (codomain of f)

$\therefore$ f is onto.

Example 4. $f: R \to R$ is defined by

$$f(x) = ax + b \text{ where } a, b, x \in R, \text{ and } a \neq 0$$

Show that f is invertible and find the inverse of f.

Solution. Firstly, we shall show that f is one to one.

Let $x_1, x_2 \in R$ such that $f(x_1) = f(x_2)$

then $$f(x_1) = f(x_2) \Rightarrow ax_1 + b = ax_2 + b$$
$$\Rightarrow ax_1 = ax_2$$
$$\Rightarrow x_1 = x_2$$

$\therefore$ f is one-one.

To show that, f is onto, let $y \in R$ such that $y = f(x)$ for some $x \in A$

then $$y = f(x)$$
$$\Rightarrow \quad y = a_x + b$$
$$\Rightarrow \quad ax = y - b$$
$$\Rightarrow \quad x = \frac{y-b}{a}$$

$\therefore$ given $y \in R$, there exists an element $x = \frac{1}{a}(y - b) \in R$ such that $f(x) = y$

This proves that f is onto

Hence f is one-one and onto

$\Rightarrow$ f is invertible and $f^{-1}(y) = \frac{1}{a}(y - b)$.

3.66 CHARACTERISTIC FUNCTION OF A SET

Definition 3.65. Let U be universal set and A be a subset of U. Then the function

$\psi(A)$: $U \to \{0, 1\}$ defined by

$$\psi_A(x) = \begin{cases} 1 \text{ if } x \in A \\ 0 \text{ if } x \notin A \end{cases}$$

is called the characteristic function of the set A.

3.66.1 Properties of Characteristic Functions

Let A and B be any two subsets of a universal set U. Then the following properties hold for all $x \in U$.

(a) $\psi_A(x) = 0 \Leftrightarrow A = \phi$

(b) $\psi_A(x) = 1 \Leftrightarrow A = U$

(c) $\psi_A(x) \leq \psi_B(x) \Leftrightarrow A \subseteq B$

(d) $\psi_A(x) = \psi_B(x) \Leftrightarrow A = B$

(e) $\psi_{A \cap B}(x) = \psi_A(x)\, \psi_B(x)$

(f) $\psi_{A\cup B}(x) = \psi_A(x) \neq \psi_B(x) - \psi_{A\cap B}(x)$

(g) $\psi_{\bar{A}}(x) = 1 - \psi_A(x)$

The operations = , +, and – used above are the usual arithmetical operations. The values of characteristic functions are always either 1 or 0. The properties (*a*) to (*g*) can easily be proved by using the definition of a characteristic function. Set identities can also be proved by using the properties of characteristic functions.

Example. Show that $(\bar{\bar{A}}) = A$ for any set A.

Solution. We have $\psi_{(\bar{\bar{A}})}(x) = 1 - \psi_{(\bar{A})}(x)$

$$= 1 - (1 - \psi_A(x))$$

$$= 1 - 1 + \psi_A(x)$$

$$= \psi_A(x)$$

$\therefore \quad (\bar{\bar{A}}) = A$

3.67 RECURSIVE FUNCTIONS

Recursion is a technique of defining a function, a set or an algorithm in terms of itself. For example, consider the set of natural numbers, we introduce the method of generating the set of natural numbers by recursion.

The natural numbers (including zero) are those objects which can be generated by staring with an initial object 0 (zero) and from any object n already generated, passing to another uniquely determined object n^+, the successor of n. The objects differently generated are always distinct. Thus the natural numbers appear as a set of objects $0, 0^+, 0^{++}, 0^{+++}, \ldots$

The transition to the usual notation is made upon introducing 1, 2, 3,..., to stand for $0^+, 0^{++}, \ldots$ and then employing the notation. The set of natural numbers is denoted by N. The set of natural numbers can also generated starting with usual null set ϕ and notion of a successor set. If A is a given set, then the successor set of A is $A \cup \{A\}$. It is denoted by A^+

$$\therefore \qquad A^+ = A \cup \{A\}$$

Let ϕ be the null set, the successor sets of ϕ are.

$$\phi^+ = \phi \cup \{\phi\},\ \phi^{++} = \phi \cup \{\phi\} \cup [\phi, \{\phi\}], \ldots$$

renaming ϕ as 0 (zero), $\phi^+ = \{\phi\} = 1,\ \phi^{++} = [\phi, \{\phi\}] = 2, \ldots$

We get the set $\{0, 1, 2, 3, \ldots\}$ in which each element is a successor set of the previous element except for the 0 element (0 is not the successor of any element).

Now, we consider recursion in terms of successor function. Let S denotes the successor function: we define

$$\text{(i) } x + 0 = x, \quad \text{(ii) } x + S(y) = S(x + y)$$

In this definition (i) is the basic, and it defines addition of 0. The recursive part defines addition of the successor of y.

Example. Consider 3 +2 We have $3 + 2 \quad = 3 + S(1)$

$$= S(3 + 1)$$
$$= S(S(3 + 0))$$
$$= S(S(3))$$
$$= S(4) = 5$$

Now, we define three functions called basis functions or initial functions:

1. Zero function: Z: $Z(x) = 0$
2. Successor function: S: $S(x) = x + 1$
3. Projection function: $\bigcup_i^n : \bigcup_i^n (x_1, x_2, \ldots, x_i, \ldots, x_n) = x_i$ (generalised identity function).

Example. $S(3) = 3 + 1 = 4,$

$$S(4) = 4 + 1 = 5,$$

$$\bigcup_1^2 (x, y) = x, \quad \bigcup_2^3 (a, b, c) = b$$

$$\bigcup_3^5 (2, 5, 8, 1, 3) = 8 \quad \text{and} \quad \bigcup_3^6 (7, 3, 1, 6, 5, 9) = 1$$

3.68 FLOOR AND CEILING FUNCTIONS

Definition 3.66. Let x be any real number, then the greatest integer that does not exceed x is called floor of x.

The FLOOR of x is denoted by $\lfloor x \rfloor$

Examples. $\lfloor 5.29 \rfloor = 5, \ \lfloor \sqrt{5} \rfloor = 2$

$$\lfloor -7.9 \rfloor = 8, \ \lfloor 4 \rfloor = 4, \ \lfloor -6 \rfloor = -6$$

3.69 CEILING FUNCTION

Definition 3.67. Let x be a real number, then the least integer that is not less then, x is called CEILING of x.

Then CEILING of x is denoted by $\lceil x \rceil$

Examples. $\lceil 2.16 \rceil = 3, \lceil \sqrt{5} \rceil = 3, \lceil -7.4 \rceil = -7$

$$\lceil -3 \rceil = -3$$

Note. If x is an integer than $\lceil x \rceil = \lfloor x \rfloor$

3.70 INTEGER VALUE FUNCTION

Definition 3.68. Let x be any real number. The integer value of x is the value of x converted into an integer by deleting the fractional part of x. It is denoted by $INT(x)$.

Example $INT(2.33) = 2$, $INT(-7.4) = -7$

$$INT(10) = 10$$

Note. If x is a positive integer then $INT(x) = \lfloor x \rfloor$ and if x is a negative integer then $INT(x) = \lceil x \rceil$.

3.71 ABSOLUTE VALUE FUNCTION

Definition 3.69. Let x be a real number then the absolute value of x denoted by $ABS(x)$ or $|x|$ is defined as follows:

$$|x| = -x \quad \text{if } x < 0$$

$$|x| = x \quad \text{if } x \geq 0$$

Example $|3.17| = 3.17$, $|-8| = -(-8) = 8$

$$|-0.74| = -(-0.74) = 0.74, \ |6| = 6$$

Note. $|x| = |-x|$

3.72 LOGARITHMIC FUNCTION

Definition 3.70. Let M be positive integer, then the Logarithm of M to the base a, written $\log_a M$ represents the exponent to which a must be raised to obtain M.

$\therefore$ $a^x = M$ and $\log_a M = x$

are equivalent statements

a is called the base.

Examples. $3^4 = 81 \Rightarrow \log_3 81 = 4$

$$\log_{10} 0.01 = -2 \text{ since } 10^{-2} = 0.01$$

Logarithms to base 10 are called common logarithm, and logarithms to the base are called natural logarithms where $e = 2.718281$

usually $\log x$ means $\log_{10} x$

3.73 PARTIAL AND TOTAL FUNCTIONS

Let x and y be sets and A be a subset of x. A function f from A to y is called a partial function from x to y. The set A is called the domain of f. If the domain of f is x, then f is called a total function from x to y.

Let f be a function from N (the set of natural numbers) to N given by $f(a) = b$ if $a = b^2$, then the domain of the function f is the set of squares and f is a partial function. We define N^n to be set of all n-tupres of elements of N. Any function $f: N^n \to N$ is a total function. If A is a subset of N^n, then the function $f: A \to N$ is a partial function. The function $f(x, y) = x + y$ for all $x, y \in N$, is total function, but the function defined by $S(x, y) = x - y$ for all $x, y \in N$ with $(x > y)$, is a partial function.

3.74 PRIMITIVE RECURSIVE FUNCTIONS

Let $g: N^n \to N$, and $h: N^{n+2} \to N$ be functions. We say that $f: N^{n+1} \to N$ defined from g and h by primitive recursion if f satisfies the conditions:

$$f(x_1, x_2, \ldots, x_n, 0) = g(x_1, x_2, \ldots, x_n)$$

and
$$f(x_1, x_2, \ldots, x_n, y + 1) = h(x_1, x_2, \ldots, x_n, y, f(x_1, x_2, \ldots, x_n, y))$$

where y is the inductive variable.

From the above definition, it is clear that primitive recursion (or simply recursion) is the operation in which a function f of $(n + 1)$ variables is defined by using two other functions g and h of n and $(n + 2)$ variables.

Definition 3.71. A function f is said to be a primitive recursive function, if it can be obtained from the initial functions by a finite number of compositions and recursion (i.e. Primitive recursion).

Example 1. Show that addition is primitive recursive

Solution: Addition is defined by

$x + 0 = x$, $x + (y + 1) = (x + y) + 1$ for all natural numbers x and y.

we define $f(x, y) = x + y$ such that

$$\begin{aligned} f(x, y + 1) &= x + y + 1 \\ &= (x + y) + 1 \\ &= f(x, y) + 1 \\ &= S(f(x, y)) \end{aligned}$$

$$f(x, 0) = x$$

More formally we define $f(x, y)$ as

$$f(x, 0) = x = U_1^1(x)$$

$$f(x, y + 1) = S(U_3^3(x, y, f(x, y)))$$

It follows that f comes from primitive recursion of U_1^1 and U_3^3 and so f is primitive recursive.

Example 2. Show that multiplication defined by

$$x * 0 = 0, x * (y + 1) = x * y + x$$

is primitive recursive.

Solution. We define

$$\mu(x, y) \text{ to be } x * y \text{ so that } \mu(x, 0) = Z(x)$$

$$\mu(x, y + 1) = \mu(x, y) + x$$

$$= f(x, \mu(x, y))$$

$$\mu(x, y + 1) = f(U_3^3 (x, y)\ \mu(x, y), (U_1^3 (x, y, \mu(x, y)))$$

where f is the addition function.

Example 3. Let x, y be position integers and suppose Q is defined recursively as follows:

$$Q(x, y) = \begin{cases} 0 \text{ if } x < y \\ Q(x - y, y) = 1 \text{ if } y \leq x \end{cases}$$

Find (i) $Q(4, 7)$, (ii) $Q(19, 6)$

Solution. (i) $Q(4, 7) = 0$ Since $(4 < 7)$

(ii) $Q(19, 6) = Q(19 - 6, 6)$

$= Q(13, 6) + 1$

$= Q(13 - 6, 6) + 1$

$= Q(7, 6) + 1 + 1$

$= Q(7 - 6, 6) + 2$

$= Q(1, 6) + 2 + 1$

$= 0 + 3$

$= 3$

Example 5. Compute (i) $A(1, 1)$ (ii) $A(1, 2)$ and (iii) $A(2, 1)$ where $A : N^2 \rightarrow N$ (called Ackermann's function) is defined by

$$A(0, y) = y + 1$$

$$A(x + 1, 0) = A(x, 1)$$

$$A(x + 1, y + 1) = A(x + 1, y)$$

Solution. (i) $A\,(1, 1) = A\,(0 + 1, 0 + 1)$

$$= A(0, A(1, 0))$$

$$= A(0, A(0, 1))$$

$$= A(0, 1 + 1) = A(0, 2)$$

$$= 2 + 1 = 3$$

(ii) $A(1, 2) = A(0 + 1, 1 + 1)$

$$= A(0, A(1, 1)$$

$$= A(0, 3) = 3 + 1 \qquad (\because \quad A(1, 1) = 3)$$

$$= 4$$

(iii) $A(2, 1) = A(1 + 1, 0 + 1)$

$$= A(\,1, \mathrm{A}\,(2 + 0))$$

$$= A(1, A(1, 1))$$

$$= A(1, 3)$$

$$= A(0, 1, 2 + 1)$$

$$= A(0, 4) = 4 + 1 = 5$$

The following are some of the primitive recursive functions

1. Sign function Sg:

$$Sg(0) = 0, Sg(y + 1) = 1$$

Sg can be expressed as

$$Sg(0) = Z(0), Sg(y + 1) = S\,(Z(\cup_2^2\ (y, Sg(y))))$$

Sign function is also called non zero test function

2. Zero test function $\overline{Sg}$

$$\overline{Sg}(0) = 1, \overline{Sg}(y + 1) = 0$$

3. Predecessor function: P

$$P(0) = 0, P(y + 1) = y\ \cup_1^2\ (y, P(y))$$

Note: $P(1) = P(0 + 1) = 0$

4. Odd and even parity function Pr:

$$Pr(0) = 0, Pr(y + 1) = \overline{Sg}(\cup_2^2(y, Pr(y)))$$

5. Proper subtraction function $\dot{-}$:

$$x \dot{-} 0 = x, x \dot{-} (y+1) = P(x \dot{-} y)$$

Note $\quad x \dot{-} y = 0 \text{ for } x < y$

$$x \dot{-} y = x - y \text{ for } x \geq y$$

6. Absolute function | |:

$$|x - y| = (x \dot{-} y) + (y \dot{-} x)$$

Let R be a relation on N^n. The characteristic function of R can be defined as follows:

$$\psi_R(x_1, x_2, \ldots, x_n) = \begin{cases} 1 \text{ if } (x_1, x_2, \ldots, x_n) \in R \\ 0 \text{ if } (x_1, x_2, \ldots, x_n) \notin R \end{cases}$$

where $(x_1, x_2, \ldots, x_n) \in N^n$

Definition 3.72. *A* relation R is said to be primitive recursive if its characteristic function is primitive recursive.

For example, consider $R = \{(x, x) : x \in N\}$

The above relation R is the relation of equality on N. We can define a characteristic function

$$f(x, y) = \overline{Sg}(|x - y|)$$

Which is such that $f(x, y) = 1$, for $x = y$

and $\quad f(x, y) = 0$ for $x \neq y$

$f(x, y)$ is required characteristic function of R which is primitive recursive. The factorial function is primitive recursive.

Regular function: Let $g(x_1, x_2, \ldots, x_n : y)$ be a total function. If there is at least one value of y say $\bar{y} \in N$, such that the function $g(x_1, x_2, \ldots, x_n, \bar{y}) = 0$ for all x-tuples $(x_1, x_2, \ldots, x_n) \in N^n$, then g is called a regular function.

Example. The function $y \dot{-} x$ is regular, such $y = 0$

$\Rightarrow \quad y \dot{-} x = 0$ for all x.

Definition 3.73. A function $f(x_1, x_2, \ldots, x_n)$ is said to be defined from a total function $g(x_1, x_2, \ldots, x_n)$ by μ-operation (or minimisation) if

$$f(x_1, x_2, \ldots, x_n) = \begin{cases} \mu_y (g(x_1, x_2, \ldots, x_n, y) = 0 \text{ if there is such } a, y \\ \text{undefined otherwise} \end{cases}$$

where μ means the least y greater than or equal to zero.

Note. If the function g is not regular, then μ operation may produce a Partial function.

3.75 RECURSIVE FUNCTION

Definition 3.74. A function f is said to be recursive if and only if it can be obtained from the initial functions by a finite number of applications of the operations of composition recursion and minimisation over regular function.

Example. The set of even natural numbers is recursive. The set of recursive functions is closed under the operations of composition, recursion and minimisation over regular functions.

Definition 3.75. A function is said to be partial recursive if it can be obtained from initial functions by a finite number of applications of the operations of composition, recursion and minimisation.

Example. The function $f(x) = x/2$ is a partial recursive function.

Example 1. The function $rm(x, y)$ defines the remainder upon division by x. Show that it is primitive recursive.

Solution. We have $rm(x, y) = 0$, if $y = 0$ and the value of $rm(x, y)$ increases by 1 when y is increased by 1, until the value becomes equal to x in which case it is equal to zero. We consider the functions.

$$S(rm\ (x, y)) \text{ and } Sg(x \mathbin{\underset{-}{\bullet}} S(rm\ (x, y)$$

where $\quad Sg(x \mathbin{\underset{-}{\bullet}} S(rm\ (x, y) = 0$ whenever $S(rm\ (x, y) = x$

and $\quad Sg(x \mathbin{\underset{-}{\bullet}} S(rm\ (x, y) = 1$ when $S(rm\ (x, y) \neq x$

and define $\quad rm\ (r, 0) = 0$

$$rm\ (x, y + 1) = S(rm\ (x, y)^* \ Sg(x \mathbin{\underset{-}{\bullet}} g(rm\ (x, y))$$

Which shows that $rm(x, y)$ is primitive recursive.

EXERCISE 3.5

1. Define the terms

   (i) function (ii) one-one function (iii) onto function

   Give examples.
2. Let $f : N \to N$ defined by $f(x) = 2x + 5, x$. Show that f is one-one.
3. If $A = \{1, 2, 3, 4\}$ and $B = \{a, b, c, d\}$ determine if the following functions are one-one or onto.

   (a) $f = \{(1, a), (2, a), (3, b), (4, d)\}$

   (b) $g = \{(1, c), (2, b), (3, a), (4, a)\}$

   (c) $h = \{(1, d), (2, b), (3, a), (4, c)\}$

4. $A = \{-1, 0, 2, 5, 6, 11\}$ and $f : A \to B$ is a function given by $f(x) = x^2 - x - 2$, $\forall x \in A$. Find the range of f, if f is onto.
5. If $A = \{1, 2, 3\}$, $B = \{p, q\}$, $C = \{a, b\}$ and $f: A \to B$, $g: B \to C$ are given by

$$f = \{(1, p), (2, q), (3, q)\}$$
$$g = \{(p, b), (q, b)\}$$

Show that $g \circ f = \{(1, b), (2, b), (3, b)\}$

6. If $A = \{1, 2, 3\}$ and f, g, h are functions on A given by

$$f = \{(1, 2), (2, 3), (3, 1)\}$$
$$g = \{(1, 2), (2, 1), (3, 3)\}$$
$$h = \{(1, 1), (2, 2), (3, 1)\}$$

find $f \circ g$, $g \circ f$, $f \circ g \circ f$, $h \circ f$, $h \circ h$, and $g \circ h$.

7. R is the set of real numbers. The functions f, g, h are defined on R as follows: $f(x) = x + 2$, $g(x) = x - 2$, and $h(x) = 3x \ \forall\, x \in R$, find $g \circ f$, $f \circ g$, $f \circ f$, $g \circ g$, $f \circ h$, $h \circ g$, $h \circ f$ and $f \circ g \circ h$.
8. Let $f: R \to R$ be defined by $f(x) = 3x + 4$. Show that f is one-one onto and find f^{-1}.
9. Let $f: A \to R$ be defined by $f(x) = \dfrac{x-2}{x-3}$, where $A = R - \{3\}$, find f^{-1}.
10. Let $A \{1, 2, \ldots, n\}$, $(n \geq 2)$, and $B = \{a, b\}$, find the number of surjections from A to B.
11. Show that the function $[x/2]$ which is equal to the greatest integer which is less than or equal to $x/2$ is primitive recursive.
12. Show that the function $f(x) = x/2$ is partial recursive.
13. Let $D(x)$ denotes "number of divisors of x" show that $D(x)$ is primitive recursive.
14. Let $\lfloor \sqrt{x} \rfloor$ be the greatest integer $\leq \sqrt{x}$ show that $\lfloor \sqrt{x} \rfloor$ is primitive recursive.
15. Show that the predicate "x is prime" is primitive recursive.
16. Show that the factorial function $n!$ is primitive recursive.

ANSWERS 3.5

4. $\{-2, 0, 18, 28, 108\}$
6. $f \circ g = \{(1, 3), (2, 2), (3, 1)\}$

$g \circ f = \{(1, 1), (2, 3), (3, 2)\}$

$f \circ g \circ f = \{(1, 2), (2, 1), (3, 3)\}$

$h \circ f = \{(1, 2), (2, 1), (3, 1)\}$

$h\ o\ h = \{(1, 1), (2, 2), (3, 11)\}$

$f\ o\ h\ o\ h = \{(1, 2), (2, 3), (3, 3)\}$

7. $g\ o\ f(x) = x, f\ o\ f(x) = x + 4, g\ o\ g = x - 4$

$h\ o\ g(x) = 3x - 6, h\ o\ f(x) = 3x + 6,$

$f\ o\ h\ o\ h\ (x) = 3x$

8. $f^{-1}(y) = \dfrac{y-4}{3}$

9. $f^{-1}(y) = \dfrac{2-3y}{1-y}$

10. $2^n - 2$

3.75 LATTICES

In this section, we introduce Lattices which have important applications in the theory and design of computers.

Definition. A Lattice is a partially ordered set $(L, \leq)$ in which every pair of elements $a, b, \in L$ has a greatest lower bound and a least upper bound.

Example. Let Z^+ denotes the set of all positive integers and let R denotes the relation "division" in Z^+; such that for any two elements $a, b \in Z^+, f\ddot{v} a\ R\ b$ if "a divides b". Then (Z^+, R) is a lattice in which the join of a and b is the least common multiple of a and b. i.e. $a \vee b = a \oplus b =$ L.C.M. of a and b. And the meet of a and b is the greatest common divisor of a and b.

i.e. $a \wedge b = a * b =$ G.C.D. of a and b

we can also write $a + b = a \vee b = a \oplus$ b = L.C.M. of a and b

$a \cdot b = a \wedge = a * b =$ G.C.D. of a and b

Example 2. Let S be a not-empty set and $L = P(S)$, Then $(P(S), \leq)$ i.e $(L, \leq)$ is a partially ordered set. If A and B are any two elements of L, then

$A \cup B = A \vee B$; and

$A \cap B = A \wedge B$

Hence $(L, \leq)$ is a Lattice.

Example 3. Let n be a positive integer and $S_{\hat{n}}$ be the set of all divisions of n. If $n = 30$, we have $S_{30} = \{1, 2, 3, 5, 6, 10, 15, 30\}$

Let R denote the relation "division". Then (S_{30}, R) is a Lattice. The Hasse diagram of (S_{30}, R) is given ahead (see Fig. 3.37).

Different lattices can be represented by the same Hasse diagram. If $(L, \leq)$ is a Lattice then $(L, \geq)$ is also a lattice. The operation of meet and join on $(L, \leq)$ become the operation of join and meet on $(L, \geq)$. The Lattices $(L, \leq)$ and $(L, \geq)$ are duals of each other.

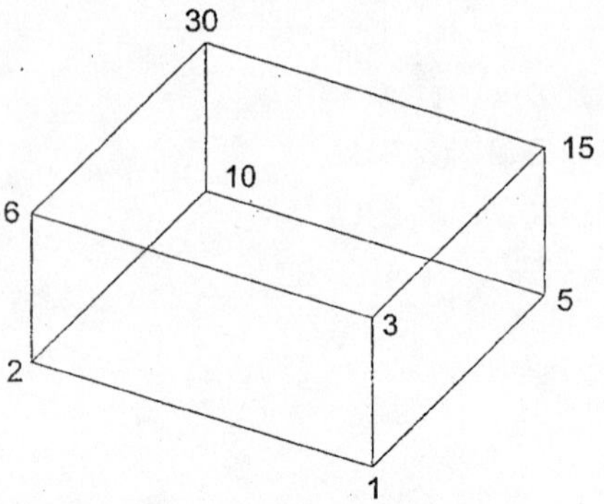

Fig. 3.37 Lattice

3.76 SOME PROPERTIES OF LATTICES

Let $(L, \leq)$ be a Lattice ' $*$ ' and '$\oplus$' denote the binary operations meet and join respectively on $(L, \leq)$. Then for any $a, b, c \in L$, we have

$(L-1) : a * a = a$; $(L-1)'$: $a \oplus a = a$ (Idempotent laws)

$(L-2) : a * b = b * a$; $(L-2)$: $a \oplus b = b \oplus a$ (Commutative laws)

$(L-3) : (a * b) * c = a * (b * c)$; $(L-3)'$ $(a \oplus b) \oplus c = a \oplus (b \oplus c)$

(Associative laws)

$(L-4) : a * (a \oplus b) = a$; $(L-4)'$: $a \oplus (a * b) = a$ (Absorption laws)

The properties $(L-1)$ to $(L-4)$ can be proved easily by using the definition of meet and join. We can apply the principle of duality and obtain $(L-1)'$ to $(L-4)'$ from $(L-1)$ to $(L-4)$.

Theorem 3.22. Let $(L, \leq)$ be a lattice in which $*$ and $\oplus$ denote the operations of meet and join respectively. Then

$$a \leq b \Leftrightarrow a * b = a \Leftrightarrow a \oplus b = b \quad \forall\ a, b, c \in L$$

Proof. Let $a \leq b$.

We know that $a \leq a$ therefore $a \leq a * b$ from the definition, we have $a * b \leq a$, therefore

$$a \leq b \Rightarrow a * b = a \qquad ...(1)$$

Now let us assume that $a * b = a$, but this is possible only

$$\text{if } a \leq b. \quad \text{i.e. } a * b = a \Rightarrow a \leq b \qquad ...(2)$$

Combining (1) and (2), we have $a \leq b \Leftrightarrow a * b = a$

We next prove that $a \leq b \Leftrightarrow a \oplus b = b$

Consider $a * b = a$, from which we have

$$b \oplus (a * b) = b \oplus a = a \oplus b$$

but

$$b \oplus (a * b) = b$$

Hence $a \oplus b = b$ follows from $a * b = a$ Similarly, by assuming that $a \oplus b = b$, we can show that

$$a * b = a$$

Hence $$a * b = a \Leftrightarrow a \oplus b = b$$

Theorem 3.23. Let $(L, \leq)$ be a lattice, then

$$b \leq c \Rightarrow \begin{cases} a * b \leq a * c \\ a \oplus b \leq a \oplus c \end{cases}$$

for all $a, d, c \in L$

(The above properties of a lattice are called properties of Isotonicity)

Proof. Using theorem 3.22, we can write

$$b \leq c \Leftrightarrow b * c = b$$

now $$(a * b) * (a * c) = (a * a) * (b * c) = a * (b * c) = a * b$$

$\therefore$ $$a * b \leq a * c \ \forall \ a, b, c * L$$

Similarly, we can pbrove that

$$a \oplus b \leq a \oplus c$$

3.77 LATTICE AS AN ALGEBRAIC SYSTEM

We now define Lacttice as an algebraic system, so that we can apply many concepts associated with algebraic system to lattices.

Definition 3.77. A lattice is an algebraic system $(L, *, \oplus)$ with two binary operations '$*$' and '$\oplus$' or L which are both commutative and associative and satisfy the absorption laws.

3.78 BOUNDED LATTICES

If L is a lattice, then every pair of elements of L has a least upper bound and a greatest lower bound. If A is a finite subset of L, then A has both least upper bound and greatest lower bounds. This properties may not hold if A is not a finite subset of L. We find the greatest lower bound and least upper bound of a subset of a lattice as follows:

Let $(L, *, \oplus)$ be a lattice and $A \subseteq L$ be a finite subset of L. The greatest and least upper bounds of A are defined as follows:

$$glb\ A = \overset{\sim}{\underset{i=1}{*}}\ a_i \text{ and } lub\ A = \overset{\sim}{\underset{i=1}{\oplus}}\ a_i$$

Where $$A = \{a_1, a_2, \ldots, a_n\}$$

Definition 3.77. A Lattice is called complete lattice if each of its non-empty subset has a least upper bound and greatest lower bound.

The least and greatest elements of lattice, if they exist are called the bounds of the lattice. They are denoted by 0 and 1 respectively. A lattice which has both 0 and 1 is called a bounded lattice. Every finite lattice must be complete and every complete lattice must have the least and greatest elements. The bounds of a lattice L satisfy the following:

$$a \oplus 0 = a, a * 1 = a$$

$$a \oplus 1 = 1, a * 0 = 0 \text{ for any } a \in L$$

If L is a bounded lattice, then the elements 0 and 1 are duals of each other. If L is a bounded lattice, then we denote it by $(L, *, \oplus, 0, 1)$.

Definition 3.78. Let $(L, *, \oplus, 0, 1)$ be a bounded lattice and $a \in L$. If there exists an element $b \in L$, such that

$$a * b = 0 \text{ and } a \oplus b = 1$$

then b is called the complement of a.

Example 1. Let A be a non-empty set and $L = P(A)$, (power set of A). Then ϕ is the lower bound of $P(A)$ and A is the upper bound of $P(A)$.

Example 2. In the lattice shown in Fig 3.38. The element a and d are complements of each other.

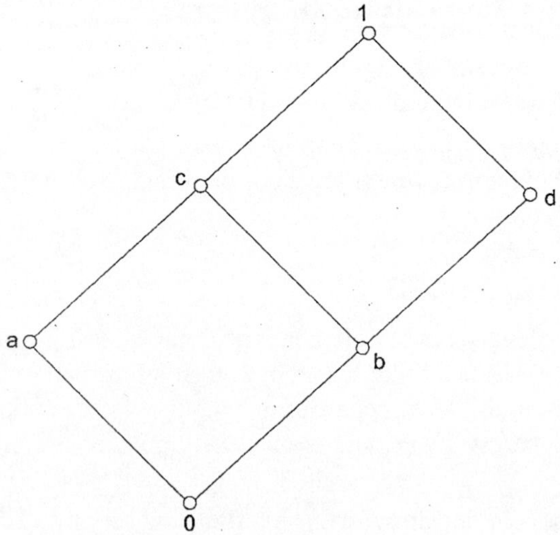

Fig. 3.38

3.79 SUB LATTICE

Definition 3.79. Let $(L, \leq)$ be a lattice. A non-empty subset A of L is called a sublattice of L if $a \oplus b \in A$ and $a * b \in A$ wherever $a \in A$ and $b \in A$.

If A is a sublattice of L, then A is closed under the operations of '* 'and /' ⊕'.

Example 1. Let Z^+ be the set of all positive integers and D denotes the relation "division" is z^+ such that for any $a, b \in Z^+$, $a \Delta b$ if a divides b. Then (Z^+, D) is a lattice in which $a \oplus b = LCM$ of a and b and $a * b = gcd$ of a and b.

Example 2. consider the lattice L shown in Fig. 3.39. The subset $A = \{a, c, d, y\}$ is a sub lattice y L.

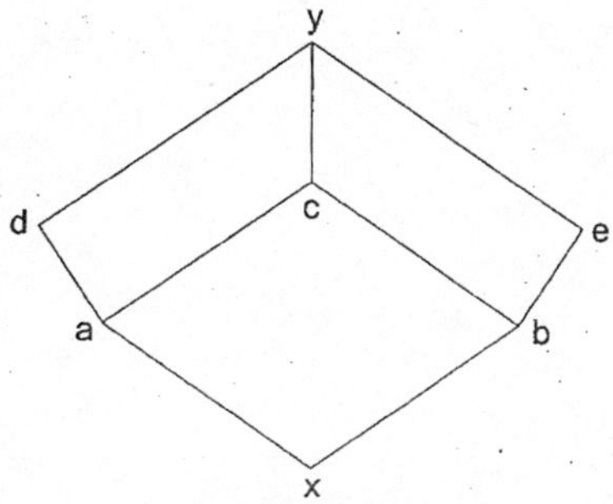

Fig. 3.39

3.80 DIRECT PRODUCTS

Let $(L_1, *, \oplus)$ and $(L_2, \wedge, \vee)$ be two Lattices. The algebraic system $(L_1 \times L_2, \bullet, +)$ in which the binary operation '+' and '•' on $L_1 \times L_2$ are defined as follows

$$(a_1, b_1)\,(a_2, b_2) = (a_1 \times a_2, b_1 \wedge b_2)$$

and $$(a_1, b_1) + (a_2, b_2) = (a_1 \oplus a_2, b_1 \vee b_2)$$

for all $$(a_1, b_1) \text{ and } (a_2, b_2) \in L_1 \times L_2$$

Example 1. Let $L_1 = \{1, 2, 4\}$ and $L_2 = \{1, 3, 5\}$ clearly L_1 and L_2 are chains and 'division' is a partial ordering on L_1 and L_2.

L_1 consists of divisors of L_1 and L_2 consists of divisors of a $L_1 \times L_2$ consists of divisors of 36. Each node in the diagram of $L_1 \times L_2$ is shown as (a, b) instead of ab

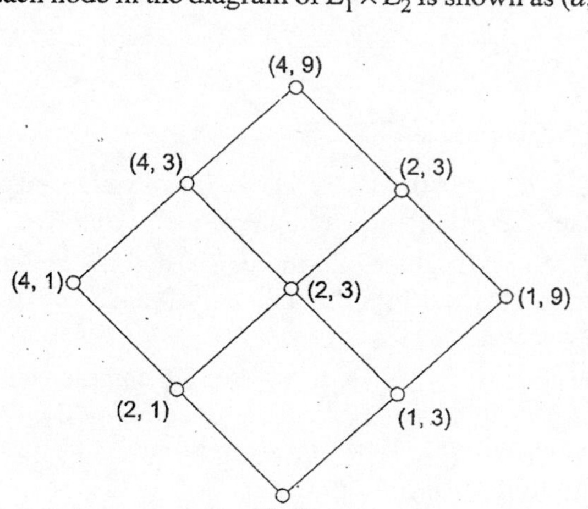

Fig. 3.40 Direct product of two lattices

Example 2. Let L_1 and L_2, be two lattices shown in Fig. 3.41, then $L_1 \times L_2$ is the lattice shown in Fig. 3.41.

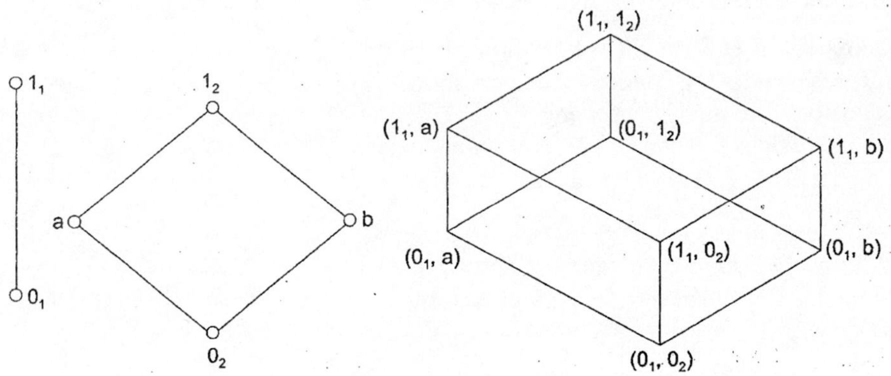

Fig. 3.41

We can use the direct product of lattices to construct larger lattices from the smaller ones. If L is a lattice, we can form the lattices

$$L \times L, L \times L \times L, \text{.... which are denoted by } L^2, L^3, \ldots$$

Example 3. If $L = \{0, 1\}$, and $(L, \leq)$ is a lattice, then $(L_n, \leq n)$ is a lattice of n-tuples of 0 and 1. Any element is the lattice $(L^n, \leq n)$ can be written as $(a_1, a_2, \ldots, a_n)$ in which a_i either 0 or 1 for $i = 1, 2, \ldots, n$.

The partial ordering relation on L^n can be defined for

$$a, b \in L^n \text{ as } a \leq_n b \Leftrightarrow a_i \leq b_i;\ i = 1, 2, \ldots, n$$

where $\quad a = (a_1, a_2, \ldots, a_n)$

and $\quad b = (b_1, b_2, \ldots, b_n)$

3.80 COMPLEMENTED LATTICE

Definition 3.80. Let $(L, *, \oplus, 0, 1)$ be a lattice. If every element of L has at least one complement then L is said to be a complemented lattice.

Example 1. Let $(L_3, \leq_3)$ be a lattice of 3-tuples of 0 and 1. The complement of an element of L_3 can be obtained by changing 1 by 0 and 0 by 1 in the 3-tuples representing the elements.

The complement of (0, 1, 1), (1, 0, ∞) and the complement of (1, 0, 1) is (0, 1, ∞) and so on. The bounds of L_3 are (0, 0, 0) and (1, 1, 1).

Example consider D_{105} (i.e. divisors of 105. If the complement of an element a is denoted by a', then we have

$$1' = 105, 3' = 35, 5' = 21$$

$$7' = 15,\ 15' = 7,\ 35' = 3$$

$$21' = 5,\ 105' = 1$$

3.81 DISTRIBUTIVE LATTICE

Definition 3.81. A laltice $(L, *, +)$ is called a distributive lattice if for all $a, b, c \in L$,

$$a * (b \oplus c) = a * b \oplus a * c$$

and
$$a \oplus (b * c) = (a \oplus b) * (a \oplus c)$$

Example 1. The lattice $(L_3, \leq_3)$ is distributive.

Example 2. The power set of a non-empty set A under the operation $\cap$ and $\cup$, is a distributive lattice.

Example 3. The lattice shown in Figure 3.42 given below is distributive.

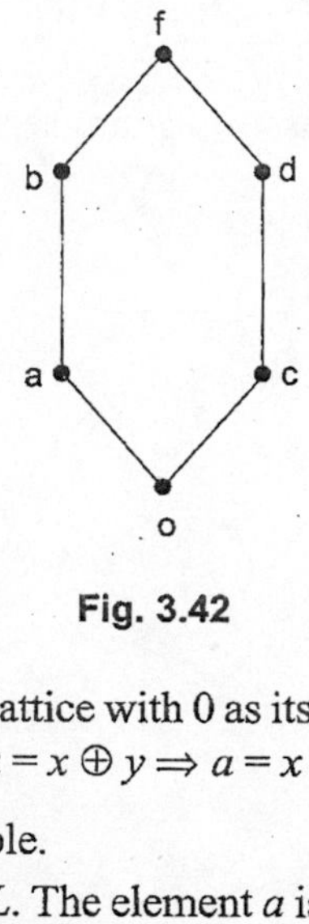

Fig. 3.42

Definition 3.82. Let $(L, \leq)$ be a lattice with 0 as its lower bound. An element $a \in L$ is said to be join irreducible if $a = x \oplus y \Rightarrow a = x$ or $a = y$.

Example $0 \in L$ is join irreducible.

Let $a \neq 0$, be an element of L. The element a is join irreducible if and only if 2 has a unique immediate predecessor.

Definition 3.83. Let $(L, \leq)$ be a lattice with an upper bound 1.

An element $a \in L$ is said to be meet irreducible if $a = x * y \Rightarrow a = x$ or $a = y$

Let $a \neq 0$ is an element of L. a is meet irreducible if a has a unique immediate successor.

In laltice of the Hasse diagram given ahead the elements x, y, z and s are join irreducible. The elements x, y, p, r, and s are meet irreducible.

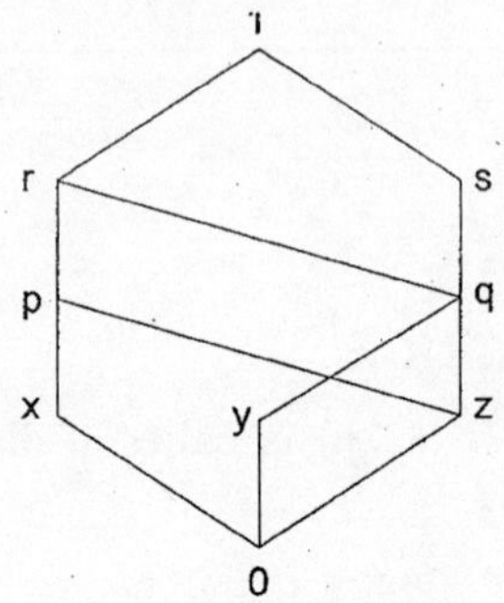

Fig. 3.43

EXERCISE 3.6

1. Define the terms
   - (a) Partially ordered set
   - (b) Linearly ordered set
   - (c) Complemented lattice
   - (d) Distributive lattice

   Give examples.
2. Determine whether the Hasse diagrams represent a lattice

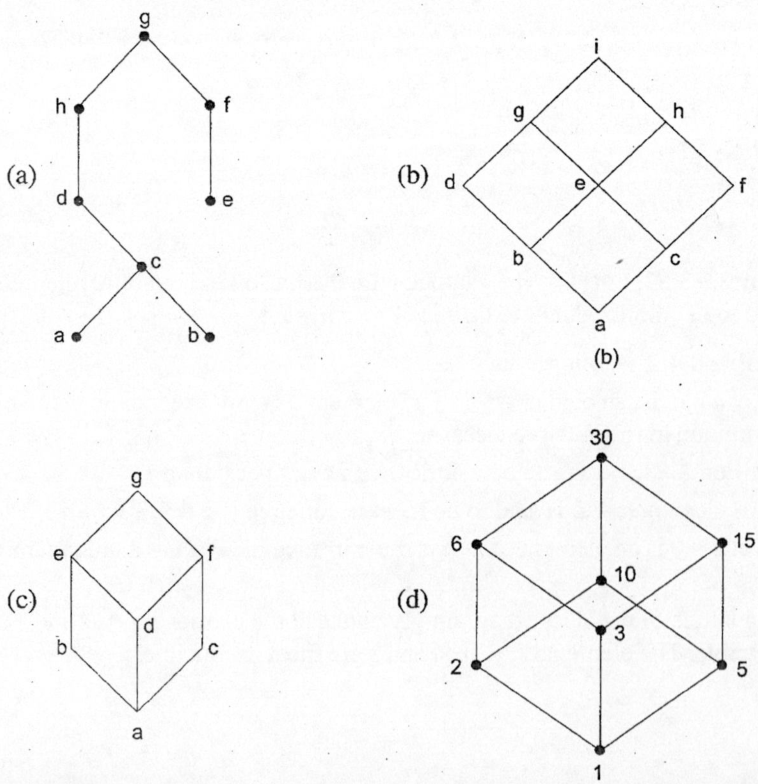

(e) 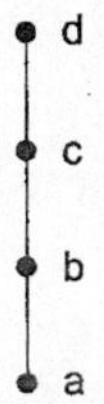

(f) 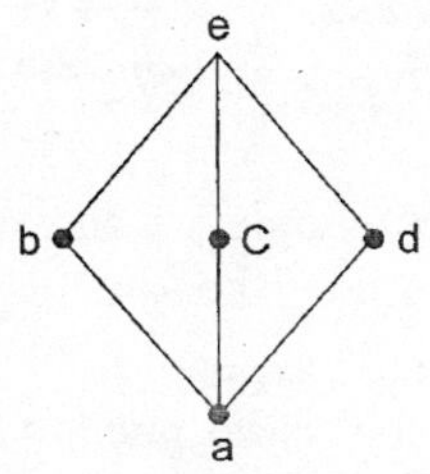

(f)

Ans. (b), (c), (d), (e) and (f) are lattices

3. Find all the upper bounds and lower bounds of the set $A = \{c. d, e\}$ for the lattice whose Hasse diagram is given below.

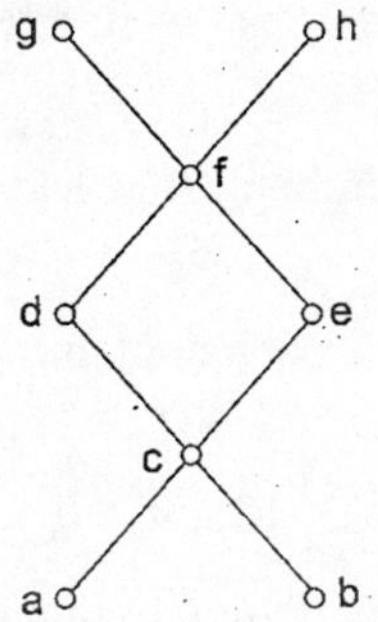

Ans. lower bounds of $A = c, a, b$

upper bounds of $A = f, g, h$

4. Define the term sub-lattice. Give examples.

5. Determine the maximal and minimal elements of

(i) $A = \{x: x$ is a real number, and $0 < x \leq 1\}$ with usual partial ordered $\leq$.

(ii) $S = \{2, 4, 6, 12, 20\}$ is ordered by divisibility.

Ans. (i) Maximal element 1, A has no minimal element.

(ii) Maximal element 20, minimal element = 2.

3.81 THE PIGEONHOLE PRINCIPLE

In this section we discuss Pigenhole Principle. It is also known as the Shoebox Principle and it can be stated as follows:

Theorem 3.24. If n pigeons are assigned to m pigeonholes and $m < n$, then some pigeonhole contains at least two pigeons.

Proof. Let $h_1, h_2,..., h_m$ demote the m pigeonholes and $p_1, p_2,..., p_m, p_{m+1},..., p_n$ denote the n pigeons. We consider the assignment of the n pigeons to m pigeonholes as follows:

Assign pigeon p_1 to the Pigeon hole h_1, pigeon p_2 to the pigeonhole h_2,..., and pigeon p_m to the pigeonhole h_m.

This assigns as many pigeons possible to individual pigeonhole. Since $m < n$. There are $(n - m)$ pigeons that have not been assigned. If all the pigeons are to be assigned to the pigeonholes, then at least one pigeonhole will be assigned a second pigeon.

Example 1. If ten people are chosen in any way. Show that at least two of them will have been born on the same day.

Solution. There are ten people and the number of days in a week is only seven. If each person (pigeon) is assigned to the day of the week (pigeonhole) on which he or she is born, the Pigeonhole Principle tells us that at least two people must be assigned to the same day of the week.

In the above example, the people are taken as the pigeons and the days of the week are taken as the pigeonholes. In general; to apply pigeonhole principle, we must identify the pigeons (i.e. the objects) and the pigeonholes (i.e categories of the desired characteristic). Also we must be able to count the number of pigeons and pigeonholes.

Example 2. Show that if seven numbers from 1 to 12 are chosen, then 2 of them will add to 13.

Solution. Construct 6 different sets each containing 2 numbers that add up to 13 as follows:

$$A_1 = \{1, 12\}, A_2 = \{2, 11\}, A_3 = \{3, 10\}, A_4 = \{4, 9\}, A_5 = \{5, 8\}, A_6 = \{6, 7\}$$

Each of the 7 numbers selected belong to one of those sets. Since there are 6 sets, the pigeonhole principle tells us that 2 of the numbers chosen belong to the same set. These numbers add up to 13.

Theorem 3.25. Extended pigeonhole principle.

If n pigeons are assigned to m pigeonholes, then one of its pigeonholes must contain at least $\lfloor n/m \rfloor + 1$ pigeon (where $\lfloor p/q \rfloor$ denotes the largest integer less than or equal to the rational number p/q).

Proof. We prove the theorem by contradiction. Let each pigeonhole contain no more than $\lfloor (n-1)/m \rfloor$ pigeons then, there are at most $m.\lfloor (n-1)/m \rfloor$ pigeon in all

but $\qquad \lfloor (n-1)/m \rfloor \leq m(n-1)/m$

$\Rightarrow \qquad \lfloor (n-1)/m \rfloor = n - 1$

i.e. there are $(n - 1)$ pigeons in all.

This contradicts or assumption. Hence one of the pigeonholes must contain at least $\lfloor (n-1)/m \rfloor + 1$ pigeons.

Example 1. Show that if any 26 people are selected, then we may choose a subset of 4 that all 4 were born on the same day of the week.

Solution. We assign each person to the week on which she or he was born. Then the number of the pigeons (people) to be assigned to 7 pigeonholes (days of the week) with $n = 26$, and $m = 7$ is at least

$$\lfloor (26 - 1)/7 \rfloor + 1 = 4$$

∴ At least 4 people must have been born on the same day of the week.

Example 2. Prove that if 30 dictionaries in a library contain a total of 61, 327 pages, then at least one of the dictionaries must have at least 2045 pages.

Solution. Here the pages of the dictionaries are pigeons and the dictionaries are pigeon holes.

Hence, by Extended pigeonhole principle, atleast one of the dictionaries must contain

$$\left\lfloor \frac{61327-1}{30} \right\rfloor + 1 = \lfloor 2044.2 \rfloor + 1$$
$$= \lfloor 2044 + 1 \rfloor = 2045 \text{ pages.}$$

EXERCISE 3.6

1. Suppose 14 students in a class appear at a university examination. Prove that there exist at least two among them whose seat numbers differ by a multiple of 13.
2. Show that if any 5 numbers from 1 to 8 are chosen, then 2 of them will add upto 9.
3. If 8 people are chosen in any way from a group. Show that, at least 2 of them will have been born on the same day of the week.
4. Show that if any 30 people are selected then we may chose, a subset of 5 so that all the 5 were born on the same day.
5. Prove that if any numbers from 1 to 25 are chosen, then one of them will be a multiple of the other.
6. Show that if any 11 numbers are chosen from the set $\{1, 2, ..., 20\}$, then one of them will be a multiple of other.
7. If 20 processors are inter-connected, show that at least 2 processors are directly connected to the same number of processors.
8. Shirts numbered consecutively from 1 to 20 one worn by 20 members of the bowling league. If any 3 of these numbers are chosen to be a team, the sum of their shirt numbers will be used as the code number of the team. Show that if any 8 of the 20 are pre-selected, then from these 8. One may form at least 2 different team having the same code number.
9. Show that if 30 dictionaries in a library contain a total of 61,327 pages, then one of the dictionaries must has at least 2045 pages.
10. If $n + 1$, numbers are chosen from $\{1, 2, ..., 2n\}$, then show that one must divide the other.
11. The members of a class of 27 pupils each 90 swimming on same of the days from Monday to Friday in a certain week. If each pupil goes at least twice, then show that there must be 2 pupils who go swimming on exactly the same day.
12. Let L be a lattice. Then prove that $a \wedge b = a$ if and only if $a \vee b = b$.

CHAPTER 4

Algebraic Structures

4.1 INTRODUCTION

In this chapter we shall deal with some important Algebraic structures (systems). We shall first explain what is meant by Algebraic structure (system) and then give examples of familiar algebraic systems namely. Semi groups, mooids and groups, which are one operational systems.

Definition 4.1. (n-ary operation): Let A be a non-empty set. A mapping $f : A^n \to A$ is called an n-ary operation on the set A.

n is called the order of the operation.

If $n = 1, f : A \to A$ is called a unary operation.

If $n = 2, f : A^2 \to A$ is called a binary operation.

If an operation (or composition) on the elements of a set A produces images which are also the elements of the same set, then the set A is said to be closed under the operation and this property is called the closure property.

4.2 ALGEBRAIC SYSTEMS (ALGEBRAIC STRUCTURES)

A system consisting of a non-empty set S and one or more n-ary operation on the set S is called an algebraic system.

If S is a non-empty set and $f_1, f_2, ..., f_n$ are n-ary operation on S, then

$(S, f_1, f_2, ..., f_n)$ is an Algebraic system (or structure)

The operation on the set S, defines a structure on the elements of S. Therefore S is called an Algebraic structure. We begin our study with the definition of an Algebraic structure.

Definition 4.2. If A is a non-empty set and $*$ is a binary operation on A, then (A_1, $*$) is called an Algebraic structure.

Example 1. Let R be the set of real numbers. Then $(R, +)$ is an Algebraic structure.

Example 2. Let Z be the set of integers. The operations addition $(+)$ and multiplication $(\times)$ are binary operations on $\geq$. Therefore $(Z, +)$ and $(Z, \times)$ are Algebraic structures.

Notation. If $*$ is a binary operation on a non-empty set A, then for the $*$ image of the ordered pair $(a, b) \in A \times A$, we write $a * b$ or $*\ (a, b)$

Multiplication is a binary operation on the sets N, R, Q and C. Union is a binary operation on the set $P(A)$, (power set of A). Therefore $(N, \bullet)$, $(R, \bullet)$, $(Q, \bullet)$ and $(C, \bullet)$ are algebraic structures. Since the sets and the operations defined on sets are abstractions, the Algebraic structures (systems) are called abstract algebras.

4.3 GENERAL PROPERTIES

We now discuss some general properties of binary operations.

Definition 4.3. Let A be a non-empty set. A binary operation $* : A \times A \rightarrow A$ is said to be commutative if for all $a, b \in A$

$$a * b = b * a$$

Example 1. Addition is a binary operation on N and is commutative.

Example 2. Let A be a non-empty set and $P(A)$ denote the power set of A. Intersection is a binary operation on $P(A)$ and is a commutative operation on $P(A)$.

Definition 4.4. Let A be a non-empty set. A binary operation $* : A \times A \rightarrow A$ is said to be associative if

$$(a * \mathrm{b}) * c = a * (b * c) \text{ for all } a, b, c \in A$$

Example 1. The operation of addition and multiplication are associative on N, the set of natural numbers.

Example 2. Addition is associative on R, Z, and C.

Definition 4.5. Let $*$ be a binary operation on a set A. If there exists on element $e_l \in A$, such that

$$e_l * a = a \quad \text{for all } a \in A$$

Then e_l is called the left identity in A with respect to the binary operation $*$.

Example. o is the left identity in Z, the set of integers with respect to the binary operation addition.

Definition 4.6. Let $*$ be a binary operation on a set A. If there exists an element $e_r \in A$, such that

$$a * e_r = a \quad \text{for every } a \in A$$

then e_r is called the right identity in A with respect to the binary operation $*$.

Example 1. o is the right identity in z, with respect to the binary operation addition.

Example 2. 1 is the right identity in the set of natural numbers with respect to the operation multiplication.

Definition 4.7. Let $*$ be a binary operation a non-empty set A. If there exists an element $e \in A$ such that

$$e * a = a * e \quad \text{for every } a \in A$$

Then e is called the $*$ identity element in A with respect to the binary operation $*$.

Example 1. o is the identity element in Z (the set of integers) with respect to addition.

Example 2. 1 is the identity element in Q (the set of rational numbers) with respect to the binary operation multiplication.

Definition 4.8. Let $*$ be a binary operation on a non-empty set A and e be the identity element in A with respect to the binary operation $*$. An element $a \in A$ is said to be invertible, if there exists an element b in A, such that

$$a * b = e = b * a$$

b is called the inverse of a in $(A, *)$

If b is the inverse of a in A, we write $b = a^{-1}$

If b is the inverse of a, then a is also inverse of b in A

i.e. a and b are inverse of each other.

Example. Consider the set of integers Z. Addition is a binary composition in Z, and – 3 is the inverse of 3 in (Z, +).

Cancellation Laws

Definition 4.9. A binary operation denoted by $*$ in a non-empty set A is said to satisfy.

(i) Left cancellation law if for all $a, b, c \in A$

$$a * b = a * c \Rightarrow b = c$$

(ii) Right cancellation law if for all $a, b, c \in A$

$$b * a = c * a \Rightarrow b = c$$

Distributive Laws

Definition 4.10. Let '$*$' and 'o' be binary operation on A. If for all a, b, c in A.

(i) $(b \, o \, c) * a = (b * a) \, o \, (c * a)$

then we say that $*$ is left distributive over 'o'.

(ii) $a * (b \, o \, c) = (a * b) \, o \, (a * c)$

then we say that $*$ is right distributive over 'o'; and

(iii) Distributive if $*$ is both left distributive and right distributive over 'o' in A.

Example. In $P(A)$, the power set of A, union of sets is distributive over intersection of sets and intersection is distributive in $P(A)$.

4.4 COMPOSITION TABLES

Composition tables for binary operation on a finite set are very useful checking various properties of the binary compositions (i.e. binary operations)

Let $A = \{a_1, a_2,..., a_n\}$ be a finite set and $*$ be a binary operation on A. To construct the composition table we write all the elements of A, horizontally in a row and vertically in a column. As shown in Table 4.1 Then $a_i * a_j$ is the element that occurs at the intersection of ith row and j^{th} column.

Table 4.1 Composition table

$*$	a_1	a_2	a_3	...	a_j	...	a_n
a_1	$a_1 * a_1$	$a_1 * a_2$	$a_1 * a_3$	...	$a_1 * a_j$	...	$a_1 * a_n$
a_2	$a_2 * a_1$	$a_2 * a_2$	$a_2 * a_3$	...	$a_2 * a_j$	...	$a_2 * a_n$
a_3	$a_3 * a_1$	$a_3 * a_2$	$a_3 * a_3$	...	$a_3 * a_j$	...	$a_3 * a_n$
⋮							
a_i	$a_i * a_1$	$a_i * a_2$	$a_i * a_3$	...	$a_i * a_j$	...	$a_i * a_n$
⋮	⋮	⋮	⋮	⋮	⋮		⋮
a_n	$a_n * a_1$	$a_n * a_2$	$a_n * a_3$	...	$a_n * a_j$	...	$a_n * a_n$

Example 1. Let $A = \{1, -1, i, -i\}$ where $i^2 = -1$, and multiplication x be a binary operation definition A. Then the binary composition table can be constructed as shown below (see Table 4.2).

Table 4.2

x	1	−1	i	$-i$
1	1	−1	i	$-i$
−1	−1	1	$-i$	i
i	i	$-i$	−1	1
$-i$	$-i$	i	1	−1

Example 2. Let $A = \{1, 2, 3, 4\}$ and $f : A \to A$ be given by

$$f = \{(1, 2), (2, 3), (3, 4), (4, 1)\}$$

Let the identity mapping on A be denoted by f_o and f by f^1, we form the composite functions

$$f o f = f^2,\ f^2 o f = f^3,\ f^4 = f^3 o f = f^4, ...$$

we observe that $f^4 = f^o, f^5 = f^1, ...$

Let $\quad F = \{f^o, f^1, f^2, f^3\}$

Then (F, o) forms an algebraic system. The composition 'o' is both commutative and associative.

f^o is the identity element (F, o). The composition table for (F, o) is given in the Table 4.3

Table 4.3

o	f^o	f^1	f^2	f^3
f^o	f^o	f^1	f^2	f^3
f^1	f^1	f^2	f^3	f^o
f^2	f^2	f^3	f^o	f^1
f3	f^3	f^o	f^1	f^2

Definition 4.10. Set (A, o) and $(B, *)$ be two algebraic systems, where o is a binary operation on A and $*$ is a binary operation on B. A mapping

$$\phi : A \rightarrow B$$

is called a homomorphism (or morphism). From A to B, if for any $a_1, a_2 \in A$.

$$\phi(a_1 \; o \; a_2) = \phi(a_1) * \phi(a_2)$$

B is called the homomorphic image of A

If such a function exists then $(B, *)$ is called the homomorphic image of (A, o).

Note. If $\phi : A \rightarrow B$ is a homomorphism, then $\phi(A) \subseteq B$

Example. Let $(R, +)$ denote the algebraic system in which R is the set of all real numbers and $+$ is a binary operation on R. Let $(R^+, \cdot)$ denote an algebraic system where R^+, denotes the set off all positive real numbers and $\cdot$ is a binary operation on it. Define a mapping $\phi : R \rightarrow R^+$ by

$$\phi(x) = 2^x \quad \forall \; x \in R$$

Then $\quad x, y \in R \Rightarrow x + y \in R$

and $\quad \phi(x + y) = 2^{x+y}$

$$= 2^x \cdot 2^y$$

$$= \phi(x) \cdot \phi(y).$$

$\therefore \phi$ is a homomorphism

Consider the relation 'congruence modulo m' defined on Z, the set of integers. The relation 'congruence modulo m' is an equivalence relation on Z, and it partitions Z into disjoint equivalence classes (called the 'residue classes modulo m' or congruence classes modulo m).

If $a \in Z$, then the equivalence class of a is denoted by $[a]$ (or $\bar{a}$)

where $[a] = \{x : x \in Z, x - a \text{ is divisible by } m\}$

Definition 4.11. Let $m \in N$, and $n \in Z$ and $[r] = \{n : x \in Z, x \equiv r \pmod{m}$ then the set

$z_m = \{[0], [1], [2], \ldots [m-1]\}$ is called the complete set of equivalence classes.

The classes (0), (1),..., $(m - 1)$ are all disjoint.

We have $(a) = (a + m) = (a + 2m)\ldots$

Example 1. If $m = 6$, then $Z_m = Z_6 = \{[0], [1], [2], [3], [4], [5]\}$

where $[0] = \{\ldots -12, -6, 0, 6, 12, \ldots\}$, $[3] = \{\ldots, -9, -3, 3, 9, 15, \ldots\}$

$[1] = \{\ldots, -11, -5, 1, 7, 13, \ldots\}$, $[4] = \{\ldots, -8, -2, 4, 10, 19, \ldots\}$

$[2] = \{\ldots, 10, -4, 2, 8, 14, \ldots\}$, $[5] = \{\ldots, -7, -1, 5, 11, 20, \ldots\}$

[0], [1], [2], [3], [4], [5] are all disjoint.

Addition of equivalence classes. We denote addition by + and define addition of equivalence classes by, $[a] + [b] = [a + b] \ \forall \ a, b \in Z_m$

Example 2. Consider $Z_4 = \{[0], [1], [2], [3]\}$

Z_4 is an algebraic system under the operation, 'addition mod 4 of equivalence classes'. The binary composition table is given in Table 4.4.

Table 4.4

+	[0]	[1]	[2]	[3]
[0]	[0]	[1]	[2]	[3]
[1]	[1]	[2]	[3]	[4]
[2]	[2]	[3]	[0]	[1]
[3]	[3]	[0]	[1]	[2]

Let $F = \{f^0, f^1, f^2, f^3\}$, where $f : A \to A$ is defined by

$$f = f^1 \approx \{(1, 2), (2, 3), (3, 4), (4, 1)\}$$

on $A = \{1, 2, 3, 4\}$

$(f, \cdot)$ is an algebraic system

Define a mapping $\phi : F \to z_4$ by

$$\phi[f^i] = [i], \quad \text{for } i = 0, 1, 2, 3$$

from the table, it is the clear that

$$\phi(f^i \cdot f^j) = \phi(f^i) +_4 \phi(f^j) \text{ for all } i = 0, 1, 2, 3, j = 0, 1, 2, 3$$

$\therefore$ ϕ is a homomorphism

Definition 4.12. Let ϕ be a homomorphism from (A, o) to $(B, *)$.

where (A, o) and $(B, *)$ are algebraic systems.

(i) If $\phi : A \to B$ is onto, then ϕ is called an epimorphism.

(ii) If $\phi : A \to B$ is one-to-one then ϕ is called a monomarphism

and (iii) if $\phi : A \to B$ is one-to-one onto then ϕ is called an isomorphism

If $\phi : A \to B$ is an isomorphism, then A and B are said to be isomorphism.

Example 1. The mapping $\phi : F \to Z_4$ defined in example 2 is an isomorphic.

Example 2. Consider the set $B = \{0, 1\}$ and + be the operation given by the Table 4.5.

Table 4.5

+	0	1
0	0	1
1	1	0

Define a mapping $\phi : Z_4 \to B$, by

$$\phi(0) = 0, \phi(1) = 1, \phi(2) = 0 \text{ and } \phi(3) = 1$$

clearly $\phi[(\text{i}) +_4 [j] = \phi(i) + \phi[j]$ for all $i, j = 0\ 1, 2, 3$

Therefore, ϕ is an epimorphism.

Definition 4.13. Let (A, o) and $(B, *)$ be two algebraic systems such that $B \subseteq A$. If $\phi: A \to B$ is a homomorphism, then ϕ is called an automorphism.

Definition 4.14. Let (A, o) be an algebraic system. If $\phi: A \to A$ is an isomorphism then ϕ is called an automorphism.

Example 1. Let (A, o) be an algebraic structure with identity e. Then show that the mapping $\phi : A \to A$, defined by

$$\phi(a) = e \quad \text{for every } a \in A,$$

is an endomorphism.

Solution. Let $a, b \in A$. Then $a \, o \, b \in A$

and $\phi(a) = e, \phi(b) = e$

now
$$\phi(a \, o \, b) = e$$
$$= e \, o \, e$$
$$= \phi(a) \, o \, \phi(b)$$

Thus, ϕ is a homomorphism of A into A

$\therefore$ ϕ is an endomorphism.

Example 2. Consider the algebraic system $(N, +)$ and $(Z_4, +_4)$

Define $\phi : N \to Z_4$, given by

$$\phi(a) = [a \text{ (mod 4)}] \text{ for any } a \in N$$

Show that ϕ is a homomorphism.

Solution. Let $a, b \in N$, Let $\phi(a) = [a \text{ (mod 4)}] = [i]$

and $\phi(b) = [b \text{ (mod 4)}] = [j]$

Then $\phi(a+b) = [(i+j) \text{ (mod 4)}] = [i] +_4 [j], = \phi(a) +_4 \phi(b)$

$\therefore$ $\phi : N \rightarrow Z_4$ is a homomorphism

Let (A, o) be an algebraic system and E be an equivalence relation on A. If for any $a_1, a_2 \in A$.

$$(a_1 \, E \, a_2') \wedge (a_2 \, E \, a_2') \Rightarrow (a_1 \circ a_2) \, E \, (a_1' \circ E_2')$$

where $a_1', a_2' \in A$

Then we say that E has substitution property. If E satisfies the substitution property then E is called a congruence relation on (A, o).

If E is an equivalence relation on an algebraic system (A, o) then E portions A in equivalence classes. The set of all equivalence classes of A is denoted by A/E. On the quotient set A/E; if we define an operation $*$, given by

$$[a_1] * [a_2] = [a_1 \circ a_2]$$

for all $a_1, a_2 \in A$

then the algebraic system $(A/E) *)$ is called Quotient algebra. We can define a function f_E from (A, o) onto the $(A/E, *)$ given by $f_E(a) = a \;\forall\; a \in A$.

clearly $f_E(a_1 \circ a_2) = [a_1 \circ a_2] = [a_1] * [a_2] = f_E(a_1) * f_E(a_2)$

for all $a_1, a_2 \in A$.

Thus f_E is a homomorphism associated with the congruence relation E

f_E is called the natural homomorphism.

EXERCISE 4.1

1. Show that addition is a binary operation on the set of m of all $m \times n$ matrices
2. Show that addition is a binary operation on the set

$$A = \{a + b\sqrt{2} : a, b \in Q\}$$

(Q is the set of rational numbers)

3. Show the additional modulo 5 $(+_5)$ is commutative and associative on $Z_5 = \{0, 1, 2, 3, 4\}$ construct composition table.
4. Show that division is a binary operation on $R - \{o\}$ and $c - \{o\}$.
5. If $*$ is a binary operation on R defined by

$$a * b = a + b - ab \;\forall\; a, b \in R$$

Then, show that the inverse of $a \neq 1$ is $\dfrac{a}{a-1}$

6. Show that the identity element of a binary operation $*$ on a set A, if it exists is unique.
7. If $*$ is an associative binary operation in A, then show that the inverse of every invertible element in $(A, *)$ is unique.
8. If $*$ is an associative binary operation a non-empty set A, such that every element is invertible, then show that left and right cancellation laws hold in $(A, *)$.

4.5 SEMI GROUP

Let A be a non-empty set and 'o' be a binary operation on A. Then the algebraic structure $(A, *)$ is called groupoid and set A is said be closed under the binary operation $*$. Addition is a binary operation on R, therefore R is closed with respect to addition. The set of all even integers is closed with respect to the binary operation addition. If binary operation $*$ on a given set A is associative, then the algebraic structure $(A, *)$ is called a semi-group.

Definition 4.14. Let S be a non-empty set and $*$ be a binary operation on S. The algebraic structure $(S, *)$ is called a semi group if the operation $*$ is associative. In other words, a non-empty set S, together with a binary operation $*$ is a semi group if

$$(a * \mathrm{b}) * c = a * (b * c) \text{ for all } a, b, c \in S$$

Example 1. Let N be the set of natural numbers. Then $(N, +)$ and $(N, \cdot)$ are semi-groups.

Example 2. Let X be a non-empty set and $P(X)$ denote the power set of X. Then $[P(X), \cup]$ and $P(X, \cap)$ are semi-groups.

Example 3. Let Z be the set of integers and Z_m be the set of equivalence classes generated by the equivalence relation "congruent modulo m" for any positive integer m The operation $+_m$ be defined on the integers of Z as follows:

For any $\quad [i], [j] \in Z_m$

$$[i] +_m [j] = [(i + j) \bmod m]$$

Then the algebraic system $(Z_m, +_m)$ is a semi-group.

4.6 HOMOMORPHISM OF SEMI-GROUPS

Definition 4.15. Let $(s, *)$ and (T, o) be any two semi-groups. A mapping $f: S \rightarrow T$, defined by

$$f(a * b) = f(a) \, o \, f(b), \quad \text{for all } a, b \in S$$

is called a semi-group homomorphism.

Example 1. Consider (*Z*, +) the semi-group under addition and

$$A = \{2^n : n \in Z\}$$

S is a semi-group under multiplication of real number

Define $f : Z \to S$ by $f(m) = 2^m \ \forall \ m \in Z$

Since $f(m + n) = 2^{m+n} = 2^m, 2^n \ \forall \ m, n \in Z$

f is a homomorphism of *Z* into *S*.

Definition 4.16. A homomorphism of a semi-group into itself is called a semi-group endomorphism.

4.7 ISOMORPHISM OF SEMI-GROUPS

Definition 4.17. Let $(S, *)$ and (T, o) be any two semi-groups. *A* homomorphism $f: S \to T$ is called a semigroup isomorphism if f is one-to-one and onto.

If $f: S \to T$ is an isomorphism then $(S, *)$ and (T, o) are said to be isomorphic.

Example. Consider the algebraic structures $(R, +)$ and $(R^+, \cdot)$ where *R* denotes the set of real numbers and R^+ denotes the set of positive real numbers, consider the mapping

$$f_a(x) = a^x \quad \forall \ n \in R$$

clearly f_a is structure preserving, and f_a is one-to-one onto therefore $f_a : R \to R^+$ is an isomorphism.

Definition 4.18. An isomorphism of a semi-group onto itself is called an automorphism.

Theorem 4.1. Let $(S, *)$, (T, o) and (V, Δ) be semi-group. If $f: S \to T$, and $g : T \to V$ be semi-group homomorphism from $(S, *)$ to (V, Δ)

Proof. Let $a, b \in S$ then

$$\begin{aligned} (g \, o f)\,(a * b) &= g[f(a * b)] \\ &= g[f(a) \, o f(b)] \\ &= g[f(a) \, \Delta \, g\,(f(b)] \\ &= (g \, o f)\, a \, \Delta \, (g \, o f)\,(b) \end{aligned}$$

$\Rightarrow g \, o f : S \to V$ is a semi-group homomorphism.

We shall state the following theorems without proof.

Theorem 4.2. The set of at endomorphism of a semi-group is a semi-group under the operation of (left) compositions.

Theorem 4.3. The set of all semi-group automorphism of a semi-group is a semi-group under the operation of (left) composition.

Theorem 4.4. Let (S, $*$) be a semi-group and (S^S, o) be the semi-group of functions from S to S, then there exists a homomorphism $\phi : S \rightarrow S^S$ under the operation of left composition.

Proof. Let $a \in A$ and let $\phi(a) = f_a$ where $f_a \in S^S$ and f_a be defined by $f_a(b) = a * b$ for any $b \in S$

$$
\begin{aligned}
\therefore \qquad f_{a*b}(c) &= (a * b) * c \\
&= a * (b * c) \quad (\therefore S \text{ is a semi-group}) \\
&= f_a f_b\,(c) \\
&= (f_a \, o \, f_b)\,(c) \quad \forall\, c \in S
\end{aligned}
$$

$$\Rightarrow \qquad f_{a*b} = \phi(a) \, o \, \phi(b) \quad \forall\, a, b \in S$$

$\therefore \phi : S \rightarrow S^S$ is a homomorphism of $(S, *)$ into (S^S, o)

Theorem 4.5. Let $(S, *)$ and (T, Δ) be two semi-groups and ϕ be a semi-group homomorphism from $(S, *)$ to (T, Δ). Corresponding to ϕ, there exists a congruence relation R on $(S, *)$ defined by

$$a\,Rb \quad \text{if} \quad \phi(a) = \phi(b) \quad \text{for all } a, b \in S$$

Proof. Clearly R is reflexive, symmetric and transitive, therefore R is an equivalence relation on S.

Let $\qquad a_1, a_2, a_1', a_2' \in S$ such that

$\qquad a_1 \, R \, a_1'$ and $a_2 \, R \, a_2'$

We have

$$
\begin{aligned}
\phi(a_1 * a_2) &= \phi(a_1) \,\Delta\, \phi(a_2) \\
&= \phi(a_1') \,\Delta\, \phi(a_2') \\
&= \phi(a_1' * a_2')
\end{aligned}
$$

i.e. $a_1 * a_2 \, R \, a_1' * a_2'$

Therefore R is a congruence relation on $(S, *)$ we now define a homomorphism from a semi-group to its Quotient semi-group, as shown in the theorem below:

Theorem 4.6. Let $(S, *)$ be a semi-group and R be a congruence relation on $(S, *)$. The quotient set S/R is a semi-group $(S/R, \oplus)$ where the operation $\oplus$ corresponds to the operation $*$ on S. Also there exists a homomorphism from $(S; *)$ onto $(S/R, \oplus)$.

Proof. Left as an exercise.

4.8 MONOID

Definition 4.19. A semi-group $(M, *)$ with an identity element with respect to the binary operation $*$ is called monoid.

In other words, an algebraic system $(M, *)$ is called a monoid if

(i) $(a * b) * c = a * (b * c)\ \forall\ a, b, c \in M$

(ii) There exists an element $e \in M$ such that $e * a = a * e\ \ \forall\ a \in M$

Example 1. Let Z be the set of integers, then $(Z, +)$ is a monoid, is the identity element with respect to the binary operation + in z.

Example 2. Let N be the set of natural numbers. (N, x) is monoid. 1 is the identity element in (N, x).

If the operation $*$ is commutative is a monoid $(M, *)$, Then we say that $(M, *)$ is commutative.

If there exists an element a in a monoid $(M, *)$ such that every element in M can be expressed as some power of a, then $(M, *)$ is called a cyclic monoid generated by a, and a is called the generator of $(M, *)$. A cyclic monoid is commutative and a cyclic monoid may have more than one generator.

Example 1. The algebraic $(Z, +)$ is a cyclic monoid $1, -1$ are its generators.

Example 2. If $M = \{-1, 1, i, -i)$ where $i = \sqrt{-1}$, then $(M, \cdot)$ is a cyclic monoid. The elements i and $-i$ are its generators.

Example 3. Consider $M = \{1, w, w^2\}$, the set of cube roots of unity (with $w^3 = 1$). $(M, \cdot)$ is a cyclic monoid generated by w. The element w^2 is also a generator of M.

4.9 MONOID HOMOMORPHISM

Definition 4.20. Let $(M, *)$ and (T, o) by any two monoids. Let e_M and e_T denote the identity element of $(M, *)$ and (T, o) respectively. A mapping

$$f : M \to T$$

is called a homomorphism, if

$$f(a * b) = f(a)\ o\ f(b) \text{ for all } a, b \in M$$

and

$$f(e_M) = e_T$$

Monoid homomorphism preserves associativity and identity. f also preserves commutativity. If $a \in M$ is invertible a^{-1} is the inverse of a in M, then $f(a^{-1})$ is the inverse of $f(a)$ i.e. $f(a^{-1}) = [f(a)]^{-1}$.

4.10 GROUPS

Definition 4.21. A group is an algebraic structure $(G, *)$ in which the operation $*$ on G satisfies the following conditions.

G_1: For all $a, b, c \in G$

$$a * (b * c) = (a * b) * c \qquad \text{(associativity)}$$

G_2: There exists an element $e \in G$, such that for any $a \in G$

$$a * e = a = e * a \qquad \text{(existence of identity)}$$

G_3: For every $a \in G$, there exists an element denoted by a^{-1} in G such that

$$a * a^{-1} = e = a^{-1} * a \qquad \text{(existence of inverse)}$$

a^{-1} is called the inverse in G.

Example 1. $(Z, +)$ is a group, where Z denotes the set of integers.

Example 2. $(R, +)$ is a group, where R denotes the set of real numbers.

4.10.1 Abelian Group

Definition 4.22. Let $(G, *)$ be a group. If $*$ is commutative

i.e. $\qquad a * b = b * a$ for all $a, b \in G$

then $(G, *)$ is called an Abelian group (or commutative group).

Example. $(Z, +)$ is an abelian group.

4.10.2 Finite Group

Definition 4.23. A group $(G, *)$ is said to be a finite group if the set G is a finite set.

Example. $G = \{-1, 1\}$ is a finite group; under the operation multiplication.

4.10.3 Infinite Group

Definition 4.24. A group $(G, *)$ which is not finite, is called an infinite group.

Example. $(Z, +)$ is an infinite group.

4.10.4 Order of a Finite Group

Definition 4.25. The order of a finite group $(G, *)$ is the number of elements in G.

The order of the group G, is denoted by $O(G)$. or by $|G|$.

Example. $G = \{1, -1, i, -i\}$ where $i = \sqrt{-1}$, is a group with respect to multiplication number of elements in G is 4

$\therefore \qquad O(G) = 4$

4.11 SOLVED EXAMPLES

Example 4.11.1. Show that the set $G = \{1, -1, i, -i\}$ is a group is an abelian group with respect to multiplication as the binary operation on G.

Solution. Let us construct the composition table.

Table. 4.6

*	1	−1	i	−i
1	1	−1	i	−i
−1	−1	1	−i	i
i	i	−i	−1	1
−i	−i	i	1	−1

From the above composition table it is clear that the algebraic structure $(G, \cdot)$ is closed and satisfies the following axioms.

Associativity: For any three elements $a, b, c \in G$,

$$(a \cdot b) \cdot c = a \cdot (b \cdot c) \quad \text{holds.}$$

Since $\quad (1 \cdot (-1)) \cdot i = -1 \cdot i = -i$

$$1 \cdot [(-1) \cdot i] = 1 - i = -i$$

$\Rightarrow \quad (1 \cdot (-1)) \cdot i = 1 \cdot (-1 \cdot i) = -i$

Similarly with any other three elements of G, we can show that the property holds.

Existence of identity: $1 \in G$, such that

$$1 \cdot a = a = a \cdot 1 \text{ for all } a \in G$$

$\therefore$ 1 is the identity in $(G, \cdot)$.

Existence of inverse: $1 \cdot 1 = 1 = 1 \cdot 1$

$$-1 \cdot -1 = 1, -i \cdot i = 1$$

Show that 1 is its an inverse

-1 is its an inverse

$-i$ is the inverse of i

and i is the inverse of $-i$

$\therefore$ Each element in G is invertible.

Commutativity: $a \cdot b = b \cdot a \ \forall \ a, b \in G$ holds in G

Example $1 \cdot 1 = 1 \cdot 1, 1 \cdot i = i \cdot 1 = i, -i \cdot 1 = 1 \cdot -i = -i$

$\therefore$ commutative law is satisfied in $(G, \cdot)$

All the axioms of an abelian group are satisfied

$\therefore$ $(G, \cdot)$ is an abelian group.

Example 4.11.2. Prove that $G = \{1, w, w^2\}$ is a group *w.r.* to multiplication where $1, w, w^2$ are cube roots of unity $(w^3 = 1)$

Solution. We construct the composition table as follows, we use

$$w^3 = 1; w^4 = w^3.w = w; \ldots$$

Table 4.7

·	1	w	w^2
1	1	w	w^2
w	w	w^2	1
w^2	w^2	1	w

The algebraic system $(G, \cdot)$ is closed under the operation multiplication.

From the composition table; it is clear that the operation is associative.

1 is the identity element in G with respect to multiplication

Since $\quad 1 \cdot a = a = a \cdot 1 \;\; \forall \; a \in G$

Each element in G is invertible : $1 \cdot 1 = 1 \Rightarrow 1$ is its own universe $w \cdot w^2 = 1 \Rightarrow w^2$ is the inverse of w and w is the inverse of the w^2.

$\therefore$ G is a group with respect to the operation multiplication

since $a \cdot b = b \cdot a$ for all $a, b \in G$, the group $(G = \{1, w, w^2\}, \bullet)$ is an abelian group.

Example 4.11.3. Prove that the set Z of all integers with the binary operation $a * b = a + b + 1 \;\; \forall \; a, b \in Z$ is a an abelian group.

Solution. Sum of two integers is an integer, therefore $a + b \in Z$ for all $a, b \in Z$

and $\quad a + b \in Z, 1 \in Z \Rightarrow a + b + 1 \in Z$

$\Rightarrow \quad a * b \in Z$ for all $a, b \in Z$

Now
$$(a * b) * c = a * b + c + 1$$
$$= a + b + 1 + c + 1$$
$$= a + b + c + 2$$

and
$$a * (b * c) = a * (b + c + 1)$$
$$= a + b + c + 1 + 1$$
$$= a + b + c + 2$$

$\therefore$
$$(a * b) * c = a * (b * c) \;\forall\; a, b, c \in Z$$

$-1 \in Z$ such that $a * (-1) = G + (-1) + 1 = a - 1 + 1 = a = (-1) * a \;\forall\; a \in Z(-1)$ is the identity element in Z with respect to $*$ a is an integer $\Rightarrow -a$ is also an integer.

$$-a \in z, -2 \in z \Rightarrow -a + (-2) = -a - 2 \in z$$
$$a * (-a - 2) = a + (-a - 2) + 1$$
$$= a - a - 2 - 1 = -1 * = (-a - 2) * a \quad \forall\; a \;\forall\; Z$$

$\therefore -a - 2$ is the inverse of a in z

i.e. each element is invertible in $(z, *)$

All the axioms of a group are satisfied

Since $a * b = a + b + 1 = b + a + 1 = b * a \;\; \forall \; a, b \in Z,$

$(Z, *)$ is commutative

$\therefore$ $(Z, *)$ is an abelian group.

Example 4.11.4. Let S be a non-empty set and $P(S)$ be the collection of all subset of S. Let the binary operation Δ called the symmetric difference of sets be defined by

$$A \Delta B = (A - B) \cup (B - A) \text{ for all } A, B \in P(S)$$

then show that $(P(S), \Delta)$ is an abelian group.

Solution. If A and B are any two subsets of S, then $A \Delta B$ is also a subset of S. Therefore $(P(S), \Delta)$ is closed with respect to operation Δ; and the following axioms are satisfied in $(P(S), \Delta)$

Associativity: For all $A, B, C \in P(S)$

$$(A \Delta B) \Delta C = A \Delta (B \Delta C)$$

can easily be verified.

Existence of identity: $\phi \in P(S)$, such that

$$A \Delta \phi = \phi \Delta A \text{ for all } A \in P(S)$$

Therefore ϕ is the identity element in $P(S)$ with respect to the composition Δ.

Existence of inverse: For all $A \in P(S)$, we have

$$A \Delta A = \phi$$

$\therefore$ A is the inverse of A

Commutative law: For all $A, B \in P(S)$

$$\begin{aligned} A \Delta B &= (A - B) \cup (B - A) \\ &= (B - A) \cup (A - B) \\ & B \Delta A \end{aligned}$$

Hence $(P(S), \Delta)$ is an abelian group.

Example 4.11.5. Show that the set of four transformations f_1, f_2, f_3 and f_4 on the set of complex numbers be defined by $f_1(z) = z, f_2(z) = -z,$

$$f_3(z) = \frac{1}{z},\ f_4(z) = \frac{-1}{z}$$

forms a finite abelian group with respect to the binary operation as the composition of product of two functions

Solution. Let $G = \{f_1, f_2, f_3, f_4\}$ and $*$ denote the composition as the composite of two functions.

We have $f_1 * f_1 = f_1,\ f_1 * f_2 = f_2 * f_1$

$$f_1 * f_2 = f_3 = f_3 * f_1,\ f_1 * f_4 = f_4 * f_1$$

$$(f_z * f_2)(z) = f_2(z) = f_2(-z) = -(-z) = z = f_1(z)$$

$\Rightarrow$ $f_2 * f_2 = f_1$

$$f_2 * f_3(z) = f_2\,(f_3(z) = f_2(1/z) = -\,1/z = f_4(z)$$

$\Rightarrow$ $\qquad f_2 * f_3 = f_4$

Similarly $\qquad f_2 * f_4 = f_3,\ f_3 * f_2 = f_4$

$$f_3 * f_3 = f_1,\ f_3 * f_4 = f_2$$

$$f_4 * f_2 = f_3,\ f_4 * f_3 = f_2 \text{ and } f_4 * f_4 = f_1$$

We have the following composition table.

Table 4.8

*	f_1	f_2	f_3	f_4
f_1	f_1	f_2	f_3	f_4
f_2	f_2	f_1	f_4	f_3
f_3	f_3	f_4	f_1	f_2
f_4	f_4	f_3	f_2	f_1

From the Table 4.8, it is clear that all the entries in the table are the elements of G. Therefore, G is closed with respect to the given operation $*$. The composition of functions is associative. The inverse elements of f_1, f_2, f_3, and f_4 are f_1, f_2, f_3 and f_4 respectively i.e. each element in G is invertible.

Since, commutative law holds good in G, the given set G is an abelian group, with respect to the operation $*$.

Definition 4.26. (Multiplication modulo p); Let p be a positive integer. Let a and be any two integers. The multiplication modulo p of a and b is defined as r, where r is the least non-negative remainder when the of a and b is divided by p. Symbolically, we write

$$a \times_p b = r,\ o \leq r c p$$

$$2 \times_5 4 = 3,\ 3x_7\, 6 = 4, \ldots$$

Example 4.11.6. Show that the set $G = \{1, 2, 3, 4)$ is an abelian group with respect to multiplication modulo 5.

Solution. The composition table for multiplication modulo 5 is

Table 4.9

$\times_5$	1	2	3	4
1	1	2	3	4
2	2	4	1	3
3	3	1	4	2
4	4	3	2	1

From the table 4.9 it is clear that G is closed with respect to the operation $\times_5$ and the binary operation $\times_5$ is associative.

1 is the identity element in $(G, \times_5)$

1 is its own inverse, 2 is the inverse of 3 and 3 is the inverse of 2.

4 is its own inverse with respect to the binary operation $\times_5$.

Commutative law, holds good in $(G, \times_5)$. Therefore $(G, \times_5)$ is an abelian group.

We shall now define a composition known as "addition modulo m" where m is a fixed integer.

Definition 4.27. (Addition modulo m). If a and b are any two integers and r is the least non-negative remainder obtained by dividing the ordinary sum of a and b by m, then the addition modulo m of a and b is r.

Symbolically, $a +_m b = r, 0 \leq r < m$

Example. $6 +_5 9 = 0, 4 +_5 8 = 2$

Since $6 + 9 = 15 = 5.3 + 0$

and $4 + 8 = 12 = 5.2 + 2$

Example 4.11.7. Show that the set $G = \{0, 1, 2, 3, 4\}$ is an abelian group with respect to addition modulo 5.

Solution: The composition table for $+_5$ is as follows.

Table 4.10

$+_5$	0	1	2	3	4
0	0	1	2	3	4
1	1	2	3	4	0
2	2	3	4	0	1
3	3	4	0	1	2
4	4	0	1	2	3

Since all the entries in the composition table are elements of G, the set G is closed with respect to addition modulo 5.

Associativity : For any $a, b, c \in G$,

$$(a +_5 b) +_5 c = a +_5 (b +_5 c) \text{ holds}$$

Example. $(1 +_5 3) +_5 4 = 3 = 1 +_5 (3 + 4)$, etc.

Existence of identity: $0 \in G$, is the identity element in $(G, +_5)$ since $o +_5 a = a = a +_5 0 \ \forall \ a \in G$.

4.12 ELEMENTARY PROPERTIES OF GROUPS

Theorem 4.7. If $(G, *)$ is a group, then the identity element in G is unique.

Proof. If possible e_1, and e_2 be identity elements in G.
e_1 is the identity element and $e_2 \in G$.

$$\Rightarrow e_1 * e_2 = e_2 = e_2 * e_1 \qquad \text{...(i)}$$

e_2 is the identity and $e_1 \in G$.

$$\Rightarrow e_2 * e_1 = e_1 = e_1 * e_2 \qquad \text{...(ii)}$$

From (i) and (ii), we get $e_1 = e_2$.
Hence proved

Theorem 4.8. The inverse of each element in a group $(G, *)$ is unique.

Proof. Let $a \in G$ and e be the identity element in G. Let $b \in G$ be an inverse of a in G, If possible let $c \in G$ be another inverse of a in G.

b is an inverse of a in G.

$$\Rightarrow a * b = b * a = e$$

Also c is an inverse of a in $G \Rightarrow a * c = c * a = e$

Now
$$\begin{aligned} b &= b * e \\ &= b * (a * c), \text{ (} e \text{ is the identity)} \\ &= (b * a) * c \\ &= e * c \text{ (by associative low)} \\ &= c \end{aligned}$$

Hence proved

Note. The identity a element is its own inverse.

Theorem 4.9. In a group (G, x)

$$(a^{-1})^{-1} = a \ \forall \ a \in G$$
$$(a^{-1} \text{ is the inverse of } a \text{ in } G)$$

Proof. G is a group.

$\therefore \ a \in G \Rightarrow a^{-1} \in G$ such that

$$a^{-1} * a = e = a * a^{-1}$$

now
$$a^{-1} \in G \Rightarrow (a^{-1})^{-1} \in G \text{ such that}$$
$$(a^{-1})\,(a^{-1})^{-1} = e = (a^{-1})^{-1} * (a^{-1})$$

Consider
$$a^{-1} * a = e$$
$$\Rightarrow \quad (a^{-1})^{-1} * (a^{-1} * a) = (a^{-1})^{-1} * e$$
$$\Rightarrow \quad ((a^{-1})^{-1} * a^{-1}) * a = (a^{-1})^{-1}$$
$$\Rightarrow \quad e * a = (a^{-1})^{-1}$$
$$\Rightarrow \quad a = (a^{-1})^{-1}$$
$$\therefore \quad (a^{-1})^{-1} = a \ \forall \ a \in G.$$

Theorem 4.10. If $(G, *)$ is a group then $(a * b)^{-1} = b^{-1} * a^{-1}$

For all $a, b \in G$ (Reversal law).

Proof. Let $a, b \in G$ and e be the identity element in G.

$$a \in G \Rightarrow a^{-1} \in G \text{ such that } a * a^{-1} = a^{-1} * a = e$$

and $b \in G \Rightarrow b^{-1} \in G$ such that $b * b^{-1} = b^{-1} * b = e$

now $a, b \in G \Rightarrow a * b \in G$ and $(a * b)^{-1} \in G$.

Consider $(b^{-1} * a^{-1}) * (a * b) = b^{-1} * (a^{-1} * a) * b$ (by associative law)

$= b^{-1} * e * b$; ($\because$ $a^{-1} * a = e$)

$= b^{-1} * b$; ($\because$ e is the identity)

$= e$ ($\because$ $b^{-1} * b = e$)

and $(a * b) * (b^{-1} * a^{-1}) = a * (b * b^{-1}) * a^{-1}$, (by associative law)

$= a * e * a^{-1}$

$= a * a^{-1}$; ($\because$ e is the identity)

$= e$

$\therefore$ $(b^{-1} * a^{-1})(a * b) = (a * b) * (b^{-1} * a^{-1}) = e$

$\Rightarrow$ $(a * b)^{-1} = b^{-1} * a^{-1}$ $\forall$ $a, b \in G$ (by the definition of inverse)

Theorem 4.11. Cancellation laws hold good in G, i.e. for all $a, b, c \in G$

$a * b = a * c \Rightarrow b = c$, (by left cancellation law)

$b * a = c * a \Rightarrow b = c$, (by right cancellation law)

Proof. $a \in G \Rightarrow a^{-1} \in G$ such that

$a * a^{-1} = a^{-1} * a = e$, where e is the identity element in G.

Consider $a * b = a * c$

$\Rightarrow$ $a^{-1} * (a * b) = a^{-1}(a * c)$

$\Rightarrow$ $(a^{-1} * a) * b = (a^{-1} * a) * c$, (by associative law)

$\Rightarrow$ $e * b = e * c$ ($\because$ a^{-1} is the inverse of a in G)

$\Rightarrow$ $b = c$ ($\because$ e is the identity element in G)

now,

$b * a = c * a$

$\Rightarrow$ $(b * a)\, a^{-1} = (c * a) * a^{-1}$

$\Rightarrow$ $b * (a * a^{-1}) = c * (a * a^{-1})$, (by associative law)

$\Rightarrow$ $b * e = c * e$ ($\because$ $a a^{-1} = e$)

$\Rightarrow$ $b = c$; (e is the identity element in G)

$\therefore$ Cancellation laws hold good in G.

Theorem 4.12. If a and b are any two elements of group $(G, *)$, then the equations $a * x = b$ and $y * a = b$ have unique solutions in G.

Proof. Let $a \in G$, then $a^{-1} \in G$. Such that $a * a^{-1} = a^{-1} * a = e$

Now $a^{-1} \in G, b \in G \Rightarrow a^{-1} * b \in G$; (closure property)

Consider,

$$a * x = b$$

$$\Rightarrow \quad a^{-1} * (a * x) = a^{-1} * b$$

$$\Rightarrow \quad (a^{-1} * a) * x = a^{-1} * b$$

$$\Rightarrow \quad e * x = a^{-1} * b$$

$$\Rightarrow \quad x = a^{-1} * b$$

$\therefore x = a^{-1} * b$ is a solution in G of the equation $a * x = b$.

Now we shall show that the solution is unique:

If possible, let there be two solutions;

Say x_1, x_2 of $a * x = b$ in G; then

$$a * x_1 = b \text{ and } a * x_2 = b$$

$$\Rightarrow \quad a * x_1 = a * x_2$$

$$\Rightarrow \quad x_1 = x_2, \text{ (by left cancellation law)}$$

$$\therefore \quad a * x = b, \text{ has unique solution in } G.$$

Similarly, we can show that

$$y * a = b, \text{ has unique solution in } G.$$

4.13 ALTERNATIVE POSTULATES FOR A GROUP

We state the following theorem (without proof). Which helps us to think of alternative postulates for a group.

Theorem 4.13. If $(G, *)$ is a semi-group such that

$$x * a = b \text{ and } a * y = b.$$

have a solutions for an $a, b \in G$ then $(G, *)$ is a group.

Theorem 4.14. If $(G, *)$ is a finite semi-group

Such that $\quad \forall\, a, b, c \in G$

$$a * b = a * c \Rightarrow b = c$$

and $\quad b * a = c * a \Rightarrow b = c$

Then $(G, *)$ is a group.

Theorem 4.15. If $(G, *)$ is a semi-group such that

(i) There exists an element $e \in G$, such that

$$a * e = a \ \forall \ a \in G.$$

(ii) $a \in G \Rightarrow$ There exists an element, $a_1 \in G$ such that

$$a * a_1 = e$$

then $(G, *)$ is a group.

Theorem 4.16. If $(G, *)$ is a finite group of orders n, then for each $a \in G$, there exists a positive integer $m \leq n$ such that $a^m = e$, e being the identity of G.

Proof. Let $a \in G$ then

$$a * a = a^2 \in a$$
$$a^2 * a = a * a * a = a^3 \in G$$
$$a^{n-1} * a = a^n \in G$$

$\therefore$ $e, a, a^2, a^3 \ldots a^n$ are all elements of G.

These elements are $(n + 1)$ in number, therefore they cannot all be distinct. In some $a^r = e$, then the theorem is proved. In the other case, we must have

$$a^r = a^s,\ 1 \leq r \leq s \leq n$$

because all the elements are not distinct.

Now, $\quad a^r = a^s$

$\Rightarrow \quad a^r * a^r = a^s * a^{-r}$

$\Rightarrow \quad a^o = a^{s-r}$

$\Rightarrow \quad e = a^{s-r}$

From our choice of r and

$$1 \leq r \leq s \leq n \ \Rightarrow \ 1 \leq s - r \leq n.$$

4.14 ORDER OF AN ELEMENT

Definition 4.25 Let $(G, *)$ be a group and $a \in G$, then the least positive integer n, if it exists such that

$a^n = e$, is called the order of a in G.

The order of an element $a \in G$ is denoted by $o(a)$.

Example 1. $G = \{1, -1, i, -i\}$ is a group with respect to multiplication 1 is the identity in G.

$$1^1 = 1^2 = 1^3 = \ldots = 1 \Rightarrow O(1) = 1$$
$$(-1)^2 = (-1)^4 = (-1)^6 = \ldots = 1 \Rightarrow O(-1) = 2$$
$$i^4 = -i^8 = i^{12} = \ldots = 1 \Rightarrow O(i) = 4$$

$$(-i)^4 = (-i)^8 = \ldots = 1 \Rightarrow O(-i) = 4$$

$\therefore$ $$O(1) = 1,\ O(-1) = 2,\ O(i) = O(-i) = 4$$

Example 2. $G = \{1, w, w^2\}$ is a group w.r. to multiplication in which we have $O(1) = 1$, $O(w) = 3$, $O(w^2) = 3$.

Example 3. In $[G = \{1, -1\}, x\}$,

The order of 1 is 1 i.e. $O(a) = 1$ and $O(e) = 2$

Note. The identity of group has order 1 i.e. $O(e) = 1$

We now prove certain properties of the order an element of a group.

Theorem 4.17. The order of every element of a finite group of order n is finite.

Proof. Let $a \in G$, then there exists a positive integer $m \leq n$ such that $a^m = e$

This proves that the order of a, can not be more than $m \leq n$;

since m is finite, $O(a)$ is finite.

Theorem 4.18. If $(G, *)$, $O(a) = n$, and m is a positive integer, then $a^m = e$, if and only if m is multiple of n,

i.e. $\quad m = nq \quad$ for some $q \in N$.

Proof. Since $O(a) = n$

There can not be another positive integer $x < n$ such that $a^x = e$

Thus, if $a^m = e$, then $m = n$

Let us assume;

$$a^m = e$$

We can find integers, q and r such that

$$m = nq + r, \quad o \leq r < n$$

We first prove that

$$a^m = e \Rightarrow m = nq, \ q \in N.$$

now $\quad a^m = e \Rightarrow e = a^{nq+r}$

$\Rightarrow \quad e = a^{nq} * a^r \qquad (\therefore a^m = e)$

$\Rightarrow \quad e = (a^n)^q * a^r$

$\Rightarrow \quad e = e * a^r$

$\Rightarrow \quad e = a^r$

$\Rightarrow \quad r = o \qquad (\because O(a) = n \text{ and } o \leq r \leq n)$

Thus we get $\quad m = nq$

Now, we prove that

$$m = nq \Rightarrow a^m = e$$

Consider,

$$m = nq \Rightarrow a^m = nq \cdot a$$

$\Rightarrow$ $$a^m = (a^n)q.$$

$\Rightarrow$ $$a^m = e^q$$

$\Rightarrow$ $$a^m = e$$

This completes the proof.

Theorem 4.19. If $(G, *)$ is a group, and

$$O(a^p) \leq O(a), \ \forall \ a \in G \text{ and } \forall \ p \in Z$$

i.e. the order of any integral power of any element of an group is less than or equal to the order of the element.

Proof. Let $O(a) = m$ and $o(a^p) = n$, where p is an integer

Since $O(a) = m$, we have $a^m = e$ and

$$(a^p)^m = a^{pm} = a^{mp} = (a^m)^p = e$$

$\Rightarrow$ $$O(a^p) \leq m$$

$\Rightarrow$ $$O(a^p) \leq O(a)$$

This completes the proof of the theorem.

Theorem 4.20. If $(G, *)$ is a group then $O(a) = O(a^{-1})$

i.e. the order of an element of a group is always equal to the order of its inverse.

Proof. By previous theorem

$$O(a^{-1}) \leq O(a) \quad \text{...(1)}$$

Also, $$a = (a^{-1})^{-1}$$

again by same property

$$O(a) \leq O(a^{-1}) \quad \text{...(2)}$$

From (1) and (2)

$$O(a) = O(a^{-1}).$$

EXERCISE 4.2

1. Define: (i) group (ii) abelian group. Give examples.
2. Show that the following algebraic structures are groups
   (a) The set of real numbers under addition
   (b) The set of rational numbers under addition
   (c) The set of non-zero real numbers under multiplication
   (d) The set of complex numbers under addition
   (e) The set of non-zero complex numbers under multiplication

(f) The set R in which $a * b = \frac{ab}{2}$ $\forall\ a, b \in R$.

3. Show that the set of all vectors in R^2 is an infinite group with vector addition as the composition.

4. Prove that the set $G = \{a + \sqrt{2}\, b : a, b \in Q\}$ is a group with respect to addition.

5. Show that the set of all vectors in R^3 with vector addition as binary composition is an infinite abelian group.

6. Show that The set of non-singular matrices of order n by m (n is fixed) with elements are rational numbers with matrix multiplication as binary composition is a group.

7. Prove that the matrices:

$$\begin{bmatrix} 1 & 0 \\ 0 & 1 \end{bmatrix}, \begin{bmatrix} 0 & 1 \\ 1 & 0 \end{bmatrix}$$

from a multiplicative group.

8. On the set of integers Z, we introduce a binary operation $*$ defined as follows. $a * b = a + b + 1$, where $+$ is ordinary addition. Show that $(Z, *)$ is a group.

9. Show that the set $G = \{-1, 1\}$, is a finite abelian group of order 2, under multiplication.

10. Prove that the set $\{1, 2, 3, 4, 5, 6\}$ of order 6 is a finite abelian group of order 6 under multiplication modulo 7.

11. Show that the set of even integers is a group under addition.

12. If Q denotes the set of rational numbers, then show that $Q - \{1\}$ is a group with respect to the operation $*$ defined by

$a * b = a + b + ab.\ \forall\ a, b \in Q - \{1\}$

13. Show that the set $G = \{6^n : n \in z\}$ is a group with respect to multiplication.

14. Show that the pair $\{(0, 4, 8, 12), +_{16}\}$ is a group.

4.15 SUB-GROUP

We now introduce the concept of a sub-group.

Definition 4.29. Let $(G, *)$ be a group and H, be a non-empty subset of G. If $(H, *)$ is itself is a group, then $(H, *)$ is called sub-group of $(G, *)$.

Example. Let $G = \{1, -1, i, -i\}$ and $H = \{1, -1\}$ G and H are groups with respect to the binary operation multiplication.

H is a subset of G, therefore $(H, \bullet)$ is a sub-group $(G, \bullet)$.

Example. Consider $(Z_6, +6)$, the group of integers modulo 6. $H = \{0, 2, 4\}$ is a subset of z_6 and $(H, +_6)$ is a group.

$\therefore$ $\{H, +_6\}$ is a sub-group.

Theorem 4.21. If $(G, *)$ is a group and $H \leq G$, then $(H, *)$ is a sub-group of $(G, *)$ if and only if

(i) $a, b \in H \Rightarrow a * b \in H$;

(ii) $a \in H \Rightarrow a^{-1} \in H$.

Proof. If $(H, *)$ is a sub-group of $(G, *)$ then conditions (i) and (ii) are obviously satisfied.

We therefore prove that if conditions (i) and (ii) are satisfied then $(H, *)$ is a sub-group of $(G, *)$.

To prove that $(H, *)$ is a sub-group of $(G, *)$ all that we are required to prove is: $*$ is associative in H and identity $e \in H$.

That $*$ is associative in H follows from the fact that $*$ is associative in G.

Also,

$$a \in H \Rightarrow a^{-1} \in H \text{ by (ii)}$$

and $$e \in H \text{ and } a^{-1} \in H \Rightarrow a * a^{-1} = e \in H \text{ by (i)}$$

Hence it is a sub-group of G.

Theorem 4.22. Let $(G, *)$ be a group and $\phi \neq H \subseteq G$.

Then $(H, *)$ is a sub-group of $(G, *)$ if and only if $a, b \in H \Rightarrow a * b^{-1} \in H$.

Proof. If $(H, *)$ is a sub-group and $a, b \in H$ then $b \in H \Rightarrow b^{-1} \in H$

$\therefore$ $$a \in H, b^{-1} \in H \Rightarrow a * b^{-1} \in H \text{ (by closure axiom).}$$

Conversely, suppose H is a non-empty subset of G which contains the element $a * b^{-1} \in H$ whenever $a, b \in H$

Then $$a, a \in H \Rightarrow a * a^{-1} = e \in H \text{ by the hypothesis.}$$

Again $$e, a \in H \Rightarrow e * a^{-1} = a^{-1} \in H.$$

Finally $$a, b \in H \Rightarrow a * b^{-1} \in H \Rightarrow a * (b^{-1})^{-1} = a * b \in H.$$

The set H "inherits" the associative law as a subject of G, so that all the groups axioms are satisfied, and $(H, *)$ is therefore a sub-group of $(G, *)$.

Theorem 4.23. Let $(G, *)$ be a finite group and $\phi \neq H \subseteq G$. Then $(H, *)$ is a sub-group $(G, *)$ if and only if

$$a, b \in H \Rightarrow a * b \in H.$$

Proof. If $(H, *)$ is a sub-group and $a, b \in H$ then $a * b \in H$ by closure axiom.

Conversely, suppose H is a non-empty subset of G and contains the element $a * b \in H$.

whenever $a = b \in H$

If $a \in H$, $a^2, a^3, a^4,\ldots$ all belongs to H. Since H is finite, the elements $a^2, a^3, a^4,\ldots$ can not be all different, that is only finite numbers of elements can be different.

Therefore some elements repeat. Let

$$a^i = a^j, i > j > o,$$

Now $\quad a^i = a^j \Rightarrow a^i * a^j = a^j * a^i.$

$\Rightarrow \quad a^{i-j} = e$

$\Rightarrow \quad a^{i-j} * a^{-1} = e * a^{-1}$

$\Rightarrow \quad a^{i-j-1} = a^{-1}$

But $\quad i - j - 1 \geq 0$

Hence $\quad a^{1-j-i} = a^{-1} \in H$. Here $i > j$

$\Rightarrow i - j \geq 0$ and $a^{i-j} = e \in H$. The set H inherits the associative law as a subset of G, so that all group axioms are satisfied and $(H, *)$ is *a* sub-group of $(G, *)$.

Theorem 4.24. If $(H_1, *)$ and $(H_2, *)$ are both sub-groups of the group $(G, *)$ then $(H_1 \cap H_2, *)$ is also a sub-group.

Proof. The set $H_1 \cap H_2 \neq \phi$, since $e \in H_1 \cap H_2$. Suppose that $a, b \in H_1 \cap H_2$, then $a, b \in H_1$ and $a, b \in H_2$, since

$(H_1, *)$ and $(H_2, *)$ are sub-groups.

$$a, b \in H_1 \Rightarrow a * b^{-1} \in H_1$$

and $\quad a, b \in H_2 \Rightarrow a * b^{-1} \in H_2$

that is $a * b^{-1} \in H_1 \cap H_2$ which implies that $(H_1 \cap H_2)$ is a sub-group of $(G, *)$.

Theorem 4.25. If $(G, *)$ is a group and $(H, *)$ is a sub-group of $(G, *)$ and $(K, *)$ is a sub-group of $(G, *)$ then $(H K, *)$ is a sub-group of $(G, *)$ if and only if $HK = KH$.

Where $HK = \{h * k/h \in H, K \in K\}$ and $KH = [k * h/h \in H, k \in K\}$

Proof. Left as an exercise to the student.

Example 1. $(\{0, 2, 4\}, +_6)$ is a sub-group of $(Z_6, +_6)$ because if $H = \{0, 2, 4\}$. Then,

$$0 +_6 0 = 0$$

$$0 +_6 2 = 2$$

$$0 +_6 4 = 4$$

$$2 +_6 2 = 4$$

$$2 +_6 4 = 0$$

$$4 +_6 4 = 2$$

i.e. $a, b \in H \Rightarrow a +_6 b \in H$.

Example 2. Let $Z = \{0, \pm 1, \pm 2, \pm 3 \ldots\}$ and let

$$H_1 = \{0, \pm 2, \pm 4, \pm 6 \ldots\}$$

and $$H_2 = \{0, \pm 3, \pm 6, \pm 9 \ldots\}$$

$$H_1 \cap H_2 = \{0, \pm 6, \pm 12 \ldots\}.$$

Here $(H_1, +)$ and $(H_2, +)$ are sub-groups of $(Z, +)$ and $(H_1 \cap H_2, +)$ is also a sub-group of $(Z, +)$...

But $(H_1 \cup H_2, +)$ is not a sub-group, since

$$H_1 \cup H_2 = \{0, \pm 2, \pm 3, \pm 4 \ldots\}$$

and $2, 3 \in H_1 \cup H_2$ but $2 + 3 = 5 \notin H_1 \cup H_2$

Hence $H_1 \cup H_2$ is not closed under +.

4.16 CENTRE OF A GROUP

Definition 4.30. Let $(G, *)$ be a group, then centre of the group G is the set of those elements of G which commute with every element of G. The centre of G is denoted by $Z(G)$

$$\therefore \quad Z(G) = \{a \in G / a * x = x * a \;\forall\; x \in G\}$$

4.17 COSETS

Let $(H, *)$ be a sub-group $(G, *)$ and $a \in G$.

Then the subsets

$$a * H = \{a * h : h \in H\}$$

is called a left coset of H in G, and subset

$$H * a = \{h * a : h \in H\}$$

is called a right coset of H in G.

Note. In general $a * H \neq H * a$, however if G is abelian then.

$$a * H = H * a \;\forall\; a \in G.$$

Example. Consider, the group $(z, +)$

Let $$H_n = \{0, \pm n, \pm 2n, \pm 3n \ldots\}$$

then the left cosets of H in z are

$$H_n, 1 + H_n, 2 + H_n, 3 + H_n \ldots (n-1) + H_n$$

and $$Z = H_n \cup 1 + H_n + 2 + H_n + \ldots + (n-1) + H_n$$

Example 2. If $H = \{1, -1\}$ $G = \{1, -1, i, -i\}$. Then H is a sub-group of G under multiplication. The left cosets of H in G are $H = \{1, -1\} \equiv -1.\ H$; $i \cdot H = \{i, -i\}$ $= (-i) \cdot H$

Theorem 4.26. If $(H, *)$ is a sub-group $(G, *)$ then $G * H = H$ if and only if $a \in H$.

Proof. Let $a * H = H$

Since $e \in h$ then $a = a * e \in a * H$

Hence $a \in H$

Conversely, let

$a \in H$

then $a * H \subseteq H$

$(H, *)$ is a sub-group.

$\therefore$ $a \in H, h \in H \Rightarrow a^{-1} * h \in H$

Now $h \in H$

$\Rightarrow$ $h = a * (a^{-1} * h) \in a * H$

$\therefore$ $h \in H \Rightarrow h \in a * H$

$\Rightarrow$ $H \subseteq a * H$

Hence $a * H = H$

We now state the following theorem without proof.

Theorem 4.27. If $(H, *)$ is a sub-group of $(G, *)$ then $a * H = b * H$ if and only if $a^{-1} * b \in H$.

Theorem 4.28. If $(H, *)$ is a sub-group of $(G, *)$ then only two left cosets of H in G are either identical or disjoint.

Proof. Let $a * H$ and $b * H$ be two left cosets of H in G. If $a * H$ and $b * H$ have no common element, then $a * H$ and $b * H$ are disjoint.

If $a * H \cap b * H \neq \phi$

Let $c \in a * H \cap b * H$

Then $c = a * h_1$ and $c = a * h_2$.

For some $h_1, h_2 \in H$

it follows that

$$a * h_1 = b * h_2$$

$\Rightarrow$ $a^{-1} * b = h_1 * h_2^{-1}$

$\Rightarrow$ $a^{-1} * b \in H \quad (\because h_1 * h_2^{-1} \in H)$

$\Rightarrow$ $a * H = b * H.$

Theorem 4.29. There exists a one to one correspondence between the elements of sub-group H and those of any coset of H in G.

Proof. Define a mapping

$$f : H \rightarrow a * H$$

by $f(h) = a * h$ for all $h \in H$

Let $h_1, h_2 \in H$ such that

$$f(h_1) = f(h_2)$$

Now, $f(h_1) = f(h_2)$

$\Rightarrow$ $a * h_1 = a * h_2$

$\Rightarrow$ $h_1 = h_2$ (by left cancellation law)

$\therefore f$ is one-one.

Every element of $a * H$ is of the form $a * h$ for same $h \in H$.

$\therefore$ f is onto.

This completes the proof of the theorem.

Note. If $(G, *)$ in a finite group, and $(H, *)$ is a sub-group of G, then any two left cosets of H in G have the same number of elements.

Theorem 4.30. If $a * H$ and $b * H$ are two distinct left cosets of H in G, then exists a one to one correspondence between $a * H$ and $b * H$.

Proof. Consider the mapping

$$f : a * H \rightarrow b * H$$

defined by

$$f(a * h) = b * h \text{ where } h \in H$$

The mapping f is one-one: Since

$$f(a * h_1) = f(a * h_2) \text{ where } h_1, h_2 \in H.$$

$\Rightarrow$ $b * h_1 = b * h_2$

$\Rightarrow$ $h_1 = h_2$ (by left cancellation law)

Thus f is one-one

f is onto mapping, since

$$b * h \in b * H \text{ is the } f\text{-image of bijective } a * h \in a * H$$

Thus f as defined above is a bijective mapping from $a * H$ to $b * H$

This completes the proof of the theorem.

4.18 LAGRANGE'S THEOREM

The theorem stated above indicate that each element of the group G belongs to one and only one left coset of H. Thus G can be partitioned by a into disjoint sets each of which has exactly as many elements as H. When the theorems are interpreted in the context of finite groups only; we obtain the following theorem known as Lagrange's theorem.

Theorem 4.31. (Lagrange's Theorems).

The order of any sub-group of a finite group divides the order of the group.

Proof. Let $(G, *)$ be a finite group of order n and $(H, *)$ be a sub-group of G of order m.

We can decompose the set G into a union of a finite number of distinct left cosets of H say K.

Let $a_1H, a_2H...a_kH$ denote the k distinct left cosets of H in G such that

$$G = a_1 * H \cup a_2 * H \cup a_3 * H \cup ... \cup a_k * H$$

Where all the k left coset. appearing on the right hand side are disjoint.

Therefore $\quad O(G) = O(a_1 * H) + O(a_2 * H) + ...O(a_k * H)$

$\Rightarrow \quad n = m + m + ...$ k terms

$\Rightarrow \quad n = km$

$\Rightarrow \quad m/n$

$\Rightarrow \quad O(H)/O(G)$

This completes the proof the theorem.

4.19 INDEX OF A SUB-GROUP

Definition 4.32. The numbers of distinct left (or right) cosets of H in G is called index of H in G.

The index of H in G is denoted by $I_G(H)$. Langrange's theorem proved above can also be stated as follows:

If $(G, *)$ is a finite group and $(H, *)$ is a sub-group of G, then

$$I_G(H) = \frac{O(G)}{O(H)}$$

Lagrange's theorem is of fundamental importance and is used to prove other important results.

Theorem 4.32. The order of every element of a finite group is a divisor of the order of the group.

Proof. Let $(G, *)$ be a finite group and $a \in G$ and let $O(a) = m$

Consider the set

$$H = \{a, a^2, a^3 ... a^m = e\}$$

Obviously $(H, +)$ is a sub-group of $(G. *)$ and $O(H) = m$. Lagrange's theorem ensures that

$$O(H)/O(G) \Rightarrow m/O(G)$$
$$\Rightarrow O(a)/O(G).$$

This completes the proof of the theorem.

We know that if H and K are two sub-groups of a group G, then HK is a sub-group of G if and only if $H * K = K * H$.

If H and K are finite sub-groups of a group G then we can find the $O(H * K)$ be making use of following theorem.

Theorem. 4.33. Let $(G, *)$ be a group $(H, *)$ and $(K, *)$ be two finite sub-groups of G such that

$$H * K = K * H \text{ then}$$

$$O(H * K) = \frac{O(H) * O(K)}{O(H \cap K)}$$

4.20 ISOMORPHISM

Let $(G, *)$ and (G', Δ) be two groups.

A bijective mapping:

$$f: G \to G'$$

Satisfying

$$f(a * b) = f(a) \Delta f(b), \forall\ a, b \in G.$$

is called an isomorphism of G to G'.

If there exists an isomorphism between two groups $(G, *)$ and (G', Δ) then G and G' are said to isomorphic to each other.

If $(G, *)$ is isomorphic to (G', Δ) then we write

$$(G, *) \cong (G', \Delta)$$

for G'.

Example 1. Consider the group $(R, +)$ and (R^+, X) where R^+ denoted the set of positive real numbers-consider the mapping

$$f_a : R \to R^+, a \in R^+$$

defined by

$$f_a(x) = a^+, \text{ for all } X \in R$$

the mapping f_a is structure preserving:

We have $\quad f_a(x + y) = a^{x+y} = a^x * a^y$ where $x, y \in R$

$\Rightarrow \quad f_a$ is structure preserving

Also $\quad f_a(x) = f_a(y)$

$\Rightarrow \quad a^x = a^y$

$\Rightarrow \quad a^{x-y} = 1$

$\Rightarrow \quad x - y = 0$

$\Rightarrow \quad x = y$

$\Rightarrow f_a$ is one-one.

From the definition:

$$y \in R^+ \Rightarrow \text{there exists a real number } x \text{ such that } y = a^x$$

$\therefore$ $\quad y \in R^+ \Rightarrow y = a^x$ for some $x \in k$

$\Rightarrow$ $\quad y = f_a(x)$

$\Rightarrow$ $\quad f_a$ is onto

$\therefore$ Thus $(R, +) \cong (R^+, x)$

Example 2. Let G be a group for a fixed $a \in G$, define

$$f_a : G \to G \text{ by } f_a(x) = axa^{-1} \; \forall x \in G$$

Show that f is an automorphism of G.

Solution. For all $x, y \in G$, we have

$$xy \in G \Rightarrow f_a(xy) = a\,xya^{-1}$$
$$= (ax^{a-1}).(aya^{-1})$$
$$= f_a(x) f_a(y)$$

and $\quad f_a(x) = f_a(y)$

$\Rightarrow$ $\quad a\,xa^{-1} = a\,ya^{-1}$

$\Rightarrow$ $\quad x = y$

$\Rightarrow$ $\quad f$ is onto

also; $\quad x \in G \Rightarrow f_a(a^{-1}xa) = a(a^{-1}xa)\,a^{-1} = x$

$\Rightarrow$ $\quad f$ is onto

f is an isomorphism on G

$\therefore$ f is an automorphism of G

Example 3. The groups $(Z, +)$ and $(nZ, +)$ $n \in Z$ are isomorphisc to each other.

4.21 PROPERTIES OF ISOMORPHISM

We now discuss some properties of isomorphic groups. Let $(G, *)$ and (G', Δ) be two groups such that $(G, *) \cong (G', \Delta)$. Also let e, e' be the identity elements of G and G' respectively, then,

(i) $f(e) = e'$

(ii) $f(a^{-1}) = [f(a)]^{-1} \; \forall \; a \in G.$

(iii) $O(a) = O[f(a)] \; \forall \; a \in G.$

(iv) $f^{-1} : G' \to G$ is an isomorphism

(v) The composite of two isomorphisms is an isomorphism.

We shall prove the following theorem, by introducing a relation in the set of all groups.

Theorem 4.34. The relation '$\cong$' in the set of all groups is an equivalence relation.

Proof. $\cong$ is reflexive:

If $(G, *)$ is a group then $(G, *) \cong (G, *)$

Since the identity mapping

$$I_G : G \to G$$

is an isomorphism from $(G, *)$ to $(G, *)$

$\cong$ is symmetric:

Let $(G, *) \cong (G', \Delta)$

then $(G', \Delta) \cong (G, *)$

Therefore $\cong$ is symmetric

$\cong$ is transitive:

Let $(G, *) \cong (G', \Delta)$ and $(G', \Delta) \cong (G'', \square)$

Then $(G, *) \cong (G''_2, \square)$

Therefore $\cong$ is transitive.

$\therefore \cong$ is reflexive, symmetric and transitive therefore $\cong$ is an equivalence relation in the set of all groups.

EXERCISE 4.3

1. Show that every group of order 2 is isomorphic to every other group of order 2.
2. Show that every group of order 3 is isomorphic to every other group of order 3.
3. Show that $(I, +)$ is isomorphic to $\{a^m : a \in z\}, +_5\ m \in Z$ is fixed.
4. Let $G = Z_6$ then show that $O(\overline{3}) = 2$.
5. Let $(G, *)$ be an infinite cyclic group then prove that G has exactly two generators.
6. If $G = <a>$ is cyclic group of order n then show that $O(a) = n$.
7. Let G be a group H be a sub-group of G such that $[G : H] = 2$ show that every left coset of H in G is also a right coset of H in G.
8. Show that any two cyclic groups of the same order are isomorphic.
9. Prove that any non-cyclic group of order 4 is isomorphic to klein 4-group.
10. If $(G, *)$ is a group of order 4, then show that G is abelian.
11. Show that the additive group $(Z, +)$ is isomorphic to the multiplicative group of non-zero elements of Z_5.
12. Find all sub-groups of a cyclic group of orders 10 and 12.

13. Let $(Z_{12}, +_{12})$ be the group of integers modulo 12 then show that the subset generated by 3.

$$<3> = \{3n \text{ (mod12)} = n \in z\}$$
$$= \{ 0, 3, 6, 9\}$$

14. Consider (G_1, I_5) the group of integers modulo 15 and the sub-group $H = \{0, 3, 6, 9, 12\}$ of G. Find all left cosets of H in G.

4.22 CYCLIC GROUPS

Let $(G, *)$ be a group. If there exists an element $a \in G$ such that

$$G = \{a^m : m \text{ is an integer}\}$$

i.e. $(G, *)$ is cyclic, if there exists an element $a \in G$ such that every element of G is a power of a and a is called the generator of the cyclic group.

If a is a generator of $(G, *)$ then we write $G = <a>$

Note. If $(G, +)$ is a cyclic group then each element of G can be expressed in the form na where n is an integer and $n \in G$.

Example 1. Let Z denote the set of integers $(Z, +)$ is an infinite cyclic group 1 and -1 are the generators of the group $(Z, +)$.

Example 2. $G = \{1, -1, i, -i\}$ is a group with respect to the binary operation '•'.

$(G, \bullet)$ is a cyclic group.

i is a generator of G.

Since $\quad (i)^4 = 1$

$$(i)^3 = -i$$
$$i^2 = -1$$
$$(i)^1 = i$$

and $\quad G = \{i^4, i^2, i, i^3\}$

$$= <i>$$

Similarly $(-i)$ is a generator.

$i, -i$ are the only generators of G.

If $(G, *)$ is an infinite group generated by an element a, then we can write,

$$G = \{\dots a^{-3}, a^{-2}, a^{-1}, a^0, a^1, a^2, a^3 \dots\}$$

And if $(G, *)$ is a finite cyclic group generated by a, then we can write,

$$G = \{a, a^2, a^3 \dots a^n = e\}$$

And if $(G, *)$ is a finite cyclic group or order n then,

$$a^i * a^j = a^{i+j} \text{ if } i + j < n$$

$$= a^0 \text{ if } i + j = 0$$

$$= a^{i+j-n} \text{ if } i + j > n$$

where a is the generator of G.

Theorem 4.35. Every cyclic group is abelian.

Proof. Let $(G, *)$ be a cyclic group generated by a

Let, $x, y \in G,$ than

$x = a^m$ and $y = a^n$ for some integers m and n.

$$x * y = a^m * a^n$$

$$= a^{m+n}$$

$$= a^{n+m}$$

$$= a^n * a^m$$

$$= y * x.$$

This completes the proof of the theorem.

Note. The converse of the theorem may not hold.

Theorem 4.36. Every group of prime order is cyclic.

Proof. Let $O(G) = p$ where 'p' is a prime number and let $a \neq e \in G$

Consider the sub group of G, generated by a

Let $H = <a>$

$\Rightarrow$ $O(H) > 1$

H is a sub group of G by Lagranges theorem $O(H)$ divides $O(G)$.

i.e. $O(H)/O(G) = 1$

$\therefore$ $O(H) = 1$ or $O(H) = p$

Since the only divisors of p are 1 and p only but $O(H) \neq 1$

$\therefore$ $O(H) = b$ but H is cyclic $\Rightarrow$ G is cyclic

Theorem 4.38. Every finite cyclic group of order n is isomorphic to $(Z_n, +)$. We leave the proof to the reader and state the following theorem without proof.

> "Every proper sub-group of an infinite cyclic group is isomorphic to the group itself".

Theorem 4.39. Every proper sub-group of finite cyclic group is a finite cyclic group.

Theorem 4.40. The only groups which do not possess proper sub-groups are the prime order finite sub-groups.

By Lagranges theorem.

$$O(H)/O(G) \Rightarrow O(H)/p$$

$$\Rightarrow \qquad O(H) = 1 \text{ or } p$$

$$\Rightarrow \qquad O(H) = p \quad \text{Since } O(H) \neq 1$$

$$O(H) = O(G)$$

But H is cyclic $\Rightarrow G$ is cyclic. Hence the theorem.

Theorem 4.37. Every infinite cyclic group is isomorphic to $(Z, +)$

Proof. Let $(G, *)$ be an infinite cyclic group generator by ϕ

Define a mapping

$$\phi : Z \to G$$

By the rule

$$\phi(x) = a^x,\ n \in Z$$

ϕ is one-one, since

$$a^x = a^y \Rightarrow x = y$$

The mapping ϕ is obviously onto.

Also, we have

$$\begin{aligned}\phi(x + y) &= a^{x+y} \\ &= a^x * a^y \\ &= \phi(x) * \phi(y)\end{aligned}$$

$\Rightarrow \phi$ is structure preserving

$\therefore\ \phi$ is an isomorphism

$$(G, *) \cong (Z, +)$$

Example. $(G, *)$ is a group order 60, find all the sub-groups of G,

Solution. We can write

$$\begin{aligned}60 &= 1 \times 2 \times 2 \times 3 \times 5 \\ &= 1 \times 2^2 \times 3 \times 5\end{aligned}$$

Therefore the factors of 60 are:

1, 2, 3, 4, 5, 6, 10, 12, 15, 20, 30, 60.

Let a be a generator of $(G, *)$ then the sub-groups of $(G, *)$ are
$\{e\}, <a>, <a^2>, <a^3>, <a^4>, <a^5>, <a^6>, <a^{10}>, <a^{12}>, <a^{15}>, <a^{20}>, <a^{30}>$.

4.23 NORMAL SUB-GROUPS

Definition 4.33. A sub-group $(H, *)$ of $(G, *)$ is called a normal sub-group of G if for all $h \in H$ and $g \in G$ $ghg^{-1} \in H$. If H is normal in G then we are write $H \not\subset G$.

Example 1. Let $(G, *)$ be a Group then $(\{e\}, *)$ is a normal sub-group in G. It is called the improper normal sub-group.

Example 2. $(G, *)$ is normal in $(G, *)$ it is called the improper normal sub group.

4.23.1 Simple Group

Definition 4.34. A group $(G, *)$ is called simple group if its only normal sub-groups are G and $\{e\}$.

Example. A group of prime order has no proper sub-groups.

Therefore, every group of prime order is simple. We now give an equivalent definition for the sub-group to be normal.

Definition 4.35. A sub-group $(H, *)$ of a group $(G, *)$ is said to be normal sub-group of $(G, *)$ is for every $g \in G, ghg^{-1} \subseteq H$.

Theorem 4.41. Every sub-group of an abelian group is normal sub-group.

Proof. Let $(G, *)$ be an abelian group and $(H, *)$ be a sub-group of G.

Now $\quad g \in G, h \in H$

$\Rightarrow \quad g * h * g^{-1} = h * g * g^{-1} \quad \therefore (G \text{ is abelian and } H \leq G)$

$$= h * e$$

$$= h$$

$\therefore \quad g * h * g^{-1} = h \in H \; \forall \; g \in G, h \in H$

$(H, *)$ is normal in G.

Theorem 4.42. A sub-group $(H, *)$ is a normal in $(G, *)$ if and only if $gh = Hg$ for each $g \in G$.

Proof. Suppose $(H, *)$ is normal in $(G, *)$. Therefore each $g \in G, g * H * g^{-1} \subseteq H$ i.e. $gH \subseteq Hg$ ...(1)

By replacing g by g^{-1} we have $g^{-1} * H$

$$g * (g^{-1} * H) * g \subseteq g * (H * g^{-1}) * g$$

$\Rightarrow \quad (g * g^{-1}) * H * g \subseteq g * H * (g^{-1} * g)$

$\Rightarrow \quad e * H * g \subseteq g * H * e$

$\Rightarrow \quad H * g \subseteq g * H$ ...(2)

Thus, from (1) and (2), $\quad Hg = gH$

Conversely assume that $\quad gH = Hg, \; \forall \; g \in G.$

Then $\quad gHg^{-1} = H \; \forall \; h \in G$

Showing that $(H, *)$ is normal in $(G, *)$.

Example. Show that every sub-group of an abelian group is normal.

Solution. Let $(G, *)$ be an abelian group and $(H, *)$ be a sub-group of G.

Let g be any element of G and $h \in H$

Then we have

$$g * h * g^{-1} = g * g^{-1} * h$$
$$= e * h$$
$$= h \in H$$

$\therefore$ Hence, H is normal in G.

4.24 QUOTIENT GROUP

Let $(H, *)$ be a normal sub-group of a group $(G, *)$. The set of all cosets of H in G is known as the quotient set G/H;

$$G/H = \{G * H : a \in G\}$$

Here $$a * H = H * a \quad \forall \ a \in G$$

Now $$(a * H) * (b * h)$$
$$(a * b * H) * H$$
$$= [(a * b) * H] * H$$
$$= (a * b) * H * H$$
$$= (a * b) * H$$

i.e. product of two left cosets of H in G is again a left coset in G. Similarly, we can show that, the product of two right cosets is again a right coset in G.

Theorem 4.43. If $(H, *)$ is a normal sub-group of the group $(G, *)$, then the system $(G/H, *)$ forms a group, known as quotient group of G by H (or factor group).

Proof. We observe the following in G/H

(i) The operation $*$ is associative in G/H for

$$\{a * H) * (b * H)\} * (c * H)$$
$$= \{(a * b) * H) (c * H)\}$$
$$= \{(a * b) * c\} * H$$
$$= a * (b * c) * H$$
$$= (a * H) * (b * c) * H$$
$$= (a * H) * \{(b * H) * (c * H)\}$$

(ii) G/H has an identity element; $H = e * H$,

For $(a * H) * (e * H) = (a * e) * H = a * H$

and $(e * H) (a * H) = (e * a) * H = a * H$

(iii) For each $a * H$ there exists $a^{-1} * H$ such that

$$(a * H) * (a^{-1} * H) = (a * a^{-1}) * H$$
$$= e * H$$
$$= (a^{-1} * a) * H$$
$$= (a^{-1} * H) * (a * H)$$

$\therefore a^{-1} * H$ is the inverse of $a * H$.

Corollary. If $(G, *)$ is a finite group and $(H, *)$ is a normal sub-group of $(G, *)$

Then $$O(G/H) = O(G)/O(H)$$

Example. If $H_3 = \{... -6, -3, 0, 3, 0, 6 ...\}$

Then $(H_3, +)$ is a normal sub-group of $(Z, +)$ of all integers. The cosets of H_3 in z are

$$H_3 = \{... -6, -3, 0, 3, 6, 9 ...\}$$
$$1 + H_3 = \{... -5, -2, 1, 4, 7, 10 ...\}$$
$$2 + H_3 = \{... -4, -1, 2, 5, 8 ...\}$$
$$\therefore \quad G/H_3 = \{H_3, 1 + H_3, 2 + H_3\}$$

Table

+	H_3	$1 + H_3$	$2 + H_3$
H_3	H_3	$1 + H_3$	$2 + H_3$
$1 + H_3$	H_3	$2 + H_3$	H_3
$2 + H_3$	$2 + H_3$	H_3	$1 + H_3$

It is clear from the table that $(Z/H_3, +)$ is a group.

4.25 PERMUTATION GROUPS

Definition 4.36. A permutation is a one-one mapping of a non-empty set onto itself.

When a set S is a finite with n elements in it, we speak of a permutation of n symbols. It is not necessary to limit our discussion with some definitions.

4.25.1 Equal Permutations

Let S be a non-empty set the permutation. The permutation f and G defined on S, are said to be aqual if $f(a) = g(a)$ for all $a \in S$.

Example. Let $$S = \{1, 2, 3, 4\}$$

and Let $$f = \begin{pmatrix} 1 & 2 & 3 & 4 \\ 3 & 1 & 2 & 4 \end{pmatrix} \quad g = \begin{pmatrix} 4 & 1 & 3 & 2 \\ 4 & 3 & 2 & 1 \end{pmatrix}$$

We have $f(1) = g(1) = 3$

$f(2) = g(2) = 1$

$f(3) = g(3) = 2$

$f(4) = g(4) = 4$

i.e. $f(a) = g(a) \ \forall \ a \in S$

therefore $f = g$

Let $S = a_1, a_2 \ldots a_n$ be a finite set. The number of permutations on S is n !. The set of all permutations on S is denoted by S_n. Where $|S_n| = n!$ If $f \in S_n$ then f is the form

$$f = \{(a\ , f(a), (a_2), f(a_2)\} \ldots (a_n, f(a_n))\}$$

It can also witten as

$$f = \begin{pmatrix} a_1 & a_2 & a_3 & \cdots & a_n \\ f(a_1) & f(a_2) & f(a_3) & \cdots & f(a_n) \end{pmatrix}$$

the images $f(a_1), f(a_2) \ldots f(a_n)$ are the elements of S arranged in some order. The order of symbols S in the first row of a permutation is immaterial but columns should not be affected and every permutation of S_n may be written in $n!$ ways.

Example. Let $S = \{1, 2, 3\}$ and $f = \begin{pmatrix} 1 & 2 & 3 \\ 2 & 3 & 1 \end{pmatrix}$.

We can write $f = \begin{pmatrix} 1 & 2 & 3 \\ 2 & 3 & 1 \end{pmatrix} = \begin{pmatrix} 2 & 3 & 1 \\ 3 & 1 & 2 \end{pmatrix} = \begin{pmatrix} 3 & 1 & 2 \\ 1 & 2 & 3 \end{pmatrix}$

$$\begin{pmatrix} 1 & 3 & 2 \\ 2 & 1 & 3 \end{pmatrix} = \begin{pmatrix} 2 & 1 & 3 \\ 3 & 2 & 1 \end{pmatrix} = \begin{pmatrix} 3 & 2 & 1 \\ 1 & 3 & 2 \end{pmatrix}$$

Hence there are $3! = 6$ ways of writing f.

4.25.2 Identity Permutation

Definition 4.37. Let S be a finite non-mpty set. An identity permutation on S denoted by I is defined as

$$I(a) = a \text{ for all } a \in S$$

Example 1. Let $S = \{a_1, a_2 \ldots a_n\}$

Then $I = \begin{pmatrix} a_1 & a_2 & \cdots & a_n \\ a_1 & a_2 & \cdots & a_n \end{pmatrix}$ is the identity permutation on S.

Example 2. Let $s = \{1, 2, 3, 4\}$ then

$$f = \begin{pmatrix} 1 & 2 & 3 & 4 \\ 1 & 2 & 3 & 4 \end{pmatrix} \text{ is the identity permutation on } S.$$

4.25.3 Product of Permutations

(or composition of permutations)

Let $s = (a_1, a_2 \dots a_n)$ and let

$$f = \begin{pmatrix} a_1 & a_2 & \dots & a_n \\ f(a_1) & f(a_2) & \dots & f(a_n) \end{pmatrix} \quad g = \begin{pmatrix} a_1 & a_2 & \dots & a_n \\ g(a_1) & g(a_2) & \dots & g(a_n) \end{pmatrix}$$

be two arbitrary permutations on S. We can find the composite of f and g as follows:

$$fog = \begin{pmatrix} a_1 & a_2 & \dots & a_n \\ f(a_1) & f(a_2) & \dots & f(a_n) \end{pmatrix} o \begin{pmatrix} a_1 & a_2 & \dots & a_n \\ g(a_1) & g(a_2) & \dots & g(a_n) \end{pmatrix}$$

$$= \begin{pmatrix} a_1 & a_2 & \dots & a_n \\ f(g(a_1)) & f(g(a_2)) & \dots & f(g(a_n)) \end{pmatrix} o \begin{pmatrix} a_1 & a_2 & \dots & a_n \\ g(a_1) & g(a_2) & \dots & g(a_n) \end{pmatrix}$$

$$= \begin{pmatrix} a_1 & a_2 & \dots & a_n \\ f(g(a_1)) & f(g(a_2)) & \dots & f(g(a_n)) \end{pmatrix}$$

Clearly fog is a permutation and $fog \in S_n$.

Note. In general $f \, o \, g \neq g \, o \, f$.

Example. Let $S = \{1, 2, 3\}$

and $$f = \begin{pmatrix} 1 & 2 & 3 \\ 2 & 3 & 1 \end{pmatrix}; g = \begin{pmatrix} 1 & 2 & 3 \\ 3 & 2 & 1 \end{pmatrix}$$

be two permutation on S, we compute fog as follows:

$$f \, o \, g = \begin{pmatrix} 1 & 2 & 3 \\ 2 & 3 & 1 \end{pmatrix} o \begin{pmatrix} 1 & 2 & 3 \\ 3 & 2 & 1 \end{pmatrix}$$

$$= \begin{pmatrix} 3 & 2 & 1 \\ 1 & 3 & 2 \end{pmatrix} o \begin{pmatrix} 1 & 2 & 3 \\ 3 & 2 & 1 \end{pmatrix}$$

$$= \begin{pmatrix} 1 & 2 & 3 \\ 1 & 3 & 2 \end{pmatrix}$$

4.25.4 Inverse of a Permutation

If f is a permutation on $s = (a_1, a_2 \ldots a_n)$ such that,

$$f = \begin{pmatrix} a_1 & a_2 & \ldots & a_n \\ b_1 & b_2 & \ldots & b_n \end{pmatrix}$$

then there exists a permutation called the inverse of f, denoted f^{-1} such that

$$f o f^{-1} = f^{-1} o f = I \qquad \text{(the identity permutation on } S\text{)}$$

where $$f^{-1} = \begin{pmatrix} b_1 & b_2 & \ldots & b_n \\ a_1 & a_2 & \ldots & a_n \end{pmatrix}$$

Theorem 4.44. If S a set of n symbols then the set S_n of all permutation on S in a group with respect to the operation product of two permutations.

Proof. We have already seen that the product (composite) of two permutations on n symbols is again a permutation. The product of permutations is associative. The identity permutation I defined on S acts as the identity element of the group. If f is a permutation on S, then f^{-1} is also a one-one mapping on S onto itself. Hence f^{-1} is also a permutation in S such that

$$f o f^{-1} = f^{-1} o f = I$$

Thus, S_n is a group.

Example. Let $S = \{1, 2, 3\}$, then the permutation on S are

$$I = \begin{pmatrix} 1 & 2 & 3 \\ 1 & 2 & 3 \end{pmatrix}, \quad \sigma_1 = \begin{pmatrix} 1 & 2 & 3 \\ 3 & 2 & 1 \end{pmatrix},$$

$$\sigma_2 = \begin{pmatrix} 1 & 2 & 3 \\ 1 & 3 & 2 \end{pmatrix}$$

$$\sigma_3 = \begin{pmatrix} 1 & 2 & 3 \\ 2 & 1 & 3 \end{pmatrix}$$

$$\rho_1 = \begin{pmatrix} 1 & 2 & 3 \\ 2 & 3 & 1 \end{pmatrix}, \quad \rho_2 = \begin{pmatrix} 1 & 2 & 3 \\ 3 & 1 & 2 \end{pmatrix}$$

$$S_3 = \{I, \sigma_1, \sigma_2, \sigma_3, \rho_1, \rho_2\}$$

is a group with respect to the operation product of permutations number of elements in $S_3 = 3! = 6$

Definition 4.38. The group S_n is called the symmetric group (or permutation group)

Note. (i) The order of S_n is $n!$

(ii) For $n \leq 2$ the group S_n is abelian and for $n \geq 3$, the group S_n is non-abelian.

If f is a permutation a finite set S we define

$$f^2 = f o f, f^3 = f o f o f \text{ ... order of a permutation.}$$

Let 'S' be a finite non-empty set with n symbols and S_n denote the set of all permutations on S. And let I denote the identity permutation on S. If $f \in S_n$ then the least positive integer k such that $f^k = I$ is called the order of f in S_n.

Example. If $f = \begin{pmatrix} 1 & 2 & 3 & 4 & 5 & 6 & 7 & 8 \\ 1 & 2 & 3 & 4 & 5 & 6 & 8 & 7 \end{pmatrix}$

Then $$f^2 = f o f = \begin{pmatrix} 1 & 2 & 3 & 4 & 5 & 6 & 7 & 8 \\ 1 & 2 & 3 & 4 & 5 & 6 & 7 & 8 \end{pmatrix}$$

$$= \text{I}$$

$\therefore$ order of f is 2.

4.25.5 Cyclic Permutation

Definition 4.39. Let $S = \{a_1, a_2, \dots a_n\}$ be a finite set of n symbols. A permutation of defined on S is said to be cyclic, if f is defined such that

$$f(a_1) = a_2, f(a_2) = a_3, \dots, f(a_{n-1}) = a_n \text{ and } f(a_n) = a_1$$

Example 1. Let $S = \{1, 2, 3, 4\}$, then

$$f = \begin{pmatrix} 1 & 2 & 3 & 4 \\ 2 & 3 & 4 & 1 \end{pmatrix} \text{ is a cyclic permutations.}$$

Example 2. Let $S = \{a, b, c\}$

Then $$f = \begin{pmatrix} a & b & c \\ b & c & a \end{pmatrix} \text{ is a cyclic permutation on } S.$$

If f is a cyclic permutation on $S = \{a_1, a_2, \dots a_n\}$,
then we can write $f = (a_1\ a_2 \dots a_n)$.

4.25.6 Transposition

Definition 4.40. A cycle of length two is called a transposition.

Example. $f = \begin{pmatrix} a & b \\ b & a \end{pmatrix} = (a\ b)$ is a transposition.

If s_n is a permutation group on n symbols, then of the $n!$ permutation in s_n, $\frac{1}{2}n!$ are even permutations and $\frac{1}{2}n!$ are add permutations. The set of all even permutations of degree n from a group under the composition of permutations. The group of even permutation is called an alternating group. It is denoted by A_n we state the following theorem without proof.

Theorem 4.45. Every permutation may be expressed as the product of transpositions in many ways.

If f is a cycle of length of n, then f can be expressed as a $(n - 1)$ transpositions.

Definition (Even Permutation). 4.41. A permutation f is said to be an even permutation, if, f can be expressed as a product of an even number of transpositions.

Definition (odd permutation) 4.42. A permutation f is said to be an odd permutation, if, f can be expressed as a product of an odd number of permutations.

(i) An identity permutation is considered an even permutation.

(ii) A transposition is always odd.

(iii) The product of two even permutations is even, and the product of two odd permutations is also even.

(iv) The product of an even and odd permutation is odd. Similarly, the product of an odd permutation and an even permutation is odd.

(v) A cyclic permutation does not change by changing the places of its elements, provided the order is not changed.

Example. $f = \begin{pmatrix} a & b & c & d \\ d & c & a & b \end{pmatrix} = \begin{pmatrix} c & d & a & b \\ a & b & d & c \end{pmatrix}$

Let $\quad S = \{a_1, a_2, a_3 \dots a_k, a_{k+1}, \dots a_n\}$

If $\quad f = \begin{pmatrix} a_1 & a_2 & a_3 & \dots & a_k & a_{k+1} & a_{k+2} & \dots & a_n \\ a_2 & a_3 & a_4 & \dots & a_1 & a_{k+1} & a_{k+2} & \dots & a_n \end{pmatrix}$

Then f is a cyclic permutation. The length of f is k and degree of f is n. We can ignore the elements $a_{k+1}, a_{k+2} \dots a_n$ which are mapped onto themselves and write

$$f = (a_1\, a_2\, a_3 \dots a_k)$$

A cyclic permutation does not change by changing the places of its elements provided their order is not changed.

Definition 4.43 (Disjoint cycles). Let $s = \{a_1, a_2, \dots a_n\}$. If f and g are two cycles on S such that they have no common elements, then f and g are said to be disjoint cycles.

Example. Let $S = \{1, 2, 3, 4, 5, 6\}$

If $f = (1\ 4\ 5)$ and $g = (2\ 3\ 6)$

then f and g are disjoint cycles on S.

The product of two disjoint cycles is always commutative.

Example. Consider the permutation $f = \begin{pmatrix} 1 & 2 & 3 & 4 & 5 & 6 & 7 \\ 2 & 3 & 4 & 5 & 1 & 7 & 6 \end{pmatrix}$

The permutation can be written as $f = (1\ 2\ 3\ 4\ 5)\ (6\ 7)$

Which is a product of two cycles. We now state the following theorem:

Theorem 4.46. Every permutation on n symbols can be expressed as a product of disjoint cycles.

Example 1. Let $A = \{1, 2, 3, 4, 5\}$

Compute $(5\ 6\ 3)\ 0\ (4\ 1\ 3\ 5)$

Solution. We have $(5\ 6\ 3) = \begin{pmatrix} 1 & 2 & 3 & 4 & 5 & 6 \\ 1 & 2 & 5 & 4 & 6 & 3 \end{pmatrix}$

and $(4\ 1\ 3\ 5) = \begin{pmatrix} 1 & 2 & 3 & 4 & 5 & 6 \\ 3 & 2 & 5 & 1 & 4 & 6 \end{pmatrix}$

$$(5\ 6\ 3)\ o\ (4\ 1\ 3\ 5) = \begin{pmatrix} 1 & 2 & 3 & 4 & 5 & 6 \\ 1 & 2 & 5 & 4 & 6 & 3 \end{pmatrix} o \begin{pmatrix} 1 & 2 & 3 & 4 & 5 & 6 \\ 3 & 2 & 5 & 1 & 4 & 6 \end{pmatrix}$$

$$= \begin{pmatrix} 3 & 2 & 5 & 1 & 4 & 6 \\ 5 & 2 & 6 & 1 & 4 & 3 \end{pmatrix} o \begin{pmatrix} 1 & 2 & 3 & 4 & 5 & 6 \\ 3 & 2 & 5 & 1 & 4 & 6 \end{pmatrix}$$

$$= \begin{pmatrix} 1 & 2 & 3 & 4 & 5 & 6 \\ 5 & 2 & 6 & 1 & 4 & 3 \end{pmatrix}$$

Example 2. Express $f = (a_1, a_2, a_3 \ldots a_n)$ as a product of transpositions.

Solution. $f = (a_1\ a_2\ a_3 \ldots a_n)$

$= (a_1\ a_2)\ o\ (a_1\ a_{n-1})\ o \ldots o\ (a_1\ a_3)\ o\ (a_1\ a_2)$

(i.e. a cycle of length can be expressed as a product of $n - 1$ transpositions)

Example 3. Show that $f = \begin{pmatrix} 1 & 2 & 3 & 4 & 5 & 6 & 7 & 8 \\ 7 & 3 & 1 & 8 & 5 & 6 & 2 & 4 \end{pmatrix}$ is even.

Solution. $f = \begin{pmatrix} 1 & 2 & 3 & 4 & 5 & 6 & 7 & 8 \\ 7 & 3 & 1 & 8 & 5 & 6 & 2 & 4 \end{pmatrix}$

$$= \begin{pmatrix} 1 & 7 & 2 & 3 & 4 & 8 & 5 & 6 \\ 7 & 2 & 3 & 1 & 8 & 4 & 5 & 6 \end{pmatrix}$$

$$= (1\,7\,2\,3)\ 0\ (4\,8)\ 0\ (5)\ 0\ (6)$$

$$= (1\,3)\ 0\ (1\,2)\ 0\ (1\,7)\ 0\ (4\,8)$$

f is expressed as a product of 4 transpositions, therefore f is even.

Example 4. Show that $f = \begin{pmatrix} 1 & 2 & 3 & 4 & 5 & 6 & 7 & 8 \\ 1 & 4 & 3 & 6 & 5 & 8 & 7 & 2 \end{pmatrix}$ is odd.

Solution. $f = \begin{pmatrix} 1 & 2 & 3 & 4 & 5 & 6 & 7 & 8 \\ 1 & 4 & 3 & 6 & 5 & 8 & 7 & 2 \end{pmatrix}$

$$= \begin{pmatrix} 1 & 2 & 4 & 6 & 8 & 3 & 5 & 7 \\ 1 & 4 & 6 & 8 & 2 & 3 & 5 & 7 \end{pmatrix}$$

$$= (1)\ o\ (2\,4\,6\,8)\ o\ (3)\ (o)\ (5)\ o\ (7)$$

$$= (2\,4\,6\,8)$$

$$= (2\,8)\ o\ (2\,6)\ o\ (2\,4)$$

f is expressed as a product of 3 transpositions. 3 is an odd number

$\therefore$ f is odd.

Example 5. Let $A = \{1, 2, 3, 4, 5, 6\}$.

Compute (i) (4135) o (563)

(ii) (563) o (4135) (VTU March 2001)

Solution. (i) Let $f = (4135)\ o\ (563)$, then

$$f = \begin{pmatrix} 1 & 2 & 3 & 4 & 5 & 6 \\ 3 & 2 & 5 & 1 & 4 & 6 \end{pmatrix} o \begin{pmatrix} 1 & 2 & 3 & 4 & 5 & 6 \\ 1 & 2 & 5 & 4 & 6 & 3 \end{pmatrix}$$

$$= \begin{pmatrix} 1 & 2 & 5 & 4 & 6 & 3 \\ 3 & 2 & 4 & 1 & 6 & 5 \end{pmatrix} o \begin{pmatrix} 1 & 2 & 3 & 4 & 5 & 6 \\ 1 & 2 & 5 & 4 & 6 & 3 \end{pmatrix}$$

$$= \begin{pmatrix} 1 & 2 & 3 & 4 & 5 & 6 \\ 3 & 2 & 4 & 1 & 6 & 5 \end{pmatrix}$$

(ii) Let $g = (563)\ o\ (4135)$, then

$$g = \begin{pmatrix} 1 & 2 & 3 & 4 & 5 & 6 \\ 1 & 2 & 5 & 4 & 6 & 3 \end{pmatrix} o \begin{pmatrix} 1 & 2 & 3 & 4 & 5 & 6 \\ 3 & 2 & 5 & 1 & 4 & 6 \end{pmatrix}$$

$$= \begin{pmatrix} 3 & 2 & 5 & 1 & 4 & 6 \\ 5 & 2 & 6 & 1 & 4 & 3 \end{pmatrix} o \begin{pmatrix} 1 & 2 & 3 & 4 & 5 & 6 \\ 3 & 2 & 5 & 1 & 4 & 6 \end{pmatrix}$$

$$= \begin{pmatrix} 1 & 2 & 3 & 4 & 5 & 6 \\ 5 & 2 & 6 & 1 & 4 & 3 \end{pmatrix}$$

Example 6. Let $S = \{1, 2, 3, 4, 5, 6\}$ and $f = \begin{pmatrix} 1 & 2 & 3 & 4 & 5 & 6 \\ 2 & 4 & 3 & 1 & 5 & 6 \end{pmatrix}$ determine the value of n such that $f^n = I_5$.

Solution. $f = \begin{pmatrix} 1 & 2 & 3 & 4 & 5 & 6 \\ 2 & 4 & 3 & 1 & 5 & 6 \end{pmatrix}$

$$\Rightarrow \quad f^2 = f\, o\, f = \begin{pmatrix} 1 & 2 & 3 & 4 & 5 & 6 \\ 2 & 4 & 3 & 1 & 5 & 6 \end{pmatrix} o \begin{pmatrix} 1 & 2 & 3 & 4 & 5 & 6 \\ 2 & 4 & 3 & 1 & 5 & 6 \end{pmatrix}$$

$$= \begin{pmatrix} 2 & 4 & 3 & 1 & 5 & 6 \\ 4 & 1 & 3 & 2 & 5 & 6 \end{pmatrix} o \begin{pmatrix} 1 & 2 & 3 & 4 & 5 & 6 \\ 2 & 4 & 3 & 1 & 5 & 6 \end{pmatrix}$$

$$= \begin{pmatrix} 1 & 2 & 3 & 4 & 5 & 6 \\ 4 & 1 & 3 & 2 & 5 & 6 \end{pmatrix}$$

$$f^3 = f^2\, o\, f = \begin{pmatrix} 1 & 2 & 3 & 4 & 5 & 6 \\ 4 & 1 & 3 & 2 & 5 & 6 \end{pmatrix} o \begin{pmatrix} 1 & 2 & 3 & 4 & 5 & 6 \\ 2 & 4 & 3 & 1 & 5 & 6 \end{pmatrix}$$

$$= \begin{pmatrix} 2 & 4 & 3 & 1 & 5 & 6 \\ 1 & 2 & 3 & 4 & 5 & 6 \end{pmatrix} o \begin{pmatrix} 1 & 2 & 3 & 4 & 5 & 6 \\ 2 & 4 & 3 & 1 & 5 & 6 \end{pmatrix}$$

$$= \begin{pmatrix} 1 & 2 & 3 & 4 & 5 & 6 \\ 1 & 2 & 3 & 4 & 5 & 6 \end{pmatrix}$$

$$= I_5$$

$\therefore \quad f^3 = I_5$ (identity permutation on s)

$\Rightarrow \quad n = 3$

Example 7. Find $\sigma^{-1}\ o\ \rho\ o\ \sigma$ where

$$\rho = (1\;3\;4)\;(5\;6)\;(2\;7\;8\;9)$$

and $$\sigma = \begin{pmatrix} 1 & 2 & 3 & 4 & 5 & 6 & 7 & 8 & 9 \\ 7 & 8 & 9 & 6 & 4 & 5 & 2 & 3 & 1 \end{pmatrix}$$

are permutations on $S = \{1, 2, 3, 4, 5, 6, 7, 8, 9\}$

Solution. We have $\rho = (1\;3\;4)\;(5\;6)\;(2\;7\;8\;9)$

$$\Rightarrow \qquad \rho = \begin{pmatrix} 1 & 2 & 3 & 4 & 5 & 6 & 7 & 8 & 9 \\ 3 & 4 & 1 & 6 & 5 & 7 & 8 & 9 & 2 \end{pmatrix}$$

and $$\sigma = \begin{pmatrix} 1 & 2 & 3 & 4 & 5 & 6 & 7 & 8 & 9 \\ 7 & 8 & 9 & 6 & 4 & 5 & 2 & 3 & 1 \end{pmatrix}$$

$$\Rightarrow \qquad \sigma^{-1} = \begin{pmatrix} 7 & 8 & 9 & 6 & 4 & 5 & 2 & 3 & 1 \\ 1 & 2 & 3 & 4 & 5 & 6 & 7 & 8 & 9 \end{pmatrix}$$

$$\sigma^{-1}\, o\, \rho\, o\, \sigma = \begin{pmatrix} 7 & 8 & 9 & 6 & 4 & 5 & 2 & 3 & 1 \\ 1 & 2 & 3 & 4 & 5 & 6 & 7 & 8 & 9 \end{pmatrix} o \begin{pmatrix} 1 & 3 & 4 & 5 & 6 & 2 & 7 & 8 & 9 \\ 3 & 4 & 1 & 6 & 5 & 7 & 8 & 9 & 2 \end{pmatrix}$$
$$o \begin{pmatrix} 1 & 2 & 3 & 4 & 5 & 6 & 7 & 8 & 9 \\ 7 & 8 & 9 & 6 & 4 & 5 & 2 & 3 & 1 \end{pmatrix}$$

$$= \begin{pmatrix} 7 & 8 & 9 & 6 & 4 & 5 & 2 & 3 & 1 \\ 1 & 2 & 3 & 4 & 5 & 6 & 7 & 8 & 9 \end{pmatrix} o \begin{pmatrix} 7 & 8 & 9 & 6 & 4 & 5 & 2 & 3 & 1 \\ 8 & 9 & 2 & 5 & 1 & 6 & 7 & 4 & 3 \end{pmatrix}$$
$$o \begin{pmatrix} 1 & 2 & 3 & 4 & 5 & 6 & 7 & 8 & 9 \\ 7 & 8 & 9 & 6 & 4 & 5 & 2 & 3 & 1 \end{pmatrix}$$

$$= \begin{pmatrix} 7 & 8 & 9 & 6 & 4 & 5 & 2 & 3 & 1 \\ 1 & 2 & 3 & 4 & 5 & 6 & 7 & 8 & 9 \end{pmatrix} o \begin{pmatrix} 1 & 2 & 3 & 4 & 5 & 6 & 7 & 8 & 9 \\ 8 & 9 & 2 & 5 & 1 & 6 & 7 & 4 & 3 \end{pmatrix}$$

$$= \begin{pmatrix} 8 & 9 & 2 & 5 & 1 & 6 & 7 & 4 & 3 \\ 2 & 3 & 7 & 6 & 9 & 4 & 1 & 5 & 8 \end{pmatrix} o \begin{pmatrix} 1 & 2 & 3 & 4 & 5 & 6 & 7 & 8 & 9 \\ 8 & 9 & 2 & 5 & 1 & 6 & 7 & 4 & 3 \end{pmatrix}$$

$$= \begin{pmatrix} 1 & 2 & 3 & 4 & 5 & 6 & 7 & 8 & 9 \\ 2 & 3 & 7 & 6 & 9 & 4 & 1 & 5 & 8 \end{pmatrix}$$

$$= \begin{pmatrix} 1 & 2 & 3 & 7 \\ 2 & 3 & 7 & 1 \end{pmatrix} \begin{pmatrix} 4 & 6 \\ 6 & 4 \end{pmatrix} \begin{pmatrix} 5 & 9 & 8 \\ 9 & 8 & 5 \end{pmatrix}$$

$$= (1\;2\;3\;7)\;(4\;6)\;(5\;9\;8)$$

Example 8. F is a permutation on $S = \{1, 2, 3, 4, 5, 6\}$ defined by

$$f = \begin{pmatrix} 1 & 2 & 3 & 4 & 5 & 6 \\ 2 & 3 & 4 & 5 & 6 & 1 \end{pmatrix}$$

compute f^2 and f^3

Solution. $f = \begin{pmatrix} 1 & 2 & 3 & 4 & 5 & 6 \\ 2 & 3 & 4 & 5 & 6 & 1 \end{pmatrix}$

$$\Rightarrow \quad f^2 = f o f = \begin{pmatrix} 1 & 2 & 3 & 4 & 5 & 6 \\ 2 & 3 & 4 & 5 & 6 & 1 \end{pmatrix} o \begin{pmatrix} 1 & 2 & 3 & 4 & 5 & 6 \\ 2 & 3 & 4 & 5 & 6 & 1 \end{pmatrix}$$

$$= \begin{pmatrix} 2 & 3 & 4 & 5 & 6 & 1 \\ 3 & 4 & 5 & 6 & 1 & 2 \end{pmatrix} o \begin{pmatrix} 1 & 2 & 3 & 4 & 5 & 6 \\ 2 & 3 & 4 & 5 & 6 & 1 \end{pmatrix}$$

$$= \begin{pmatrix} 1 & 2 & 3 & 4 & 5 & 6 \\ 3 & 4 & 5 & 6 & 1 & 2 \end{pmatrix}$$

and $$f^3 = f o f^2 = \begin{pmatrix} 1 & 2 & 3 & 4 & 5 & 6 \\ 2 & 3 & 4 & 5 & 6 & 1 \end{pmatrix} o \begin{pmatrix} 1 & 2 & 3 & 4 & 5 & 6 \\ 3 & 4 & 5 & 6 & 1 & 2 \end{pmatrix}$$

$$= \begin{pmatrix} 3 & 4 & 5 & 6 & 1 & 2 \\ 4 & 5 & 6 & 1 & 2 & 3 \end{pmatrix} o \begin{pmatrix} 1 & 2 & 3 & 4 & 5 & 6 \\ 3 & 4 & 5 & 6 & 1 & 2 \end{pmatrix}$$

$$= \begin{pmatrix} 1 & 2 & 3 & 4 & 5 & 6 \\ 4 & 5 & 6 & 1 & 2 & 3 \end{pmatrix}$$

$$= \begin{pmatrix} 1 & 4 & 2 & 5 & 3 & 6 \\ 4 & 1 & 5 & 2 & 6 & 3 \end{pmatrix}$$

$$= (1\ 4)\ o\ (2\ 5)\ o\ (3\ 6)$$

4.26 GROUP HOMOMORPHISM

The concept of group isomorphism was already introduced. A more general concept is that a group homomorphism with which are concern ourselves in this section.

Definition 4.44. Let $(G, *)$ and $(\overline{G}, \Delta)$ be any two groups. A mapping $f: G \to \overline{G}$ is called a homomorphism of G to $\overline{G}$ if $f(a * b) = f(a)\, \Delta\, f(b)\ \forall\ a, b \in G$.

Example 1. Let $G = R$, $\overline{G} = R - \{o\}$

$(G, \cdot)$ and $(\overline{G}, \cdot)$ are groups and G and G' one group with respect to multiplication.

Then the mapping $f: G \to \overline{G}$ defined by $f(a) = a^n \ \forall \ a \in G$, where $n \in Z$ is fixed, is a homomorphism.

Since $\qquad f(ab) = (ab)^n = a^n \cdot b^n = f(a) f(b)$.

Example 2. Let $G = R$, the set of real numbers and $\overline{G} = R - \{o\}$

$(G, +)$ and $(\overline{G}, \cdot)$ are groups. Define a mapping $f: G \to \overline{G}$ by

$$f(a) = 2^a \ \forall \ a \in G.$$

clearly $f(a + b) = 2^{a+b} = 2^a \, 2^b = f(a) f(b)$

$\Rightarrow f$ is a homomorphism of G into $\overline{G}$.

Definition. 4.45. Let $(G, *)$ and $(\overline{G}, \Delta)$ be two groups. A mapping $f: G \to \overline{G}$ defined by $f(a * b) = f(a) \Delta f(b)$ is called isomorphism, if f is a one-one onto mapping.

From the above definition, it is clear that a one-one onto homomorphism is an isomorphism. Thus every isomorphism is necessarily a homomorphism, whereas the converse need not be true. If $f: G \to \overline{G}$ is onto homomorphism, then $\overline{G}$ is called homomorphic image of G. Every quotent group is a homomorphic image of the group.

Definition 4.46. Let $(G, *)$ be a group. A mapping f: $G \to G$ is called, endomorphism of G. If $f(a * b) = f(a) * f(b)$.

Definition 4.47. Let $(G, *)$ be a group. A mapping $f: G \to G$ defined by $f(a * b) = f(a) * f(b)$ is called an automorphism if 'f' is a onto mapping.

4.27 KERNEL OF A HOMOMORPHISM

Let G and $\overline{G}$ be any two groups and $f: G \to \overline{G}$ be a homomorphism. Then Kernel of f denoted by ker f and kerf $= k = a \in G : f(a) = \overline{e}$.

Where $\overline{e}$ is the identity of $\overline{G}$. Ker f is called the Kernel of homomorphism f. Thus the Kernel of f is the set of all those elements of the domain set which are mapped onto the identity of the range set.

Theorem 4.47. Let $(G, *)$ and $(\overline{G}, \cdot)$ be two groups and f be a homomorphism then

(i) $f(e) = \overline{e}$, where e is the identity of G and $\overline{e}$ is the identity of $\overline{G}$.

(ii) $f(a^{-1}) = [f(a)]^{-1} \ \forall \ a \in G$.

(iii) $f(a^n) = [f(a)]^n \ \forall \ a \in G$ and $n \in Z$.

Proof. (i) $a \in G \Rightarrow a = a * e$

$$\Rightarrow \qquad f(a) = f(a * e)$$
$$= f(a) \cdot (e) \quad (\because f \text{ is a homomorphism})$$

Now $f(a) \in \overline{G}$ and $\bar{e}$ is the identity of $\overline{G}$

$$\Rightarrow \qquad f(a) \cdot \bar{e} = f(a)$$
$$\therefore \qquad f(a) \cdot f(e) = f(a) = f(a) \cdot \bar{e}$$
$$\Rightarrow \qquad f(a) \cdot f(e) = f(a) \cdot \bar{e}$$
$$\Rightarrow \qquad f(e) = \bar{e} \text{ (by left cancellation how)}$$

(ii) $a \in G \Rightarrow a^{-1} \in G \ \forall \in G.$

$$\Rightarrow a * a^{-1} \in G$$
$$\Rightarrow e \in G.$$
$$\therefore \qquad \Rightarrow f(a * a^{-1}) = f(e)$$
$$\Rightarrow f(a) \, o \, f(a^{-1}) = f(e) = \bar{e}$$
$$\Rightarrow [f(a)]^{-1} = f(a^{-1}) \ \forall \ a \in G.$$

(iii) Left as an exercise to the student.

Theorem 4.48. Let f be a homomorphism of $(G, *)$ into $(\overline{G}, \cdot)$ with Kernel K, then K is a normal sub-group of G.

Proof. We know that $f(e) = \bar{e} \quad \therefore \quad \hat{e} \in K$

Thus Ker $f = K$ is a non-empty sub-set of G.

K is a sub-group of G.

$$a, a, b \in K \Rightarrow f(a) = \bar{e} \text{ and } f(b) = \bar{e}$$

now $\qquad f(a * b) = f(a) \cdot f(b)$

$$= \bar{e} \cdot \bar{e} \quad (\therefore f \text{ is a homomorphism})$$
$$= \bar{e}$$

$$\therefore \qquad a \in K, b \in K \Rightarrow a * b \in K$$

Now $\qquad a \in K \Rightarrow a \in G \Rightarrow a^{-1} \in G$

Also $\qquad f(a^{-1}) = [f(a)]^{-1}$

$$= (\bar{e})^{-1}$$
$$= \bar{e}$$
$$\Rightarrow a^{-1} \in K \ \forall \ a \in K$$

Thus K is a sub-group of G.

We now prove that K is a normal sub-group G.

$$k \in K, a \in G.$$

$$\Rightarrow \quad f(a * k * a^{-1})$$

$$\Rightarrow \quad f(a * k) \cdot f(a^{-1})$$

$$= f(a) \cdot f(k) \cdot f(a^{-1})$$

$$= f(a) \cdot \bar{e} \cdot f(a^{-1})$$

$$= f(a) \cdot f(a^{-1})$$

$$= f(a * a^{-1})$$

$$= f(e) = \bar{e}$$

$$\Rightarrow \quad a * k * a^{-1} \in K \ \forall \ k \in K, a \in G.$$

Hence, K is a normal sub-group of G. Now, we state the following theorem without proof.

Theorem 4.49. Let f be a homomorphism of $(G, *)$ into $(\overline{G}, \cdot)$ with Kernel K, then $G/k \cong \overline{G}$.

4.28 SOLVED EXAMPLES

Example 1. Let G be $(Z, +)$ *i.e.* the group of integers under addition and let f: $G \to G$ defined by $\phi(x) = 3x \ \forall \ x \in G$. Prove that f is a homomorphism, determine its Kernel.

Solution. We have $(x) = 3x \ \forall \ x \in G$. $\forall \ x, y \in G \Rightarrow x + y \in G$ ($\therefore$ G is a group of under addition)

Now, $\quad f(x + y) = 3\,(x + y)$

$$= 3x + 3y$$

$$= f(x) + f(y)$$

Hence f is homomorphism

Kernel of homomorphism consists of all the integers whose Image is zero.

Thus $\quad K = \{o\}$

Example 2. Let $\phi : (G, *) \to (\overline{G}, o)$ defined by $\phi(a) = \bar{e}$. Prove that $\forall \ a \in G$ is homomorphism.

Solution. $\forall \ a, b \in G \Rightarrow a * b$ ($\because$ G is a Group)

Now $\quad \phi(a * b) = \bar{e}$

$$= \bar{e} \ o \ \bar{e} = \phi(a) \ o \ \phi(b)$$

Hence ϕ is homomorphism.

Example 3. Consider two groups G and $\overline{G}$ where $G = (Z, +)$ and

$$\overline{G} = \{2^m : m \in Z\}.$$

Let $\phi = Z \rightarrow \{2^m/m \text{ is an integer}\}$. Defined by $f(m) = 2^m$, where $m \in Z$. Prove that ϕ is homomorphism.

Example 4. Prove that the mapping ϕ is an automorphism where,

(a) G, the group of positive integers under addition $\phi(x) = -x$.

(b) G, the group of positive reals under multiplication $\phi(x) = x^2$.

Solution. (*a*) Let $x, y \in G$ then $x + y \in G$

$$(\because G \text{ is a group under addition})$$

Now $\quad \phi(x + y) = -(x + y) = -x - y = \phi(x) + \phi(y)$

$\therefore \quad \phi$ is homomorphism

Now $\quad y \in G \Rightarrow y \in G \quad \{\because G \text{ is a group under addition}\}$

Now $\quad \phi(-y) = -(-y) = y \in G.$

$\therefore$ ϕ is auto.

Hence ϕ is an automorphism.

(b) Let $x, y \in G$. Then $x * y \in G$.

($\because G$ is a group under multiplication.)

Now $\quad \phi(x * y) = (xy)^2 = x^2 * y^2 = \phi(x) * \phi(y)$

$\therefore \quad \phi$ is homomorphism

Now $y \in G$ there exists $y^{1/2} \in G$ such that

$$\phi(y^{1/2}) = (y^{1/2})^2 = y$$

$\therefore \quad \phi$ is automorphism.

$$\phi(x) = \phi(y) = x^2 = y^2 \Rightarrow x = y.$$

$\therefore \quad \phi$ is one-one.

Example 5. If for a group $G, f : G \rightarrow G$ is given by $f(x) = x^2$ for all $x \in G$ is a homomorphism, prove that G is abelian.

Solution. $a, b \in G \Rightarrow a\, b \in G$

$$f : G \rightarrow G \text{ is a homomrphism}$$

$\therefore \quad f(a) = a^2, f(b) = b^2$ and

$$f(ab) = (ab)^2$$

Now $\quad f(ab) = (ab)^2$

$\Rightarrow \quad f(a) = f(b) = (ab)(ab)$

$\Rightarrow \quad a^2 b^2 = (ab)(ab)$

$\Rightarrow \quad (aa)(bb) = (ab)(ab)$

$\Rightarrow \quad a(ab)b = a(ba)b$

$\Rightarrow \quad (ab)b = (ba)b$

$\Rightarrow \quad ab = ba$ (cancellation law hold good in G)

$\Rightarrow \quad G$ is abelian.

Example 6. Let G be the group of non zero real numbers and multiplication $\overline{G} = \{1, -1\}$; be the group under multiplication. Define a mapping

$$\phi : G \to \overline{G}$$

by $\quad \phi(x) = 1, \quad$ if x is positive,

and $\quad \phi(x) = -1, \quad$ if x is negative.

where $x \in G$. Then ϕ is a Homomorphism

Example 7. Let G be the group all real 2×2 matrices $\begin{bmatrix} a & b \\ c & d \end{bmatrix}$ such that $ad - bc \neq o$. Let $\overline{G}$ be the group of all non zero real number under multiplication. Define

$$\phi : G \to \overline{G} \text{ by}$$

$$\phi \begin{bmatrix} a & b \\ c & d \end{bmatrix} = ad - bc$$

clearly ϕ is a homomorphism of G onto $\overline{G}$

Example 8. Let G be the group of positive real numbers under multiplication and let $\overline{G}$ be the group of all real numbers under addition. Define a mapping

$$f : G \to \overline{G} \text{ by}$$

$$f(x) = \log_{10} x$$

Then $\quad x \in G,\ y \in G \Rightarrow xy \in G$

$\Rightarrow \quad f(xy) = \log_{10} xy = \log_{10} x + \log_{10} y$

$\quad = f(x) + f(y)$

for all $\quad x, y \in G.$

$\therefore \quad f$ is a homomorphism of G into $\overline{G}$

Example 9. Let $G = \{1, -1\}$, G is a group a under multiplication. Define $f: Z \to G$ by

$$f(n) = 1, \text{ if } n \text{ is even}$$

and $\quad f(n) = -1, \text{ if } n \text{ is odd}$

show that G is a homomorphic image of Z.

Solution. Consider any two integers. m, n. The following cases arise

Case I. Both m, n are even

Now m, n even $\Rightarrow m + n$ is even and $m + n \in Z$

$\Rightarrow \quad f(m+n) = 1 = 1.1 = f(m).f(n)$

$\Rightarrow \quad f(m+n) = f(m)f(n)\ \forall m, n \in Z$

$\Rightarrow \quad f$ is a homomorphism

Case II. One of m and n even and the other is odd.

Let m be an odd integer and n be an odd integer, then $m + n$ is odd

$\therefore \quad f(m+n) = -1 = -1.1 = f(m)f(n)$

Similarly when m is even and n is odd, we have $f(m+n) = f(m)f(n)$

Case III. When m and n are odd

Now m is odd, n is odd $\Rightarrow m + n$ is even

$\Rightarrow \quad f(m+n) = 1 = (-1).(-1)$

$\quad = f(m)f(n)$

$\therefore$ In all the cases given above, we have

$$f(m+n) = f(m)f(n)\ \forall m, n \in Z$$

$\therefore \quad f: Z \to G$ is a homomorphism

further, we see that f is onto

$\therefore \quad f: Z \to G$ is an epimorphism,

Consequently G is a homomorphic image of Z.

4.28 CAYLEY'S THEOREM

Every finite group G is isomorphic to a parmutation group.

4.29 REGULAR PARMUTATION GROUP

The parmutation group to which it is isomorphic is called a regular parmutation group.

Example. Find a regular parmutatation group isomorphic to the group $G = \{1, -1, i, -i\}$.

Solution. Let G' denote the regular permutation group of G which isomorphic to G.

By Cayley's theorem G' consisting the four parmutations f_1, f_2, f_3, f_4 where

$$f_1 = \begin{pmatrix} 1 & -1 & i & -i \\ 1.1 & 1-1 & 1-i & 1-i \end{pmatrix} = \begin{pmatrix} 1 & -1 & i & -i \\ 1 & -1 & i & -i \end{pmatrix} = I$$

$$f_2 = \begin{pmatrix} 1 & -1 & i & -i \\ 1-1 & -1-1 & i-1 & -i-1 \end{pmatrix} = \begin{pmatrix} 1 & -1 & i & -i \\ -1 & 1 & -i & i \end{pmatrix}$$

$$f_3 = \begin{pmatrix} 1 & -1 & i & -i \\ 1-i & -1-i & i.i & -i.i \end{pmatrix} = \begin{pmatrix} 1 & -1 & i & -i \\ i & -i & -1 & 1 \end{pmatrix}$$

$$f_4 = \begin{pmatrix} 1 & -1 & i & -i \\ 1-i & -1.-i & i.-i & -i.-i \end{pmatrix} = \begin{pmatrix} 1 & -1 & i & -i \\ -i & i & 1 & -1 \end{pmatrix}$$

$\therefore \qquad G' = \{I, f_2, f_3, f_4\}$

EXERCISE. 4.4

1. Define
   (a) Sub-group
   (b) Order of a group
   (c) Finite group
   (d) Order of an element
2. Show that the set of all elements a of an abelian group G which satisfy $a^2 = e$ forms a sub-group of G.
3. If G is a sub-group and $N(a) = \{x \in g : ax = xa\}$ for $a \in G$, then prove that $N(a)$ is a sub-group of G.
4. If $G = \{1, -1, i, -i\}$ is a group under multiplication then write all sub-groups of G.
5. Let (G, o) be a group and $a \in G$. If $N(G)$ denotes the normalizer of a in G then show that $(N(G), o)$ is a sub-group of (G, o).
6. Define a cyclic group and give examples.
7. Show that the group $\{a, a^2, a^3 = e, o\}$ is a cyclic group.
8. Let p be the collection of sub-groups of $(G, *)$ prove that the intersection of sub-group in p is a sub-group of $(G, *)$.
9. Find all the sub-groups of a cyclic group of order 60.
10. Find all the sub-groups of a cyclic group of order 10.

11. Show that every sub-group of a cyclic group is cyclic.
12. List all sub-groups of a cyclic group of order 12.
13. Show that group $\{(0, 6), +_{12}\}$ is a sub-group of $(Z_{12}, +_{12})$ of integers modulo12.
14. Show that every cyclic group is commutative.
15. State and prove Lagranges theorem.
16. If $(G, *)$ is a finite group of composite order then show that $(G, *)$ has non-trivial sub-groups.
17. Show that every group of prime order is cyclic.
18. Prove that sub-group of an abelian group is normal.
19. Prove that every quotient group of an abelian group is abelian.
20. If $(G, *)$ is a finite group of order n, with generator a. Then show that a^m is also a generator of G if and only if $m < n$ and $(m, n) = 1$.
21. If (H_1), and (H_2) are two normal sub-groups of (G) then show that $H_1 \cap H_2$ is also a normal sub-group of G.
22. Show that the intersection of any collection of normal sub-groups itself a normal sub-group.
23. Prove that every sub-group of index 2 is a normal sub-group.
24. Prove that the symmetric group of two symbols (S_2, o) is commutative.
25. Determine whether the following permutations all odd or even.

(a) $\begin{pmatrix} a & b & c \\ c & a & b \end{pmatrix}$

(b) $\begin{pmatrix} 1 & 2 & 3 & 4 & 5 & 6 & 7 \\ 3 & 2 & 4 & 5 & 6 & 7 & 1 \end{pmatrix}$

(c) $\begin{pmatrix} 1 & 2 & 3 & 4 & 5 & 6 & 7 & 8 & 9 \\ 6 & 1 & 4 & 3 & 2 & 5 & 7 & 9 & 8 \end{pmatrix}$

(d) $\begin{pmatrix} 1 & 2 & 3 & 4 & 5 \\ 3 & 2 & 4 & 1 & 5 \end{pmatrix}$

26. Express the following permutations as the product of transpositions.

(a) $\begin{pmatrix} 1 & 2 & 3 & 4 & 5 & 6 & 7 \\ 5 & 7 & 6 & 2 & 1 & 3 & 4 \end{pmatrix}$

(b) $(1\ 2\ 3)\ o\ (4\ 5)\ o\ (6\ 7\ 8)$

27. $G = \{f_1, f_2 f_3 f_4\}$ where

$$f_1 = \begin{pmatrix} 1 & 2 & 3 & 4 \\ 1 & 2 & 3 & 4 \end{pmatrix}, f_2 = \begin{pmatrix} 1 & 2 & 3 & 4 \\ 2 & 1 & 4 & 3 \end{pmatrix}, f_3 = \begin{pmatrix} 1 & 2 & 3 & 4 \\ 3 & 4 & 1 & 2 \end{pmatrix},$$

$$f_4 = \begin{pmatrix} 1 & 2 & 3 & 4 \\ 4 & 3 & 2 & 1 \end{pmatrix}$$

Show that $(G, \cdot)$ is an abelian group.

28. $\sigma, \tau \in S_6$, where $\sigma = \begin{pmatrix} 1 & 2 & 3 & 4 & 5 & 6 \\ 3 & 1 & 4 & 5 & 6 & 2 \end{pmatrix}$

$$\tau = \begin{pmatrix} 1 & 2 & 3 & 4 & 5 & 6 \\ 2 & 4 & 1 & 3 & 6 & 5 \end{pmatrix}$$

find (a) $\sigma\tau$ (b) $\sigma\tau^2$ (c) $\sigma\,\tau\,\sigma^{-1}$

Ans. (a) (2 5) (b) $\begin{pmatrix} 1 & 2 & 3 & 4 & 5 & 6 \\ 5 & 4 & 1 & 3 & 6 & 2 \end{pmatrix}$ (c) $\begin{pmatrix} 3 & 1 & 4 & 5 & 6 & 2 \\ 1 & 5 & 3 & 4 & 2 & 6 \end{pmatrix}$

CHAPTER 5

Elementary Combinatorics

5.1 INTRODUCTION

In this chapter we study permutations, combinations and binomial theorem. A permutation is an ordered arrangement of a finite number of objects among themselves. It may be regarded as a function of the set of objects onto itself. A combination is an unordered selection of elements of a given set. The study of arrangements objects is called combinatorics. It deals with the counting of arrangements of objects. In this chapter we discuss basic counting techniques: Known as sum and product rules of combinatorics. We also discuss circular permutations and multinomial theorem. We begin our study with basic counting techniques.

5.2 BASIC OF COUNTING

5.2.1 Sum Rule (Principle of Disjunctive Counting)

Let S be a set and $|S|$ denote the number of elements in S. If S is a union of disjoint non-empty subsets $A_1, A_2 \ldots A_n$ then

$$|S| = |A_1| + |A_2| + \ldots + |A_n|.$$

In the above statement the subsets A_i of S are all disjoint i.e. they have no element in common. If A_i and A_j are two subsets of S, then

$$A_i \cap A_j = \phi \text{ (for } i \neq j)$$

and we have

$$S = A_1 \cup A_2 \cup A_3 \cup \ldots \cup A_n$$

that is each element of S is exactly in one of the subsets A_i. In other words, the subsets $A_1, A_2 \ldots A_n$ form a partition of S.

We now state the sum rule for counting events.

Let S be a sample space. Two events E_1 and E_2 of X are said to be mutually exclusive if the events have no elements in common. If $E_1, E_2 \ldots E_n$ are mutually exclusive events of S, then we can state sum rule for counting events as follows:

If $E_1, E_2 \ldots E_n$ are mutually exclusive events of a sample space S, and F_1 can happen in m_1 ways... E_2 can happen in m_2 ways ... E_n can happen in m_n ways then E_1 or ... E_2 or E_n can happen $m_1 + m_2 + \ldots m_n$ ways.

Example 1. How many ways can we get a sum of 7 or 11 when two distinguishable dice are rolled?

Solution. The two dice are distinguishable, therefore the ordered pairs (a, b) and (b, a) are distinct when $a \neq b$ i.e. $(a, b) \neq (b, a)$ for $a \neq b$.

The ordered pairs in which the sum is 7 are: (1, 6), (2, 5), (3, 4), (4, 3), (5, 2), (6, 1). These ordered pairs are distinct.

$\therefore$ There are 6 ways to obtain the sum 7 similarly the ordered pairs: (5, 6), (6, 5) are distinct.

$\therefore$ The number of ways in which we get a sum 11 with the two dice is 2.

Therefore, we canget a sum 7 to 11 with two distinguishable dice in $6 + 2 = 8$ ways.

Example 2. How many ways can we draw a club or a diamond from a pack of cards?

Solution. There are 13 clubs and 13 diamonds in a pack of cards.

The number of ways a club or a diamon may be drawn $13 + 13 = 26$.

Example 3. In how ways can be drawn an ace or a king from an ordinary deck of playing cards?

Solution. Number of Aces in a pack = 4

Number of kings in a pack = 4

Number of ways of an Ace or king can be drawn from the pack $= 4 + 4 = 8$.

5.2.2 Product Rule (The Principle of Sequential Counting)

If $A_1, A_2 \ldots A_n$ are non-empty sets, then the number of elements in the Cartesian products $A_1 \times A_2 \times \ldots \times A_n$ is the product $\prod_{i=1}^{n} |A_i|$

i.e., $$|A_1 \times A_2 \times \ldots \times A_n| = \prod_{i=1}^{n} |A_i|$$

Product rule in terms of events

If $E_1, E_2, \ldots E_n$ are events and E_1 can happen in m_1 ways E_2 can happen m_2 ways ... E_n can happen in m_n ways, then the sequence of events E_1 first followed by E_2 followed by E_3 ... followed by E_n ... can happen in $m_1 \times m_2 \times \ldots \times m_n$ ways.

The product rule is also known as the multiplication rule of combinatorics.

Example 1. How many possible out comes are there when we roll a pair of dice, one red and one green?

Solution. The red die can land in any of six ways and for each of these six ways, the green die can also land in six ways.

The number of possible out comes when two dice are rolled $= 6 \times 6 = 36$.

Example 2. In how many different ways one can answer all the questions of a true-false test consisting of 4 questions.

Solution. There are two ways of answering each of the 4 questions. Therefore by product rule the number of ways in which all the 4 questions can be answered

$$= 2 \times 2 \times 2 \times 2 = 16.$$

Example 3. Find the number of n license plates that can be made where each plate contains two distinct letters followed by three different digits.

Solution. First letter can be printed in 26 different ways. Since the second letter must be different from the first, we have 25 chances for the second letter. Similarly the first digit can be printed in 10 ways, the second digit in license plate can be printed in 9 ways and the third in 8 ways.

Therefore, the number of license plates that can be printed, so that each plate contains two distinct letters followed by three different digits 26.25.10.9.8 = 4,68,000.

EXERCISE 5.1

1. In how many ways can we draw a king or a queen from ordinary deck of playing cards?
2. How many ways can we draw a club or a spade from a pack of cards?
3. How many ways can we get a sum 6 or 9 when two distinguishable dice are rolled?
4. How many possible out comes are there when we roll a pair of dice, one yellow and one red?
5. In how many different ways are can answer all the questions of a true-false test consisting of 5 questions?
6. How many ways can we get a sum of 8 when two in distinguishable dice are rolled?
7. How many ways can we get a sum of 4 or 8 when two distinguishable dice are rolled? How many ways can we get an even sum?
8. In how many ways can an organisations containing 26 members elect a president, a treasurer and a secretary (assuming no person is elected to more than one position)?

9. In a railway compartment 6 seats are vacant on a bench. In how many ways can 3 passengers can sit on them?
10. There are 10 buses plying between on a bench. In how many ways can 3 passengers can sit on them?
11. Suppose a license plate contains two letters followed by three digits with the first digit not zero. How many different plates can be printed?
12. Suppose a license plate contains 3 English letters followed by 4 digits.
    (a) How many different license plates can be manufactured if repetition of letter and digits are allowed?
    (b) How many plates are possible if only the letters are repeated?
    (c) How letters are possible if only the digits can be repeated?
13. How many many different license plates are there that involve 1, 2, or 3 letters followed by 4 digits?
14. There are 10 true-false questions on an examinations. How many sequences are possible?
15. Suppose that a states license plates consists of three letters followed by three digits. How many different plates can be manufactures (if repetitions are allowed)?
16. If there are 12 boys and 16 girls in a class find the number of ways of selecting one student as class representative.

ANSWER 5.1

1. 8 **2.** 26 **3.** 9 **4.** 36
5. 32 **6.** 12 **7.** 8, 18 **8.** 15,600
9. 120 **10.** 90 **11.** 6,08,600
12. $26^3 \times 10^4$, $26^3 \times 10 \times 9 \times 8 \times 7$, $26 \times 25 \times 24 \times 10^4$.
13. $(26 + 26^2 + 26^3)10^4$. **14.** 1024 **15.** $26^3 \times 10^3$
16. 28.

5.3 PERMUTATIONS

Definition 5.1. A permutation of n objects taken r at a time is an arrangement of r of the objects $(r \leq n)$.

A permutation of n objects taken r at a time is also called r-permutation or an r-arrangement. Various symbols are used to indicate the number of permutations of n things taken r at a time then we shall use in this text is $P(n, r)$, $(r \leq n)$. The symbols ${}^np_r$ or $P\binom{n}{r}$ or $n_{(r)}$ are also used to denote the number of permutations of n objects taken r at a time.

Example 1. Consider the three letters a, b, c. The arrangements of the letters a, b, c taken two at a time are

$$ab, ba, ac, ca, bc, cb$$

∴ The numbers of 2-arrangements are 6. i.e. the number of permutation of 3 letters taken 2 at a time

$$= {}^3P_2 = P(3, 2) = 6.$$

5.3.1 Notation

In this chapter we use $\{p_1a, p_2b, p_3c, p_4d\}$ to indicate either:

(i) That we have $p_1 + p_2 + p_3 + p_4$ objects which include p_1a's, p_2b's, p_3c's, p_4d's.

(ii) That we have 4 objects a, b, c and d with the condition that
a can be chosen at most p_1 times
b can be chosen at most p_3 times
c can be chosen at most p_3 times
d can be chosen at most p_4 times

The numbers p_1, p_2, p_3 and p_4 are called repetition numbers.

Example. The there permutations of $\{3a, 1b, 1c\}$ are *aaa, aab, baa, aac, aca, abc, acb, bac, bca, cab, cba, aba*.

5.4 FACTORIAL FUNCTION

If n is a natural number, then the product of all the natural numbers from 1 to n is called "n-factorial". It is denoted by the symbol $n!$. The symbols $\underline{|n}$ and $n_{(n)}$ are also used.

From the definition

$$n! = n(n-1)(n-2) \ldots 3 \cdot 2 \cdot 1.$$

The factorial function $n!$ can also be defined recursively as follows:

$$0! = 1$$

$$(n+1)! = n!\,(n+1),\ n \geq 0.$$

From the above recursive definition, we get

$$1! = 0!\,(1) = 1$$

$$2! = 1!\,(2) = 1.2$$

$$3! = 2!(3) = 1 \cdot 2 \cdot 3$$

Theorem 5.1. (Number of r-permutations without repetition). The number of r-permutations of n objects is $P(n, r) = n(n-1)(n-2) \ldots (n-r+1) = \dfrac{n!}{(n-r)!}$.

Proof. The first element in an r-permutations of n-objects can be chosen in n different ways. Once the first element has been selected, the second element can be selected in n–1 ways. We continue selecting the elements, having selected $(r-1)^{\text{th}}$ element.

We select rth element i.e. the last element. The rth element in the permutation can be selected in $n - r + 1$ ways. By product rule, the number of r-permutations of a set of n distinct objects is

$$P(n, r) = n(n-1)(n-2) \ldots (n-r+1),$$

We may also write $P(n, r)$ in factorial

$$P(n, r) = n(n-1)(n-2) \ldots (n-r+1) = \frac{n(n-1)(n-r+1)\,(n-r) \ldots 2.1}{(n-r) \ldots 2.1} = \frac{n!}{(n-r)!}$$

5.5 SOLVED EXAMPLES

Example 5.5.1. Find 8!

Solution. For $n \geq 0$, we have $(n + 1) = n!\,(n + 1)$

Hence,

$$\begin{aligned} 8! &= (7+1)! \\ &= 7!\,8 \\ &= 6!\,7{\cdot}8 \\ &= 5!\,6{\cdot}7{\cdot}8 \\ &= 4!\,5{\cdot}6{\cdot}7{\cdot}8 \\ &= 3!{\cdot}4{\cdot}5{\cdot}6{\cdot}7{\cdot}8 \\ &= 2!{\cdot}3{\cdot}4{\cdot}5{\cdot}6{\cdot}7{\cdot}8 \\ &= 1!{\cdot}2{\cdot}3{\cdot}4{\cdot}5{\cdot}6{\cdot}7{\cdot}8 \\ &= 1{\cdot}2{\cdot}3{\cdot}4{\cdot}5{\cdot}6{\cdot}7{\cdot}8 \\ &= 40,320. \end{aligned}$$

Example 5.5.2. Simplify $\dfrac{(n+1)!}{n!}$

Solution. $$\frac{(n+1)!}{n!} = \frac{(n+1)\,n(n-1) \ldots 3.2.1}{n(n-1) \ldots 3.2.1} = n+1$$

or $$\frac{(n+1)!}{n!} = \frac{(n+1)\,n!}{n!} = (n+1)$$

Example 5.5.3. Prove that $2P.(n, n-2) = P(n, n)$

Solution. $= 2p\,(n, n-2) = 2\dfrac{n!}{(n-(n-2)!}$

$$= \frac{2.n!}{n-n+2!} = \frac{2.n!}{2!}$$

$$= n! = P(n, n)$$

Example 5.5.4. How many words of three distinct letters can be formed from the letters of the word LAND?

Solution. The number of three distinct letter words that can be formed from the 4 letters of the word LAND is $p(4, 3) = \dfrac{4!}{(4-3)} = \dfrac{4!}{1!} = 4! = 24.$

5.6 PERMUTATIONS WITH LIKE ELEMENTS

Theorem 5.2. The number of permutations of n objects of which are q_1 are alike q_2 are alike .... q_r are alike is

$$P(n_1, q_1, q_2, \ldots q_r) = \frac{n!}{q_1!\,q_2!\ldots q_r!}$$

where $$n = q_1 + q_2 + \ldots + q_r$$

Proof. Let the number of permutations be x. If the q_1 like objects are unlike, then for each of these x arrangements the q_1 like objects could be rearranged among themselves in $q_1!$ ways without altering the positions of the other objects.

∴ The number of permutations would be $x\,q_1!$

Similarly, if all the objects were unlike numbers of permutations would be $xq_1!\,q_2! \ldots q_r!$

But if all the objects were unlike, the number of permutations with the n objects would be $n!$ Hence

$$xq_1!\,q_2! \ldots q_r! = n!$$

∴ $$x = \frac{n!}{q_1!\,q_2!\ldots q_r!}$$

i.e. $$P(n\,q_1q_2 \ldots q_r) = \frac{n!}{q_1!\,q_2!\ldots q_r!}$$

Example. There are 4 black, 3 green and 5 red balls. In how many ways can they be arranged in a row?

Solution. Total number of balls 4 black + 3 green + 5 red = 12

The black balls are alike.

The green balls are, and the red balls are unlike.

∴ The number of ways in which the balls can be arranges in a row

$$= \frac{12!}{4!\,3!\,5!} = 27{,}720.$$

5.7 ORDERED SAMPLES

In combinational analysis many problems are concerned with choosing a ball from an *urn* containing n balls or a card from a deck. When we choose one ball after the other from the *urn* (or a card from a deck) say r times, we call each choice an ordered sample of size r. We have two cases.

(i) **Sampling with replacement**

The drawn ball may be replaced in the *urn* before the next ball is drawn. There are n different ways to choose each ball, therefore by fundamental principle of counting (product rule): there are n^r different ordered sample with replacement of size r.

(ii) **Sampling without replacement**

When the ball drawn is not replaced in the *urn* after it is drawn; repetitions do not occur, in the ordered sample.

In this case, an ordered sample of size r without replacement, is an r-permutation of the objects in the *urn*.

∴ There are $\dfrac{n!}{(n-r)!}$ different ordered samples of size r without replacement.

Example 1. Suppose an *urn* contains 5 balls. Find the number of ordered sample of size 2.

(i) With replacement (ii) without replacement.

Solution. (i) There are 5 balls, and each ball in the ordered sample can be chosen in 5 ways.

Hence, there are $5 \cdot 5 = 25$; samples with replacement.

(ii) The first ball in the orders sample can be chosen in 5 ways and the next ball in the ordered sample can be chosen in 4 ways (when the first drawn ball is not replaced.).

∴ There are $5 \times 4 = 20$, samples without replacement.

Example 2. In how many ways can one choose two cards in succession from a deck of 52 cards, such that first chosen card is not replaced.

Solution. There are 52 cards in the deck of cards since the chosen card is not replaced the first can be chosen in 52 different ways and the second can be in 51 different ways.

∴ The number of ways in which the 2 cards are chosen $= 52 \times 51 = 2{,}652$.

Example 3. A box contains 10 light bulbs. Find the number of n ordered samples of

(a) Size 3 with replacement, and

(b) Size 3 without replacement.

Solution. (a) $n = 10^r$

$$= 10^3 = 10 \times 10 \times 10 = 1000$$

and (b) $P(10, 3) = 10 \times 9 \times 8 = 720.$

5.8 CIRCULAR PERMUTATIONS

Definition 5.2 A circular permutations of n objects is an arrangement of the objects around a circle.

In circular arrangements, we have to consider the relative positions of the different things. The circular permutations are different only when the relative order of the objects is changed otherwise they are same.

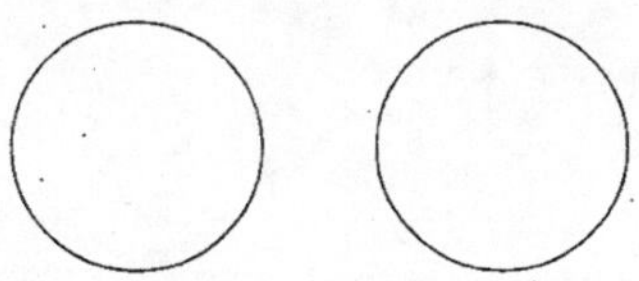

5.8.1 Number of Circular Permutations

Let n distinct be given. If the n objects are to be arranged round a circle we take an objects and fix it in one position.

Now the remaining $(n - 1)$ objects can be arranged to fill the $(n - 1)$ positions in the circle $(n - 1)!$ ways.

∴ The number of circular permutations of n different objects $= (n - 1)!$

5.8.2 Number of Different Circular Permutations

We consider the order, clockwise or anti-clockwise of objects around a circle as the same circular permutation. Every arrangement with n objects round a circle is counted twice in $(n - 1)!$

Circular permutations.

∴ The total number of different permutations of n distinct objects is $= \dfrac{(n-1)!}{2}$

Example 1. In how many ways can a party of 9 persons arrange themselves around a circular table?

Solution. One person can sit at any place in the circular table. The other 8 persons can arrange themselves in 8! ways i.e. the 9 persons can be arranged among themselves round the table in $(9 - 1)! = 8!$ ways.

Example 2. In how many ways 5 gents and 4 ladies dine at a round table, if no two ladies are to sit together?

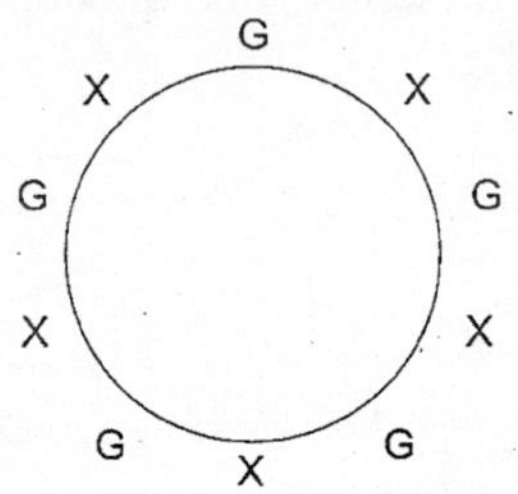

Solution. Since no two ladies are to sit together they should seat themselves in between gents (i.e. a lady is to be seated in between two gents). The 5 gents can sit round the circular table in 5 positions (marked *G* in the Fig.) they can be arranged in $(5 - 1)! = 4!$ ways. The ladies can sit in the 4 out of 5 seats (marked *X* in the Fig.). This can be done in $P(5, 4)$ ways.

∴ The required number of ways in which 5 gents and 4 ladies can sit round a table

$$= 4!\ P(5, 4)$$

$$= (4.3.2.1) \times (5.4.3.2)$$

$$= 2{,}880.$$

Example 3. Twelve persons are made to sit around a round table. Find the number of ways they can sit such that 2 specified are not together.

Solution. 12 persons can sit round a table in $(12 - 1)! = 11!$ ways.

The total number of ways in which 2 specified persons are together in $2!.10!$

∴ The required number of seating arrangements in which 2 specified persons are not together

$$= 11! - 2! \cdot 10!$$
$$= 11 \cdot 10! = 2! \cdot 10!$$
$$= 10! \cdot (11 - 2)$$
$$= 9 \cdot 10!$$

EXERCISE 5.2

1. In how many ways can 5 Telugu, 3 English, 2 Hindi books arranged on a shelf, if the books of each language are to be together?

2. Suppose a license plate contains two letters followed by three digits with the first digits not zero. How many different license plates can be printed?
3. How many license plates can be formed involving 3 English letters and 4 non-zero digits, if all the letters must appear either in the beginning or in the end.
4. From the digits 1, 2, 3, 4, 5, 6 how many three digits odd numbers can be formed, if the repetition of digits is not allowed?
5. There are 10 true-false questions on an examinations. How many sequences are possible?
6. Suppose that a states 11 cases plates consists of three letter followed by three digits. How many plates can be manufactured (if repetitions are allowed)?
7. Suppose a license plate contains 2 letters followed by 3 digits. How many different license plates can be printed?
8. Suppose a license plate contains 3 English letters followed by 4 digits.
   (a) How many different plates can be manufactured if repetition of letters and digits are allowed?
   (b) How many plates are possible if only the letters are repeated?
   (c) How many plates are possible if only the digits can be repeated?
9. Solve for n in $\dfrac{(n-2)!}{n!} = 56.$
10. Find n if $\dfrac{(n-1)!}{(n-2)!} = 12.$
11. A man wished to travel from one point in a city to a second point, which is five blocks south and six blocks east of his starting point. In how many ways he make the journey if he always travels either south or east?
12. Find the number of ways of arranging the letter of the word TENNESSEE all at a time (a) if there no restriction (b) if the first two letters must be *E*.
13. Suppose a license plate contains three distinct letters followed by four digits with first digit not zero. How many different license plate can be printed?
14. In how many ways can 52 playing cards be distributed to four digits giving 13 cards each?
15. Find the number of arrangements that can be made out of the letters of the following words.
   (i) Accountant
   (ii) Independent
   (iii) Assassination
16. How many five digit numbers can be formed with the digits 2, 3, 5, 7, 9 which are:

(a) Odd
(b) Even
(c) Greater than 30,000
(d) Lie between 30,000 and 90,000

17. In how many ways 6 rings be worn on 4 finger when (i) there can be only one ring on each other (ii) there can be any number of rings on each finger.
18. In how many ways 6 gents and 5 ladies dine at a round table, if no two ladies are to sit together?
19. If $P(n, 4) : P(n, 3) = 9{:}1$ find n.
20. In how many ways 7 ladies and 7 gents can be seated at a round table: if no two ladies are to sit together.
21. In how many ways can 12 beads of which 3 are alike of one kind, 2 are alike of the another kind and the rest are different be formed into a ring.
22. In how many ways can 4 men and 3 ladies be arrangesd at a round table if the 3 ladies
    (i) never sit together
    (ii) always sit together?
23. How many words can be formed out of the letters of the word DAUGHTER so that vowels always occur together?
24. Suppose repetitions are not permitted.
    (a) How many ways three digit numbers can be formed from the six digits 2, 3, 4, 5, 7 and 9?
    (b) How many of these numbers are less than 400?
    (c) How many are even?
    (d) How many are odd?
    (e) How many are multiples of 5?
25. Find the number of ways that three Americans, four French men, four Danes and two Italians can be seated in a row so that those of the same nationality sit together to solve the problem if they sit at a round table.
26. Find the number of n permutations that can be formal from all the letters of the word MISSISSIPPI.
27. Find n if $2P(n \cdot 2) + 50 = P(2n, 2)$
28. How many routes are there from the lower left corner of an $n \times n$ squares gird to the upper right corner if we are restricted to travelling only to the right or upward?
29. How many of the different permutations of the letters of word ENGLISH will
    (i) Start with E
    (ii) Start with E and end with N?

30. How many different numbers of six digits can be formed using the digits 0, 1, 2, 3, 4 and 5. How many of these are divisible by 5?

31. Twelve persons are made to sit around a table. Find the numbers of ways they can sit such that 2 specified persons are not together.

ANSWERS 5.2

1. 5! 3! 3! 2! **2.** 608400 **3.** $2 \times 26^3 \times 9^4$

4. 60 **5.** 1024 **6.** $26^3 \times 10^3$

7. $(26 + 26^2)10^3$

8. $26^3 \times 10^4, 26^3 \times 10 \times 9 \times 8 \times 7, 26 \times 25 \times 24 \times 10^4$

9. 6 **10.** 3 **11.** 924

12. (a) 3780 (b) 630 **13.** 14,04,00,000 **14.** $52/(131)^4$

15. (i) 2,26,800 (ii) 7560 (iii) 10810800

16. (a) 96 (b) 24 (c) 96 (d) 72 **17.** 86,200

18. 9 **19.** 3628800 **20.** 1663200

21. (i) 4320 (ii) 720 **22.** 4320

23. (a) 120 (b) 40 (c) 40 (d) 80 (e) 20.

5.9 COMBINATIONS

Definition 5.3. A combination of n objects taken at a time is an unordered selection of r of the n objects $(r \leq n)$.

A combination of n objects taken r at a time is also called r-combination of n objects.

Example 1. The two combinations of a, b, c, d taken two at a time are ab, ac, ad, bc, bd, and cd.

Example 2. Consider the objects a, b, c from which the selections are to be made by taking 2 objects at a time.

Solution. The 2 combinations are objects ab ac, bc.

Example 3. The 3 combinations of four objects a, b, c, d taken 3 at a time are abc, abd, acd, bcd.

The number of combinations of n objects taken r at a time is denoted by $C(n, r)$. The symbols $n_{(r)}$, ${}^nC_r$; $\binom{n}{r}$ and $C_{n,r}$ are also used to denote r-combinations of n objects.

Theorem 5.3. (Number of r-combinations without repetition). The number of r-combinations of n objects taken r at a time is

$$(n, r) = \frac{p(n, r)}{r!} = \frac{n!}{r!(n-r)!} : (1 \le r \le n)$$

Proof. r-combination means a selection of r-objects from the n-objects, in which order of the objects does not matter. Each r-combination contains r-objects and these r-objects can be arranged among themselves in $r!$ ways. Hence each r-combination gives rise to $r!$ permutations.

Therefore $C(n, r)$ combinations will give rise to $c(n, r).\ r!$ permutations. But the number of r-permutation of n objects is $P(n, r)$

$\therefore \quad C(n, r).\, r! = P(n, r)$

$$= \frac{n!}{(n-r)!r!} = \frac{n}{r!(n-r)!} = c(n, r)$$

Corollary 4. If $C(n, x) = C(n, y)$ then either

$$x = y \text{ or } x + y = n.$$

Proof. $\quad C(n, x) = C(n, y)$

$\Rightarrow \quad x = y \quad$ ...(i)

and $\quad C(n, x) = C(n, y)$

$\Rightarrow \quad C(n, x) = C(n, y)$

$\Rightarrow \quad C(n, x) = C(n, n-y)$

$\Rightarrow \quad x = n - y$

$\Rightarrow \quad x + y = n \quad$ ...(ii)

From (i) and (ii) either

$$x = y \text{ or } x + y = n.$$

$$\Rightarrow \quad C(n, r) = \frac{P(n, r)}{r!}$$

$$= \frac{n!}{r!(n-r)!} \quad \left(\therefore P(n,r) = \frac{n!}{(n-r)!}\right) \quad (1 \le r \le n)$$

Corollary 1

$$C(n, n) = \frac{n!}{n!(n-n)!} = \frac{n!}{n!0!} = 1 \quad \text{(Boundary condition)}$$

Corollary 2

$$C(n, 0) = \frac{n!}{0!(n-0)!} = \frac{n!}{0!n!} = 1 \quad \text{(Boundary condition)}$$

Corollary 3

$$C(n, r) = C(n, n-r) \qquad \text{(symmetric property)}$$

Proof. $C(n, n-r) = \dfrac{n!}{(n-r)!\,n-(n-r)!} = \dfrac{n!}{(n-r)!\,(n-n+r)!}$.

5.10 r-SUBSET

Definition 5.4. Let A be a set containing n elements. An r-subset of A is a selection of r elements of A without regard to the order.

Each r-subset is a r-combination and the number of r subsets of A is

$$C(n, r) = \frac{p(n, r)}{r!}$$

Combinationally $C(n, r)$ represents the number of ways of choosing r objects from n distinct objects (i.e. number of ways of selecting r elements from the n elements of the set A).

5.11 NUMBER OF SUBSETS OF SET

Let A be a non-empty set with n distinct objects: Each element of the set A either is a member of a particular subset of A or is not a member. The first element may either selected or not selects as a member of subset. It can therefore be treated in two ways, selected or rejected. For each of these, the second element of the set may treates in two ways selected or rejected. Therefore, the first two elements may be treated in $2 \times 2 = 2^2$ ways. Similarly the third element may be either selected or rejected in the two ways for each of the 2^2 ways. Hence the first three elements, may be treated in 2^3 ways. By continuing the argument we can show the n elements of the set A can be treated in 2n ways. Therefore, total number of subsets of A is 2^n.

i.e. if $\qquad |A| = n,$

then $\qquad |P(A)| = 2^n.$

It may happen on certain occasions that we require the number of subsets of a set which contain repetitions of certain elements, for example, consider the set $A = \{a, a, b, b, b, c\}$. A has seven elements of which 2 are alike of one kind and 4 are alike of a second kind.

For any subset we may select no, one or two. Hence may be selected in a's 3 ways. Similarly, for each fo these ways, the four b's may be selected in 5 ways. Therefore the a's and b's may be selected in $3 \times 5 = 15$ ways. Each of these 15 ways the remaining element C can be selected in 2 ways. Therefore the number of subsets of A is

$$3 \times 5 \times 2 = 30$$

In the same way, we may prove that the number of subsets of a set A of n elements of which p are alike of one kind, q are alike of a second kind and r are alike of a third kind is

$$(p + 1)(q + 1)(r + 1)\, 2^{n-p-q-r}$$

where $\quad (n \geq p + q + r)$

Example 1. Find the number of subsets of

$$A = \{2, 2, 2, 3, 3, 5, 11\}$$

Solution. There are three 2's two 3's in the set. The remaining two elements i.e. 5 and 11 are distinct we have $p = 3$, $q = 2$

∴ The number of subsets of A is

$$(3 + 1)(2 + 1)\, 2^{7-3-2}$$
$$= 4 \times 3 \times 2^2$$
$$= 48.$$

Example 2. How many selections any number at a time, may be made from three white balls, four green balls, one red ball and are black ball, if at least one must be chosen.

Solution. Total number of balls

$$= 3 \text{ white} + 4 \text{ green} + 1 \text{ red} + 1 \text{ black} = 9.$$

We have $\quad p = 3,\ q = 4$

∴ The number of selections that can be made is

$$(3 + 1)(4 + 1)\, 2^{9-7}$$
$$= 4 \times 5 \times 2^2$$
$$= 80 \text{ (This includes the null set)}$$

If at least are ball is to be chosen then the number of selections

$$= 80 - 1$$
$$= 79.$$

5.11 COMBINATIONS OF *n* THINGS TAKING ANY NUMBER OF THEM AT A TIME WHEN ALL THE THINGS ARE DIFFERENT

Each one of the n different things may either be selected or not selected i.e. there are two ways of selecting for each one of the n things.

∴ The total number of combinations of n things taking any number of things is 2^n.

But this includes the case in which all the things are rejected.

Hence the total number of ways in which one or more things are taken.

Corollary 1. The total number of combinations of n things taken 1, 2, 3,..., n at a time.

i.e. $\quad C(n, 1) + C(n, 1) + ... + C(n, n) = 2^n - 1.$

Example 1. In how many ways can a person invite one or more of his 5 friends to a party?

Solution. For every friend there are two possibilities i.e. he may be invited or he may not be invited to attend the party.

∴ The number of ways in which we can invite his five friends at a party is 2^5.

But this includes the case when none of his friend is invited.

Hence, the number of ways in which he invites one or more of his five friends to a party is

$$2^5 - 1 = 32 - 1 = 31.$$

5.12 COMBINATIONS WHEN ALL THE GIVEN THINGS ARE NOT DIFFERENT

Suppose that out of $m + n + p + ...$ things m are alike and of one kind, n are alike and of second kind and the rest are different, say they are k in number.

Out of m things we may take 0, 1, 2 ... or m. Hence there $(m + 1)$ ways of selecting the m things similarly n things which are alike may be selected in $(n + 1)$ ways and p things which are alike may be selected in $(p + 1)$ ways the remaining k different things may be selected in 2^k ways. These include the case in which all rejected.

∴ The total number of combinations

$$= (m + 1)(n + 1)(p + 1)2^k - 1.$$

5.13 BASIC IDENTITIES

Example 1. Prove that $C(n, r) = C(n, n - r)$

Solution. $C(n, r)$ denotes the number of selections of n objects when r-things are selected, and the remaining $(n - r)$ objects are not selected. Therefore selecting of r-things is the same as discarding $(n - r)$ objects. For each selection of r-things i.e. for each r-combination there is a discarding of $n - r$ remaining objects. The number of ways in which we discard $(n - r)$ objects is $C(n, n - r)$.

Since the set of r-combinations and the set of discarding of $(n - r)$ objects all in one-one corresponding we get,

$$C(n - r) = C(n, n - r)$$

Example 2. (Pascals Identity)

Show that $\quad C(n - r) = C(n - 1, r - 1) + C(n - 1, r).$

Solution. Method 1: $C(n - r)$ is the number of r-combinations of n objects. Let n be any one of the n objects. Fix the object x the $C(n, r)$ combinations can be grouped into:

(i) Those r-combinations that contain the object x.

(ii) Those r-combinations that do not contain the object x.

Since the objects x is in all r-combinations of (i) We are to choose $(r-1)$ objects in each combinations of (i) from the remaining $(n-1)$ objects, this can be done in $C(n-1, r-1)$ ways.

$\therefore$ The number of combinations that contain x, is

$$C(n-1, r-1)$$

To count the number of r-combinations of (ii) i.e. the number of r-combinations which do not contain x, we omit the objects x from n objects and choose r objects from the remaining $(n-1)$ objects.

Therefore, the number r-combinations that do not contain x is $C(n-1, r)$

$$C(n, r) = C(n-1, r-1) + C(n-1, \text{r})$$

Method 2: Consider R.H.S

$$C(n-1, r-1) + C(n-1, \text{r})$$

$$= \frac{(n-1)!}{(r-1)!(r-1-r+1)!} + \frac{(n-1)!}{r!(n-1-r)!}$$

$$= \frac{(n-1)!}{(r-1)!(n-r)!} + \frac{(n-1)!}{r!(n-1-r)!}$$

$$= \frac{(n-1)!}{(r-1)!(n-r)(n-r-1)!} + \frac{(n-1)!}{r(r-1)!(n-1-r)!}$$

$$= \frac{(n-1)!}{(r-1)!(n-r-1)}\left[\frac{1}{n-r} + \frac{1}{r}\right]$$

$$= \frac{(n-1)!}{(r-1)!(n-r-1)!}\left[\frac{r+n-r}{r-(n-r)}\right]$$

$$= \frac{n\,(n-1)!}{r(r-1)!(n-r)(n-r-1)!}$$

$$= \frac{n!}{r!(n-r)!} = C(n, r)$$

Example 3. (Pascal's row sum identity) Show that $C(n, 0) + C(n, 1) + \ldots + C(n, n) = 2^n$.

Solution. This is equivalent to finding the number of subsets of a set A with n elements. The number of subsets of A can be considered follows:

The number of subsets having no elements (i.e. subsets with 0 element is $C(n, 0)$.

The number of subsets having 1 element is $C(n, 1)$...

The number of subsets with n elements is $C(n, n)$. Adding we get the number of elements in the power set of A.

i.e. $$|(A)| = C(n, 0) + C(n, 1) + ... + C(n, n)$$

but the number of elements in the power set of A is 2^n.

$\therefore$ $$C(n, 0) + C(n, 1) + ... + C(n, n) = 2^n.$$

5.13 PARTITION AND CROSS PARTITIONS

Let S be a non-empty set. The collection $\{A_1, A_2 ... A_r\}$ of subsets of S is a partition of S if and only if

(i) $S = A_1 \cup A_2 \cup ... \cup A_r$

(ii) For any $A_i, A_j \in S$

either $A_i = A_j$ or $A_i \cap A_j = \phi$.

$A_1, A_2 .... A_r$ are called cells of the partition and each A_i a non-empty subset of S.

If $\{A_1, A_2,..., A_r\}$ and $\{B_1, B_2,..., B_r\}$ are both partitions of a non-empty set S, then $\{A_i \cap B_j\}$ forms a partition of $S\{A_i \cap B_j\}$ is called a cross partition of S.

Example. Let $S = \{a, b, c, d, e, f, g, h\}$ and $A_1 = \{a, d, e\}, A_2 = \{b, c\}, A_3 = \{f\}$, $A_4 = \{g, h\}$ then $\{A_1, A_2, A_3, A_4\}$ forms a partition of S. We observe that

(i) $A_1 \cup A_2 \cup A_3 \cup A_4 = S$ and

(ii) $A_i \cap A_j = \phi$ for all $A_i, A_j \in S$.

5.3.1 Ordered Partition

A partition a non-empty set is called an ordered partition of S if there is a specified order on the subsets of S.

An ordered t-tuple of sets $\{A_1, A_2,..., A_t\}$ is called a t-part ordered partition if the sets $A_1, A_2,..., A_t$ form a partition of S.

Example. Let $S = \{a, b, c, d, e\}$ and $A_1 = \{a, b\}$

$A_2 = \{c\}, A_3 = \{d\}, A_4 = \{e\}$; then $\{A_1, A_2, A_3, A_4\}$

is an ordered partition of S. If is a 4-part ordered partition.

Theorem 5.4. Let S contain n distinct objects. Then the number of ways S can be partitions in r subsets $A_1, A_2,... A_r$ with p_1 objects p_2 objects in A_2 ... and p_r objects in A_r is

$$P(n; p_1 \, p_2 \cdots p_r) \quad \frac{n!}{p_1! p_2! \ldots p_r!}$$

Proof. Since the p_1 objects going into A_1 can be chosen in $c(n_1, p_1)$ ways the p_2 objects going into A_2 can be chosen in $(n - p_1, p_2)$ ways, the p_3 objects going into A_3 can be chosen in $(n - p_1, n - p_2, p_3)$ ways ... and p_r objects going into A_r can chosen in $(n - p_1, - p_2 \cdots p_{r-1}, p_r)$ ways the total number of partitions is

$$C(n, p_1)\, C(n - p_1, p_2)\, C(n - p_1, - p_2, p_3)\, C(n - p_1 - p_2 \dots p_{r-1}, p_r)$$

$$= \frac{n!}{p!(n-p_1)!} \cdot \frac{(n-p_1)!}{p_2!(n-p_2)!} \cdot \frac{(n-p_1-p_2)!}{p_3!(n-p_3)!}$$

$$\dots \frac{(n - p_1 - p_2 - \dots - p_{r-1})!}{p_r!\, 0!}$$

$$= \frac{n!}{p_1!\, p_2! \dots p_r!}$$

Example. A farmer buys 4. cows, 2 goats, and 5 ducks from a person who has 7 cows, 5 goats and 8 ducks. How many choices the farmer have.

Solution. The farmer can choose 4 cows from 7 cows in $C(7, 4)$ ways, 2 goats from 5 goats in $C(5, 2)$ ways and 5 ducks from 8 ducks in $C(8, 5)$ ways.

∴ The farmer can choose 4 cows, 2 goats and 5 ducks in

$$C(7, 4)\, C(5, 2)\, C(8, 5).$$

$$= \frac{7 \cdot 6 \cdot 5 \cdot 4}{1 \cdot 2 \cdot 3 \cdot 4} \cdot \frac{5 \cdot 4}{1.2} \cdot \frac{8 \cdot 7 \cdot 6}{1 \cdot 2 \cdot 3}$$

$= 19{,}600$ ways.

Theorem 5.5. The number of combinations of $p_1 + p_2 + \dots + p_r$ objects if p_1 are alike of one kind, p_2 are alike of another kind ... p_r are alike of rth kind is

$$(p_1 + 1)(p_2 + 1) \dots (p_r + 1) - 1.$$

Proof. Out of $(p_1 + p_2 + \dots + p_r)$ objects any number of objects can be selected and each combination may contain any number of objects from 1 to $(p_1 + p_2 + \dots + p_r)$ objects.

Consider p_1 objects out of the p_1 like things none or 1 or 2 or 3 ... all of the p_1 may be selected. Therefore, the number of combinations with p_1 like objects in $(p_1 + 1)$.

Similarly, the number of possible combinations with p_2 like object is $(p_2 + 1)$ ... the number of combinations of p_r objects is $(p_r + 1)$.

∴ The total number of combinations with $p_1 + p_2 + \dots + p_r$ objects is

$$(p_1 + 1)\, (p_2 + 1)\, (p_r + 1)$$

but these combinations include the case in which all objects are rejected.

∴ The total number of combinations is

$$(p_1 + 1)\, (p_2 + 1) \dots (p_r + 1) - 1.$$

Example 1. There are twelve students in a class. Find the number of ways that the twelve students take three different tests if four students are to take cash test.

Solution. We find the number of ordered partitions of twelve students into cells containing for students each. There are $\dfrac{12!}{4!4!4!} = 34{,}650$ such petitions.

The required number of ways, the students can write the tests is 34,650.

Example 2. Find the number m of ways that a set X containing ten elements can be partitioned into two cells.

Solution. Let A denote a subset of X each A divides X into two disjoint sets A and $\overline{A}$ (complement of A). Thus there are 2^{10} such divisions of X. This number includes the case in which $A = \phi$ and $A = X$ (i.e. improper subsets of X).

$\therefore$ The number of partitions $[A, \overline{A}\,]$ of X is $2^{10} - 2 = 1024 - 2 = 1022$ since each unordered partition determines two ordered partitions.

We, have,

$$m = \frac{1022}{2} = 511$$

$\therefore$ Thus there are 511 partitions of X into two non-empty disjoint cells.

5.14 PERMUTATIONS AND COMBINATIONS WITH UNLIMITED REPETITIONS

Let $U(n, r)$ denote r-permutations of n-objects with unlimited repetitions and $V(n, r)$ denote the number of r-combinations with unlimited repetitions, then

$$U(n, r) = n^r \text{ and } V(n, r) = C(n - 1 + r, n - 1).$$

Consider the set $\{\infty \cdot a_1, \infty \cdot a_2 \ldots \infty \cdot a_n\}$ where $a_1, a_2 \ldots a_n$ are all distinct. Any r-combination is of the form$\{x_1 \cdot a_1,\ x_2 \cdot a_2 \ldots\ x_n \cdot a_n\ \}$ where each x; is a non-negative integer and

$$x_1 + x_2 + \ldots + x_n = r.$$

The numbers $x_1 + x_2 + \ldots + x_n$ are called repetition numbers. Conversely any sequence of non-negative integers

$$x_1 + x_2 + \ldots + x_n \text{ where } \sum_{i=1}^{n} x_i = r$$

Corresponds to a r-combination $\{x_1 \cdot a_1,\ x_2 \cdot a_2\ \ldots\ x_n \cdot a_n\}$. The Following observations are made:

The number of r-combinations of

$$\{\infty \cdot a_1,\ \infty \cdot a_2 \ldots \infty \cdot a_n\}$$

= The number of non-negative integers solution of

$x_1 + x_2 + \ldots + x_n = r.$

= The number of ways of playing r indistinguishable balls in n numbered boxes

= The number of binary numbers with $n - 1$ one's and r zeros.

$= C(n - 1 + r, r)$

$= C(n - 1 + r, n - 1)$

Example 1. Find the 4 combinations of {1, 2, 3, 4 ,5, 6}

Solution. The 4 combinations are

1234, 1235, 1236, 1245, 1246, 1256, 1345, 1346, 1356, 1456, 2345, 2346, 2456, 3456.

Example 2. Find the number of 3-combinations of $\{\infty \cdot a_1, \infty \cdot a_2, \infty \cdot a_3, \infty \cdot a_4\}$

Solution. We have $n = 4$, $r = 3$

$\therefore$ The number of 3-combinations of the given set is $C(4 - 1 + 3, 3)$

$= C(6, 3)$

$= 20.$

Example 3. Find the number of non-negative integral solutions to $n_1 + n_2 + n_3 + n_4 = 20$

Solution. We have $n = 4$, $r = 20$

$\therefore$ The number of non-negative integral solution

$= C(4 - 1 + 20, 20)$

$= C(23, 20) = 1{,}771.$

Example 4. Find the number of 4-combinations of 5 objects with unlimited repetitions.

Solution. We have $n = 5$, $r = 4$

$\therefore$ The number of 4-combinations

$= C(5 - 1 + 4, 4)$

$= C(8, 4) = 70.$

Example 5. Find the number of ways of placing 8 similar balls in 5 numbered boxes.

Solution. The number of ways of placing similar balls in 5 numbered boxes is

$= C(5 - 1 + 8, 8)$

$= C(12, 8) = 495.$

Example 6. Find the number binary numbers with five and three 0's.

Solution. The number of binary numbers with five is and 0's is

$= C(5 + 3, 3)$

$= C(8, 3)$

$= 56.$

Example 7. How many integral solutions are there to $x_1 + x_2 + x_3 + x_4 + x_5 = 16$, where each x: ≥ 2?

Solution. Let $x_i = y_i + 2$ where $y_i \geq 0$.

$\therefore \quad x_1 + x_2 + x_3 + x_4 + x_5 = 16$

$\Rightarrow \quad y_1 + 2 + y_2 + 2 + y_3 + 2 + y_4 + 2 + y_5 + 2 = 16$

$\Rightarrow \quad y_1 + y_2 + y_3 + y_4 + y_5 = 6.$

$\therefore$ The number of integral solutions of $x_1 + x_2 + x_3 + x_4 + x_5 = 16$ is the same as the numbers of integral solution of $y_1 + y_2 + y_3 + y_4 + y_5 = 6$.

$\therefore$ There are $C(5 - 1 + 6, 6) = C(10, 6)$ such solutions.

Example 8. How many integral solutions are there of $x_1 + x_2 + x_3 + x_4 + x_5 = 30$ where $x_1 \geq 2, x_2 \geq 3, x_3 \geq 4, x_4 \geq 2, x_5 \geq 0$?

Solution. Let $x_1 = y_2 + 2, x_2 = y_2 + 3, x_3 = y_3 + 4$

$x_4 = y_4 + 2, x_5 = y_5 + 0.$

$x_1 + x_2 + x_3 + x_4 + x_5 = 30$

$\Rightarrow \quad y_1 + 2 + y_2 + 3 + y_3 + 4 + y_4 + 2 + y_5 + 0 = 30$

$\Rightarrow \quad y_1 + y_2 + y_3 + y_4 + y_5 = 19.$

The number of integral solutions of $x_1 + x_2 + x_3 + x_4 + x_5 = 30$ where $x_1 \geq 2, x_2 \geq 3, x_3 \geq 4, x_4 \geq 2, x_5 \geq 0$ is the same as the number of integral solutions of

$$y_1 + y_2 + y_3 + y_4 + y_5 = 19$$

There are $C(5 - 1 + 19, 19) = C(23, 19)$ such solutions.

Example 9. How many outcomes are possible by casting a 6 faced die 10 times.

Solution. This is same as placing 10 similar balls into 6 numbered boxes.

$\therefore$ There are $C(6 - 1 + 10, 10)$

$= C(15, 10)$ possible outcomes.

Example 10. Enumerate the number of non-negative integral solutions to the in equality

$$x_1 + x_2 + x_3 + x_4 + x_5 \leq 19.$$

Solution. We can express the problem as follows

$$x_1 + x_2 + x_3 + x_4 + x_5 = 0$$
$$x_1 + x_2 + x_3 + x_4 + x_5 = 1$$
$$x_1 + x_2 + x_3 + x_4 + x_5 = 2$$
$$\ldots$$
$$x_1 + x_2 + x_3 + x_4 + x_5 = 19.$$

The number of non-negative integral solutions of

$$x_1 + x_2 + x_3 + x_4 + x_5 = 0$$

is $C(5 - 1 + 0, 0)$

The number of non-negative integral solutions of

$$x_1 + x_2 + x_3 + x_4 + x_5 = 1$$

is $C(5 - 1 + 1, 1)$

The number of non-negative integral solutions of

$$x_1 + x_2 + x_3 + x_4 + x_5 = 19$$

is $C(5 - 1 + 19, 19)$

∴ The number of non-negative integral solutions of

$$x_1 + x_2 + x_3 + x_4 + x_5 \leq 19$$

is $C(5 - 1 + 0, 0) + C(5 - 1 + 1, 1) + \ldots + C(5 - 119, 19)$.

EXERCISE 5.3

1. If $C(n + 1, r + 1)$: $C(n, r)$: $c(n - 1, r - 1)$: 11: 6: 3. Find n and r.
2. A committee of seven is to be formed from a boys and 5 girls. In how many ways can this be done, when the committee contains:
   (i) Exactly three girls
   (ii) At least three girls.
3. Among 20 members to a team, there are two wicket keepers and five bowlers. In how many ways can eleven persons be chosen to include only one wicket keeper and at least three bowlers?
4. How many different seven person can be formed each containing women from the available set of 20 women and 4 men from an available set of 30 men?
5. In how many ways can a persons invite one or more of his seven friends to a party.
6. How many triangles can be formed by joining 12 points, 7 of which are in the same straight line?
7. In how many ways 5 white balls and 3 black balls can be arranged in a row so that no two black balls may be together?
8. In how many ways can a committee of 6 men and 2 women be formed out of 10 men and 5 women?
9. A student is to answer out of 10 questions in an examination.
   (i) How many choices he has?
   (ii) How many if he must answer the first three question?

(iii) How many if he must answer at least four of the five questions?

10. Out of 4 officers and 10 clerks in an office a committee conststing of 2 officers ane 3 clerks is to be formed. In how many ways committee be done if:

    (i) Any officers and any clerk can be included?
    (ii) Are particular clerk must be on the committee?
    (iii) Are particular officer cannot be an the committee?

11. If $C(n, 15) = C(n, 27)$, find the value of $C(n, 30)$.
12. If $P(n, r) = 5040$ and $C(n, r) = 210$, find n and r.
13. From 6 consonants and 4 vowels how many words can be made, each containing 4 consonants and 3 vowels?
14. A committee of 8 is to be selected out of 6 males and 8 females. In how many ways can it be formed so that the male way not be out numbered?
15. There are twelve points in a given plane, no three of them on the same time. How many lines are determined by these points?
16. In how many ways can three or more persons be selected from twelve persons?
17. Which regular polygon has the same number of diagonals as sides?
18. Find the regular polygon which has twice as many diagonals as sides.
19. Find the number of triangles that can be formed by the vertices of an octagon.
20. Find the number of ways in which 9 toys can be divided evenly among three children.
21. Find the number of ways in which fourteen people can be partitioned into six committees, where two of the committee contain three people each and the remaining four committees contain two people each.
22. In how many ways can a cricket eleven be selected out of 14 players when the captain is always to be included?
23. A box contains 2 white 3 black and 4 red balls. In how many ways three balls can be drawn from the box if at least are black ball is to be included in the draw.
24. How many different possible outcomes are possible by tossing 6 similar coins?
25. There are 5 true and false questions in an examination. How many sequences of answers are possible?
26. Find the 3 combinations of $\{3a, 2b, 2c, 1d\}$
27. There are 25 true or false questions on an examination. How many different ways can a students to the examination, if he or she can also choose to leave the answer blank?
28. Find the number of 4 combinations of

$$\{\infty \cdot a_1, \infty \cdot a_2, \infty \cdot a_3, \infty \cdot a_4, \infty \cdot a_5\}.$$

29. Find the number of 3 combinations of 5 objects with unlimited repetitions.
30. Find the number of non-negative integral solutions to
$$x_1 + x_2 + x_3 + x_4 + x_5 = 15.$$
31. How many integral solutions are there to $x_1 + x_2 + x_3 + x_4 + x_5 = 20$ where each $x_i \geq 2$.
32. Find the number of integral solutions to $x_1 + x_2 + x_3 + x_4 + x_5 = 20$ where
$$x_1 \geq 3, x_2 \geq 2, x_3 \geq 4, x_4 \geq 6, x_5 \geq 0.$$
33. Enumerate the number of placing 20 indistinguishable balls into 5 boxes where each box is non-empty.
34. How many different outcomes are possible from tossing 10 similar dice.
35. Enumerate the number of non-negative integral solutions to the in equality
$$x_1 + x_2 + x_3 + x_4 + x_5 \geq 19.$$
36. Find the number of integral solutions to $x_1 + x_2 + x_3 + x_4 + x_5 = 50$, where $x_1 \geq 4, x_2 \geq 7, x_3 \geq -14, x_4 \geq 10$.
37. Find the number of 10-permutations of $\{3a, 4b, 2c, 1d\}$.

ANSWER 5.3

1. $n = 10, r = 5$ **2.** (i) 1260 (ii) 1716 **3.** 54054
4. 31,241,700 **5.** 127 **6.** 185
7. 20 **8.** 2100 **9.** (i) 45 (ii) 21 (iii) 35
10. (i) 720 (ii) 210 (iii) 360 **11.** 496
12. $n = 10, r = 4$ **13.** 302400 **14.** 1414
15. 66 **16.** 4017 **17.** Pantagon
18. 7 Sides (Heptagon) **19.** 56 **20.** 1680
21. 3153150 **22.** $c(13, 10)$ **23.** 64
24. 7 **25.** 32
26. *aaa, aab, aac, aad, bba, bbd, cca, ccb, ccd, abc, abd, acd, bcd.*
27. 3^{25} **28.** 70 **29.** 35
30. 316,251 **31.** $c(14, 10)$ **32.** $c(9, 5)$
33. $c(9, 15)$ **34.** 3003
35. $c(5 - 1 + 10, 4) + c(5 - 1 + 1, 4) + \ldots + c(5 - 1 + 19, 4)$
36. $c(54, 3)$ **37.** 12,600

5.15 BINOMIAL THEOREM

In this section we obtain the formula for the expansion of $(a + b)^n$. The Binomial theorem gives a formula for the coefficients in the expansion of $(a + b)^n$. Since

$$(a+b)^n = \underbrace{(a+b)(a+b)\dots(a+b)}_{n-\text{factors}}$$

The expansion results from selecting either a or a from each of the n-factors. Multiplying the selection together, and then summing all such products obtained.

Let us consider the expansion $(a + b)^3$ as an example.

$$\begin{aligned}(a+b)^3 &= (a+b)(a+b)(a+b)\\ &= (a+b)[(a+b)(a+b)]\\ &= (a+b)[(a+b)a+(a+b)b]\\ &= (a+b)\ (aa+ba+ab+bb)\\ &= a[aa+ba+ab+bb]+b[aa+ba+ab+bb]\\ &= aaa+aba+aab+abb+baa+bba+bab+bbb.\end{aligned}$$

There are eight terms in the expansions of which some are alike. Each term is a product three factors, The first being obtained from the first binomial, the second from the second binomial, and the third form the third binomial in $(a + b)(a + b)$ $(a + b)$. All such possible selections are included in the eight terms. Each of the eight terms is a product of three letters are selected from each of the binomial factors. In the expansion a^3(i.e. aaa) appears only once a^2b(i.e. aba, aab, bba) appears there times ab^2(abb, bba, bab) appears three times and b^3 (i.e. bbb) appears only once.

Therefore we can write

$$(a+b)^3 = a^3 + 3a^2b + 3ab^2 + b^3.$$

We observe that each term of this expansion is formed by selecting one and only one term from each of the three binomial factors, multiplying these together and adding like products.

We can select either none, one, two or three b's . If we select one from each of factor. This can be done in only $C(3, 0)$ one way. This gives the first term aaa i.e. a^3. If we select one b, we can select it from anyone of the three factors. This can be done in $C(3, 1) = 3$ ways. For each of these selection, we must select one from each of the other two factors and this can be done in only one way. Hence, we have three terms each equal to a^2b giving the second term $3a^2b$. We may how select two b's from the three factors in $C(3, 2) = 3$ ways and an a must be selected in one way from the remaining factor. This gives the term $3ab^2$. Finally, we select three b's in $C(3, 3) = 1$ way and obtain the term $bbb = b^3$. Therefore, we can write

$$\begin{aligned}(a+b)^3 &= C(3,0)a^3 + C(3,1)a^2b + C(3,2)ab^2 + C(3,3)b^3\\ &= a^3 + 3a^2b + 3ab^2 + b^3\\ &= \sum_{r=0}^{3} C(3,r)\,a^{3-r}.\,b^r\end{aligned}$$

We now consider the expansion of $(a + b)^n$ where a and b are real numbers and n is a positive integer.

Theorem 5.6. (The Binomial theorem)

Let n be a positive integer then for all a and b

$$(a+b)^n = C(n, 0)a^n + C(n, 1)a^{n-1} b + C(n, 2)a^{n-2} b$$
$$+ \ldots + C(n, r)\, a^{n-r}\, b^r + \ldots + C(n, n)b^n.$$
$$= \sum_{r=0}^{n} C(n, r)\, a^{n-r} \cdot b^r \cdot$$

Proof. First proof

$$(a, b)^n = (a + b)\,(a + b) \ldots n \text{ factors}$$

Each term in the expansion is obtained by forming the product of one term each binomial factors.

Therefore, each term in the expansion is of the form

$$a^{n-r}\, b^r \;(r = 0, 1, 2 \ldots n)$$

The term $a^{n-r} b^r$ known as the general term is formed by selecting one b from each of r of the n factors. This can be done in $C(n, r)$ ways. For each of these ways the $(n - r)$ a's can be selected one from each of the remaining $(n - r)$ binomial factors, in only one way.

Hence, coefficient of $a^{n-r} b^r$ is $C(n, r)$. Therefore,

$$\therefore \quad (a+b)^n = \sum_{r=0}^{n} C(n, r)\, a^{n-r} .\, b^r$$
$$= C(n, 0)a^n + C(n, 1)b + C(n, 2)a^{n-2}b + \ldots + C(n, r)\, a^{n-r}b^r$$
$$+ \ldots + C(n, n)b^n.$$

Second Proof. (Proof by mathematical induction)

Let r denote the set of all positive integers we use the identities

$$C(x, r) + C(x, r + 1) = C(x + 1, r + 1) \cdot C(k, 0)$$
$$= C(k + 1, 0)$$

and $\quad C(k, k) = C(k +1, k + 1) = 1$

Proof. Let S be the set of positive integers for which

$$(a + b)^n = \sum_{r=0}^{n} C(n, r)\, a^{n-r} \cdot b^r \cdot$$

Taking $n = 1$ we obtaining

$$\sum_{r=0}^{1} C(1, r)\, a^{1-r}\, b^r = C(1, 0)a^1 \cdot b^0 + C(1, 1)\, a^0 \cdot b^1$$

$$= a + b$$

$\therefore$ The theorem holds for $x = 1$ i.e. $1 \in S$.

Assume that the theorem is true for $n = k$ i.e. $k \in S$. Hence

$$(a + b)^k = \sum_{r=0}^{k} C(k, r)\, a^{k-r} \cdot b^r$$

$$= C(k, 0)a^k + C(k, 1)\, a^{k-1}b + \ldots + C(k, r-1)a^{k-r+1}b$$

$$+ C(k, r)a^{k-r}\, b^r + \ldots + C(k, k)b^k.$$

Then

$$(a + b)^{k+1} = (a + b)\,(a + b)^k$$

$$= a(a + b)^k + b(a + b)^k$$

$$= C(k, 0)a^{k+1} + C(k, 1)a^k\, b + C(k, 2)a^{k-1}\, b^r + \ldots$$

$$+ C(k, r-1)a^{k-r+1}\, b^{r-1} + C(k, r)\, a^{k-r+1}\, b^r + \ldots + C(k, k)\, ab^k.$$

$$= C(k, 0)a^k\, b + C(k, 1)a^{k-1}\, b^2 + c(k, 2)a^{k-2}\, b^3 + \ldots$$

$$+ c(k, r-1)a^{k-r+1}\, b^r + c(k, r)a^{k-r+1}\, b^{r+1} + \ldots + c(k, k)b^{k+1}.$$

$$= C(k, 0)a^{k+1} + [C(k, 0) + C(k, 1)a^k\, b + [C(k, 1) + C(k, r)]a^{k-1}b^r + \ldots$$

$$+ [c(k, r-1) + c(k, r)]a^{k-r+1}\, b^r + \ldots + c(k, k)b^{k+1}.$$

$$= C(k+1, 0)a^{k+1} + C(k+1, 1)a^k\, b + C(k+1, 2)a^{k-1}b^r + \ldots$$

$$+ C(k+1, r-1)a^{k-r+1}\, b^r + \ldots + C(k+1, k+1)b^{k+1}.$$

$$= \sum_{r=0}^{k+1} C(k+1, r)\, a^{k+1-r} \cdot b^r.$$

Therefore, the theorem is true for $x = k + 1$ also i.e. $k + 1 \in S$ if $k \in S$

Hence, $S = N$.

i.e. by the principle of Mathematical induction the theorem is true for all positive integers and

$$(a + b)^n = \sum_{r=0}^{n} C(n, r)a^{n-r} \cdot b^r \text{ for all } n \in N$$

We note the following in the expansion

(i) The number of terms in the expansion $(a + b)^n$ is $n + 1$

(ii) The general term is $c(x, r)a^{n-r}b^r$ (i.e. $(r + 1)$ the term is the general term). It is denoted by T_{r+1}.

(iii) The coefficients of terms equidistant from ends of the expansion are equal.

5.15.1 Pascal's Triangle

Consider the following expansions

$$(a + b)^0 = 1$$

$$(a + b)^1 = a + b$$

$$(a + b)^2 = a + 2ab + b^2$$

$$(a + b)^3 = a^3 + 3a^2b + 3ab^2 + b^2$$

$$(a + b)^4 = a^4 + 4a^3b + 6a^2b^2 + 4ab^3 + b^4.$$

$$(a + b)^5 = a^5 + 5a^4b + 10a^3b^2\ 10a^2b^3 + 5ab^4 + b^5.$$

It can be can be seen that the coefficients follow a definite pattern which can be most easily demonstrated by means of triangle known as Pascals triangle.

1
1 1
1 2 1
1 3 3 1
1 4 6 4 1
1 5 10 10 5 1
....

We can form the lines of the triangle by using the identity

$$C(x, r) + C(x, r + 1) = C(x + 1, r + 1).$$

Example 1. Expand $(2a + b)^4$

Solution. $(2a + b)^4 = C(4, 0)\ (2a)^4 + C(4, 1)\ (2a)^3b + C(4, 2)\ (2a)^2\ b^2$

$$= C(4, 3)\ (2a)b^3 + C(4, 4)$$

$$= 16a^4 + 32a^3b + 24a^2b^2 + 8ab^3 + b^4$$

Example 2. Find the general term in the expansion of $\left(x^3 + \frac{1}{x}\right)^{10}$

Solution. $\left(x^3 + \frac{1}{x}\right)^{10} = (x^3 + x^{-1})^{10}$

The general term in the expansion is

$$T_{r+1} = C(10, r)\,(x^3)^{10-r}\,(x-1)^r$$
$$= C(10, r)x^{30-3r}\,x^{-r}$$
$$= C(10, r)\,x^{30-4r}$$

Example 3. Find the term containing a^4 in the expansion of $\left(a - \frac{2}{a}\right)^8$.

Solution. The general term is

$$T_{r+1} = C(8, r)\,a^{8-r}\left(\frac{-2}{a}\right)^r$$
$$= C(8, r)a^{8-r}\,(-2a^{-1})^r$$
$$= C(8, r)a^{8-r}\,(-2)^r(a^{-1})^r$$
$$= C(8, r)\,(-2)^r\,a^{8-r}\,a^r$$
$$= C(8, r)\,(-2)^r\,a^{8-2r}$$

Putting $8 - 2r = 4$, we get $r = 2$

$\therefore$ The term containing a^4 is

$$T_{r+1} = T_3 = C(8, r)\,(-r)^4\,a^4.$$
$$= C(8, r)^4\,a^4$$
$$= 284\,a^4$$
$$= 112\,a^4$$

Example 4. Find the term independent of x in the expansion of $\left(x^2 + \frac{1}{x}\right)^{12}$

Solution. $\left(x^2 + \frac{1}{x}\right)^{12} = (x^2 + x^{-1})^{12}$

$\therefore$ The general term in the expansion of $(x^2 + x^{-1})^{12}$ is

$$T_{r+1} = C(12, r)(x^2)^{12-r}(x^{-1})^r$$
$$= C(12, r)\,x^{24-2r}\,x^{-r}$$
$$= C(12, r)\,x^{24-3r}$$

if $\quad 24 - 3r = 0$

$\quad 3r = 24$

$\Rightarrow \quad r = 8$

Hence the coefficient x^0 is $C(12, 8)$

$\therefore$ The term is dependent of x is $C(12, 8) = 495$

Replacing a by x and b by 1 we can write binomial theorem as:

$$(x+1)^n = \sum_{r=0}^{n} C(n, 1)\, x^{n-r}$$

$$= \sum_{r=0}^{n} C(n, n-r)\, x^{n-r}$$

$$= \sum_{r=0}^{n} C(n, 1) x^r = (1+x)^n$$

Consider $(1 = x)^n = \sum_{r=0}^{n} C(n, 1)\, x^r$.

replacing x by $-x$, we have

$$(1-x)^n = \sum_{r=0}^{n} C(n, 1)\,(-x)^r = \sum_{r=0}^{n} C(n, 1)\,(-1)^r x^r.$$

taking $a = b = 1$ binomial theorem can be written as

$$(1+1)^n = C(n, 0) + C(n, 1) + \ldots + C(n, n)$$

$$2^n = C(n, 0) + C(n, 1) + \ldots + C(n, 1)$$

If we set $a = 1$ and $b = -1$ in the binomial theorem, we get

$$(1+1)^n = C(n, 0) - C(n, 1) + C(n+2) + \ldots + (-1)^n\, C(n, n)$$

$$\Rightarrow \qquad C(n, 0) - C(n, 1) + \ldots + (-1)^n\, C(n, n) = 0$$

which shows that the alternating sum of the member of any row of Pascals triangle is zero. This can also be written as

$$C(n, 0) + C(n+2) + C(n, 4) + \ldots$$

$$= C(n, 1) + C(n+3) + \ldots$$

The coefficients

$$C(n, 0),\ C(n, 1)\ C(n+2) \ldots = C(n, n)$$

are called Binomial coefficients. These coefficients can also be written as

$$C_0, C_1, C_2, \ldots, C_n$$

5.17 MULTINOMIAL THEOREM

The binomial theorem can be extented to give a formula power of multinomial $(x_1 + x_2 + \ldots x_t)^n$ as follows

Theorem The multinomial theorem

Let n be a positive integer then
for all $x_1 + x_2 + ... + x_t$

$$(x_1 + x_2 + ... x_t)^n = \Sigma_p(n; q_1, q_2 ... q_t)\, x_1^{q_1}\, x_2^{q_2} ... x_t^{q_t}.$$

where the summation extends over all sets of non-negative integers $q_1, q_2 ... q_t$.

Proof. The coefficient of $x_1^{q_1}\, x_2^{q_2} ... x_t^{q_t}$ is the number of ways of arranging n letters

$$\{q_1x_1, q_2x_2 ... q_tx_t\}$$

$\therefore$ The coefficients of $x_1^{q_2}, x_2^{q_2} ... x_t^{q_t}$ is $(n; q_1, q_2 ... q_t)$

Hence, $(x_1 + x_2 ... + x_t)^n = \Sigma P(x_i\, q_1, q_2 ... q_t)\, x_1^{q_1}\, x_2^{q_2} ... x_t^{q_t}$.

In the above expansion $x_1^{q_1}\, x_2^{q_2} ... x_t^{q_t}$ is a selecting of n objects with repetition from t distinct types. The number of ways of selecting the objects is $C(n + t - 1, n)$

Therefore, the number of terms in the expansion of $(x_1 + x_2 + ... n_t)^n$ is $C(n + t - 1, n)$.

Example 1. Find the coefficient of $x^5\, y^{10} z^5\, w^5$ in the expansion of $(x + 7y + 3z + w)^{25}$.

Solution. We have $n = 25, x_1 = x_1, x_2 = 74\, x_3 = 3z$ and $x_4 = w$
$\therefore$ The coefficients of $x^5\, y^{10} z^5\, w^5$ in the expansion of $(x + 7y + 3z + w)^{25}$ is

$$P(25, 5, 10, 5, 5) = \frac{25! \cdot 1^5 \cdot 7^{10} \cdot 3^5 \cdot 1^5}{5!\,10!\,5!\,5!}$$

Example. Find the number of terms in the expansion of $(2x + 3y - 5z)^8$
Solution. We have $n = 8, t = 3$
$\therefore$ The number of terms is

$$\begin{aligned} C(n + t - 1, n) &= C(8 + 3 - 1, 8) \\ &= C(10, 8) \\ &= 45. \end{aligned}$$

5.18 SOLVED EXAMPLES

Example 5.18.1. Prove that $C(n, 1) + C(n, 3) + ...$

$$= C(n, 0) + C(n, 2) + ... = 2^{n-1}.$$

Solution. We know that

$$n(C, 1) + n(C, 3) + n(C, 5) + ...$$
$$= n(C, 0) + n(C, 2) + n(C, 4)$$

Let S denote the common total of these sums. Adding right hand side to the left, we get

$$C(n, 0) + C(n, 1) + C(n, 2) + C(n, 4) + \ldots + C(n, n) = 2S.$$

$$\Rightarrow 2^n = 2S$$

$$\Rightarrow S = 2^{n-1}.$$

Example 5.18.2. In the expansion of $(1 + x)^n$. Prove that

$$C_0^2 + C_1^2 + C_2^2 + \ldots + C_n^2 = \frac{2n!}{(n!)^2}$$

Solution. We have,

$$(1 + x)^n = C(n, 0) = C(n, 1)x + \ldots + C(n, n)x^n \quad \ldots(i)$$

We know that the coefficients of the terms equivalent form the beginning and end are equal, therefore, we can write

$$(1 + x)^n = C(n, n) = C(n, 1)x + \ldots + C(n, 0)x^n \quad \ldots(ii)$$

Multiplying (i) and (ii) we, get,

$$(1 + x)^{2n} = [C(n, 0) = C(n, 1)x + \ldots + C(n, n)x^n]$$

$$x[c(n, n) = c(n, n-1)x + \ldots + c(n, 0)]x^n.$$

From the right hand side, we see that the coefficients of x^n is

$$C(n, 0)^2 = C(n, 1)^2 + \ldots + C(n, n)^2$$

But the coefficient of x^n in $(1 + x)^{2n}$ is given by

$$C(2n, n) = \frac{2n!}{n! \cdot n!} = \frac{2n!}{(n!)^2}$$

$$C_0^2 + C_1^2 + \ldots + C_n^2 = \frac{2n!}{(n!)^2}$$

Example 5.18.3. If $(1 + x)^n = C(n, 0) + C(n, 1)x + C(n, 2)x^2 + \ldots + C(n, n)x^n$

Prove that

$$C(n, 1) + 2C(n, 2) + 3C(n, 3) + \ldots + nC(n, n) = n2^{n-1}$$

Solution.

$$C(n, 1)x + 2C(n, 2)x^2 + \ldots + nC(n, n)x^n$$

$$= nx + 2 \cdot \frac{2(n-1)}{2!}x^2 + \ldots + n \cdot x^n$$

$$= nx + n(n-1)x^2 + \ldots + nx^n$$

$$= nx[1 + (n-1)x + \ldots + x^{n-1}]$$

$$= nx\,[C(n-1, 0) + C(n-1, 1)\,x + \ldots + C(n-1, n-1)\,x^{n-1}]$$

$$= nx\,(1 + x)^{n-1}$$

Putting $x = 1$ on both sides, we get,

$$C(n, 1) + 2 \cdot C(n, 2) + ... + n \cdot C(n, n) = n \cdot 2^{n-1}.$$

Example 5.18.4. If $C(n, 0)$ $C(n, 1)$ $C(n, 2)$ ... $C(n, n)$ are the coefficients in the expansion of $(1 + x)^n$ prove that:

$$C(n, 0)\ C(n, r) + C(n, 1)\ C(n, r + 1) + ...$$
$$+ C(n, n - r)\ C(n, n) = \frac{2n!}{(n - r)!\,(n + r)!}$$

Solution. We know that

$$(1 + x)^n = C(n, 0) + C(n, 1)x + C(n, 2)x^2$$
$$+ ... + C(n, r)x^r + ... + C(n, n-1)x^{n-1}$$
$$+ C(n, n)x^n \qquad ...(i)$$

Writing in two reverse order we get

$$(1 + x)^n = C(n, n)x^2 + C(n, n - 1)x^{n-1} + ...$$
$$+ C(n - 1)x + C(n, 0) \qquad ...(ii)$$

Since $\quad C(n, r) = C(n, n - r)$

We have $C(n, n) = C(n, 0)$ $C(n, n - 1) = C(n, 1)$

We can therefore write

$$(1 + x)^n = C(n, 0)x^n + C(n - 1)x^{n-1} + C(n, 2)x^{n-2}$$
$$+ ... + C(n, n - 1)x + C(n, n) \qquad ...(iii)$$

Multiplying (i) and (ii) and equating the coefficient of x^{n+r} on both sides, we get,

$$C(n, 0) + C(n, r) + C(n, 1) + C(n, r + 1)$$
$$+ C(n, 2)\ C(n, r + 2) + ... + C(n, n - r)\ C(n, n)$$
$$= \text{coefficient of } x^{n+r} \text{ in } (1 + x)^{2n}$$
$$= C(2n, n + r)$$
$$= \frac{2n!}{(n + r)!\,(n - r)!}$$

Example 5.18.5. Prove that

$$\frac{C(n, 1)}{C(n, 0)} + 2 \cdot \frac{C(n, 2)}{C(n, 1)} + 3 \cdot \frac{C(n, 3)}{C(n, 2)} + ... + n\frac{C(n, n)}{C(n, n - 1)} = \frac{n(n + 1)}{2}$$

Solution. We know that,

$$\frac{C(n, 1)}{C(n, 0)} = \frac{n}{1} = n$$

2. $$\frac{C(n,2)}{C(n,1)} = \frac{\frac{2(n-1)}{1.2}}{\frac{n}{1}} = \frac{n(n-1)}{n} = n-1$$

3. $$\frac{C(n,3)}{C(n,2)} = \frac{\frac{3.n(n-1)(n-2)}{1.2.3}}{\frac{n(n-1)}{1.2}} = n-2$$

$$n \cdot \frac{C(n,n)}{C(n,n-1)} = n \cdot \frac{1}{n} = 1$$

Adding we get,

$$\frac{C(n,1)}{C(n,0)} + 2 \cdot \frac{C(n,2)}{C(n,1)} + 3 \cdot \frac{C(n,3)}{C(n,2)} + \ldots + n \cdot \frac{C(n,n)}{C(n,n-1)}$$

$$= n + (n-1) + (n-2) + \ldots + 1 \leq 1 + 2 + 3 + \ldots + (n-1) + n$$

$$= \frac{n(n+1)}{2}$$

Example 5.18.6. How many non-negative integral solutions are arise for the system: $x_1 + x_2 + x_3 + x_4 + x_5 = 0$.

Solution. We have $n = 5$, $r = 40$

The number of ways of keeping r similar balls in n boxes is $C(n - 1 + r; r)$ but

$$C(n - 1 + r; r) = C(n - 1 + r, n - 1)$$

$\therefore$ The number of non-negative integral solutions

$$= C(n - 1 + r; n - 1)$$

$$= C(5 - 1 + 40, 5 - 1)$$

$$= C(44, 4)$$

Example 5.18.7. (Pascals's diagonal sum identity)

$$C(n, 0) + C(n + 1, 1) + C(n + 2, 2) + \ldots + C(n + r, r) = C(n + 4 + 1, r)$$

Solution. The number of wap in which r distinguishable balls can be distributed into $n + 2$ boxes is

$$C(n + 2 - 1 + r, r) = C(n + r + 1, r)$$

But the balls may also be distributed as follows: for each $0 \leq k \leq r$, we can distributek of the balls into the first $n + 1$ boxes and the remaining in the last box.

This can be done is $\sum_{k=0}^{n} C(n+k, k)$ ways

$$\therefore \quad \sum_{k=0}^{n} C(n+k, k) = C(n+r+1, r)$$

Example 5.18.8 Using the diagonal sum identity of Pascal show that $1 + 2 + ... + n = n(n + 1)/2$.

Solution: Consider L.H.S.

$$\begin{aligned} 1 + 2 + ... + n &= C(1, 0) + C(2, 1) + ... + C(n, 1) \\ &= C(1 + 0, 0) + C(1 + 1, 1) + ... (C(1 + (n-1), n-1) \\ &= C(1 + (n-1) + 1, n-1) \\ &= C(n + 1, n-1) \\ &= C(n + 1, 2) \\ &= n(n + 1)/2 \\ &= \text{L.H.S.} \end{aligned}$$

Example 5.18.9. (Pascal's column identity)

Prove: $\quad C(r, r) + C(r + 1, r) + ... + C(n, r) = C(n + 1, r + 1)$.

algebrically.

Solution. We have $C(n + 1, r + 1) = C(n, r + 1) + C(n, r)$

$$\begin{aligned} C(n; r + 1) &= C(n-1, r + 1) + C(n-1, r) \\ C(n-1, r + 1) &= C(n-2, r + 1) + C(n-2, r) \\ &\vdots \\ C(r + 2, r + 1) &= C(r + 1, r + 1) + C(r + 1, r) \\ C(r + 1, r + 1) &= C(r, r) \end{aligned}$$

Adding the above termss and cancelling the like terms on L.H.S. with those in R.H.S. we get

$$C(r, r) + C(r + 1, r) + ... + C(n, r) = C(n + 1, r + 1)$$

Example 5.18.10. Using Pascal's column identity show that

$C(n + 1, r) + C(n + 2, r) + ... + C(n + m, r) = C(n + m + 1, r + 1) - C(n + 1, r + 1)$

Solution. From Pascal's column sum identity we have

$$C(r, r) + C(r + 1, r) + ... + C(n, r) = C(n + 1, r + 1) \qquad ...(1)$$

replacing n by $n + m$, we get

$$C(r, r) + C(r+1, r) + ... + C(n+1, r) + ... + C(n+m, r)$$
$$= C(n+m+1, r+1) \quad ...(2)$$

(2) – (1) gives

$$C(n+1, r) + C(n+1, r) + ... + C(n+m, r)$$
$$= C(n+m+1, r+1) - C(n+1, r+1)$$

Example 5.18.11. Show that $C_0 + C_{2/3} + C_{4/5} + ... = 2^n/n + 1$.

Solution. We have

$$(1+x)^n = C(n, 0) + C(n\ , 1)x + ... + C(n, r)x^n$$
$$(1-x)^n = C(n, 0) + C(n, 1)^x + ... + (-1)^n\ C(n, r)x^n$$

$\therefore \quad (1+x)^n + (1-x)^n = 2[C(n, 0) + C(n, 2)x^2 + ...]$

$\Rightarrow \quad (1+x)^n + (1-x)^n = 2[C_0 + C_2x^2 + ...]$

Integrating on both sides we get

$$\frac{(1+x)^{n+1}}{n+1} - \frac{(1-x)^{n+1}}{n+1} = 2(C_0x + C_2x^3/3 + ...) + k \quad ...(1)$$

where k is the constant of integration

Putting $x = 0$ in (1) we get

$$0 = k$$

$$\frac{2^{n+1}}{n+1} - 0 = 2(C_0 + C_{2/3} + C_{4/5} + ...) + k$$

i.e. $\quad 2(C_0 + C_{2/3} + C_{4/5} + ...) = \dfrac{2^{n+1}}{n+1} \quad (\because k = 0)$

i.e. $\quad C_0 + C_{2/3} + C_{4/5} + ... = \dfrac{2^{n+1}}{n+1}$

Hence proved.

Example 5.18.12. Show that

$$\frac{(1+x)^n + (1-x)^n}{2} = C(n, 0) + C(n, 2)x^2 + ... + C(n, q)x^q$$

where $q = \begin{cases} n & \text{if } n \text{ is even} \\ n-1 & \text{if } n \text{ is odd} \end{cases}$

and verify that

$$C(n, 0) + C(n, 2) + ... + C(n, q) = \begin{cases} 2^{n-1} & \text{for } n > 0 \\ 1 & \text{for } n = 0 \end{cases}$$

for $q = \begin{cases} n & \text{if } n \text{ is even} \\ n-1 & \text{if } n \text{ is odd} \end{cases}$

Solution. We have

$$(1+x)^n = C(n, 0) + C(n, 1)x + C(n, 2)x^2 + ... + C(n, n)x^n$$

$$(1-x)^n = C(n, 0) - C(n, 1)x + C(n, 2)x^n + ... + (-1)^2 C(n, n)x^n$$

Adding we get

$$(1+x)^n + (1-x)^n = 2[C(n, 0) + C(n, 2)x^2 + ... + C(n, q)x^q]$$

where $q = \begin{cases} n & \text{if } n \text{ is even} \\ n-1 & \text{if } n \text{ is odd} \end{cases}$

Taking $x = 1$ in the above expression we get

$$(1+1)^n + (1-1)^n = 2[C(n, 0) + C(n, 2) + ... + C(n, q)]$$

$$\Rightarrow \quad 2^n = 2[C(n, 0 + C(n, 2) + ... + C(n, q)]$$

$$\Rightarrow \quad C(n, 0) + C(n, 2) + ... + C(n, q) = 2^{n-1} \quad \text{for } n > 0$$

$$= 0 \quad \text{for } n = 0$$

Hence proved.

Example 5.18.13. Determine the coefficient of $x^5\, y^{10}\, z^5\, w^5$ in the expansion of

$$(x - 7y + 3z - w)^{25}$$

Solution. The coefficient of $x^5\, y^{10}\, z^5\, w^5$ in the expansion of $(x - 7y + 3z - w)^{20}$ is

$$\frac{25!(1)^5(-7)^{10} \cdot 3^5 \cdot (-1)^5}{5!\ 10!\ 5!\ 5!}$$

$$= \frac{25! 7^5\ 3^5}{5!\ 5!\ 5!\ 10!}$$

Example 5.18.14. Find the coefficient of $a^6\, b^3\, c^3$ in the expansion of $(a+b+2c)^{12}$

Solution. The coefficient of $a^6\, b^3\, c^3$ in the expansion of $(a + b + 2c)^{12}$ is

$$\frac{(12)! \cdot 2^3}{6!\ 3!\ 3!}$$

Example 10. Show that $k^3 = C(k, 1) + 6C(k, 2) + 6C(k, 3)$.

Solution. R.H.S. $C(k, 1) + 6C(k, 2) + 6C(k, 3)$

$$= k + 6\frac{k(k-1)}{2!} + 6\frac{k(k-1)(k-2)}{3!}$$

$$= k + 3k^2 - 3k + k^3 - 3k^2 + 2k$$

$$= k^3 = \text{L.H.S.}$$

EXERCISE 5.4

1. Expend $(1 + x + x^2)^4$ in assending powers of x.
2. Find the 10th term of $\left(2x^2 + \frac{1}{x}\right)^{12}$.
3. Find the term independent of x in the expansion of
$$\left(x - \frac{1}{x^2}\right)^{3n} \text{ is } (-1)^n \frac{3n!}{n!\,(2n)!}$$
4. Prove that the term independent of x in the expansion of
$$\left(x - \frac{1}{x^2}\right)^{3n} \text{ is } (-1)^n \frac{3n!}{n!\,(2n)!}$$
5. Give a combinational proof of pascal's column identify given by
$$C\cdot(r, r) + C(r + 1, r) + ... + C(n, r) = C(n + 1, r + 1)$$
6. Prove the pascal's diagonal sum identity
$$C(n + 1, r) = C(n + 1, 1) + C(n + 2, 2) + ... C(n + r, r)$$
7. Use the diagonal sum identity of pascal to show that
$1.2 + 2.3 + ... + n(n + 1) = n(n + 1)(n + 2)/3$
8. Prove that following by combinational argument
   (i) $C(2n, 2) = 2C(n, 2) + n^2$
   (ii) $(n - r)\, C(n, r) = n(n - 1, r)$.
9. Show that
$$C(n, 0) + C(n + 1, 1) + C(n + 2, 2) + ... + C(n + r, r)$$
$$= C(n + r + 1, r).$$
10. Prove that
$$C(n, 0) + 2C(n, 1) + 2^2C(n, 2) + ... + 2^nC((n, n) = 3^n.$$
11. Show that $\frac{(1+x)^n + (1-x)^n}{2} = C(n, 0) + C(n, 2)x^2 + ... + C(n, q)x^q$

   where $q = n$ if n is even $= n - 1$ if n is odd.

12. Show that the product of r consecutive integers is divisible by $r!$
13. State and prove Binomial theorem.
14. State multinomial theorem.
15. In the expansion of $(1 + px)^n$, the first three terms are $1, 5x, \frac{45x^2}{4}$. Find the values of n and p.
16. Find the coefficient of $x_1^2\, x_3\, x_4\, x_4^5$ in the expansion of $(x_1 + x_2 + x_3 + x_4 + x_5)^{10}$. Also find the number of terms in the expansion.
17. Find the coefficient of $x^3y^3z^2$ in the expansion of $(2x - 3y + 5z)^8$.
18. In the expansion of $\left(a^2 - \frac{1}{a^3}\right)^n$ the fifth term in $10a^{-4}$. Find the value of n.
19. In the expansion of $(ax + by)^n$ the coefficients of the first three terms are 729, 486, and 135 respectively if $a > 0$. Find the values of a, b and n.
20. Find the coefficient of $x^3y^3z^2$ in $(x + y + z)^7$.
21. Write the expansion of $(x + y)^4$.
22. Write the expansion of $(x + y)^6$.
23. What is the coefficient of x^9 in $(2 - x)^{19}$?

ANSWERS 5.4

1. $1 - 4x + 10x^2 - 16x^3 + 19x^4 - 16x^5 + 10x^6 - 4x^7 + x^8$.
2. $\frac{1760}{x^2}$
3. 7920

15. $P = \frac{1}{2}, n = 10$
16. 12,600,1001
17. $(2^3)\cdot(-3)^3\,(5)^2\,(560)$
18. $n = 8$
19. $a = 3, b = \frac{1}{3}, n = 6$
20. 210
21. $x^4 + 4x^3y + 6x^2y^2 + 4xy^3 + y^4$
22. $x^6 + 6x^5y + x^4\,y^2 + 20x^3y^3 + 15x^2y^4 + 6xy^5 + y^6$
23. $-2^{10}\; c(19, 9)$.

CHAPTER 6

Recurrence Relations

6.1 INTRODUCTION

In chapter 3, we have defined a set of recursive functions and recurrence relations. In this chapter we first explain what in meant by a numeric function and then explain the methods of solving recurrence relations.

A function whose domain is the set of natural numbers and whose range is the set of real numbers is called a numeric function.

For example, consider the function defined by

$$a_r = \begin{cases} 0, & \text{for } 0 \le r \le r \\ 2^{-r} + 5, & \text{for } \ge m \end{cases}$$

is a numeric function. Lower case letters are used to denote numeric functions.

Example. $a = \{a_0, a_1, a_2, a_3, \ldots, a_r, \ldots\}$

is a numeric function, where $a_0, a_1, a_2, a_3, \ldots, a_r, \ldots$ denote of a at $0, 1, 2, 3, \ldots, r, \ldots$

If a and b denote two numeric functions, then their sum and product are also numeric functions. If k is a scalar and a is a numeric function, then $k\,a$ is also a numeric function.

If $a = \{a_1, a_2, \ldots, a_r, \ldots\}$

is a numeric function $a_{r+1} - a_r$ is called the forward difference at r. It is denoted by Δa_r. The backward difference at r is $a_r - a_{r-1}$. It is denoted by ∇a_r.

In general Δa denotes the forward difference of a and ∇a denotes the backward difference of a.

The convolution of two numeric functions a and b is a numeric function c. where,

$$c_r = a_0\, b_r + a_1\, b_{r-1} + \ldots + a_{r-1}\, b_1 + a_r\, b_0$$

$$= \sum_{i=0}^{r} a_i \, b_{r-i}$$

The convolution of a and b is denoted by $a * b$.

6.2 GENERATING FUNCTIONS

In this section, we introduce an alternate way to represent numeric functions. For the numeric function

$$a = \{a_0, a_1, a_2, \ldots a_r, \ldots\}$$

We define an infinite series.

$$a_0 + a_1 z + a_2 z^2 + \ldots + a_r z^r + \ldots$$

which is called the generating function of a. It is denoted by $A(z)$ or by $G(z)$. The coefficient of z^r in $A(z)$ is the value of the numeric function a at r. For example, consider

$$a = (2^0, 2^1, 2^2, 2^3, \ldots 2^r \ldots)$$

The generating function a is

$$2^0 + 2^1 z + 2^2 z^2 + \ldots + 2^r z^r + \ldots$$

The above infinite series can be written in closed form as

$$A(z) = \frac{1}{1 - 2z}$$

In the generating function $A(z) = a_0 z^0 + a_1 z^1 + a_2 z^2 + \ldots a_r z^r + \ldots = \sum_{r=0}^{\infty} a_r z^r$

The term $a_0 z^0 = a_0$ is called the constant term, and the term $a_r z^r$ is called the term of degree r. Note that $A(z)$, generates its coefficients. If all the coefficient are zero from the some point on, $A(z)$ is just a polynomial of degree r, if $a_r \neq 0$ and $a_s = 0$ for $s \geq r + 1$.

The generating functions $A(z) = a_0 z^0 + a_1 z^1 + \ldots + a_r z^r + \ldots = \sum_{r=0}^{\infty} a_r z^r$

and $B(z) = b_0 z^0 + b_1 z^1 + \ldots + b_r z^r + \ldots = \sum_{r=0}^{\infty} b_r z^r$

are said to be equal if $a_r = b^r$ for each $r \geq 0$ and $A(z) + B(z) = \sum_{r=0}^{\infty} (a_r + b_r) z^r$.

If k is a scalar, then

$$kA(z) = k(a_0 z^0 + a_1 z^1 + ... + a_r z^r + ...)$$

$$= k \sum_{r=o}^{\infty} a_r z^r$$

The product $A(z)\,B(z)$ is defined as

$$A(z)\,B(z) = a_0 b_0 z^0 + (a_0 b_1 + a_1 b_0) z^1 + (a_0 b_2 + a_1 b_1 + a_2 b_0) z^2 + ...$$
$$+ (a_0 b_r + a_1 b_{r-1} + ... + a_r b_0) z^r + ...$$

$$= \sum_{k=0}^{\infty} \left(\sum_{j=0}^{k} a_j b_{k-j} \right) z^k$$

Shifting Properties of Generating Functions

If $A(z)$ is the generating function for the sequence $(a_0, a_1, a_2, ...)$ then multiplying $A(z)$ by z amounts to shifting the sequence to one place to the right and inserting a zero in front. Multiplying $A(z)$ by z^2 amounts to shifting the sequence two positions to the right and inserting 2 zeros in front and multiplying $A(z)$ *by* z^k amounts to shifting the sequence k positions to the right and inserting k zeros in front.

Thus

$$zA(z) = (0, a_0, a_1, a_2, ...)$$
$$z^2 A(z) = (0, 0, a_0, a_1, a_2, ...)$$
$$z^3 A(z) = (0, 0, 0, a_0, a_1, a_2, ...)$$

and

$$A(z) = \sum_{r=0}^{\infty} a_r z^r \Rightarrow z^k A(z) = \sum_{r=0}^{\infty} a_r z^{r+k}$$

If $(a_0, a_1, a_2, ...) = (1, 1, 1, ...)$

Then $(1, 1, 1, ...)$ is generated by $\dfrac{1}{1-z} = \sum_{r=0}^{\infty} z^r$

i.e. the sequence $\{a_n\}$ where $a_n =$ for all $n \geq 0$, is generated by $\sum_{r=0}^{\infty} z^r$

If $\dfrac{1}{1-z} = \sum_{r=0}^{\infty} z^r$, then $\dfrac{z}{1-z} = \sum_{r=0}^{\infty} z^{r+1} = \sum_{r=1}^{\infty} z^r$

$\dfrac{z}{1-z} = \sum_{r=1}^{\infty} z^r$ generates the sequence $(0, 1, 1, 1, ...)$

and $\frac{z^2}{1-z} = \sum_{r=2}^{\infty} z^r$, generates the sequence (0, 0, 1, 1, 1,...)

Similarly, we have $\frac{1}{(1-z)^2} = \sum_{r=0}^{\infty} (r+1)z^r$

$$\frac{z}{(1-z)^2} = \sum_{r=1}^{\infty} r\,z^r$$

where $\frac{1}{(1-z)^2}$ generates (1, 2, 3, 4,...) and $\frac{z}{(1-z)^2}$ generates (0, 1, 2, 3, 4,...) applying these shifting properties, we get different generating functions which are given below.

Sequence	Generating function	
$C(k, r)$	$(1+z)^k$	
1^n	$\frac{1}{1-z}$	
a^n	$\frac{1}{1-az}$	
$(-1)^n$	$\frac{1}{1+z}$	
$(-a)^n$	$\frac{1}{1+az}$	
$C(k-1+n, n)$	$\frac{1}{(1-z)^k}$	(k is a fixed integer)
$C(k-1+n, n)a^n$	$\frac{1}{(1-az)^k}$	(k is a fixed integer)

Example 1. The generating function of $a_0, a_1,..., a_r,...$

where $\quad a_r = 5z^r, \quad (r \geq 0)$

is $\quad A(z) = \frac{5}{1-2z}$

Example 2. The generating function of the sequence $a_0, a_1, \ldots, a_r, \ldots$

where $$a_r = 5^{r+z}.\ r \geq 0$$

is $$A(z) = \frac{25}{1-5z}$$

Example 3. The generating function of the sequence $\{a_r\}$ with

$a_r = 8$, is

$$A(z) = \sum_{r=0}^{\infty} 8\, z^r$$

Example 4. The generating function for the sequence 1, 1, 1, 1, 1, 1 is

$$A(z) = 1 + z + z^2 + z^3 + z^4 + z^5 = \frac{z^6 - 1}{z - 1}$$

Example 5. If $G(z)$ denotes the generating function for the sequence {1, 1, 1, 1,....,] then

$$G(z) = 1 + z + z^2 + z^3 + \ldots = \frac{1}{1-z};\ 1 = 1 < 1$$

Example 6. Let m be a positive integer, and let $a_r = c(m, r)$ for $r = 0, 1, 2, \ldots, m$. Then the generating function for the sequence $a_0, a_1, a_2, \ldots, a_m$ is

$$A(z) = \{(m, 0) + C(m, 1)\, z + \ldots + C(m, m) z^m$$
$$= (1 + z)^m$$

Example 7. If $G(x)$ denotes the generating function for the sequence

$$\{a_0, a_1, a_2, \ldots, a_r \ldots\}$$

where $a_r = C(n, r)$ then the generating function for the sequence is

$$G(x) = 1 + C(n, 1)x + C(n, 2)x^2 + \ldots + C(n, r)\, x^n + \ldots$$

If $$G(x) = 1 + C(n, 1)n + \ldots + x^n$$

then $$G(x) = (1 + x)^n$$

Similarly we have

$$G(x) = (1 + x^x)^n = \sum_{k=0}^{n} C(n, k) x^{rk}$$

$$G(x) = (1 + ax)^n = \sum_{r=0}^{n} C(n, r) a^r\, x^r$$

$$= 1 + C(n, 1)ax + C(n, 2)\, a^2\, x^2 + \ldots + \ldots + a^n\, x^n$$

The generating function for the sequence.

$$1 + x^r + x^{2r} + \ldots \text{ is } \frac{1}{1-x^r}$$

Example 8. Find the sequence generated by

$$5x^2(1-x)^{-1}$$

Solution: We have

$$\begin{aligned} 5x^2(1-x)^{-1} &= 5x^2(1 + x + x^2 + x^3 + \ldots) \\ &= 5x^2 + 5x^3 + 5x^4 + \ldots \\ &= 0 + 0.x + 5x^2 + 5x^3 + 5x^4 + \ldots \end{aligned}$$

The sequence generated by $5x^2(1-x)^{-1}$ is

$$0, 0, 5, 5, 5, \ldots$$

Example 9: Find the sequence generated

$$\text{by } 4x^2 + e^{2x}$$

Solution: We have $4x^2 + e^{2x}$

$$= 4x^2 + 1\left[1 + \frac{(2x)}{1!} + \frac{(2x)^2}{2!} + \frac{(2x)^3}{3!} + \ldots\right]$$

$$= 1 + \frac{2x}{1!} + \left(4x^2 + \frac{4x^2}{2!}\right) + \frac{(2x)^3}{3!} + \ldots$$

$$= 1 + \frac{2x}{1!} + \left(4 + \frac{2^2}{2!}\right)x^2 + \frac{2^3}{3!}x^3 + \ldots$$

$\therefore$ The sequence generated by $4x^2 + e^{2x}$ is

$$1, \frac{2x}{1!}, 4 + \frac{4x^2}{2!}, \frac{2^3}{3!}, \ldots$$

Example 10: Find the sequence generated by

$$\frac{1}{1-z} + 5z^3$$

Solution: We have

$$\begin{aligned} \frac{1}{1-z} + 5z^3 &= (1-z)^{-1} + 5z^3 \\ &= (1 + z + z^2 + z^3 + \ldots) + 5z^3 \end{aligned}$$

$$= 1 + z + z^2 + 5z^3 + ...$$

$\therefore$ The sequence generated by $\frac{1}{1-z} + 5z^3$ is

$$1, 1, 1, 6, 1, 1, ...$$

Example 11. Find the generating function for the sequence

$$0, 1, -2, 3, -4$$

Solution. We can write the sequence as follows

$$0 + 1.z + (-2)\cdot z^2 + 3\cdot z^3 + (-4)\cdot z^4 + ...$$
$$= z(1 - 2z + 3z^2 - 4z^3 + ...)$$
$$= z(1 + z)^{-2}$$

$\therefore$ The generating for the given sequence is

$$z(1 + z)^{-2}$$

Ecxample 12. Find the generating function for the sequence

$$1^2, 2^2, 3^2, ...$$

Solution. We have

$$0 + 1.z + 2.z^2 + 3.z^3 + ... = z(1 - z)^{-2} \qquad ...(1)$$

differentiating (1) w.r.t. z, we get

$$1 + 2(2z) + 3.(z^2) + 4.(4z^3) + ... = \frac{d}{dz}\left(\frac{z}{(0-z)^2}\right)$$

$$\Rightarrow \qquad 1^2 + 2^2z + 3^2.z^2 + ... = \frac{1+z}{(1-z)^3}$$

Hence, the generating function for the given sequenceis $\frac{1+z}{(1-z)^3}$

Example 13. Find the generating function for the sequence

$$0^2, 1^2, 2^2, 3^2, ...$$

Solution. We have

$$0^2 + 1^2.z + 2^2.z^2 + 3^2.z^2 + ... = z(1^2 + 2^2z + 3^2z^2 + ...)$$
$$= \frac{z(1+z)}{(1-z)^3}.$$

Thus the generating functions for the given sequence is $\frac{z(1+z)}{(1-z)^3}$.

Example 14. Find the generated sequence for the sequence

$$1^2, 2^2, 3^2, \ldots$$

Solution. We know that the generating function for the sequence

$$0^2 + 1^2.z + 2^2.z^2 + 3^2.z^3 + \ldots = \frac{z(1-z)}{(1-z)^3} \qquad \ldots(1)$$

differentiating (1) with respect to z, we get

$$1 + 2^3.z + 3^3.z^2 + \ldots = \frac{d}{dz}\left(\frac{z(1+z)}{(1-z)^3}\right)$$

$$\Rightarrow \qquad 1^3 + 2^3.z + 3^3.z^2 + \ldots = \frac{z^2+4z+1}{(1-z)^4}$$

The generating function for the sequence is $1^3, 2^3, 3^3, \ldots$ is $\dfrac{z^2+4z+1}{(1-z)^4}$

Example 15. Determine a generating function for the numeric function

$$a_r = \begin{cases} 2^r & \text{if } r \text{ is even} \\ -2^r & \text{if } r \text{ is odd} \end{cases}$$

Solution. We have $a_r = (-1)^r 2^r$

$$\therefore \qquad \sum_{r=0}^{\infty} a_r z^r = \sum_{r=0}^{\infty} (-1)^r 2^r . z^r$$

$$= \sum_{r=0}^{\infty} (-2)^r z^r = [1-(-2z)]^{-1}$$

$$= (1+2z)^{-1}$$

is the required generating funciton.

Example 16. If n is a positive integer, prove that

$$(1-z)^{-n} = \sum_{r=0}^{\infty} C(n+r-1, r)z^r$$

$$\therefore \qquad \sum_{r=0}^{\infty} C(n+r-1, r)z^r = \sum_{r=0}^{\infty} (-1)r\, C(-n, r)z^r$$

$$= \sum_{r=0}^{\infty} C(-n1r)\,(-z)r$$

$$= (1-z)^{-n}$$

Hence proved.

Example 17. Obtain a generating function for the sequence

1, 1, 1, 0, 1, 1, ...

Solution. Here we have $a_0 = 1, a_1 = 1, a_2 = 1, a_3 = 0, a_4 = 1, ...$

∴ The generating function for the sequence is

$$A(z) = a_0 + a_1 z + a_2 z_2 + a_3 z^3 + a_4 z^4 + ...$$
$$= 1 + 1.z + 1.z^2 + 1.z^3 + 1.z^4 + ...$$
$$= 1 + z + z^2 + z^4 + ...$$
$$= (1 + z + z^2 + z^3 + z^4 + ...) - z^3$$
$$= (1-z)^{-1} - z^3.$$

Example 18. Find the term independent of x (i.e. constant term) in the expansion of

$$\left(3x^2 - \frac{2}{x}\right)^{15}$$

Solution. We have $\left(3x^2 - \frac{2}{x}\right)^{15}$

$$= (3x^2)^{15}\left(1 - \frac{2}{3x^3}\right)^{15}$$

$$= 3^{15}(x^2)^{15} \sum_{r=0}^{15} C(15, r)\left(\frac{-2}{3}\right)^r x^{-3x}$$

$$= 315 \sum_{r=0}^{15} C(15, r)\left(\frac{-2}{3}\right)^r x^{30-3r}$$

The term independent of x is x^0

taking $30 - 3r = 0$, we get

$$3r = 30 \Rightarrow r = 10$$

∴ The coefficient of x^0 is the expansion is

$$3^{15}\, C(15, 10) \left(\frac{-2}{3}\right)^{10}$$

$$= 3^{15}\, C(15, 10) \cdot \frac{(-2)^{10}}{3^{10}}$$

∴ The constant term is $3^5.2^{10}\, C(15, 10)$.

Example 19. Find the coefficient x^{27} in

$$(x^4 + x^5 + x^6 + ...)^5$$

Solution. We have $(x^4 + x^5 + x^6 + ...)^5$

$$= (x^4)^5(1 + x + x^2 + ...)^5$$
$$= x^{20}(1 + x + x^2 + ...)^5$$
$$= x^{20}[(1 - x)^{-1}]^5$$
$$= x^{20}(1 - x)^{-5}$$
$$= x^{20}\, \Sigma\, C(4 + r, r)x^r$$
$$= \Sigma\, C(4 + r, r)x^{20 + r}$$

taking $20 + r = 27$ we get

$$r = 7$$

∴ The coefficient x^{27} is $C(4 + 7, 7) = C(11, 7)$

$$= \frac{11!}{7!\ 4!} = 330$$

Example 20. Using generating function find the number of

(i) non-negative and

(ii) positive integral solutions of the equation

$$x_1 + x_2 + x_3 + x_4 = 25$$

Solution. (i) There are no constants for x_i, $i = 1, 2, 3, 4$ therefore x_i's can take the values 0, 1, 2, 3, 4, ...

We can write

$$f_i(x) = 2^0 + x^1 + x^2 + x^3 + x^4 + ... \quad i = 1, 2, 3, 4$$

The generating function for the number of non-negative integer solution is

$$f(x) = f_1(x)\, f_2(x)\, f_3(x)\, f_4(x)$$
$$= (1 + x + x^2 + ...)\,(1 + x + x^2 + ...)\,(1 + x + x^2 + ..)\,(1 + x + x^2 + ...)$$
$$= (1 + x + x^2 + ...)^4$$

$$= [(1-x)^{-1}]^4 = (1-x)^{-4}$$
$$= \Sigma\, C(3+r, r)x^r$$

taking $x^r = x^{25}$ we get $r = 25$

$\therefore$ The coefficient of x^{25} is $C(3 + 25, 25) = C(28, 25)$

$$= \frac{28}{25!\ 3!} = 3276$$

(ii) In this case x_i's can take values 1, 2, 3, 4, ...

we can write

$$\therefore \qquad f_i(x) = x_1 + x_2 + x_3 + ..., \quad i = 1, 2, 3, 4$$
$$= x(1 + x + x^2 + ...) \quad i = 1, 2, 3, 4$$

The generating function for the number of positive integer solution is

$$f(x) = f_1(x) + f_2(x) + f_3(x) + f_4(x)$$
$$= x^4(1 + x + x^2 + ...)^4 = x^4[(1-x)^{-1}]^4$$
$$= x^4(1-x)^{-4}$$
$$= x^4 \sum_{r=0}^{\infty} C(3r, r)x^r = \sum_{r=0}^{\infty} C(3+r, r)x^{4+r}$$

We have $25 = 4 + r \Rightarrow r = 21$

$\therefore$ The coefficient of x^{25} is $C(3 + 21, 31)$

$$= C(24, 21)$$
$$= \frac{24!}{21!\ 4!} = 2024$$

EXERCISE 6.1

1. Find the sequencer generated by
$$(3 + x)^3$$
2. Find the sequence generated by
$$3x^3 + e^{2x}$$
3. Find the sequencer generated by
   (a) $x^2(1-x)^{-1}$ (b) $2x^2(1-x)^{-1}$
4. Find a generating function for the sequence $0^3, 1^3, 2^3, ...$
5. Find a generating function for the sequence $1, -2, 3, -4, ...$

6. Find a generating function for the sequencer

$$1, 1, 0, 1, 1, ...$$

7. Find a generating function for the sequencer

$$a_r = 0 \text{ for } 0 \leq r \leq n-1$$

and $$a_r = 0\ C(r-1, n-1) \text{ for } r \geq n$$

8. Find the coefficient of x^5 in $(1 - 2x)^{-7}$.

9. Find the coefficient of x^8 in $\dfrac{1}{(x-3)(x-2)^2}$.

10. Show that the coefficient x^n is the sequence

$$(x^2 + x^3 + x^4 + ...\ 1^4\ C(n-5, 3)$$

11. Determine the coefficient of x^{18} in the product

$$(x + x^2 + x^3 + x^4 + x^5)\ (x^2 + x^3 + x^4 + ...)^5$$

12. Obtain the generating function fo the sequencer

$$1, \frac{1}{1!}, \frac{1}{2!}, \frac{1}{3!}, ...$$

13. In how many ways can 2 oranges be distributed among the children A, B, C so that A gets atleast four, B and C get atleast two, but C gets no more than five.

Hint. Let x_1, x_2, x_3 denote the number of oranges which A, B and C get respectively then we have

$$x_1 + x_2 + x_3 = 12 \quad \text{(number of oranges)}$$

where $x_1 \geq 4$, $x_2 \geq 2$, $x_3 \leq 5$.

14. Using generating function, prove that the number of ways of choosing with repetitions, r of n objects in $C(n + r - 1, r)$.

15. In how many ways can we distribute 24 pencils to 4 children so that each child gets at least 3 pencils but not more than eight.

Hint. We have $x_1 + x_2 + x_3 + x_4 = 24$, $3 \leq x_i \leq 8$, $i = 1, 2, 3, 4$ where x_i, $i = 1, 2, 3, 4$ denotes the number of pencils got by each child.

ANSWERS 6.1

1. 27, 27, 9, 1, 0, 0, ...

2. $1, \dfrac{2}{1!}, \dfrac{2^2}{2!}, 3 + \dfrac{2^3}{3!}, \dfrac{2^4}{4!}, ...$

3. (a) 0, 0, 1, 1, 1, ... (b) 0, 0, 2, 2, ...

4. $\dfrac{x(x^2+4x+1)}{(1-x)^4}$

5. $(1+x)^{-2}$

6. $(1-x)^{-1}-x^2$

7. $\left(\dfrac{z}{1-z}\right)^n$

8. $25.\dfrac{11!}{5!\,6!}$

9. $\dfrac{-1}{12}\sum_{r=0}^{\infty}\left(\dfrac{1}{3}\right)^r\left(\dfrac{1}{2}\right)^{8-r} C(9-r, 8-r)$

11. $C(11, 7) + C(10, 6) + C(9, 5) + C(8, 4) + C(7, 3)$

12. e^x

13. 14 ways

14. 125 ways

6.3 RECURRENCE RELATIONS

In chapter 3, we have discussed recursive definition of a sequence. The recursive formula for defining a sequence (or a numeric function) is called a recurrence relation. If $a = \{a_0, a_1, a_2,...a_r,...\}$ is a numeric function then the recurrence relation for a is an equation relating a_r for any r; to one or more a_is $(i < r)$. Every recursive formula includes a starting point. The information a accompanying a recursive formula about beginning of the sequence (or numeric function) is called initial condition. A recurrence relation is also called a difference equation.

Example 1. The recurrence relation

$$a_n = a_{n-1} + 5, \text{ with } a_1 = 2$$

recursively defines the sequence 7, 12, 17,...

$$a_1 = 2 \text{ is the initial condition.}$$

Example 2. The recurrence relation

$$a_n = a_{n-1} + a_{n-2}, \text{ with } a_1 = a_2 = 1$$

defines the Fibonacci sequence,

$$1, 1, 2, 3, 5, 8, 13,...$$

$a_0 = 1$, $a_1 = 1$ are called the initial conditions.

Example 3. Fibonacci sequence and Rabits: consider the problem in which a young pair of rabbits (one of each sex) is placed on an island. A pair of rabbits does not bread until they are 2 months old. After they are 2 months old; each pair of rabbits produces another pair each month. Find a recurrence relation for the number of pairs of rabbits on the island after n months; assuming that no rabbits ever die.

Solution. Let a_n denote the number of pairs of rabbits on the island after n months.

At the end of first month, the number of pairs of rabbits on the island is $a_1 = 1$. Since this pair does not bread during the second month we have $a_2 = 1$. To find the number of pairs of rabbits after n months, add the number on the island the previous month, a_{n-1}, and the number of new-born pairs which equal a_{n-2}, since each new-born pair comes from a pair at least 2 month old.

Consequently, the sequence $\{a_n\}$ satisfies, the recurrence relation (i.e. the Fibonacci sequence).

$$a_n = a_{n-1} + a_{n-2}, \quad n \geq 3$$

with $a_1 = 1$, $a_2 = 1$, as the initial conditions.

6.3.1 Linear Recurrence Relations

Let r and k be non-negative integers, then a recurrence relation of the form

$$C_0(r)\, a_r + C_1(r)\, a_{r-1} + \ldots + C_k(r)\, a_{r-k} = f(r) \text{ for } r \geq k$$

where $C_0(r)$, $C_1(r)$..., $C_k(r)$ and $f(r)$ are functions of r is said to be a recurrence relation of degree k.

If $C_0(r) > 0$, and $C_k(r) > 0$, then it is said to be a linear recurrence relation of degree k. If $C_0(r)$, $C_1(r)$,..., $C_k(r)$ are constants then the recurrence relation said to be a linear recurrence relation with constant coefficients.

If $f(r) = 0$, then the recurrence relation is called a linear homogeneous relation. A recurrence relation which is not homogeneous is said to be inhomogeneous relation. It is of the form $C_0\, a_r + C_1\, a_{r-1} + \ldots + C_k\, a_{r-k} = f(r)$. The recurrence relations:

$$a_n - 6a_{n-1} + 11a_{n-2} + 6a_{n-3} = 2n$$

and

$$a_n - 9a_{n-1} + 26\, a_{n-2} - 24a_{n-3} = 5n$$

are linear recurrence relations with constant coefficients. The relation:

$$a_n - 6a_{n-1} + 11a_{n-2} + 6a_{n-3} = 0$$

is a homogeneous recurrence relation.

The relation: $a_r + 9a_{r-2} = 0$

is a second order recurrence relation with constant coefficients.

6.4 SOLVING RECURRENCE RELATIONS

We shall now describe some methods of solving recurrence relation in this section. We know that very recurrence relation is accompanied by conditions. Any numeric function that can be described by a recurrence relation together with an appropriate set of boundary conditions is called a solution of the recurrence relation. If $a = (a_0, a_1, a_2, ..., a_r...)$ is a solution to a recurrence relation, then it is said to satisfy the recurrence relation. A given recurrence relation may or may not have a solution. We shall now consider the methods of solving homogeneous recurrence relations.

6.4.1 Solving Recurrence Relations by Substitution

In this method, the recurrence relation for a_r is used repeatedly to solve. A general expression for a_r in terms of r is obtained by this method.

Example 1. Solve the recurrence relation

$$a_{n+1} = 5a_n, \text{ for } n \geq 0, \text{ given taht } a_0 = 2.$$

Solution. We have $a_{n+1} = 5a_n$

$\therefore$
$$a_1 = 5a_0$$
$$a_2 = 5a_1 = 5.5.a_0 = 5^2.a_0$$
$$a_3 = 5a_2 = 5^3a_0$$
$$a_4 = 5a_3 = 5^4a_0$$

The general solution is $a_n = 5^n a_0$

The initiated condition is $a_0 = 2$

$\therefore$ The particular solution is $a_n = 5^n.2$

i.e. $a_n = 2.5^n$

Example 2. Solve the recurrence relation

$$4a_r - 5a_{r-1} = 0, r \geq 1, a_0 = 1$$

Solution. We have $4a_r - 5a_{r-1} = 0 \Rightarrow 4a_r = 5a_{r-1}$

$\Rightarrow$
$$a_r = (5/4)a_{r-1}$$

$\therefore$ The general solution is ar = $(5/4)^r$

$$a_0 = 1 \text{ the initial condition}$$

$\therefore$ The particular solution is ar = $(5/4)^r$

Example 3. If an is a solution of $a_{n+1} = ka_n$ for $n \geq 0$ and $a_3 = 153/49$ and $a_5 = \dfrac{1377}{2401}$ find the value of k.

Solution. The general solution of $a_{n+1} = ka_n$ is

$$a_n = k^n a_0, \quad n \geq 1$$

which gives $\quad a_3 = k^3 a_0 \Rightarrow k^3 a_0 = 153/49 \qquad (1)$

and $\quad a_5 = k^5 a_0 \Rightarrow k^5 a_0 = 1377/2401 \qquad (2)$

$\therefore$ from (1) and (2) $\dfrac{k^5 a_0}{k^3 a_0} = \dfrac{1377}{2401} \cdots \dfrac{49}{153} \Rightarrow ak^2 = 9/49$

$\Rightarrow \qquad k \pm 3/7$

Example 4. Solve the recurrence relation.

$a_r = 2a_{r-1} + 1$ with $a_1 = 7$, for $r > 1$, by substitution method.

Solution. The initial condition is $a_1 = 7$

we have $\quad a_2 = 2a_1 + 1 = 2 - 7 + 1$

$$a_3 = 2a_2 + 1 = 2 - (2 - 7 + 1) + 1 = 2^2 \cdot 7 + 2 + 1$$

$$a_4 = 2a_3 + 1 = 2(2^2 \cdot 7 + 2 + 1) + 1 = 2^3 \cdot 7 + 2^2 + 2 + 1$$

$$a_r = 2^{r-1} \cdot 7 + 2^{r-2} + 2^{r-3} + \ldots + 2 + 1$$

$$\Rightarrow \qquad a_r = 7 \cdot 2^{r-1} + \left(\frac{2^{r-1} - 1}{2 - 1}\right) \quad \left(\because 1 + 2 + \ldots 2^{r-2} = \frac{2^{r-1} - 1}{2 - 1}\right)$$

$$\Rightarrow \qquad a_r = 7 \cdot 2^{r-1} + 2^{r-1} - 1, \text{ for } r \geq 1$$

$\therefore$ The solution is $a_r = 7 \cdot 2^{r-1} + 2^{r-1} - 1$

Example 5. Solve the recurrence relation $a_r = a_{r-1} + f(r)$ for $r \geq 1$, by substitution.

Solution. We have,

$$a_1 = a_0 + f(1)$$

$$a_2 = a_1 + f(2) = a_0 + f(1) + f(2)$$

$$a_3 = a_2 + f(3) = a_0 + f(3) + f(1) + f(a)$$

$$= a_0 + f(1) + f(2) + f(3)$$

$$\ldots$$

$$a_r = a_0 + f_1(1) + f(2) + f(3) + \ldots + f(r)$$

$$\Rightarrow \qquad a_r = a_0 + \sum_{n=1}^{r} f(n)$$

is the required solution.

Example 6. Solve the recurrence relation $a_n = a_{n-1} + n$, $a_0 = 2$ by substitution.

Solution. We have,

$$a_n = a_{n-1} + n, \quad a_0 = 2$$

$$\therefore \quad a_1 = a_0 + 1$$

$$a_2 = a_1 + 2 = a_0 + 1 + 2 = a_0 + \sum_{1}^{2} r$$

$$a_3 = a_2 + 3 = a_0 + 1 + 2 + 3 = a_0 + \sum_{1}^{3} r$$

$$\vdots$$

$$a_{n-1} = a_{n-2} + (n-1) = a_0 + \sum_{1}^{n-1} r$$

$$a_n = a_{n-1} + (n-1) = a_0 + \sum_{r=1}^{n} r = a_o + \frac{n(n+1)}{2}$$

$$\therefore \quad a_n = 2 + \frac{n(n+1)}{2} \text{ is the solution.}$$

Example 7. Solve $a_r = a_{r-1} + r^2$; $a_0 = 7$ by substitution method.

Solution. We have, $a_r = a_{r-1} + r^2$; $a_0 = 7$

$$\therefore \quad a_1 = a_0 + 1^2 x$$

$$a_2 = a_1 + 2^2 = a_0 + 1^2 + 2^2$$

$$a_3 = a_2 + 3^2 = a_0 + 1^2 + 2^2 + 3^2$$

$$\vdots$$

$$a_r = a_{r-1} + (r-1)^2 = a_0 + 1^2 + 2^2 + 3^2 + \dots + r^2$$

$$\therefore \quad a_r = a_0 + \frac{r(r+1)(2r+1)}{6}$$

$\therefore$ The required solution is $a_r = 7 + \dfrac{r(r+1)(2r+1)}{6}$

Example 8. Solve $a_n = a_{n-1} + 5^n$, $a_0 = 1$

Solution. $a_n = a_{n-1} + 5^n$, $a_0 = 1$

$$\Rightarrow \quad a_1 = a_0 + 5^1$$

$$a_2 = a_1 + 5^2 = a_0 + 5^1 + 5^2$$

$$a_3 = a_2 + 5^3 = a_0 + 5^1 + 5^2 + 5^3$$

$$\vdots$$

$$a_n = a_{n-1} + 5^n = a_0 + 5^1 + 5^2 + ... + 5^n$$

Now $\quad a_0 = 1$

$\Rightarrow \quad a_n = 1 + 5^1 + 5^2 + 5^3 + ... + 5^n = 1 + 5\,(1 + 5 + ... + 5^{n-1})$

$$= 1 + \frac{5\,(5^n - 1)}{5 - 1} = 1 + \frac{5^{n+1} - 5}{4} = \frac{5^{n+1} - 1}{4}$$

Example 9. Solve $a_r = a_{r-1} + r \cdot 2^r$, given $a_0 = 1$

Solution: We have $a_r = a_{r-1} + r \cdot 2^r$, $a_0 = 1$

$\therefore \quad a_1 = a_0 + 1 \cdot 2^1$

$$a_2 = a_1 + 2 \cdot 2^2 = a_0 + 1 \cdot 2^1 + 2 \cdot 2^2$$

$$a_3 = a_2 + 3 \cdot 2^3 = a_0 + 1 \cdot 2^1 + 2 \cdot 2^2 + 3 \cdot 2^3$$

$$\vdots$$

$$a_r = a_{r-1} + (r - 1)\, 2^{r-1} = a_0 + 1 \cdot 2^1 + 2 \cdot 2^2 + ... + (r - 1) \cdot 2^{r-1}$$

but $\quad a_0 = 1$

$\therefore \quad a_r = 1 + \sum r \cdot 2^r$ is the solution.

EXERCISE 6.2

1. Solve the recurrence relation by substitution.

(a) $a_r = a_{r-1} + r$, $\quad a_0 = 5$

(b) $a_r = a_{r-1} + r^2$, $\quad a_0 = 3$

(c) $a_n = a_{n-1} + n^3$, $\quad a_0 = 5$

2. Solve the recurrence relation by substitution.

(a) $a_n = a_{n-1} + n\,(n - 1)$, $\quad a_0 = 1$

(b) $a_n = a_{n-1} + n \dfrac{1}{n(n+1)}$, $\quad a_0 = 1$

3. Solve $a_n = a_{n-1} + 3^n$, $\quad a_0 = 1$

4. Solve $a_r = a_{r-1} + r \cdot 3^r$, $\quad a_0 = 1$

5. Solve $a_r = a_{r-1} + (2r + 1)$, $\quad a_0 = 1$

6. Solve $a_n = n\, a_{n-1}$, $n \geq 1$, $a_0 = 1$.

7. Solve $3a_n + 1 - 4a_n = 0$, $n \geq 0$, $a_1 = 5$.

8. Solve $a_{n+1} - a_n = 3n^2 - n$, $n \geq 1$, $a_0 = 3$.

9. Solve $a_n - a_{n-1} = 3n^2, n \geq 1, a_0 = 7$.

10. Solve $a_n - a_{n-1} = n^3$, n^3 1, $a_0 = 25$.

ANSWER

1. (a) $a_r = 5 + \dfrac{r(r+1)}{2}$

(b) $a_r = 3 + \dfrac{r(r+1)(2r+1)}{6}$

(c) $a_n = 5 + \dfrac{n^2(n+1)^2}{4}$

2. (a) $a_n = 1 + \dfrac{n(n+1)(2n+1)}{6} - \dfrac{n(n+1)}{2}$

(b) $a_n = 2 - \dfrac{1}{n+1}$

3. $a_n = \dfrac{3^{n+1} - 1}{2}$

4. $a_r = 1 + \sum r \cdot 3^r$

5. $a_r = (r+1)^2$

6. $a_n = n!$

7. $a_n = 15/4 \cdot \left(\dfrac{4}{3}\right)^n$

8. $a_n = 3 + n(n-1)^2$

9. $a_n = (n+1)^3$

10. $a_n = \dfrac{1}{4} n^2 (n+1)^2 + 5$

6.4.2 Method of Generating Functions

Recurrence relations can also be solved by using generating functions. Some equivalent expressions used are given below.

If $A(z) = \sum_{r=0}^{\infty} a_r z^r$,

then $\sum_{r=k}^{\infty} a_r z^r = A(z) - a_0 - a_1 z - ...a_{k-1} z^{k-1}$

$$\sum_{r=k}^{\infty} a_{r-1} z^r = z\,(A(z) - a_0 - a_1 z - ... - a_{k-2} z^{k-2})$$

$$\sum_{r=k}^{\infty} a_{r-2} z^r = z^2\,(A(z) - a_0 - a_1 z - ... - a_{k-3} z^{k-3})$$

$$\vdots$$

$$\sum_{r=k}^{\infty} a_{r-k} z^r = z^k A(z)$$

where $A(z)$ is called a generating function for a given recurrence relation

Example 1. Solve the recurrence relation $a_r - 7a_{r-1} + 10\, a_{r-2} = 0, r \geq 2$

Solution. Multiplying each term of the recurrence relation by z^r,

we get $$a_r z^r - 7a_{r-1} z^r + 10\, a_{r-2} z^r = 0$$

Taking the sum 2 to ∞, we get

$$\Sigma a_r z^r - 7\sum_{r=2}^{\infty} a_{r-1} z^r + 10 \sum_{r=2}^{\infty} a_{r-2} z^r = 0$$

replacing each infinite sum by the corresponding equivalent expression give above we get

$$(A(z) - a_0 - a_1 z) - 7 - (A(z) - a_0) - 10z^2 A\,(z) = 0$$

Simplifying, we get

$$A(z)\,(1 - 7z + 10z^2) = a_0 + a_1 z - 7a_0 z$$

$$\Rightarrow \qquad A(z) = \frac{a_0 + (a_1 - 7a_0)z}{1 - 7z + 10z^2} = \frac{a_0 + (a_1 - 7a_0)z}{(1 - 2z)\,(1 - 5z)}$$

Decomposing $A(z)$ as a sum of Partial fractions, we can write

$$A(z) = \frac{c_1}{1 - 2z} + \frac{c_2}{1 - 5z}$$

$$= c_1 \sum_{r=0}^{\infty} 2^r z^r + c_2 \sum_{r=0}^{\infty} 5^r z^r$$

The solution is

$$\therefore \qquad a_r = c_1\, 2^r + c_2\, 5^r.$$

Example 2. Solve the recurrence relation

$$a_n - 9a_{n-1} + 26\, a_{n-2} - 24\, a_{n-3} = 0 \quad \text{for } n \geq 3$$

Solution. Let $A(z) = \sum_{n=0}^{\infty} a_n z^n$.

Consider $a_n - 9a_{n-1} + 26\, a_{n-2} - 24\, a_{n-3} = 0$ ...(1)

Multiplying each term by z^n and taking the sum for $n = 3$ to ∞ b. We get

$$\sum_{n=3}^{\infty} a_n z^n - 9\sum_{n=3}^{\infty} a_{n-1} z^n + 26\sum_{n=3}^{\infty} a_{n-2} z^n - 24\sum_{n=3}^{\infty} a_{n-3} z^n = 0$$

replacing the infinite sums by equivalent expressions, we get

$$(A(z) - a_0 - a_1 z - a_2 z^2) - 9z\,(A(z) - a_0 - a_1 z) + 26z^2(A(z) - a_0) - 24z^3 A(z) = 0$$

Simplifying, we get

$$A(z)\,[1 - 9z + 26z^2 - 24z^3] = a_0 + a_1 z + a_2 z^2 - 9a_0 z - 9a_1 z^2 + 26a_0 z^2$$

$$\Rightarrow \; A(z)\,(1 - 2z)\,(1 - 3z)\,(1 - 4z) = a_0 + (a_1 - 9a_0)z + (a_2 - 9a_1 + 26a_0)z^2$$

$$\therefore \qquad A(z) = \frac{a_0 + (a_1 - 9a_0)z + (a_2 - 9a_1 + 26a_0)z^2}{(1 - 2z)\,(1 - 3z)\,(1 - 4z)}$$

Decomposing $A(z)$, as a sum of partial fraction we can write

$$A(z) = \frac{c_1}{1 - 2z} + \frac{c_2}{1 - 3z} + \frac{c_3}{1 - 4z} = c_1 \sum_{n=3}^{\infty} 2^n z^n + c_2 \sum_{n=3}^{\infty} 3^n z^n + c_3 \sum_{n=3}^{\infty} 4^n z^n$$

$\Rightarrow \quad a_n = c_1 \cdot 2^n + c_2 \cdot 3^n + c_3 \cdot 4^n$ is the required solution

Example 3. Solve $a_n - 6a_{n-1} + 12a_{n-2} - 8a_{n-3} = 0$

Solution. Let $\qquad A(z) = \sum_{n=0}^{\infty} a_n z^n$

Consider $\qquad a_n - 6a_{n-1} + 12a_{n-2} - 8a_{n-3} = 0$ ...(1)

multiplying each term by z^n and taking the sum for $n = 3$, we get

$$\sum_{n=3}^{\infty} a_n z^n - 6\sum_{n=3}^{\infty} a_{n-1} z^n + 12\sum_{n=3}^{\infty} a_{n-2} z^n - 8\sum_{n=3}^{\infty} a_{n-3} z^n = 0$$

replacing infinite sums by equivalent expression we get

$$(A(z) - a_0 - a_1 z - a_2 z^2) - 6 = (A(z) - a_0 - a_1 z) + 12z_2(A(z) - a_0) - 8z^3 A(z) = 0$$

Simplifying we get

$$A(z) = \frac{a_0 + (a_1 - 6a_0)z + (a_2 - 6a_1 + 12a_0)z^2}{1 - 6z + 12z^2 - 8z^3}$$

$$= \frac{a_0 + (a_1 - 6a_0)z + (a_2 - 6a_1 + 12a_0)z^2}{(1-2z)^3}$$

$$= \frac{C_1}{1-2z} + \frac{C_2}{(1-2z)^2} + \frac{C_3}{(1-2z)^3} \quad \text{(say)}$$

$$= C_1 \sum_{n=0}^{\infty} 2^n z^n + C_2 \sum_{n=0}^{\infty} C(n+1, n) 2^n z^n + C_3 \sum_{n=0}^{\infty} C(n+2, n) 2^n z^n$$

$$\Rightarrow \quad \sum_{n=0}^{\infty} a_n z^n = \sum_{n=0}^{\infty} \left(C_1 z^n + C_2 (n+1) 2^n + C_3 \frac{(n+2)(n+1)}{2} . 2^n \right) . 2^n$$

$$\therefore \quad a_n = C_1 2^n + C_2 (n+1) 2^n + C_3 \frac{(n+2)(n+1)}{2} . 2^n$$

is the required solution.

6.4.3 Generating Functions for the First Order Recurrence Relations

Consider the equation

$$a_r = Ca_{r-1} + f(r), \quad r \geq 1 \qquad \text{...(1)}$$

where C is constant

The equation (1) can be written as

$$a_{r+1} = Ca_r + \phi(r) \quad \text{for } r \geq 0 \qquad \text{...(2)}$$

where $\phi(r) = f(r+1)$

multiplying (2) by 2^{r+1} we get

$$a_{r+1} 2^{r+1} - Ca_r z^n + 1 = f(r+1) z^{r+1} \qquad \text{...(3)}$$

Taking the sum of all relation obtained from (3) for $n = 0, 1, 2, ...$ we get

$$\sum_{n=0}^{\infty} a_{r+1} z^{r+1} - C \sum_{n=0}^{\infty} a_r . z^r . z = \sum_{n=0}^{\infty} f(r+1) z^{r+1}$$

$$\Rightarrow \quad \sum_{n=0}^{\infty} a_{r+1} z^{r+1} - C.Z \sum_{n=0}^{\infty} a_r z^r = z \sum_{n=0}^{\infty} \phi(r) z^r \qquad \text{...(4)}$$

Equation (4) can be written as

$$\sum_{r=0}^{\infty} a_r z^r - Cz \sum_{r=0}^{\infty} a_r z^r = z \sum_{r=0}^{\infty} \phi(r) z^r$$

$$\Rightarrow \quad \left(\sum_{r=0}^{\infty} a_r z^r - a_0\right) - Cz \sum_{r=0}^{\infty} a_r z^r = z \sum_{r=0}^{\infty} \phi(r) z^r$$

$$\Rightarrow \quad (1 - Cz) \sum_{r=0}^{\infty} a_r z^r = a_0 + z \sum_{r=0}^{\infty} \phi(r) z^r$$

$$\Rightarrow \quad (1 - Cz)\, A(z) = a_0 + g(z) \qquad \text{...(5)}$$

where $g(z)$ dentoes the generating function for the sequencer $\phi(0), \phi(1), \ldots$ from equation (5), we get

$$A(z) = \frac{a_0 + zg(z)}{1 - Cz} \qquad \text{...(6)}$$

If $\phi(r) = 0$, then $g(z) = 0$, in which case

we have $$A(z) = \frac{a_0}{1 - Cz}$$

$A(z)$ is the generating function for the recurrence relatin given by (1).

Generating function for the second order recurrence relations: the relations of the form.

$$a_r + Aa_{r-1} + \beta a_{r-2} = f(x) \quad r \geq 2 \qquad (1)$$

are called second order recurrence relation.

Applying the method explained above, we get the generaing function for the equation (1), which is of that form

$$A(z) = \frac{a_0 + (a_1 + a_0 A)z + z^2 g(z)}{1 + Az + \beta z^2} \qquad (2)$$

If $g(z) = 0$, then the generating function is

$$A(z) = \frac{a_0 + (a_1 + a_0 A)}{1 + Az + \beta z^2}$$

Example 1. Consider the recurrence relation

$$a_n - 4a_{n-1} = 0, \quad n \geq 1, a_0 = 1 \qquad (1)$$

the equation (1) can be written as

$$a_{n+1} - 4a_n = 0$$

we have $a_0 = 1$, $C = 4$

$\therefore$ The generating function for (1) is

$$A(z) = \frac{a_0 + zg(z)}{1 - Cz} = \frac{1+0}{1-4z} = \frac{1}{1-4z}$$

now $$A(z) = \sum_{r=0}^{\infty} a_r z^r = \frac{1}{4-4z} = \sum_{r=0}^{\infty} 4^r z^r$$

$\Rightarrow$ $$a_r = 4^r$$

$\therefore$ The generating function for the given recurrence relation is $\frac{1}{4-4z}$ and the solution of the recurrence relation is $a_r = 4^r$.

Example 2. Consider the recurrence relation

$$a_{n+1} - a_n = 5^n, \quad n \geq 0 \text{ and } a_0 = 1 \qquad \text{...(1)}$$

In this recurrence relation we have

$$a_0 = 1, C = 1, \phi(n) = 5^n$$

The generaing function is

$$A(z) = \frac{a_0 + z\,g(z)}{1 - Cz} = \frac{1 + zg(z)}{1 - z} \qquad \text{...(2)}$$

where $g(z) = \sum_{r=0}^{\infty} \phi(n) z^n = \sum_{r=0}^{\infty} g^n z^n = (1 - 5z)^{-1}$ ...(3)

$\therefore$ from (2) and (3) we get

$$A(z) = \frac{1 + 2(1-5z)^{-1}}{1-z} = \frac{1 + \frac{z}{1-5z}}{1-z}$$

$$= \frac{1 - 5z + z}{(1-z)(1-5z)} = \frac{1-4z}{(1-2((1-5z)} \qquad \text{...(4)}$$

Consider $\frac{1-4z}{(1-z)(1-5z)}$

let $$\frac{1-4z}{(1-z)(1-5z)} = \frac{A}{1-z} = \frac{B}{1-5z}$$

$$\Rightarrow \qquad \frac{1-4z}{(1-z)(1-5z)} = \frac{A-5Az+B-5z}{(1-2)(1-5z)} = \frac{(A+B)-(5A+B)z}{(1-z)(1-1-5z)}$$

Equation the coefficients of like terms we get

$$\therefore \qquad A + B = 1 \qquad \text{...(i)}$$

$$5A + B = 4 \qquad \text{...(ii)}$$

Solving (i) and (ii) we get

$$A = \frac{3}{4} \quad \text{and} \quad B = \frac{1}{4}$$

$$\therefore \qquad A(z) = \frac{3/4}{1-z} + \frac{1/4}{1-5z} = \frac{3}{4}(1-z)^{-1} + \frac{1}{4}(1-5z)^{-1}$$

$$= \frac{3}{4}\Sigma z^n + \frac{1}{4}\Sigma n.2^n$$

$$\therefore \qquad \sum_{n=0}^{\infty} a_n z^n = \Sigma\left(\frac{3}{4} + \frac{1}{4}.5^n\right).z^n$$

$$\Rightarrow \qquad a_n = \frac{1}{4}(3 + 5^n)$$

is the required solution.

6.4.4 Method of Characteristic Roots

In this section, we explain the method of solving linear recurrence relation by the method characteristic roots.

Consider,

$$c_0\, a_r + c_1\, a_{r-1} + c_2\, a_{r-2} + \ldots + c_k a_{r-k} = 0 \qquad \text{...(1)}$$

Substituting $A\, \alpha^r$ for a_r in (1), we get

$$c_0\, A\, \alpha^r + c_1\, A\, \alpha^{r-1} + \ldots + c_k\, A\alpha^{r-k} = 0 \qquad \text{...(2)}$$

where the constant A is determined by the boundary conditions.

(2) Can be simplified and determined and written as follows:

$$c_0\alpha^r + c_1\alpha^{k-1} + \ldots + c_k = 0 \qquad \text{...(3)}$$

The equation (3) is called the characteristic equation of the difference equation (i.e. recurrence relation).

Equation (3) is an equation kth degree, therefore it has k roots. Let α_1, α_2,..., α_k be the roots of the equation (3). The roots α_1, α_2,..., α_k are called characteristic roots and

$$a_r = A_1\, \alpha_1{}^r + A_2\, \alpha_2{}^r + \dots + A_k\, \alpha^k$$

is a solution of the difference equation given by (1).

The constant $A_1, A_2, \dots, A_k$ are determined by boundary conditions.

Some roots of the characteristic equation may be repeated (i.e. multiple notes). Let α_1 be a roots of multiplicity p, then the corresponding solution is given by

$$[A_1\, r^{p-1} + A_2\, r^{p-2} + \dots + A_r]\alpha_1^r$$

where the constant A_i's are to be determined by boundary conditions.

Example 1. Solve the recurrence relation:

$$a_r - 7a_{r-1} + 12a_{r-2} = 0$$

Solution. The characteristic equation of the given equation is

$$\alpha^2 - 7\alpha + 12 = 0$$

$$\Rightarrow \qquad (\alpha - 3)(\alpha - 4) = 0$$

$\therefore$ The characteristics roots are $\alpha = 3, \alpha = 4$

and the solution of the given recurrence relation is

$$a_r = A_1\, 3^r + A_2 \cdot 4^r$$

Example 2. Solve the recurrence relation

$$a_r + 9a_{r-1} + 27a_{r-2} + 27a_{r-3} = 0$$

Solution. The characteristic equation of the given equation is

$$\alpha^3 + 9\alpha^2 + 27\alpha + 27 = 0$$

$$\Rightarrow \qquad (\alpha + 3)^3 = 0$$

$$\Rightarrow \qquad \alpha = -3, -3, -3$$

i.e. the characteristic roots are $-3, -3, -3$

$$\therefore \qquad a_r = (A_1\, r^2 + A_2\, r + A_3)(-3)^r$$

is the solution of the given equation.

Example 3. Solve $a_r - 7a_{r-1} + 10a_{r-2} = 0, r \geq 2$ (J.N.T.U. Dec. 2002)

Solution. We have $a_r - 7a_{r-1} + 10a_{r-2} = 0, r \geq 2$,

the characteristic equation is

$$\alpha^2 - 7\alpha + 10 = 0$$

$$\Rightarrow \qquad (\alpha - 2)(\alpha - 5) = 0$$

$$\therefore \qquad \alpha = 2 \text{ or } 5$$

The general solution is

$$a_r = A_1\, 2^r + A_2\, 5^r$$

Example 4. Solve the recurrence relation

$$a_n - 9a_{n-1} + 26a_{n-2} - 24a_{n-3} = 0, \text{ for } n \geq 3$$

Solution. We have $a_n - 9a_{n-1} + 26a_{n-2} - 24a_{n-3} = 0, n \geq 3$

The characteristic equation is

$$\alpha^3 - 9\alpha^2 + 26\alpha - 24 = 0$$

$\Rightarrow$ $$(\alpha - 2)(\alpha - 3)(\alpha - 4) = 0$$

$\Rightarrow$ $$\alpha = 2, \text{ or } \alpha = 3, \text{ or } \alpha = 4$$

$\therefore$ The general solution of the given equation is

$$A_n = A_1 \cdot 2^n + A_2 \cdot 3^n + A_3 \cdot 4^n$$

Example 5. Solve

$$a_{r+4} - 4a_{r+3} + 6a_{r+2} - 4a_{r+1} + 4a_r = 0$$

Solution. The characteristic equation of the given equation is

$$\alpha^4 - 4\alpha^3 + 6\alpha^2 - 4\alpha + 4 = 0$$

$\Rightarrow$ $$(\alpha - 1)^4 = 0$$

$\Rightarrow$ $$\alpha = 1, 1, 1, 1$$

$\therefore$ The general solution is

$$a_r = (A_1 + A_2 r + A_3 r^2 + A_4 r^3)\, 1^r$$

i.e. $$a_r = A_1 + A_2 r + A_3 r^2 + A_4 r^3$$

Example 6. Consider the Fibonacci sequence, defined by

$$a_n = a_{n-1} + a_{n-2}, \text{ with } a_0 = 0, a_1 = 1$$

as the initial conditions.

The characteristic equation of the recurrence relation is

$$\alpha^2 = \alpha + 1 \qquad \text{...(1)}$$

i.e. $$\alpha^2 - \alpha - 1 = 0$$

$\Rightarrow$ $$\alpha = \frac{1 \pm \sqrt{5}}{2}$$

i.e. the characteristic roots of (1) are $\frac{1+\sqrt{5}}{2}, \frac{1-\sqrt{5}}{2}$

The complete solution is

$$a_n = A_1\left(\frac{1+\sqrt{5}}{2}\right)^n + A_2\left(\frac{1-\sqrt{5}}{2}\right)^n \qquad \text{...(2)}$$

$$a_0 = 0 \Rightarrow A_1\left(\frac{1+\sqrt{5}}{2}\right)^0 + A_2\left(\frac{1-\sqrt{5}}{2}\right)^0 = 0 \qquad ...(3)$$

$$\Rightarrow A_1 + A_2 = 0$$

and $$a_1 = 1 \Rightarrow A_1\left(\frac{1+\sqrt{5}}{2}\right) + A_2\left(\frac{1-\sqrt{5}}{2}\right) = 0 \qquad ...(4)$$

Solving (3) and (4) we get

$$A_1 = 1/\sqrt{5} \text{ and } A_2 = -\,1/\sqrt{5}$$

The complete solution of the Fibonacci sequence is given by

$$a_n = \frac{1}{\sqrt{5}}\left(\frac{1+\sqrt{5}}{2}\right)^n - \frac{1}{\sqrt{5}}\left(\frac{1-\sqrt{5}}{2}\right)^n$$

Example 7. Solve the recurrence relation

$$S(k) - 9S(k-1) + 18S(k-2) = 0, S(0) = 1, S(1) = 4$$

Solution. Consider

$$S(k) - 9S(k-1) + 18S(k-2) = 0 \qquad ...(1)$$

Let $S(k) = A\alpha^k$, then

$$S(k-1) = A\alpha^{k-1}$$

$$S(k-2) = A\alpha^{k-2}$$

$\therefore$ from (1), we get

$$A\alpha^k - 9\cdot\alpha^{k-1} + 18\alpha^{(k-2)} = 0$$

$\Rightarrow$ $$A\alpha^{k-2}(\alpha^2 - 9\alpha + 18) = 0$$

$\Rightarrow$ $$\alpha^2 - 9\alpha + 18 = 0$$

$\Rightarrow$ $$(\alpha - 3)(\alpha - 6) = 0$$

$\Rightarrow$ $$\alpha = 3 \text{ or } \alpha = 6$$

$\therefore$ $S(k) = A_1 3^k + A_2 6^k$ is the solution of (1)

Now, $$S(0) = 1 \Rightarrow A_1 + A_2 = 1 \qquad ...(2)$$

$$S(1) = 4 \Rightarrow 3A_1 + 6A_2 = 4 \qquad ...(3)$$

Solving (2) and (3) we get $A_1 = \frac{2}{3}, A_2 = \frac{1}{3}$

$\therefore$ The general solution is

$$S(k) = \frac{2}{3}\cdot 3k + \frac{1}{3}\cdot 6^k$$

Example 8. Solve the recurrence relation

$$a_n = 2(a_{n-1} - a_{n-2}), \quad \text{for } n \geq 2$$

given that $a_0 = 1$, and $a_1 = 2$.

Solution. The characteristic equation is

$$\alpha^2 - 2\alpha + 2 = 0$$

$$\Rightarrow \qquad a = 1 \pm i$$

The general solution is

$$a_n = r^n[A_1 \cos n\theta + A_2 \sin n\theta] \qquad \text{...(1)}$$

where A_1 and A_2 are arbitrary constants

$$r = |1 \pm i| = \sqrt{2}\,,\ \theta = \tan^{-1}\frac{1}{1} = \frac{\pi}{4}$$

$$\therefore \qquad a_n = \left(\sqrt{2}\right)^n \left| A_1 \cos\frac{n\pi}{4} + A_2 \sin\frac{n\pi}{4} \right| \qquad \text{...(2)}$$

The initial conditions are $a_0 = 1$, and $a_1 = 2$

$\therefore$ from (2), we get

$$\therefore \qquad a_0 = \left(\sqrt{2}\right)^0 \left[A_1 \cos\frac{\pi}{4} + A_2 \sin 0\right] = 1 \Rightarrow A_1 = 1$$

$$a_1 = \left(\sqrt{2}\right)^1 \left[A_1 \cos\frac{\pi}{4} + A_2 \sin\frac{\pi}{4}\right]$$

$$\Rightarrow \qquad a_1 = \sqrt{2}\left[1.\frac{1}{\sqrt{2}} + A_2\frac{1}{\sqrt{2}}\right]$$

$$\Rightarrow \qquad 2 = 1 + A_2 \quad \Rightarrow \quad A_2 = 1$$

$$\therefore \qquad A_1 = 1, A_2 = 1$$

The general solution is

$$a_n = \left(\sqrt{2}\right)^n \left[\cos\frac{n\pi}{4} + \sin\frac{n\pi}{4}\right], \quad n \geq 2$$

(with initial conditions $a_0 = 1$ and $a_1 = 2$)

Example 9. Solve $a_{n+2}^2 - 5a_{n+1} + 4a_n^2 = 0, \quad n \geq = 0$

given $a_0 = 4$ and $a_1 = 13$.

Solution. Consider

$$a_{n+2}^2 - 5a_{n+1} + 4a_n^2 = 0, \quad n \geq = 0 \qquad ...(1)$$

Taking $b_n = a_n^2$, the relation 91) can be written as

$$b_{n+2} - 5bn + 4b_n = 0 \qquad ...(2)$$

(2) can be written as

$$b_n - 5b_{n-1} + 4b_{n-2} = 0, \quad n \geq 2 \qquad ...(3)$$

The characteristic equation for the relation (3) is

$$\alpha^2 - 5\alpha + 4 = 0$$

$$\Rightarrow \quad (\alpha - 1)(\alpha - 4) = 0$$

$$\therefore \quad \alpha = 1, \alpha = 4$$

The general solution for (3) is

$$b_n = A.1^n + B.4^n = A + B.4^n \qquad ...(4)$$

where A and B are arbitrary constants

It is given that $a_0 = 4$ and $a_1 = 13 \Rightarrow a_0^2 = 16,\ a_1^2 = 169 \Rightarrow b_0 = 16, b_1 = 169$

from (4) we get

$$b_0 = A + B$$

i.e. $$16 = A + B \qquad ...(i)$$

and $$b_1 = A + 4B \qquad ...(ii)$$

Solving (i) and (ii) we get A = – 35, B = 51

$$\therefore \quad b_n = -35 + 51.4^n$$

which gives $$a_n^2 = -35 + 51.4^n \Rightarrow a_n = \pm\sqrt{51.4^n - 35}$$

This is the required solution for (1).

6.5 SOLUTION OF INHOMOGENEOUS LINEAR RECURRENCE RELATIONS

In this section we learn how to solve the inhomogeneous recurrence relations. Any solution to the recurrence relation is the sum of two parts.

The homogeneous solution and the particular solution when the right hand side of the recurrence relation is set to zero. The solution is called a homogeneous solution.

The solution which satisfies the recurrence relation with $f(r)$ on the right hand side is called the particular solution. The homogeneous solution is denoted by $a_r^{(h)}$

and the particular solution is denoted by $a_r^{(p)}$. We follow the same procedure as in solving homogeneous recurrence relations for determining the homogeneous solution, but there is no general procedure for determining the particular solution. It depends on the nature of $f(r)$.

The methods are described in terms of examples. In all these methods (explained below), we guess the general form of the particular solution.

6.5.1 When $f(r)$ is a Constant

In this case the particular solution is also a constant.

Example. Find the particular solution of the recurrence relation

$$a_r - 7a_{r-1} + 12a_{r-2} = 1 \qquad ...(1)$$

Solution. $f(r)$ (i.e. R.H.S. of (1)) is a constant, therefore the particular solution is a constant. Substituting p into the given equation. We get

$$p - 7p + 12p = 1 \Rightarrow 6p = 1$$

$$\Rightarrow p = 1/6$$

The particular solution is $a_r^{(p)} = 1/6$.

Example 2. Solve the recurrence relation

$$a_r - 6a_{r-1} + 8a_{r-2} = 9 \qquad \text{where } a_0 = 10 \text{ and } a_1 = 25$$

Solution. Here $f(r) = 9$, is a constant

The characteristic equation (polynomial) of the given relation is

$$\alpha^2 - 6\alpha + 8 = 0$$

$$\Rightarrow \qquad (\alpha - 2)(\alpha - 4) = 0$$

$$\therefore \qquad \alpha = 2 \text{ or } \alpha = 4$$

$$a_r^{(h)} = c_1 \cdot 2^n + c_2 \cdot 4^n.$$

$$f(r) = a = a(1)^n$$

$f(r)$ is a constant. The particular solution is of the form p. Substituting $a_r^p = p$ into the given recurrence relation, we get,

$$p - 6p + 8p = 9 \Rightarrow 3p = 9$$

$$\text{or} \qquad p = 3 \quad \therefore a_r^{(p)} = 3$$

$\therefore$ The general solution of the recurrence relation is

$$a_r = c_1\, 2^n + c_2\, 4^n + 3$$

Now using the initial conditions $a_0 = 10$, $a_1 = 25$, we get

$$a_0 = c_1 + c_2 + 3 = 10 \qquad ...(i)$$

$$a_1 = 2c_1 + 4c_2 + 3 = 25 \qquad ...(ii)$$

Solving (i) and (ii) we get $c_1 = 3$, $c_2 = 4$

$\therefore$ The general solution of the given recurrence relation is

$$a_r = 3 \cdot 2^n + 4 \cdot 4^n + 3$$

6.5.2 When *f*(*r*) is a polynomial in *r* and is of degree *m*.

In this case the particular integral is of the form

$$p_1 r^m + p_2 r^{m-1} + \ldots + p_{m+1}$$

Example 1. Find the particular solution of $a_r + 5a_{r-1} + 6a_{r-2} = 3r^2$

Solution. Let the form of the particular solution be

$$p_1 r^2 + p_2 r + p_3$$

where p_1, p_2 and p_3 are constants to be determined.

Substituting the expression into the left hand side of the given recurrence relation we get

$$p_1 r^2 + p_2 + p_3 + 5p_1(r-1)^2 + 5p_2(r-1) + 5p_3$$
$$+ 6p_1 (R-2)^2 + 6p_2 (r-1) + 6p_3 = 3r_2$$

Simplifying, we get

$$12p_1 r^2 - (34p_1 - 12p_2)\, r + (29p_1 - 17p_2 + 12p_3) = 3r^2$$

$\therefore$ We have $\quad 12p_1 = 3 \qquad \ldots(1)$

$$34p_1 - 12p_2 = 0 \qquad \ldots(2)$$

$$29p_1 - 17p_2 + 12p_3 = 0 \qquad \ldots(3)$$

Solving the equation (1), (2) and we get

$$p_1 = \frac{1}{4},\ p_2 = \frac{17}{24},\ p_3 = \frac{115}{288}$$

$$a_r^{(p)} = \frac{1}{4} r^2 + \frac{17}{24} r + \frac{115}{288}$$

6.5.3 When *f*(*r*) is of the form λ^r and λ is not a Characteristic Root of the Recurrence Relation

In this case the particular solution is of the form $p\lambda^r$ further more, when $f(r)$ is of the form

$$(p_1 r^m + p_2 r^{m-1} + \ldots + p_{m+1})\lambda^r$$

The corresponding particular solution is of the form

$$(p_1 r^m + p_2 r^{m-1} + \ldots + p_{m+1})\lambda^r$$

where λ is not a characteristic root of the recurrence relation.

When λ is characteristic root with multiplicity p and $f(r)$ is of the form $(b_1 r^m + b_2 r^{m-1} + ... + b_{m+1})\lambda^r$, then the particular solution is of the form

$$r^{p-1}(p_1 r^m + p_2 r^{m-1} + ... + p_{m+1})\lambda^r$$

Example 4. Find the particular integral of

$$a_r + 3a_{r-1} - 10a_{r-2} = r + 1$$

Solution. $f(r)$ (i.e. R.H.S.) is a polynomial of degree 1. Let particular integral be of the form $p_1 r + p_2$ where p_1, p_2 are constants to be determined.

From the given equation we have

$$p_1 r + p_2 + 3[p_1(r-1) + p_2] - 10\,[p_1(r-2) + p_2] = r + 1$$

$$\Rightarrow \quad p_1 r + p_2 + 3p_1 r - 3p_1 + 3p_2 - 10p_1 r + 20p_1 - 10p_2 = r + 1$$

$$\Rightarrow \quad (p_1 + 3p_1 - 10p_1)r + (p_2 - 3p_1 + 3p_2 + 20p_1 - 10p_2) = r + 1$$

$$\Rightarrow \quad -6p_1 r + (-6p_2 + 17p_1) = r + 1$$

Equating the coefficient of like terms, we get

$$-6p_1 = 1 \qquad ...(1)$$

$$-6p_2 + 17p_1 = 1 \qquad ...(2)$$

Solving (1) and (2), we get $p_1 = -1/6, \quad p_2 = -23/6$

$$\therefore \quad a_r^{(p)} = -\frac{1}{6}r + \left(-\frac{23}{6}\right)$$

Example 2. Find the general solution of

$$a_r - 7a_{r-1} + 10_{r-2} = 7.3^r, r \geq 2$$

Solution. The characteristic equation is

$$\alpha^2 - 7\alpha + 10 = 0$$

$$\Rightarrow \quad (\alpha - 2)(\alpha - 5) = 0$$

$$\therefore \quad \alpha = 2, \text{ or } \alpha = 5$$

The homogeneous solution is

$$a_r^{(h)} = A_1 2^r + A_2 5^r$$

Let $P \cdot 3^r$ be the particular solution of the given recurrence relation. Substituting $P \cdot 3^r$ for a_r in the given recurrence relation, we get

$$P \cdot 3^r - 7P \cdot 3^{r-1} + 10P \cdot 3^{r-2} = 7 \cdot 3^r$$

$$\Rightarrow \quad -2 \cdot p^r = 7 \cdot 3^2$$

$$\Rightarrow \quad p = -63/2$$

$\therefore$ The particular solution is $a_r^{(p)} = (-63/2)3^r$

$\therefore$ The general solution of the given recurrence relation is

$$a_r = a_r^{(h)} + a_r^{(p)} = A_1 2^r + A_2 5^r - (-63/2)\cdot 3^r$$

Example 3. Solve $a_r - 10a_{r-1} + 25a_{r-2} = 2^n$

Solution. The homogeneous equation is

$$a_r - 10a_{r-1} + 25a_{r-2} = 0$$

The characteristic equation is

$$\alpha^2 - 10\alpha + 25 = 0$$

$$\Rightarrow \quad (\alpha - 5)^2 = 0$$

$$\therefore \quad \alpha = 5, 5$$

The homogeneous solution is

$$a_r^{(h)} = (A_1 + A_2 r)\cdot 5^r$$

The multiplicity of 5 is 2, and the right hand side of the given recurrence relation is 2^r.

2 is not a characteristic root of the given relation.

Let $\quad a_r^{(p)} = p\cdot 2^r$

Substituting; into the given recurrence relation we get

$$p\cdot 2^r - 10p\cdot 2^{r-1} + 25\,p2^{r-2} = 2^r$$

$$\Rightarrow \quad 2^r\left(p - \frac{10p}{2} + \frac{25p}{4}\right) = 2^r$$

$$\Rightarrow \quad \frac{4p - 20p + 25p}{4} = 1 \Rightarrow 9p = 4$$

$$\Rightarrow \quad = 4/9$$

$$\therefore \quad a_r^{(p)} = \frac{4}{9}\cdot 2^r$$

$\therefore$ The general solution of the given recurrence relation is

$$a_r = a_r^{(h)} + a_r^{(p)} = (A_1 + A_2 r)5^r + \frac{4}{9}\cdot 2^r$$

Example 4. Solve $a_r - 6a_{r-1} + 8a_{r-2} = r\cdot 4^r$, where $a_0 = 8$, and $a_1 = 22$

Solution. The given recurrence relation is

$$a_r - 6a_{r-1} + 8a_{r-2} = r - 4^r \qquad ...(1)$$

The characteristic equation is

$$\alpha^2 - 6\alpha + 8 = 0$$

$$\Rightarrow \qquad (\alpha - 2)(\alpha - 4) = 0$$

$\therefore$ The characteristic roots are 2, 4.

The homogeneous solution is $a_r^{(h)} = A_1 \cdot 2^n + A_2 \cdot 4^r$

Hence 4 is characteristic root with multiplicity 1, the particular solution takes the form

$$r(p_1 + p_2 r) \cdot 4^r$$

substituting this expression into (1) we get

$$16r(p_1 + p_2 r) - 24(r-1)(p_1 + p_2(r-1)] + 8(r-2)(p_1 + p_2(r-2) = 16r \quad ...(2)$$

The above expression holds for all values of r and in particular when $r = 0$

Substituting $r = 0$ into (2) we get,

$$p_1 + p_2 = 0 \qquad ...(3)$$

Putting $r = 1$into (2) we get,

$$P_1 + 3p_2 = 2 \qquad ...(4)$$

Solving (3) and (4) we obtain

$$p_1 = -1, p_2 = 1$$

$\therefore$ The particular solution is $a_r^{(p)} = r(-1 + r) \cdot 4^r$

or $\qquad a_r^{(p)} = r(r-1) \cdot 4^r$

$\therefore$ The general solution of (1) is

$$a_r = a_r^{(h)} + a_r^{(p)} = A_1 \cdot 2^r + A_2 \cdot 4^r + r(r-1) \cdot 4^r \qquad ...(5)$$

The initial condition $a_0 = 8$, $a_1 = 22$, give $A_1 = 5$ and $A_2 = 3$

$\therefore$ The general solution is

$$a_r = r(r-1) \cdot 4^r + 5 \cdot 2^r + 3 \cdot 4^r$$

Example 5. Find the particular integral of

$$a_n + a_{n-1} = 3n \cdot 2^n$$

Solution. The general form of the particular solution is $(p_1\ n + p_2)2^r$. Substituting: into

$$a_n + a_{n-1} = 3n \cdot 2n, \text{ we get}$$

$$(p_1 n + p_2)2^n + (p_1(n-1) + p_2)\, 2^{n-1} = 3n \cdot 2^n$$

Simplifying, we get

$$\left[\frac{3}{2}p_1 n + \left(-\frac{1}{2}p_1 + \frac{3}{2}p_2\right)\right]2^n = 3n\,2^n$$

$$\Rightarrow \qquad \frac{3}{2}p_1 n + \left(-\frac{1}{2}p_1 + \frac{3}{2}p_2\right) = 3n$$

Comparing, we get $\frac{3}{2}\,p_1 = 3$ ...(1)

and $$-\frac{1}{2}p_1 + \frac{3}{2}p_2 = 0$$ ...(2)

Solving the above equations, we get

$$p_1 = 2,\ p_2 = \frac{2}{3}$$

∴ The particular solution is

$$a_n^{(p)} = \left(2n + \frac{2}{3}\right)2^n$$

EXERCISE 6.2

1. Solve the recurrence relations
   (a) $a_r = 7a_{r-1} - 10a_{r-2}$, $a_0 = 4$, $a_1 = 17$
   (b) $a_r - 4a_{r-1} - 11a_{r-2} + 30a_{r-3} = 0$
   given initial conditions,
   $a_0 = 0$, $a_1 = -35$, $a_2 = -85$
   (c) $a_n - a_{n-1} - 6a_{n-2} = -30$, with $a_0 = 20$, $a_1 = -5$
2. Solve $s_k - 2s_{k-1} + s_{k-2} = 2$
3. Solve $a_n - 9a_{n-1} + 26a_{n-2} - 24a_{n-3} = 0$ for $n \geq 3$
4. Solve $a_r - 6a_{r-1} + 12a_{r-2} - 8a_{r-3} = 0$ by generating functions.
5. Write a general expression for $A(z) = {}^{P(z)}\!/_{Q(z)}$ specifying coefficient for $P(z)$ and $Q(z)$, where $A(z)$ generates the sequence $\{a_r\}$ and a_r satisfies the following recurrence relations
   (i) $a_r - 5a_{r-1} + 8a_{r-2} - 4a_{r-3} = 0$, $r \geq 3$
   (ii) $a_r + 5a_{r-1} + 3a_{r-2} = 0$, for $n \geq 2$, $a_0 = 1$, $a = 2$
6. Solve $a_n - 9a_{n-1} + 20a_{n-2} = 0$, $n \geq 2$, $a_0 = -3$
   $a_1 = -10$

7. Solve $a_r - 7a_{r-1} - 12a_{r-2} = 0, n \geq 2$
8. Solve $a_n - 7a_{n-1} + 16a_{n-2} - 12a_{n-3} = 0, n \geq 3$
9. Solve
   (a) $a_n - 5a_{n-1} + 6a_{n-2} = 0, n \geq 2, a_0 = 1, a_1 = -2$
   (b) $a_n - 3a_{n-2} + 2a_{n-3} = 0$, for $n \geq 3$, $a_0 = 1, a_1 = 0, a_2 = 0$
10. Solve $a_r - 3a_{r-1} - 4a_{r-2} = 0$, for $r \geq 2$, $a_0 = a_1 = 1$
11. Solve $a_r - 4a_{r-1} - 12a_{r-2} = 0$ for $r \geq 2$, $a_0 = 5/2$, $a_1 = 8$
12. Solve $a_n - 5a_{n-1} + 6a_{n-2} = 2^n$
13. Solve $a_n - 6a_{n-1} + 8a_{n-2} = 3^n$, where $a_0 = 3$, and $a_1 = 7$
14. Solve $a_r - 5a_{r-1} = r$ for $r \geq 1$, $a_0 = 0$
15. Solve $5(n) - 6S(n-1) + 9S(n-2) = 3^{n+1}$
16. Solve $5(k) - 3S(k-1) - 4S(k-2) = 4^k$
17. Find the general solution of:
    $T(k) - 7T(k-1) + 10T(k-2) = K^2 + 1$
    where $T(0) = 4$, $T(1) = 17$

ANSWERS 6.2

1. (a) $a_r = 7.2^r + 3.5^r$
   (b) $a_r = 4(-3)^r + 1 \cdot 2^r + (-5) \cdot 5^r$
   (c) $a_n = 11(-2)^r + 4 \cdot 3^r + 5$
2. $S_k = 25 - 10k + k^2$
3. $a_n = A_1 \cdot 2^n + A_2 \cdot 3^n + A_3 \cdot 4^n$
4. $a_r = A_1 \cdot 2^r + A_2\,(r+1)2^r + A_3\,\dfrac{(r+2)\,(r+1)}{2} \cdot 2^r$
5. (a) $A(z) = \dfrac{a_0 + (a_1 - 5a_0)\,z + (a_2 - 5a_1 + 8a_0)z^2}{1 - 5z + 8z^2 - 4z^3}$

   (b) $A(z) = \dfrac{1 + 72}{1 + 52 + 3z^2}$
6. $a_n = 5 \cdot 4^n + 2 \cdot 5^n$
7. $A(z) = \dfrac{a_0 + (a_1 - 7a_0)\,z}{1 - 7z + 12z^2}$
8. $A(z) = \dfrac{a_0 + (a_1 - 7a_0)z + (a_2 - 7a_1 + 16a_0)z^2}{1 - 7z + 16z^2 - 12z^3}$

9. (a) $A(z) = \dfrac{1-7z}{1-5z+6z^2} = \dfrac{5}{1-2z} - \dfrac{4}{1-32}$, $a_n = 5\cdot 2^n - 4\cdot 3^n$

(b) $a_n = \dfrac{1}{12}\{-3 + 4\cdot 3^n - (-3)^n\}$

10. $a_r = \dfrac{2}{5}\cdot 4^r + \dfrac{3}{5}(-1)^r$

11. $a_r = \dfrac{5}{3}6^r + \dfrac{7}{3}(-2)^r$

12. $a_n = A_1\, 3^n + A_2\, 2^n + A_3\,(n+1)\cdot 2^n$

13. $a_n = (-9)3^n + 5\cdot 4^n + 7\cdot 2^n$

14. $a_r = \dfrac{(r+1)r}{2}$

15. $S(n) = (A_1 + A_2 n)\, 3^n + 1.5.n^2 3^n$

16. $S(k) = A_1(-1)k + A_2\, 4^k + 0.8k\, 4^k$

17. $T(k) = (-3)2^k + (3.5)\, 5^k + 3.468 + 1.625k + 0.25k^2$

CHAPTER 7

Graph Theory

7.1 INTRODUCTION

Graph theory is intimately related to many branches of mathematics. It is widely applied in subjects like computer technology, Communication Science, Electrical Engineering, Architecture, Physics, Operations Research etc. In the earlier stages it was called Slum Topology. Graph theory was born in 1736, with Euler's paper on Konigsberg bridge problem. Konigsberg bridge is the best known example in graph theory. It was a long pending problem. Euler solved this problem by means of a graph. Euler become father of graph theory. Kirchoff, Mobius, DeMorgan and Hamilton have laid strong foundations and contributed much to the development of the subject. In this chapter basic concepts and terms of graph theory have been introduced.

7.2 BASIC CONCEPTS

7.2.1 Graph

Definition 7.1. A graph G is a pair of sets (V, E); where V is a non-empty set. The set V is called the set of vertices and the set E is called the set of edges (or lines).

A graph may be represented by a diagram in which each vertex is represented by a point in the plane and each edge is represented by a straight line (or curve) joining the points. The objects shown in Fig. 7.1 are graphs.

In a graph $G = (V, E)$, the sets V and E are assumed to be finite sets. A vertex in a graph G is called a node, a point, or a junction. An edge in a graph G is called a line; a branch or an arc. If $G = (V, E)$ is a graph, then the number of vertices in G is denoted by $|V|$ (or $|V(G)|$) and the number of edges in G is denoted by $|E|$ (or $|E(G)|$). If $\{u, v\}$ is an edge in a graph G, then the vertices u and v are called adjacent vertices in G.

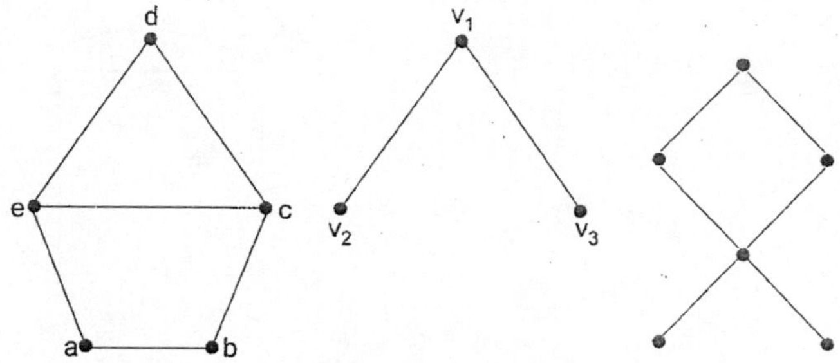

Fig. 7.1 Graph

7.2.2 Non-directed graph

Definition 7.2. Let $G = (V, E)$ be a graph. If the elements of E are unordered pairs of vertices of G, then G is called a non-directed graph.

The Graphs in Fig. 7.2 are non-directed graphs.

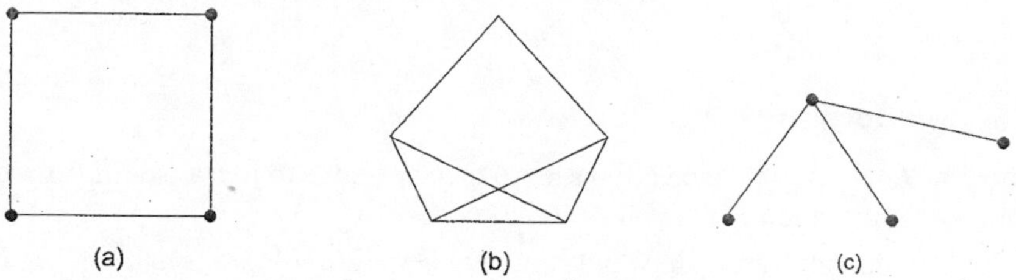

(a) (b) (c)

Fig. 7.2 Non-directed graphs

7.2.3 Directed Graph

Definition 7.3. Let $G = (V, E)$ be a graph. If the elements of E are ordered pairs of vertices of G, then G is called a directed graph (or a digraph).

If e is an edge in a directed graph G, joining the vertices u, v then $e = (u, v)$ is an ordered pair and is a directed edge in G. The edge e begins at u and ends at v. The vertex u is called the origin or the initial point of the directed edge e and v is called the destination or terminal point of e.

The graphs in Fig. 7.3 are directed graphs.

7.2.4 Self-Loop

Definition 7.3. An edge associated with the unordered pair $\{v_i, v_i\}$ where $v_i \in V$, of a graph $G = (V, E)$ is called a self-loop in the graph G.

i.e. A self-loop in the graph G is an edge joining a vertex to itself.

In Fig. 7.4, there is a self-loop on the vertex v.

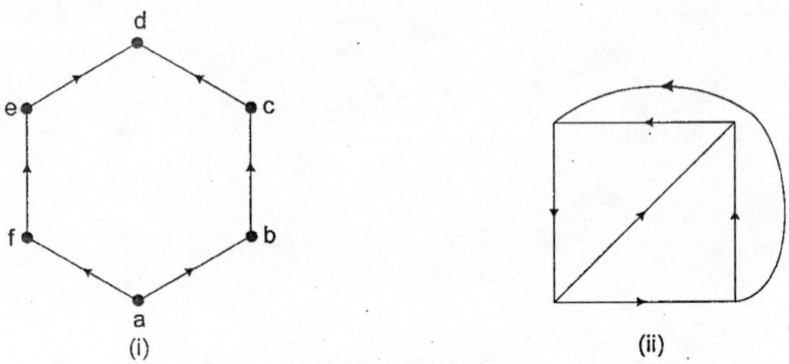

Fig. 7.3 Directed graph

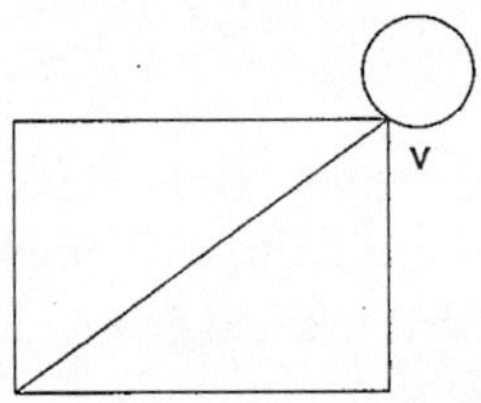

Fig. 7.4

7.2.5 Multigraph

Definition 7.4. A graph which allows more than one edge to join a pair of vertices is called a multigraph.

Fig. 7.5 is a multigraph in which we have $G = \{2(a, b), 3(b, c), (a, c), (a, d), (e, d)\}$.

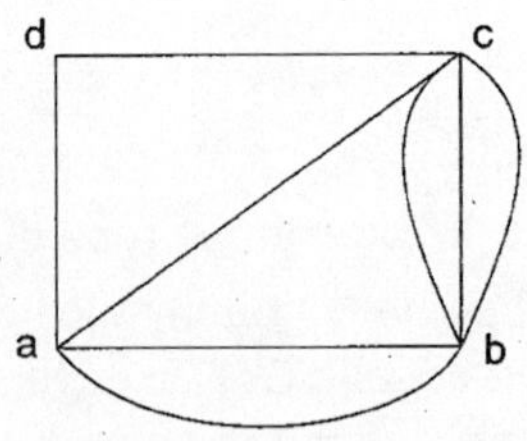

Fig. 7.5 A multigraph

7.2.6 Simple Graph (Loop Free Graph)

Definition 7.5. A group G with no self-loops is called a simple graph.

The graphs in Fig. 7.6 are simple graph.

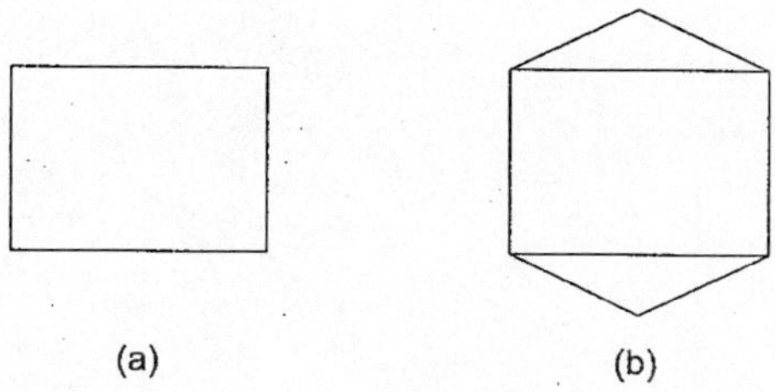

Fig. 7.6 Simple graph

7.2.7 Non-simple Graph

Definition 7.6. If a graph *G*, is not a simple graph, then *G* is called a non-simple graph (see Fig. 7.7).

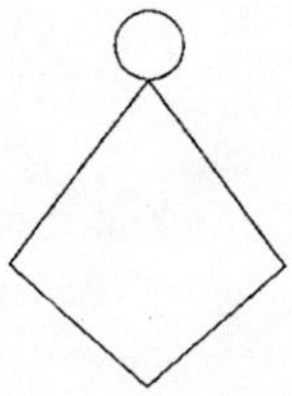

Fig. 7.7 Non-simple graph

7.2.8 Null Graph

Definition 7.8. A graph $G = (V, E)$ in which the set of edges *E* in empty is called a Null graph (see Fig. 7.8).

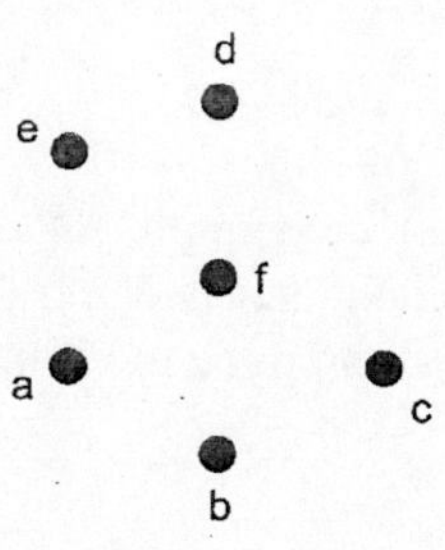

Fig. 7.8 Null graph

7.2.9 Weighted Graph

Definition 7.9. A graph *G* in which weights are assigned to every edge is called a weighted graph (Fig. 7.9).

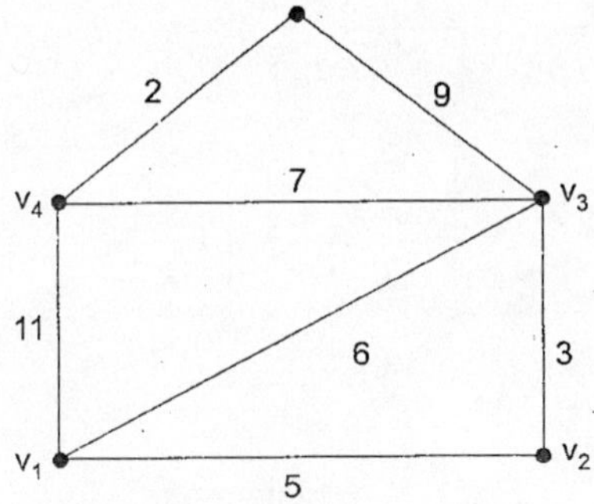

Fig. 7.9 Weighted graph

7.2.10 Finite Graph

Definition 7.10. A graph $G = (V, E)$ in which both $V(G)$ and $E(G)$ are finite sets, is called a finite graph.

7.2.11 Pseudograph

Definition 7.11. A graph having loops but no multiple edges is called a Pseudograph.

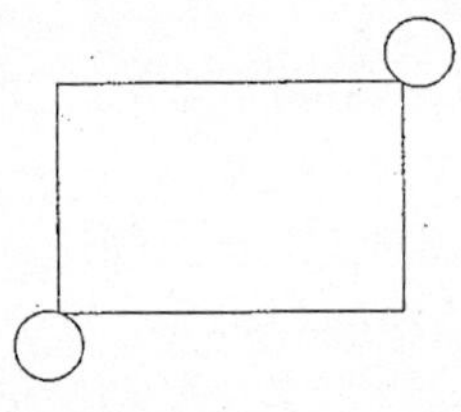

Fig. 7.10 Pseudograph

7.3 INCIDENCE AND DEGREE

Definition 7.12. Let G be a non-directed graph. An unordered pair $\{u, v\}$ where u, v are vertices of G, is an edge incident on u and v. If G is a directed graph, an edge (u, v) is said to be incident from u, and to be incident to v.

7.3.1 In-degree

Definition 7.13. In a directed graph, the number of edges ending at a vertex v of G is called the in-degree of the vertex v.

The in-degree of a vertex V in G is denoted by $\deg_G^+(v)$. (or by indeg (v)). The indegree of the vertex v_1 in the graph given below is 3. (see Fig. 7.11)

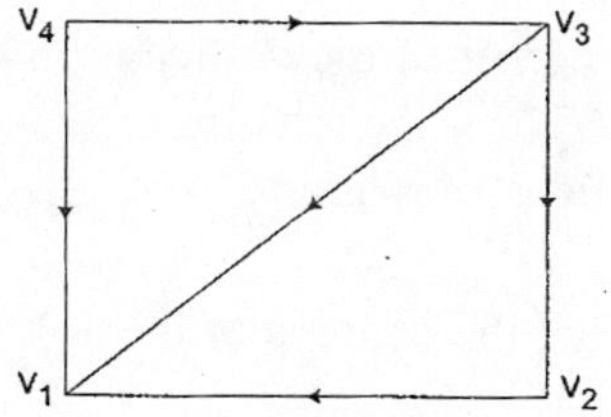

Fig. 7.11

7.3.2 Out-degree

Definition 7.14. Let *G* be a directed group and *v* be a vertex of *G*. The out-degree of *v* is the number of edges begining at *v* is called the out-degree of *v*.

The out-degree of the vertex *v* in *G* is denoted by $\deg_G^-(V)$ (or by out deg (*v*).

In Fig. 7.12 the out-degree of the vertex v_1 is 2.

Fig. 7.12

7.33 DEGREE OF A VERTEX

In a non-directed graph *G*, the degree of a vertex *v* is determined by counting each loop incident on *v* twice and each other edge once. The degree of *v* in *G* is denoted by deg(*v*) (or by *d*(*v*)) and is defined as follows.

Definition 7.15. The number of edges incident with a vertex *v* of a graph with self-loops counted twice is called the degree of a vertex *v*.

The degree of a vertex in a graph is sometimes referred to as its vabency.

Note. A vertex of odd degree in a graph is called odd vertex, and a vertex of even degree in a graph is called even vertex.

Example. In the Fig. 7.13, $\deg(v_1) = 3$, $\deg(v_2) = 1$, $\deg(v_3) = 1$, $\deg(v_4) = 1$

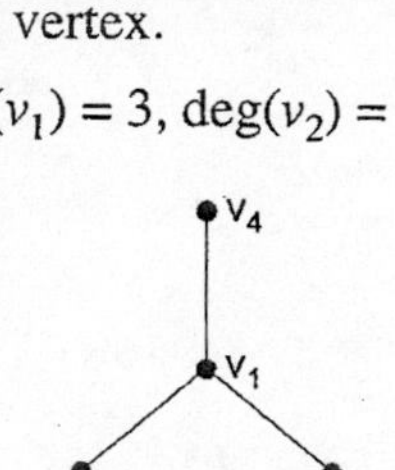

Fig. 7.13

Example. In the graph of Fig. 7.14, we have $\deg(v_1) = 3$, $\deg(v_2) = 4$, $\deg(v_3) = 2$, $\deg(v_4) = 3$.

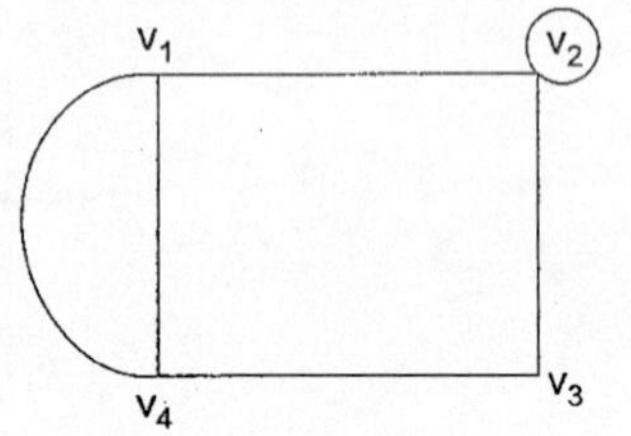

Fig. 7.14

The minimum of all the degrees or the vertices of a graph G is denoted by $\delta(G)$, and the maximum of all the degrees of the vertices of a graph G is denoted by $\Delta(G)$.

$\therefore$ $\delta(G)$ = minimum of all the degrees of the vertices of $G = (V, E)$

$= \min \{\deg (v_i) : v_i \in V\}$

and $\Delta(G)$ = maximum of all the degree of the vertices of $G = (V, \in)$

$= \max \{\deg(v_i) : v_i \in V\}$

7.3.4 Isolated Vertex

Definition 7.16. A vertex of degree zero in a graph is called an isolated vertex.

An Isolated vertex, vertex in a graph G has no edges incident with it.

The vertex v_1 is an isolated vertex in Fig. 7.15.

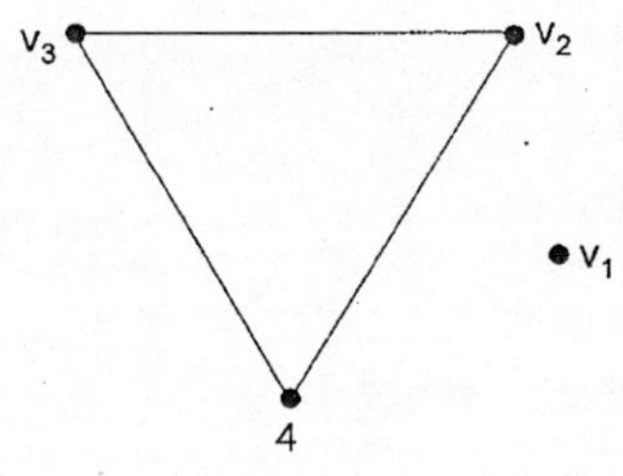

Fig. 7.15 Isolated vertex in a graph

7.3.5 Pendant Vertex

Definition 7.17. A vertex of a graph with degree one is called a pendant vertex (or an end vertex). In the graph shown in Fig. 7.16, the vertices v_1 and v_4 are pendant vertices.

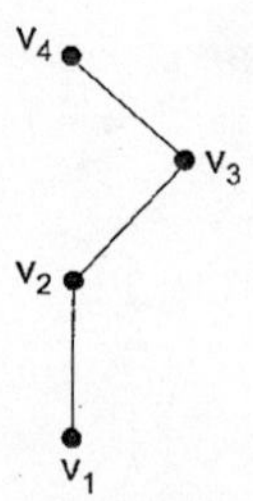

Fig. 7.16

7.3.5 k-regular Graph

Definition 7.18. A graph G is said to be k-regular, if every vertex of G has degree k.

For a k-regular graph, we have

$$\delta(G) = \Delta(G) = k$$

i.e., all the vertices of G have the same degree k.

A regular graph degree zero has no edges. In a regular graph of degree 1, every component contains exactly one line. If G is a 2-regular graph, then every component of G has a cycle. If G is a 3-regular graph, then G is called a cubic graph.

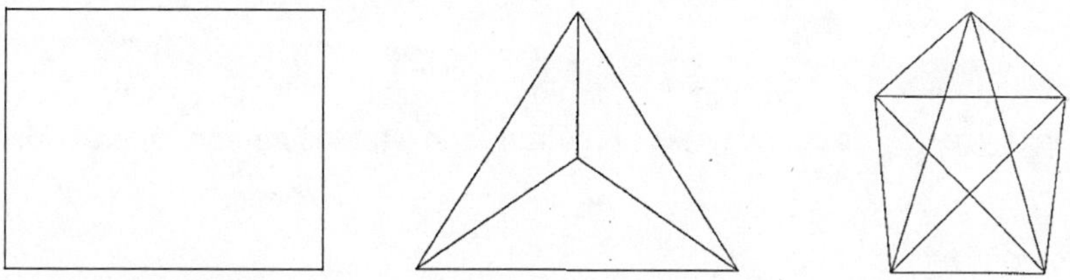

(a) 2-regular graph (b) 3-regular graph (c) 4-regular graph

Fig. 7.17

7.4 LABELED GRAPH

7.4.1 Vertex Labeling

Definition 7.19. A vertex labeling of a graph $G = (V, E)$, is a function

F: $V(G) \to D$, where D is some domain of labels.

7.4.2 Edge labeling

Definition 7.20. An edge labeling of a graph G is a function f: $E(G) \to D$, where D is some domain of labels.

If G is a labeled graph, then each vertex of G is assigned a unique label.

The graph in Fig. 7.18 is a labeled graph.

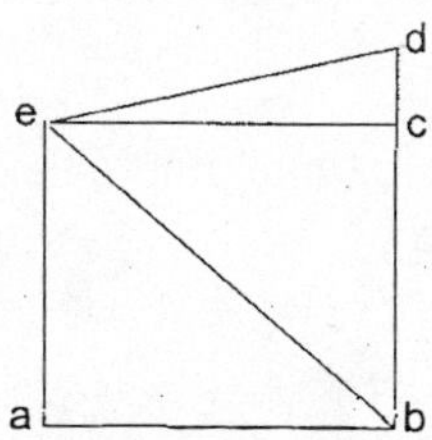

Fig. 7.18

7.5 COMPLETE GRAPH

Definition 7.21. A simple graph G, in which every pair of distinct vertices are adjacent is called a complete graph. If G is a complete graph with n vertices, then it is denoted by k_n. Fig. 7.19 shows k_3, k_4 and k_5.

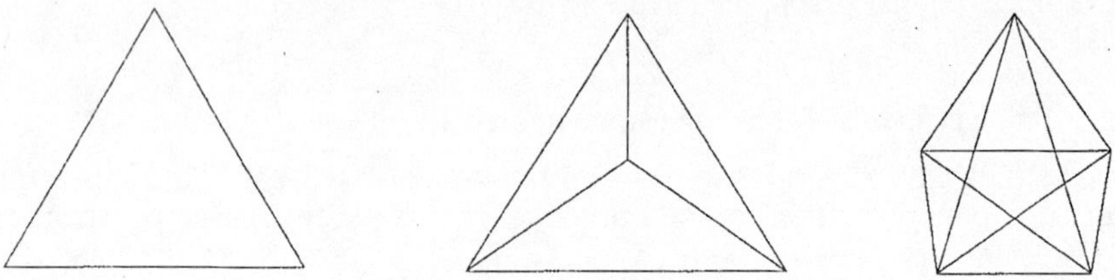

(a) k_3 (b) k_4 (c) k_5

Fig. 7.19 Complete graphs

In a complete graph, there is an edge between every pair of vertices. The complete graph k_n is called $(n - 1)$ regular graph, and the number of edges in k_n is $\frac{n(n-1)}{2}$.

7.6 Order of a Graph

Definition 7.22. If $G = (V, E)$ is a finite graph, then the number of vertices denoted by $|V|$ is called the order of G.

Thus, the cardinality of the vertex set V is called the order of G.

Example. The graph shown in Fig. 7.20 is of order 6.

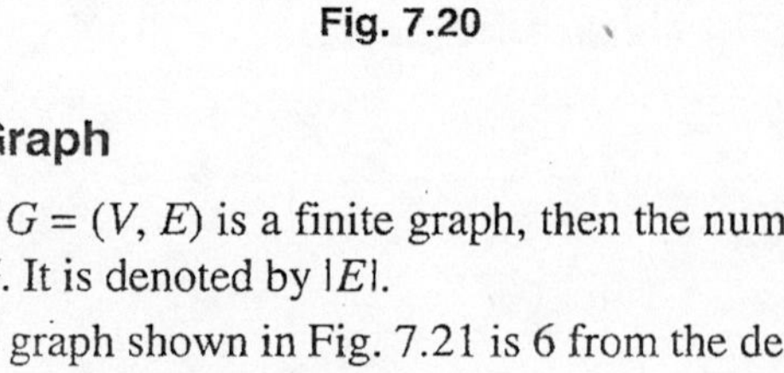

Fig. 7.20

7.7 Size of a Graph

Definition 7.23. If $G = (V, E)$ is a finite graph, then the number of edges in G is called the size of G. It is denoted by $|E|$.

The size of the graph shown in Fig. 7.21 is 6 from the definition the size of a graph is the cardinality of the edge set E.

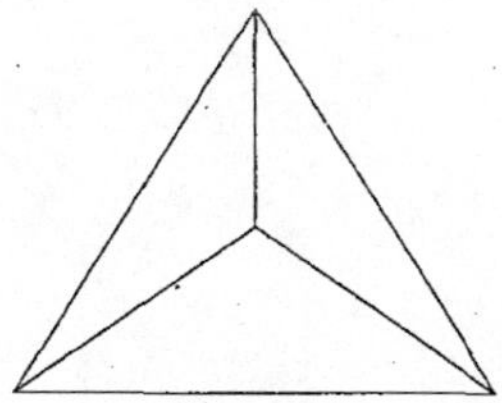

Fig. 7.21

We shall refer to a graph order m and size n as an (m, n) graph.

If G is (p, q) graph, then has p vertices and q edges.

Example. Let $V = \{v_1, v_2, v_3, v_4\}$

and $E = \{(v_1, v_2), (v_1, v_3), (v_1, v_4)\}$

Then $G = (V, E)$ is a (4, 3) graph. It can be represented as shown in Fig. 7.22.

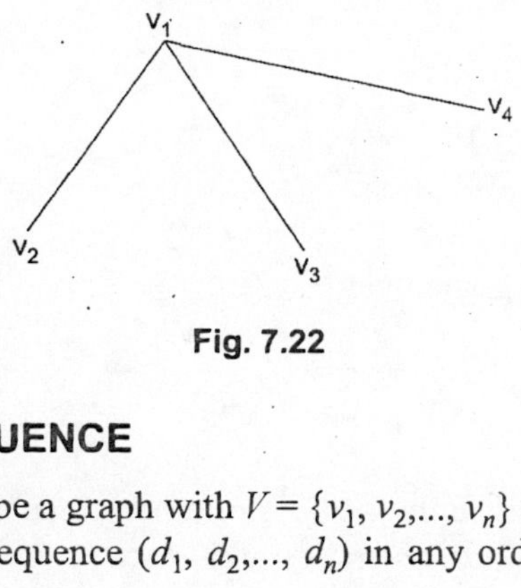

Fig. 7.22

7.8 DEGREE SEQUENCE

Definition 7.24. Let G be a graph with $V = \{v_1, v_2, ..., v_n\}$ as the vertex set and let $d_i = \deg(v_i)$, then the sequence $(d_1, d_2, ..., d_n)$ in any order is called the degree sequence of G.

The vertices of a graph G can be ordered so that the degree sequence is monotonically increasing, and

$$\delta(G) = d_1 \leq d_2 \leq d_3 \leq ... \leq d_n = \Delta(G)$$

The set of distinct non-negative integers occurring in a degree sequence of a graph G is called its degree set. If G_1 and G_2 are graphs having the same degree sequence, then G_1 and G_2 are said to be equivalent. It is customary to denote the degree sequence in power notation. If $(Z, 2, 2, 3, 3, 4, 5, 5, 6)$ is the degree sequence of a graph G, then it is represented as $2^3\ 3^2\ 4^1\ 5^2\ 6^1$ the degree set being $\{1, 2, 3\}$.

Example. Consider the graph shown in Fig. 7.23

We have $\deg(v_1) = d_1 = 0$, $\deg(v_2) = d_2 = 2$, $\deg(v_3) = d_3 = 1$

$\deg(v_4) = d_4 = 2$, $\deg(v_5) = d_5 = 4$, $\deg(v_6) = d_6 = 1$

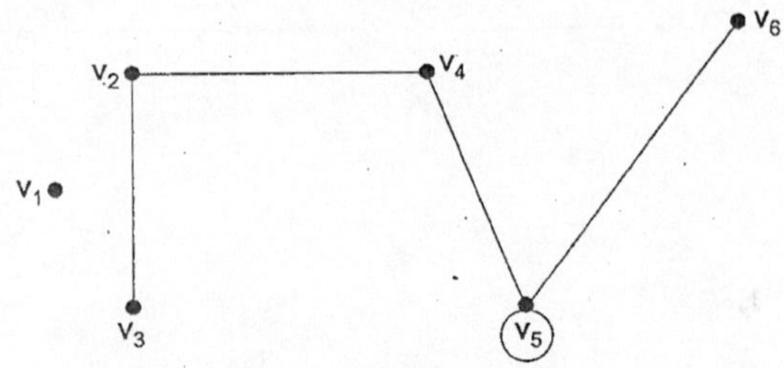

Fig. 7.23

the vertex v_1, is not adjacent to any other vertex of G. Therefore v_1 is an isolated vertex in G.

The degree sequence of G is (0, 1, 1, 2, 2, 4).

There are two vertices of odd degree in G.

Theorem 7.1. (The first Theorem of graph theory or the handshaking theorem)

(i) The sum of degree of the vertices of a nondirected graph $G = (V, E)$ is twice the number of edges in G.

i.e. $$\sum_{i=1}^{n} \deg(v_i) = 2\,|E|$$

where $v_1, v_2, \ldots, v_n \in V$

and (ii) if G is a directed graph, then

$$\sum_{i=1}^{n} \deg^{+}_{G}(v_i) = \sum_{i=1}^{n} \deg^{-}_{G}(v_i)$$

Solution. Let $v_1, v_2, \ldots, v_n$ be the vertices of a non-directed graph G. Then each edge of G is incident with two vertices and hence 'contributes 2 to the sum of degree of all the vertices of G.

∴ The sum of degree of all the vertices in G is twice the number of edges in G.

i.e. $$\sum_{i=1}^{n} \deg_G(v_i) = 2\,|E|$$

(ii) Let G be directed graph and e be an edge associated with a vertex pair (v_p, v_q). The edge e contributes one to the out-degree of v_p and one to the in-degree of v_q. This is true for all the edges in G. Hence

$$\sum_{i=1}^{n} \deg^{+}_{G}(v_i) = \sum_{i=}^{n} \deg^{-}_{G}(v_i) = |E|$$

Corollary 7.1.1. In a non-directed graph the number of odd degree vertices is even.

Proof. Let $G = (V, E)$ be a non-directed graph. Let w denote the set of odd vertices in G and U denote the set of even vertices in G.

Then $$\sum_{v_i \in V} \deg_G(v_i) = \sum_{v_i \in w} \deg(v_i) + \sum_{v_i \in U} \deg(v_i)$$

$$\Rightarrow \quad \sum_{v_i \in V} \deg(v_i) - \sum_{v_i \in U} \deg(v_i) = \sum_{v_i \in w} \deg(v_i) \qquad \text{...(i)}$$

$$\sum_{v_i \in V} \deg(v_n) \text{ is even and } \sum_{v_i \in U} \deg(v_i) \text{ is even, therefore}$$

$$\sum_{v_i \in V} \deg(v_i) - \sum_{v_i \in U} \deg(v_i) \text{ is even}$$

i.e. L.H.S. of (1) is even, therefore R.H.S. of (1) is even.

i.e. the number of odd degree vertices in G is even.

Corollary 7.1.2. If $k = \delta(G)$, is the minimum degree of the vertices of a non-directed graph $G = (V, E)$, then

$$k\,|V| \leq 2\,|E|$$

in particular, if G is a k-regular graph, then

$$k\,|V| = 2\,|E|$$

If G is a multigraph then the multiple edges joining any two vertices of G are called parallel edges. If G is a simple graph, then G is without parallel edges and self-loops.

Let G be a simple graph with one vertex. The maximum number of edges in G is zero i.e. $(1 - 1)$. Therefore the maximum degree of the vertex in a simple graph G with one vertex is zero. If G is a simple graph with 2 vertices then the maximum degree of a vertex in G is one i.e. $(2 - 1)$. In general, it can be shown that the maximum degree of any vertex in a simple group G with n vertices is $(n - 1)$. This can be stated in the form of a theorem as follows.

Theorem 7.2. The maximum degree of any vertex in a simple graph with n vertices is $(n - 1)$.

7.9 SOLVED EXAMPLES

Example 7.9.1 Draw a graph representing the problem of three houses and three utilities say water, gas and electricity.

Solution. Let H_1, H_2 and H_3 denote the houses. The utilities water, gas and electricity be denoted by W, G and E respectively.

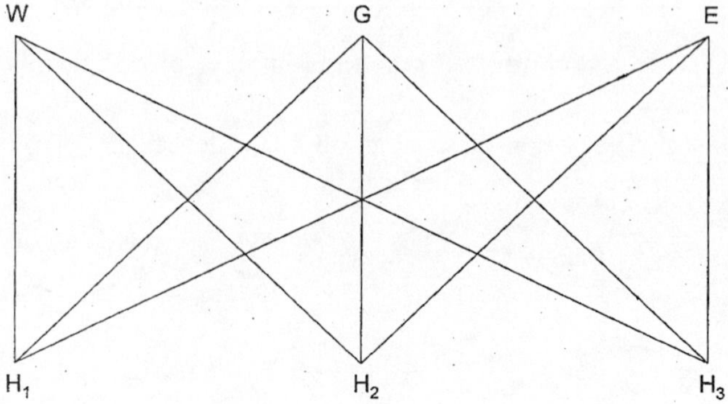

Fig. 7.24

Example 7.9.2. Draw the graphs of the chemical compounds
(a) C_2H_6 and (b) C_4H_{10}

Solution. (a) The graph of C_2H_6 is shown in 7.25(a)
(b) The graph of C_4H_{10} is shown in 7.25(b).

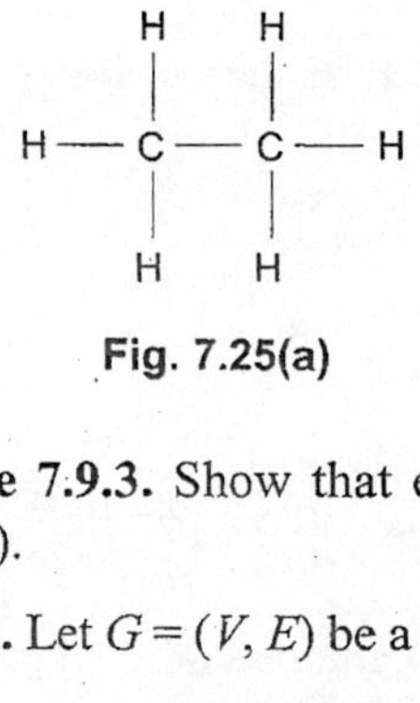

Fig. 7.25(a)

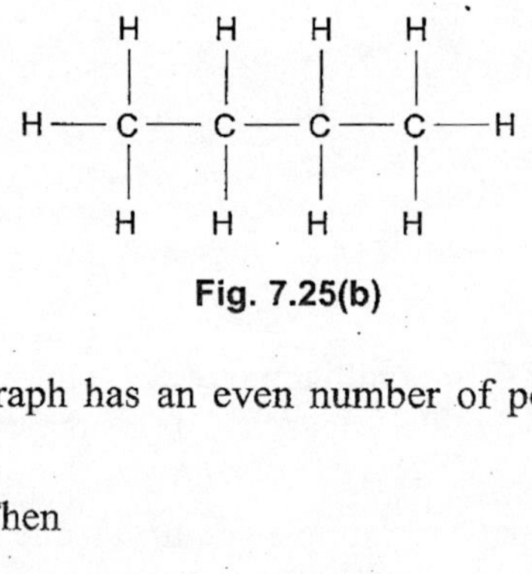

Fig. 7.25(b)

Example 7.9.3. Show that every cubic graph has an even number of points (vertices).

Solution. Let $G = (V, E)$ be a cubic graph. Then

$$\sum_{v_i \in V} \deg(v_i) = 3\,|V| = 2 \text{ (number of edges)}$$

$\Rightarrow$ $3\,|V|$ is even

$\Rightarrow$ $|V|$ is even

Example 7.9.4. Represent the graph

$$G = \{(1, 2, 3, 4), |x-4| : |x-4| \leq 1\}$$

Solution. We have

$$V = \{1, 2, 3, 4\}$$

$$E = \{(1, 1), (1, 2), (2, 2), (2, 3), (3, 3), (3, 4)\}$$

and $G = (V, E)$ is a non-directed graph. It can be represented as shown in Fig. 7.26.

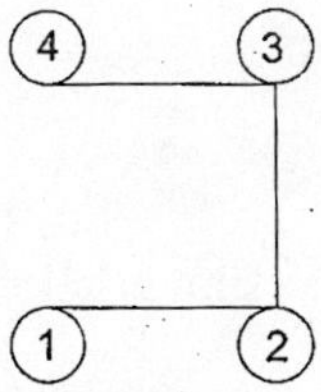

Fig. 7.26

Example 7.9.5. For the graph shown in Fig. 7.27, verify

$$\sum \deg(v_i) = 2|E|$$

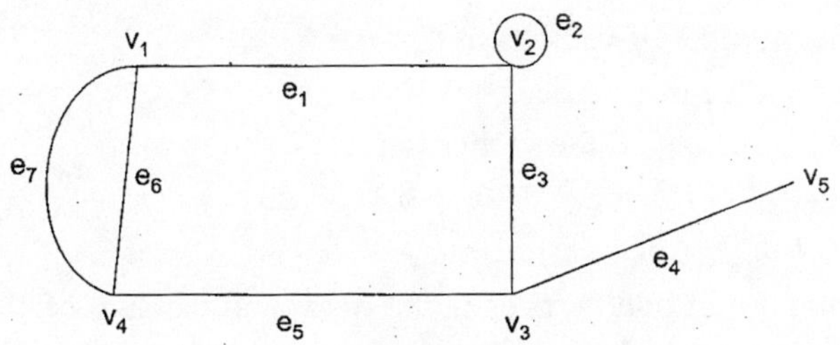

Fig. 7.27

Solution. We have, $V = \{v_1, v_2, v_3, v_4, v_5\}$

$$E = \{e_1, e_2, e_3, e_4, e_5, e_6, e_7\}$$

$$d_1 = \deg(v_1) = 3, \quad d_2 = \deg(v_2) = 4, \quad d_3 = \deg(v_3) = 3$$

$$d_4 = \deg(v_4) = 3, \quad d_5 = \deg(v_5) = 1.$$

$$\therefore \qquad \sum \deg(v_i) = d_1 + d_2 + d_3 + d_4 + d_5 = 3 + 4 + 3 + 3 + 1 = 14$$

and $|E|$ = number of edges in $G = 7$

$$\therefore \qquad \sum \deg(v_i) = 14 = 2\,|E|$$

verified.

Example 7.9.6. G is a graph with p vertices and q edges.

Show that $\delta \le \dfrac{2q}{p} \le \Delta$

where $\qquad \delta = \delta(a) = \min\{\deg(v_i) : v_i \in V(G)\}.$

$$\Delta = \Delta(G) = \max\{\deg(v_i) = v_i \in V(G)\}$$

Solution. Let $V = \{v_1, v_2, \ldots, v_p\}$

Then, we have, $\delta \leq \deg(v_i) \leq \Delta$ for all i

Hence $$p\delta \leq \sum_{i=1}^{p} \deg(v_i) \leq p\,\Delta$$

$\Rightarrow$ $$p\delta \leq 2q \leq p\,\Delta \qquad (\because \text{ sum of all degrees of vertices in } G \text{ is even})$$

$\Rightarrow$ $$\delta \leq \frac{2q}{p} \leq \Delta$$

Example 7.9.7. A sequence $d = (d_1, d_2, ..., d_n)$ is graphic, if there is a simple non-directed graph with the degree sequence d. Show that the following sequences are not graphic

(a) (2, 3, 4, 5, 6, 7) (b) (2, 2, 4)

Solution. (a) The number of vertices in $G = n = 6$

The number odd degree vertices = 3

The sum of degrees of all the vertices in G

$$= 2 + 3 + 5 + 6 + 7 = 23$$

and the maximum degree of a vertex in $G = 7$

Since the number of odd vertices is not even, and the sum of all degrees of vertices in G is not even, the given degree sequence is not graphic.

Also we have line maximum degree of v_i in $G = 7 \neq 6 - 1 = 5$

$\therefore$ The given degree sequence is not graphic.

(b) The graph has 3 vertices. The maximum degree in the graph must be 3 – 1 = 2. But the maximum degree in the given degree sequence of the graph is 4. Hence the given degree sequence is not graphic.

Example 7.9.8. Draw a picture of the following graph and state whether it is directed or non-directed and whether it is simple.

$$G = (V, E) \text{ where } V = \{a, b, c, d, e\} \text{ and}$$
$$E = \{(a, b), (a, c), (a, d), (a, e), (e, e), (c, a), (a, a), (b, c), (c, c)\}$$

Solution. The graph G is directed and is not simple since it has multiple edges.

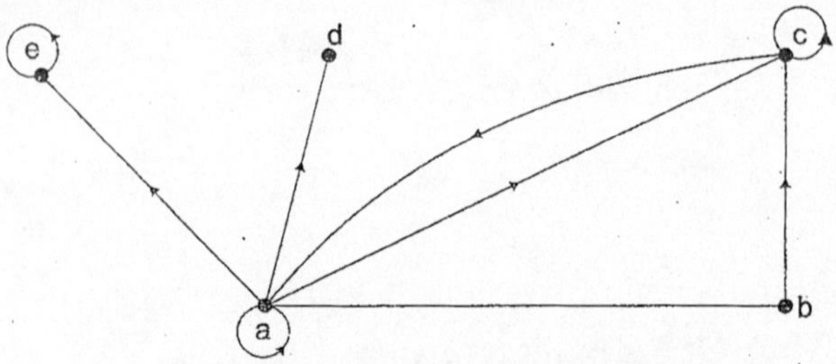

Fig. 7.28

Example 7.9.9. Find the order and size of the graph shown in Fig. 7.29.

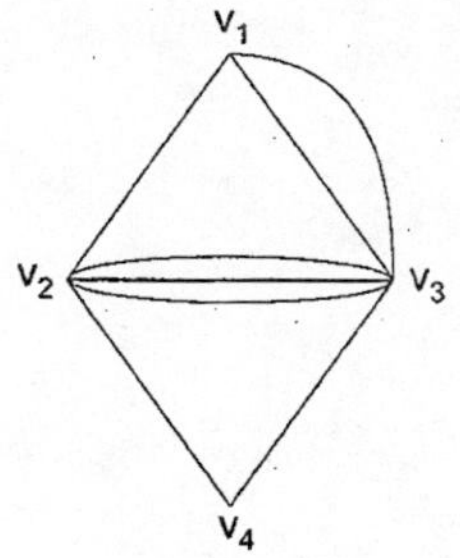

Fig. 7.29

Solution. Order of $G = |V| = 4$

Size of the graph $= |E| = 8$

Example 7.9.10. Given an example of (a) A simple graph (b) A pseudo graph (c) A multigraph.

Solution. (a) Fig 7.30 (a) is a simple graph (b) Fig. 7.30 (b) is a pseudo graph and (c) Fig 7.30 (c) is a multigraph.

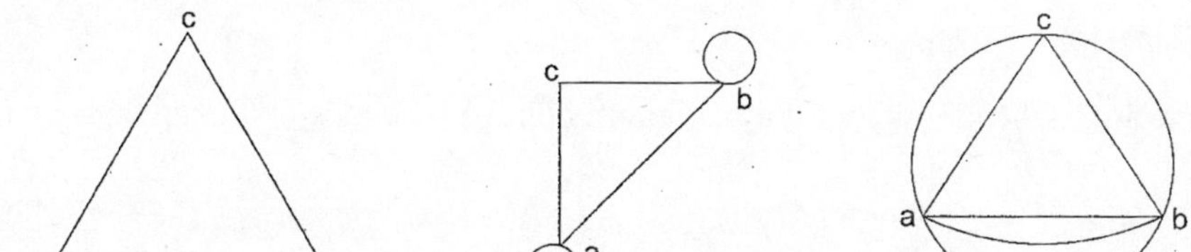

(a) A Simple graph (b) A pseudo graph (c) A multigraph

Fig. 7.30

Example 7.9.11. Draw a non-simple graph with degree sequence (1, 1, 3, 3, 3, 4, 6, 7).

Solution. G is non-directed; therefore G presents self loops in it. And it can be drown as shown in Fig. 7.31.

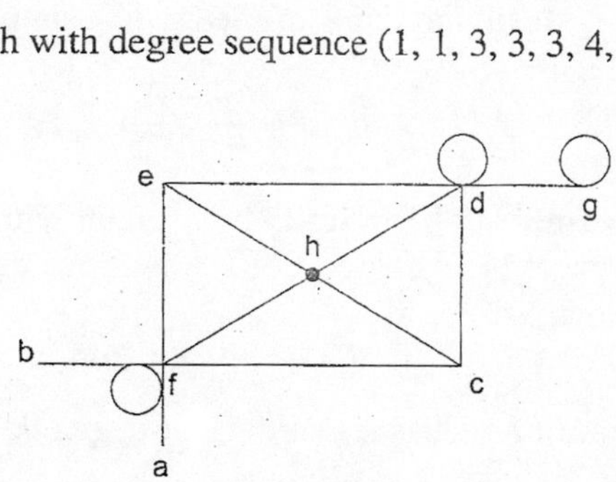

Fig. 7.31

Example 7.9.12. Suppose G is a non-directed graph with 12 edges. If G has 6 vertices each of degree 3 and the rest have degree less than 3. What is the minimum number of vertices G can have?

Solution. Number of edges in $G = 12$

$$\therefore \qquad \sum \deg(v_i) = 2|E| = 2 \times 12 = 24$$

We have 6 vertices of degree 3. Let x denotes the number of vertices each of whose degree is less than 3. Then,

$$\sum \deg(v_i) < 6.3 + 3x$$

$$\Rightarrow \qquad 24 < 18 + 3x$$

$$\Rightarrow \qquad 3x > 6$$

$$\Rightarrow \qquad x > 2$$

The least positive integer for which the inequality holds is $x = 3$

$\therefore$ Hence, the minimum number of vertices G can have is $3 + 6 = 9$.

Example 7.9.13. A non-directed graph G has 8 edges. Find the number of vertices, if the degree of each vertex in G is 2.

Solution. Given $|E| = 8$. Let n denotes the number of vertices in G.

we have $$\sum_{v_i \in V} \deg(v_i) = 2|E|$$

$$\Rightarrow \qquad 2n = 2 \times 8 \qquad \text{(Since degree of each vertex is 2)}$$

$$\Rightarrow \qquad n = 8$$

Example 7.9.14. Show that the maximum number of edges in a simple graph is $\dfrac{n(n-1)}{2}$.

Solution. Let G be a simple graph with n vertices. Since G is simple, G has no self-loops and parallel (multiple) edges. If G has maximum number of edges then there will be an edge between every pair of vertices in G. We can draw edge in G in n_{c_2} ways.

$\therefore$ Maximum number of edge in a simple graph G is

$$n_{c_2} = \frac{n(n-1)}{2}$$

Example 7.9.15. Set G be a graph with n vertices and m edges, such that the vertices have degree k or $k + 1$. Prove that if G has N_k vertices of degree k, then show that

$$N_k = (k+1)\,n - 2m$$

Solution. G has N_k vertices of degree k, therefore the remaining $n - N_k$ vertices are of degree $k + 1$.

Hence $\sum_{v_i \in V} \deg(v_i) = k\,N_k + (n - N_k)\,(k + 1)$

$$\Rightarrow \qquad 2m = k\,N_k + nk + n - k\,N_k - N_k$$

$$\Rightarrow \qquad N_k = (k + 1)\,n - 2m$$

Example 7.9.16. Construct two cubic graph on 10 vertices.

Solution. The following graphs are examples of cubic graphs.

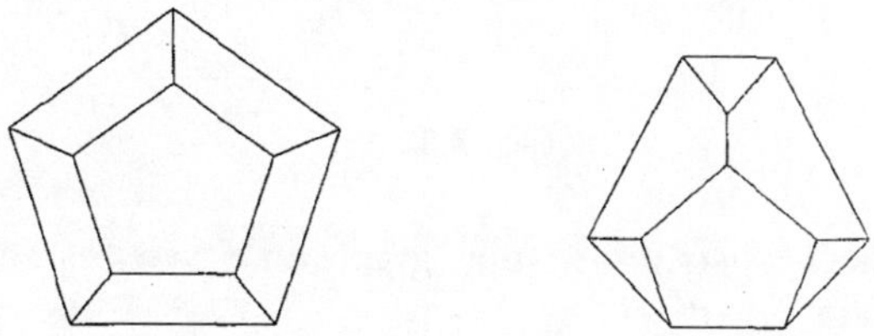

(a) (b)

Fig. 7.32

Example 7.9.17. Find the size of an r-regular (p, q) graph.

Solution. G is a r-regular graph, therefore degree of each vertex in G is r and number of edges in G is q.

$$\therefore \qquad \sum_{1}^{p} \deg(v_i) = 2q$$

$$\Rightarrow \qquad \sum_{r=1}^{p} r = 2q$$

$$\Rightarrow \qquad pr = 2q$$

$$\Rightarrow \qquad q = \frac{pr}{2}$$

$\therefore$ The size of an r-regular (p, q)-graph is $\dfrac{p.r}{2}$.

EXERCISE 7.1

1. Define the terms:

   (a) Graph (b) Self loop (c) Digraph
   (d) Multigraph (e) Pseudo graph (f) Order of graph
   (g) Size of a graph

   Give examples.

2. Draw a diagram for each of the following graphics
   (a) $E = \{(a, d), (a, f), (b, c), (b, f), (c, e)\}$, $V = \{a, b, c, d, e\}$
   (b) $V = \{v_1, v_2, v_3, v_4, v_5\}$, $E = \{(v_1, v_2), (v_2, v_3), (v_2, v_4), (v_4, v_5)\}$
3. Find the degrees of the vertices in the graphs G_1, G_2 and G_3 shown in Fig. 7.32

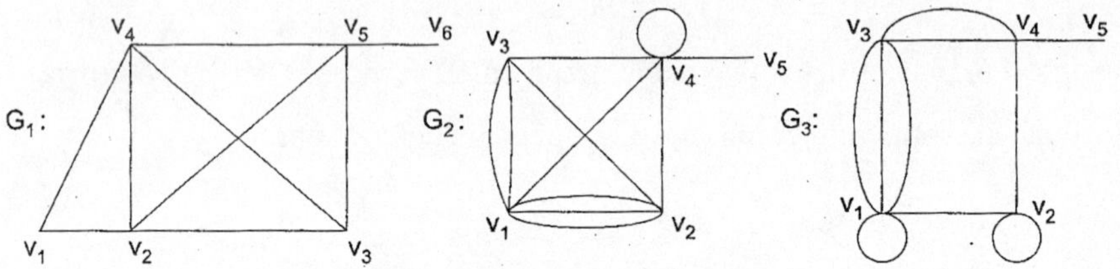

Fig. 7.32

4. Show that there exists a 4-regular graph on 6 vertices and construct 4-regular graph on 6 vertices.
5. Find the in-degree and out-degree of each vertex in the graph with directed edges shown in Fig. 7.33.

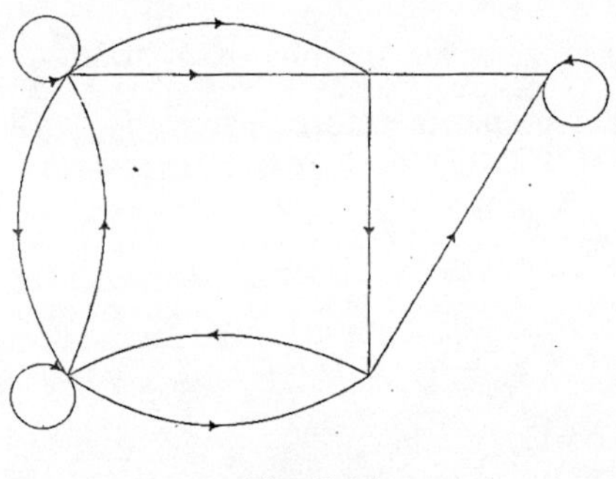

Fig. 7.33

6. Find the in-degree and out-degree of each vertex in the graph G, shown in Fig. 7.34.

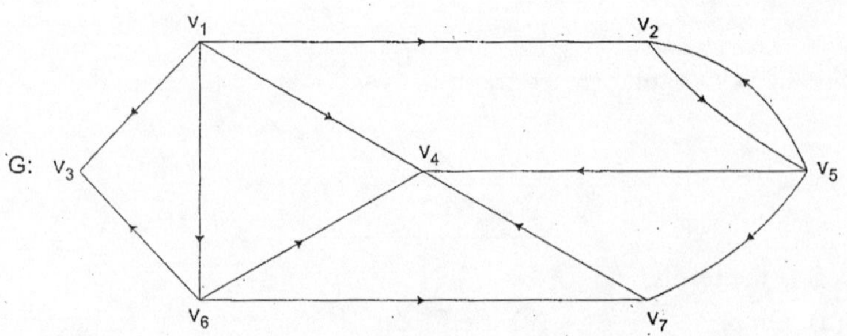

Fig. 7.34

7. Find the order and size of the graph G (Fig. 7.35).

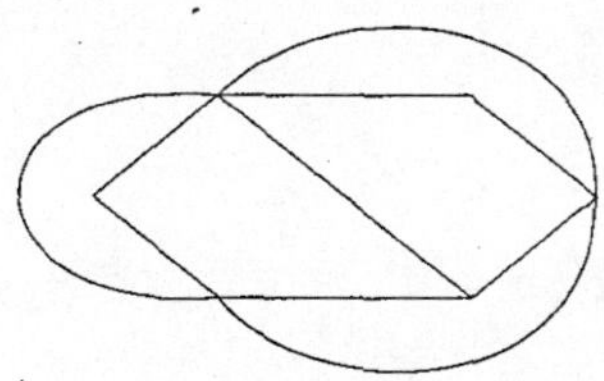

Fig. 7.35

8. Construct a cubic graph on 10 vertices. Draw three different representations.
9. Suppose you are married and you and your husband visited a party with 3 other married couples. If no one shakes hands with him-self or his wife, find how many hands have you shaken and how many did your husband shake. Give limits.
10. What is the largest number of vertices in a graph with 35 edges, if all vertices are of degree at least 3.
11. How many vertices do the following graphs have if they contain
    (i) 16 edges and all vertices of degree 2.
    (ii) 21 edges, 3 vertices of degree 4 and the remaining vertices are such that each is of degree 3.
    (iii) 24 edges and all vertices of same degree.
12. Suppose all vertices of a graph G have degree k where k is an odd number. Show that the number of edges in G is a number of k.
13. Let G be a (p, q) graph, all of whose points have degree k of $k + 1$. If G has t points ($t > 0$), of degree k, show that $t = p(k + 1) - 2q$.
14. Is there a simple graph with the degree sequence
    (a) (1, 1, 3, 4, 6, 7)
    (b) (1, 2, 3, 3, 3, 4, 6, 7)
    (c) (1, 1, 2, 3)
    (d) (2, 3, 3, 4, 5, 6)
    (e) (2, 2, 4, 8)
    (f) (6, 6, 6, 6, 4, 3, 3, 0)
    (g) (6, 6, 5, 4, 3, 3, 1)
15. Which of the following sequence are graphical
    (a) (2, 2, 2, 2)
    (b) (3, 3, 3, 3)
    (c) (2, 2, 2, 2, 2)
16. (Havel-Hakimi). Prove that there exists a simple graph with degree sequence $(d_1, d_2,..., d_n)$ where $d_1 \le d_2 \le ... < d_n$, if and only if there exists are with degree sequence

$$(d'_1, d'_2,..., d'_{n-1})$$

where

$$d'k = \begin{cases} d_k & \text{for } k = 1, 2, \ldots, n - d_n - 1 \\ d_{k-1} & \text{for } k = n - d_n, \ldots n - 1 \end{cases}$$

17. Which of the following sequence are graphical

(a) 6, 5, 5, 4, 3, 3, 2, 2, 2

(b) 5, 5, 4, 4, 3, 2, 2, 1, 1

(c) 7, 6, 5, 4, 4, 3, 2, 1

(d) 4, 4, 4, 4, 3, 3

(e) 3, 3, 3, 3, 3, 3, 3, 3

18. A graph G has 21 edges, 3 vertices of degree 4, and other vertices are of degree 3. Find the number of vertices in G.

ANSWERS 7.1

9.

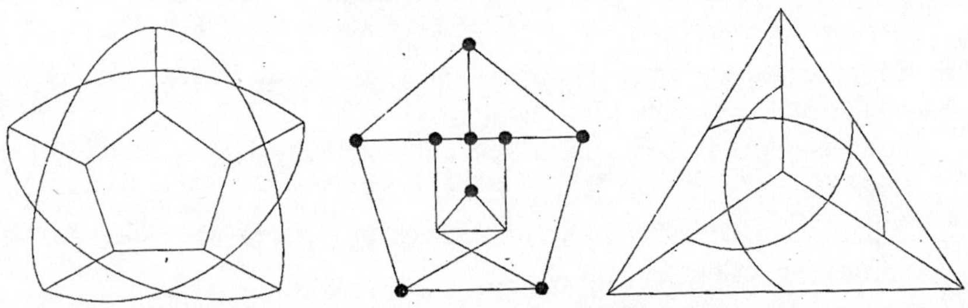

10. $\delta = 0$, and $\Delta = 36$

11. 23

12. (i) 16 (ii) 13 (iii) $\dfrac{48}{x}$; where x is the degree of v_i for all i.

15. (a) to (g) No

16. (a), (b) and (c) are graphics

17. (a) Yes (b) No (c) No (d) Yes (e) Yes

18. 13

7.10 EDGES IN SERIES

In a graph G two edges are said to be in series if they have exactly one vertex in common.

In Fig. 7.36, the edges e_1, and e_2 are in series.

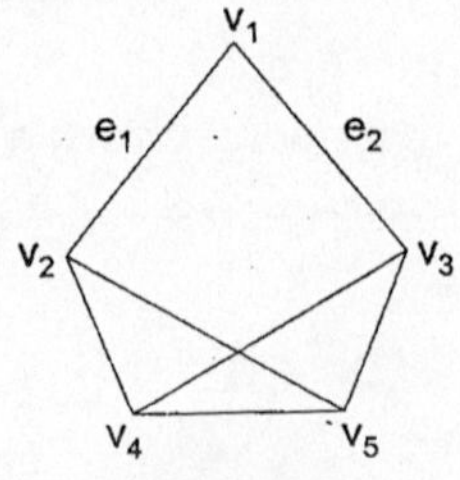

Fig. 7.36

7.11 ADJACENCY

In a graph, if there is an edge e incident from the vertex u to the vertex v, or incident on u and v, then the vertices said to be adjacent.

In the graph 7.37, the vertices v_1 and v_2 are adjacent.

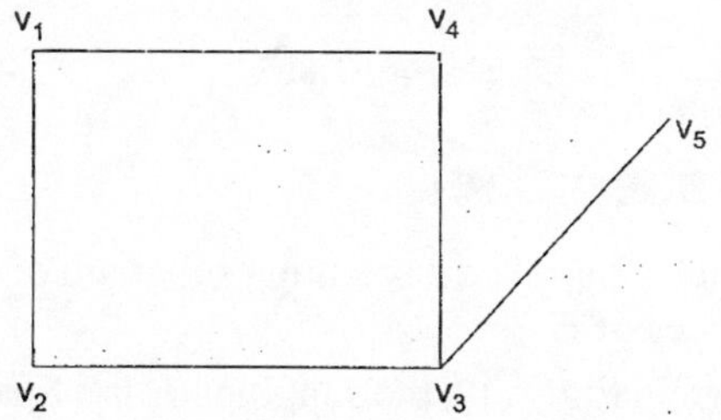

Fig. 7.37

Two non-parallel edges in a graph G are said to be adjacent, if they are incident on the same vertex.

In the graph of Fig. 7.38, the edges e_1 and e_2 are adjacent.

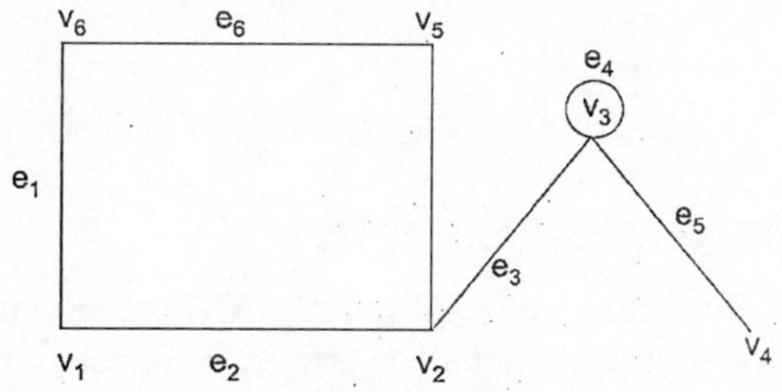

Fig. 7.38

7.12 SUBGRAPH

Let G be a graph. A graph H is called a subgraph of G if $V(H)$ is a subset of $U(G)$ and $E(H)$ is a subset of $E(G)$.

If H is a subgraph of G then

(i) All the vertices of H are in G

(ii) All the edges of H are in G and

(iii) Each edge of H has the same end point in H as in G

In Fig. 7.39, the graph H is sub-graph of the graph G.

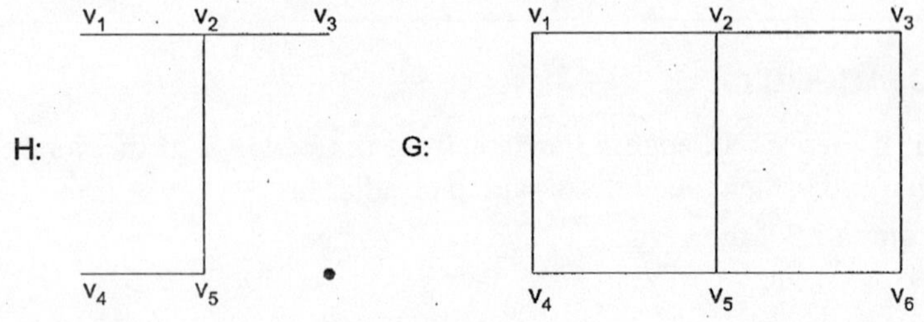

Fig. 7.39

7.13 SPANNING SUBGRAPH

A subgraph H of a graph G is called a spanning subgraph of G if $V(H) = V(G)$, i.e., H contain all the vertices of G.

In the graph shown in Fig. 7.40, the subgraph H is a Spanning subgraph of G.

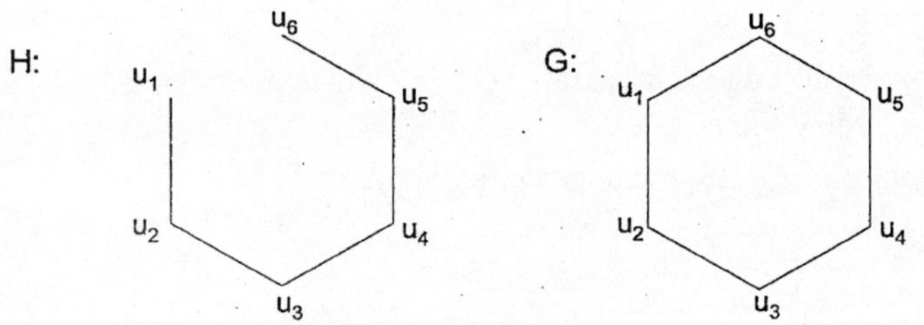

Fig. 7.40

7.14 EDGE DISJOINT SUBGRAPHS

Let G be a graph. Two subgraphs H_1 and H_2 of G are said to be edge disjoint subgraphs of G if H_1 and H_2 do not share any edges in common.

7.15 REMOVAL OF VERTICES AND EDGES FROM A GRAPH

The removal of a vertex v_i from a graph G, results in a subgraph of G, consisting of all points of G except v_i and all edges of G not incident with v_i. The graph obtained is denoted by $G - v_i$, and is the maximal subgraph of G not containing v_i (see Fig. 7.41).

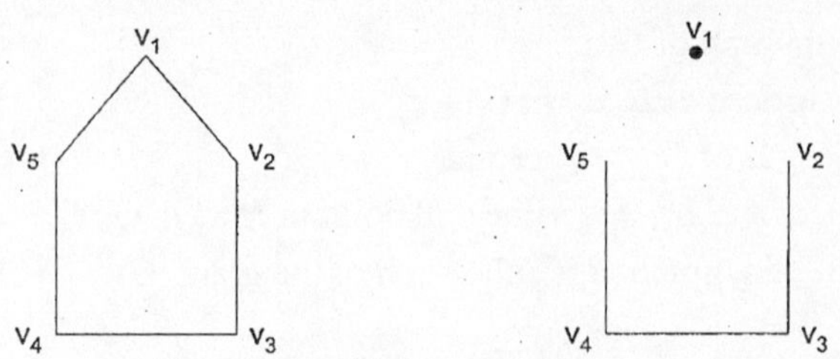

Fig. 7.41 Graph minus a vertex v_1

The removal of an edge e_j from a graph results in the spanning subgraph of G, which contains all the edges of G which contains all the edges of G except the edge e_j. It is denoted by $G - e_j$ and is the maximal subgraph of G not containing e_j.

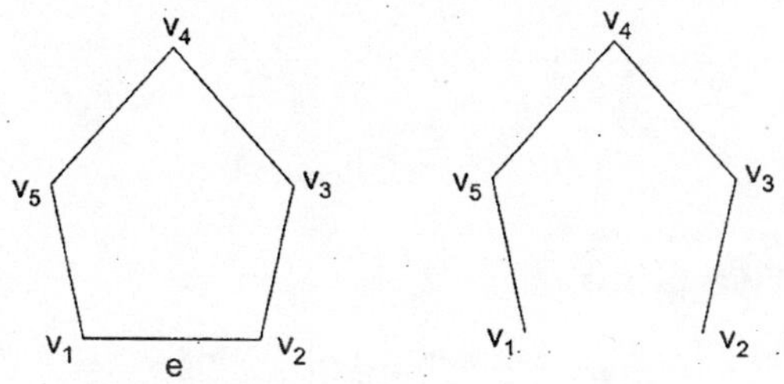

Fig. 7.42 Graph minus an edge

7.16 ADDITION OF A VERTEX

Let G be a graph and v be a vertex, which is not in G, then the graph obtained by joining v with each vertex of G, is the sum graph which is denoted by $G + v$.

7.17 OPERATIONS ON GRAPHS

7.17.1 Union of Graphs

Let $G = (V_1, E_1)$ and $G_2 = (V_2, E_2)$ be two graphs. Then the union of G_1 and G_2 denoted by $G_1 \cup G_2$ is defined as the graph $G = (V, E)$ where

$$V(G) = V(G_1) \cup V(G_2)$$

and $$E(G) = E(G_1) \cup E(G_2)$$

Consider the graphs G_1 and G_2 shown in Fig. 7.43.

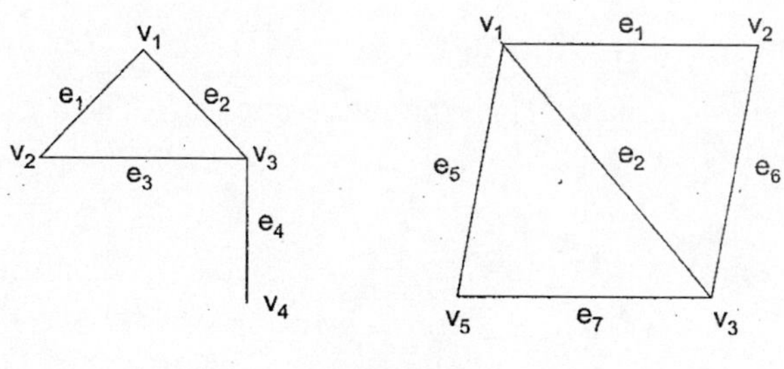

Fig. 7.43

The union of G_1 and G_2 is the graph shown in Fig. 7.44.

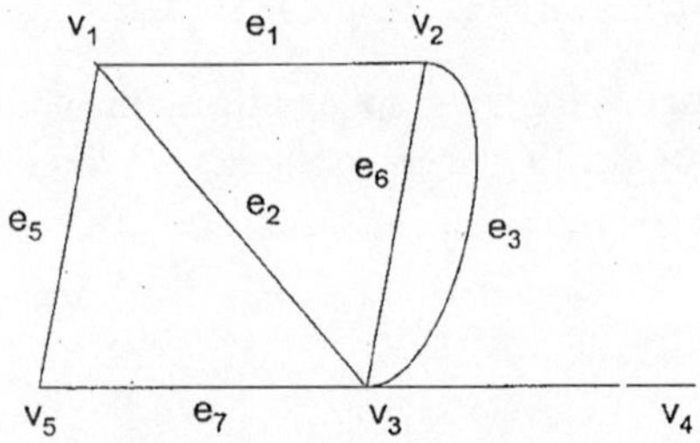

Fig. 7.44 Union of graphs

7.17.2 Sum of Two Graphs

Let $G_1 = (V_1, E_1)$ and $G_2 = (V_2, E_2)$ denote two vertex disjoint graphs. Then the sum of G_1 and G_2 denoted by $G_1 + G_2$ is defined as $G_1 \cup G_2$ together with all edges joining the vertices of V_1 to vertices of V_2.

Example. If G_1 is the graph 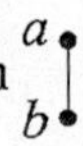

[Fig. 7.45(a)] and G_2 is the graph 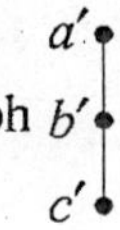

[Fig. 7.45(b)]

Then $G_1 + G_2$ is the graph given below.

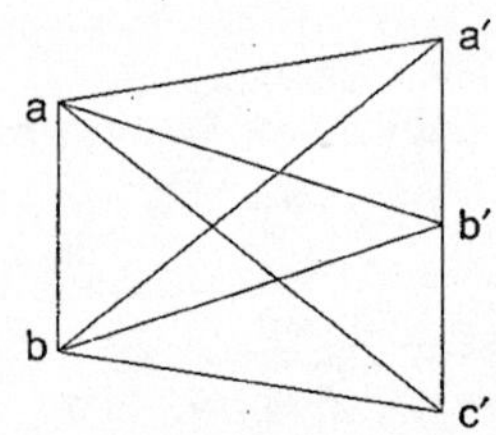

Fig. 7.45

7.17.3 Intersection of Graphs

Let G_1 and G_2 be two graphs. Then the intersection of G, and G_2 denoted by $G_1 \cap G_2$ is defined as the graph G.

Such that

(i) $V(G) = V(G_1) \cap V(G_2)$

(ii) $E(G) = E(G_1) \cap E(G_2)$

Example.

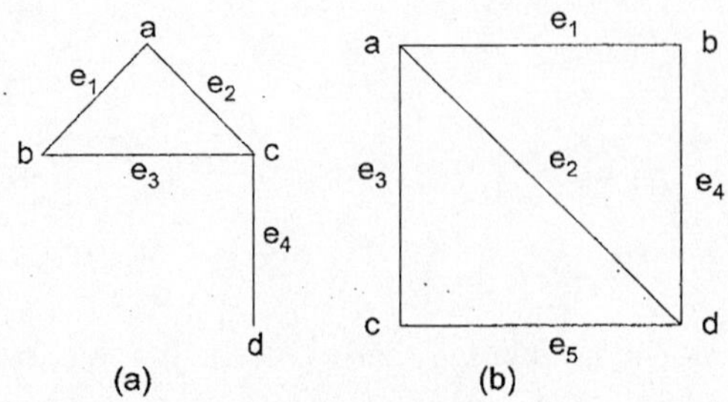

Fig. 7.46

Then $G_1 \cap G_2$ is the graph

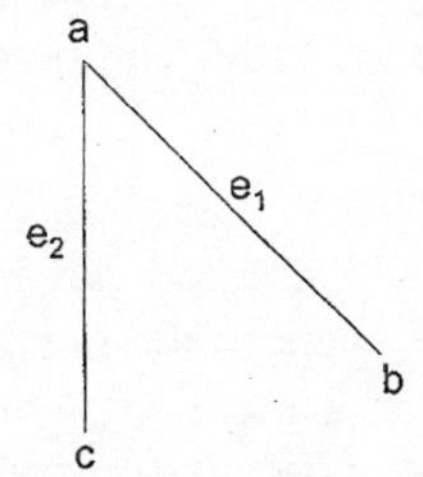

Fig. 7.46 Intersection of two graphs

7.18 COMPLEMENT OF A GRAPH

Let G be a graph with n vertices, then $k_n - G$ is called the complement of G. It is denoted by $\overline{G}$. A graph and its complement are shown in Fig. 7.47.

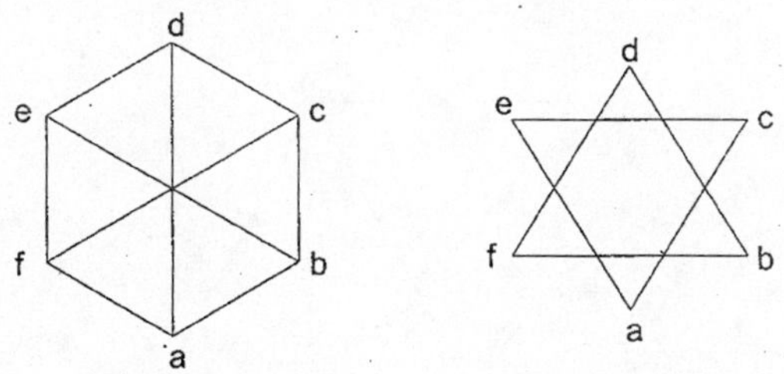

(a) Graph G (b) Complement of G

Fig. 7.47

7.18.1 Complement of a Subgraph

Let G be a graph and H be a subgraph of G. The complement of H in G is the graph obtained by deleting the edges of H from those of G.

The complement of H in G is denoted by $\overline{H}$(G).

In Fig. 7.48, a subgraph H of G and its complement $\overline{H}$ are shown.

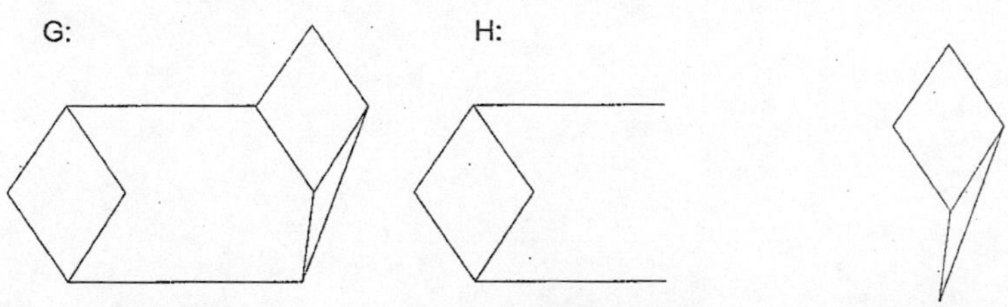

(a) Graph G (b) Subgraph H of G (b) Complement $\bar{H}$ of H

Fig. 7.48

7.19 WALKS, PATHS AND CIRCUITS

Definition 7.25. A walk in a graph is defined as a finite alternating sequence of vertices and edges $v_0, e_1, v_1, e_2\ v_2, ..., u_{n-1}, e_n\ v_n$, beginning and ending with points (vertices) in which edge is incident with two points immediately preceding and following it.

A walk is a graph is denoted by W.

The walk $W = v_0, e_1, v_1, e_2, ..., v_{n-1}, e_n, v_n$ may be written as $W : v_0, - v_1 - v_2 - ... - v_{n-1} - v_n$ and is called $v_0 - v_n$ walk in the graph. The vertices v_0 and v_n with which the walk begins and ends are called the initial and terminal vertices respectively. If $v_0 \neq v_n$, then the walk is called an open walk and if $v_0 = v_n$, the walk is called a closed walk. A walk is also called a **chain.**

The number of occurrence of edges in a walk is called the **length of walk**. It is denoted by $l(W)$.

Example. Consider the graph G shown in Fig. 7.49.

$$W: v_1, e_1, v_2, e_2, v_3, e_3, v_4, e_4, v_5$$

is a walk in the graph G.

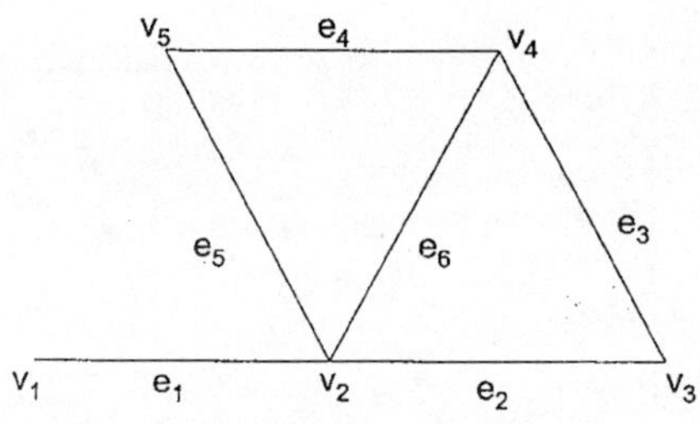

Fig. 7.49

The length of the walk is 4

i.e. $l(W) = 4$

Definition 7.26. A walk in which no edges are repeated is called trial.

Consider the graph shown in Fig. 7.50.

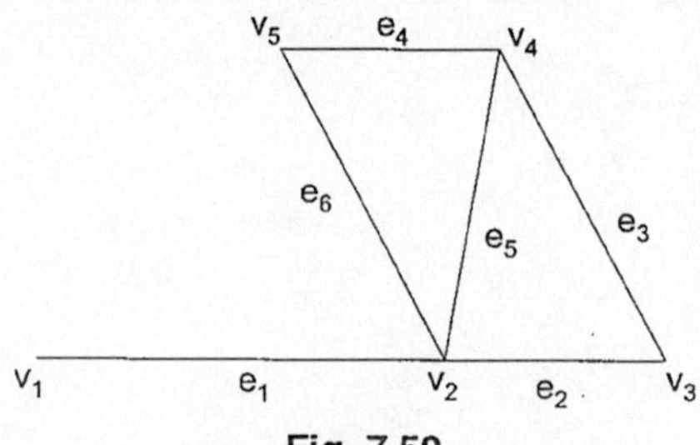

Fig. 7.50

$$W = v_1\, e_1\, v_2\, e_2\, v_3\, e_3\, v_4$$

is a walk.

Since, no edges are repeated in W, it is called trail. The length of the trail is 3. $v_1 \neq v_4$, in W, therefore W is an open walk in G. It is an open walk, which is also an open trail.

Definition 7.27. Let G be a non-directed graph. A trail in G in which no vertices are repeated is called a **path**.

A sequence P of zero or more edges (in G) of the form $\{v_0, v_1\}$, $|v_1, v_2|$, $\{v_2, v_3\}, \ldots, \{v_{n-1}, v_n\}$ where $v_0, v_1, v_2, \ldots, v_n$ are the vertices of G is a path in G.

The vertex v_0 is called the initial vertex and the vertex v_n is called the terminal vertex of the path P.

The path P can also be written as

$$P : v_0 - v_1 - v_2 - \ldots - v_{n-1} - v_n$$

It is called $v_0 - v_n$ path.

If $v_0 \neq v_n$, then the path P is called an open path. And if $v_0 = v_n$, then the path P is called a closed path.

The number of occurrence of edges in a path p, is called the length of the path. A path of length zero is called a trivial path.

The terminal vertices of an open path are of degree one. A path is simple if all edges and vertices on the path are distinct, except that the end vertices v_0 and v_n may be equal.

If $v_0\, e_1\, v_1\, e_2\, v_2, \ldots v_{n-1}\, e_n\, v_n$ is a path in a graph in $G = (V, E)$.

Then, it is a sequence of edges $e_1\, e_2 \ldots e_n$ in E; for which there is a sequence of vertices $v_0, v_1, \ldots, v_n$ such that $e_i = (v_{i-1}, v_i)$ or $e_i = (v_i, v_{i-1})$ for each $1 \leq i \leq n$. If P is a simple path of length n then P has $n + 1$, distinct vertices and if P is closed path of length n, then p has n distinct vertices and n distinct edges. Consider the graph shown in Fig. 7.51.

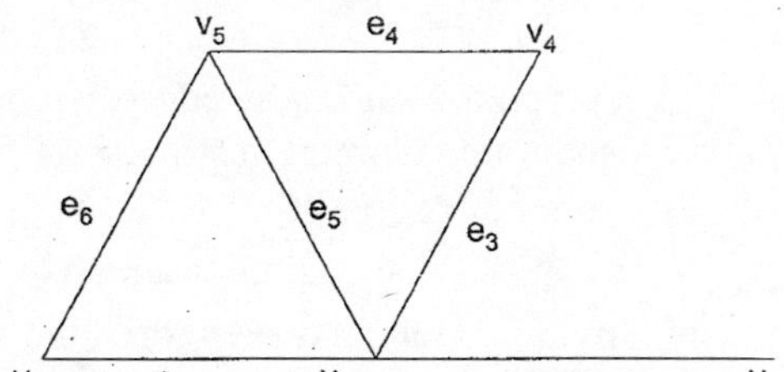

Fig. 7.51

$v_1\, e_1\, v_2\, e_3\, v_4\, e_4\, v_5$ is a path in G.

In the graph shown in Fig. 7.52, $a - e - c - b - d$ is a path of length of 4.

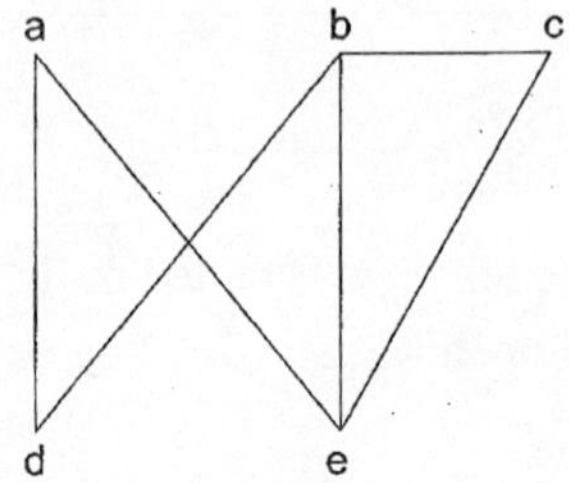

Fig. 7.52

Note. If P is a path in a graph $G = (V, E)$, then $V(P) \subseteq V(G)$ and $E(P) \subseteq E(G)$

Also $$1 \leq |V(P)| \leq |E(P)| \leq n$$

A path $v_1 - e_1 - v_2 - e_2 - ... - v_{n-1} - e_n - v_n$, is said to traverse a vertex v if one or more of the edges e_i's is to or from v and v is not serving as one of the end points of the path.

Definition 7.28. (Circuit): A path of length $n \geq 1$, with no repeated edges and whose end points are equal is called a **circuit**.

A circuit in a graph G, is a closed non-intersecting work in which every vertex is of degree two. a circuit no other repeated vertex except its end points is called a cycle. It is a simple circuit in G. The terms a circuit and cycle are synonymous. The definitions of simple, circuit and cycle apply equally to directed and non-directed paths in a graph. In the directed graph shown in Fig. 7.53. $a - b - c - d - e$ is a directed path, $a - b - c - d - e - a$ is directed cycle.

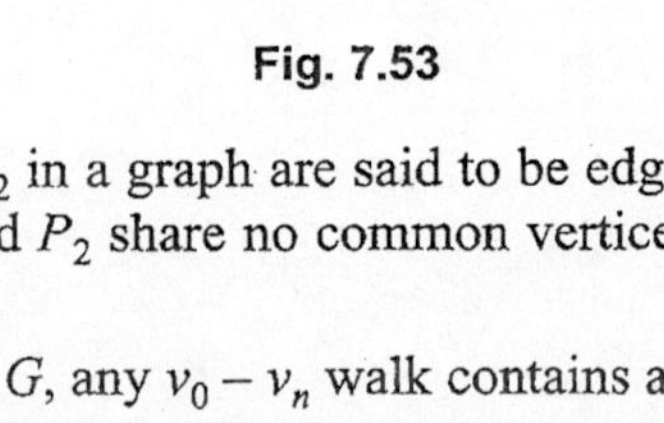

Fig. 7.53

Two paths P_1 and P_2 in a graph are said to be edge-disjoint if they share no common edges. If P_1 and P_2 share no common vertices then they are said to be vertex disjoint.

Theorem 7.3 In a graph G, any $v_0 - v_n$ walk contains a path.

Proof. We prove the theorem by induction on the length of the walk. If the length of $v_0 - v_n$ walk is 0 or 1, then the walk is obviously a path. Now, let us assume that the result holds for all walks of length less than n.

Let $v_0, v_1, v_2,..., v_n$ be a walk of length n. If all the vertices v_i, $1 \leq i \leq n$ are distinct then the walk is a path. If not there exist i and j such that $v_i = u_j$ for same i, j such that $1 \leq i \leq j \leq n$.

Now the walk $v_0 - v_1 - v_2 - \ldots\ v_i, v_{i+1},\ldots, v_n$ is a walk whose length is less than n, which by induction hypothesis contains a $v_0 - v_n$ path.

Theorem 7.4. If $\delta(G) \geq k$, then the graph G has a path of length k.

Proof. Let v_1 be an arbitrary vertex in G. Choose a vertex say v_2, which is adjacent to v_1. Since $\delta(G) \geq k$, there exists at least $k - 1$, vertices other than v_1 which are adjacent to v_2. Choose another vertex $v_3 \neq v_1$ such that v_3 is adjacent to v_2. In this way, we can find $v_4, v_5,\ldots, v_i$ where $1 \leq i \leq \delta(G)$. Having choosen the vertices v_1, $v_2,\ldots, v_i$, where $1 \leq i \leq \delta(G)$, we can find another vertex v_{i+1} which is different from the vertices $v_1, v_2,\ldots, v_i$, such that v_{i+1} is adjacent to v_i proceeding in this way, we can find a path of length k in G.

7.20 CONNECTED GRAPH

A graph G is said to be connected if every pair of points in G are joined by a path.

If G is not connected then G is said to be disconnected.

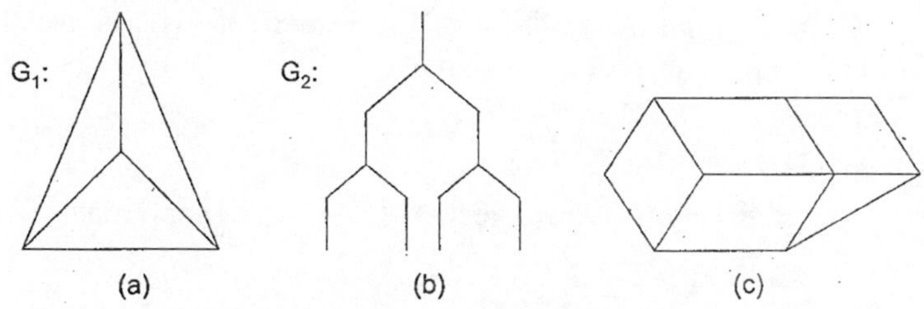

Fig. 7.54 A connected graphs

7.20.1 Component of a Graph

A maximal connected subgraph of a graph G is called a component of G. If G is disconnected then G has at least two components clearly a graph G is connected if and only if has exactly one component.

The number of components of a graph G is denoted by $k(G)$.

Theorem 7.5. If G is a graph with n points (vertices) and $\delta(G) \geq \dfrac{n-1}{2}$, then G is connected.

Proof. Let is assume that G is not connected. Then G has more than one component. Consider any component $G_1 = (V_1, E_1)$ of G.

Let $v_1 \in V_1$. Since $\delta(G) \geq \dfrac{n-1}{2}$, there exist at least $\dfrac{n-1}{2}$ points in G_1 which are adjacent to v_1 in G_1.

$\therefore \qquad |V_1| \geq \frac{n-1}{2} + 1$

$\Rightarrow \qquad |V_1| \geq \frac{n+1}{2}$

Thus, each component of G has at least $\frac{n+1}{2}$ points and G has at least two components.

Hence, the number of points (vertices) in $G \geq \frac{2 \cdot (n+1)}{2}$

i.e. $\qquad |V(G)| \geq (n+1)$

which is a contradiction.

$\therefore$ G is connected.

7.21 PARTITIONS

Let $G = (V, E)$ be a graph. A partition of the vertex set $V(G)$ is a collection $\{V_i\}_{1 \leq i \leq \alpha}$ of non-empty subsets of V such that

(i) $V_1 \cup V_2 \cup \ldots \cup V_\alpha = V, \quad (\alpha \neq 1)$

and (ii) $V_i \cap V_j \neq \phi$, whenever $i \neq j$.

A partition of edge set $E(G)$ is a collection $\{E_i\}_{1 \leq i \leq \beta}$ of non-empty subsets of E such that

(i) $E_1 \cup E_2 \cup \ldots E_\beta = E \ (\beta \neq 1)$

and (ii) $E_i \cap E_j = \phi$, whenever $i \neq Ji$.

A partition of the edge set E is also called the Edge decomposition of G.

Example. Consider the graph shown in Fig. 7.55.

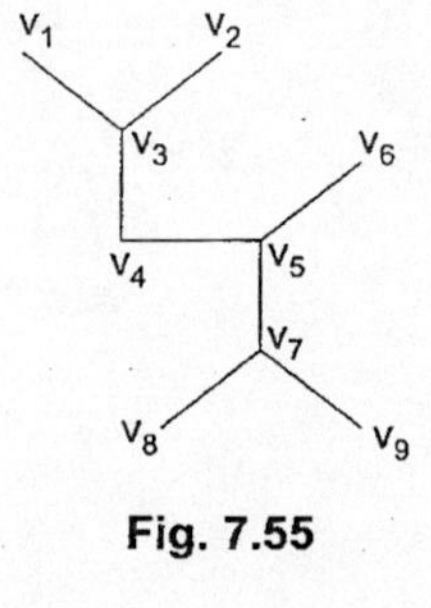

Fig. 7.55

$$V_1 = \{v_1, v_2, v_3\}, V_2 = \{v_4, v_5, v_6, v_7\}, V_3 = \{v_8, v_9\}$$

is a vertex partition of G

and $\quad E_1 = \{\{v_1, v_3\}, \{v_3, v_2\}\}, E_2 = \{v_3, v_4\}, \{v_4, v_5\}, \{v_5, v_6\}\}$

$E_3 = \{v_5, v_7\}\}, E_4 = \{v_7, v_8\}\}, E_5 = \{\{v_7, v_9\}\}$

is a partition of edge set $E(G)$ i.e. Edge decomposition of G.

Theorem 7.6. A graph G is connected if and only if for any partition of V into subsets V_1 and V_2, there is an edge joining a vertex of V_1 to a vertex of V_2.

Proof. Let G be a connected graph and $V = V_1 \cup V_2$ be a partition of V into two subsets.

Let $\quad u_1 \in V_1$ and $u_2 \in V_2$

Since the graph G is connected there exists a $u - v$ path in G. Say $u = v_0, v_1, v_2,..., v_n = v$. Let i be the least positive integer such that $v_i \in V_2$. The $v_{i-1} \in V_1$ and the vertices v_{i-1}, v_i are adjacent. Thus there is an edge joining $v_{i-1} \in V_1$ and $v_i \in V_2$.

Conversely, let G be a disconnected graph. Then G contains at least two components. Let V_1 be the set of all vertices of one component and V_2 be the set of remaining vertices of G. Clearly $V_1 \cup V_2 = V$ and $V_1 \cap V_2 = \phi$. The collection $\{V_1, V_2\}$ is a partition of V and there is no edge joining any vertex of V_1 to any vertex of V_2.

Hence the theorem.

Theorem 7.7. If G is a simple graph with n vertices and k components, then G can have at most $(n - k)(n - k + 1)/2$ edges.

Proof. Let G be a simple group with n vertex and $G_1,..., G_2,..., G_k$ be the k components of G. Let the number of vertices in the ith component G_i be n_i.

Then, $$|V(G_1)| + |V(G_2)| + ... + |V(G_k)|$$
$$= n_1 + n_2 + ... + n_k$$
$$= |V(G)| = n$$

and $$\max |E(G_i)| \leq \frac{n_i(n_i - 1)}{2}, \text{ where } n_i > 1$$

$\therefore$ $$E(G) \leq \sum_{i=1}^{k} \max |E(G_i)| = \sum_{i=1}^{k} \frac{n_i(n_i - 1)}{2}$$

$\Rightarrow$ $$E(G) \leq \frac{1}{2}\left[\sum_{i=1}^{k} n_i^2 - \sum_{i=1}^{k} n_i\right] = \frac{1}{2}\left[\sum_{i=1}^{k} n_i^2 - n\right] \qquad ...(i)$$

now $$\sum_{i=1}^{k} (n_{i-1}) = (n_{i-1}) + (n_{2-1}) + ... + (n_{k-1})$$
$$= (n_1 + n_2 + ... + n_k) - (1 + 1 + ... k \text{ times})$$
$$= n - k$$

Squaring on both sides, we get

$$\left[\sum_{i=1}^{k}(n_i - 1)\right]^2 = (n-k)^2 = n^2 + k^2 - 2nk$$

$$\Rightarrow \quad \sum_{i=1}^{k} (n_i - 1)^2 + 2 \text{ (non-negative terms)} = n^2 + k^2 - 2nk$$

$$\Rightarrow \quad \sum_{i=1}^{k} (n_i - 1)^2 = n^2 + k^2 - 2nk - 2 \text{ (non-negative terms)}$$

$$\Rightarrow \quad \sum_{i=1}^{k} (n_i - 1)^2 \le n^2 + k^2 - 2nk$$

$$\Rightarrow \quad \sum_{i=1}^{k} n_i^2 + \sum_{i=1}^{k} 1 - 2\sum_{i=1}^{k} n_i \le n^2 + k^2 - 2nk$$

$$\Rightarrow \quad \sum_{i=1}^{k} n_i^2 + k - 2n \le n^2 - k^2 - 2nk$$

$$\Rightarrow \quad \sum_{i=1}^{k} n_i^2 - n \le n^2 - nk + n - nk + k^2 - k$$

$$\Rightarrow \quad \sum_{i=1}^{k} n_i^2 - n \le n(n-k+1) - k(n-k+1) = (n-k)(n-k+1)$$

$$\therefore \quad \sum_{i=1}^{k} n_i^2 - n \le (n-k)(n-k+1) \qquad \ldots(2)$$

From (1) and (2) we obtain

$$E(G) \le (n-k)(n-k+1)/2$$

hence proved.

Theorem 7.8. If G is not connected then $\overline{G}$ is connected.

Proof. Let G be a disconnected graph. Then G has more than one component.

Let u, v be any two vertices of G. The theorem is proved if we show that there is a $u-v$ path in $\overline{G}$.

If u, v are in different components of G, then u, v are not adjacent in G. Hence thus are adjacent in $\overline{G}$.

If u, v are in the same components of G, choose a vertex w, in a different component of G. Then $u-w-v$ is a $u-v$ path in $\overline{G}$. Hence $\overline{G}$ is connected.

7.22 CYCLE GRAPH

Definition 7.29. A cycle graph of order n is a connected graph whose edges from a cycle of length n.

A cycle graph of order n is denoted by C_n. The graph shown in Fig. 7.56(a) is a cycle graph of order 5 and the graph sown in Fig. 7.56(b) is a cycle graph of order 6.

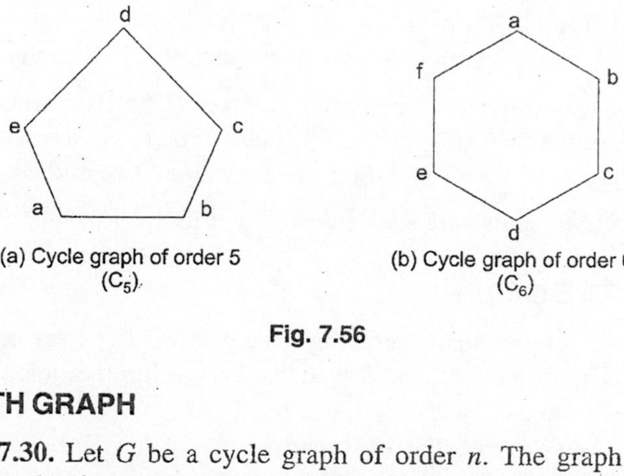

(a) Cycle graph of order 5 (C_5)

(b) Cycle graph of order 6 (C_6)

Fig. 7.56

7.23 PATH GRAPH

Definition 7.30. Let G be a cycle graph of order n. The graph obtained by remaining an edge from G is called a path graph of order of order n. It is denoted by P_n.

The graph shown in Fig. 7.57, is a path graph of order 5.

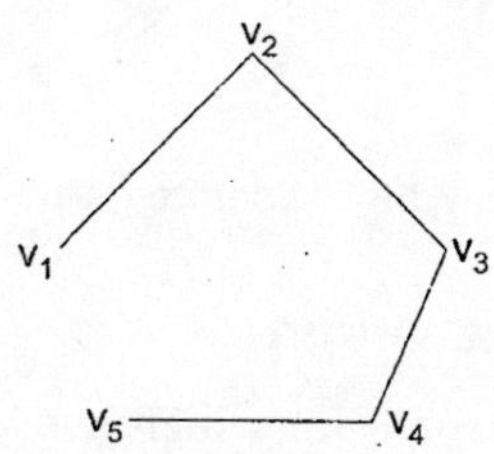

Fig. 7.57 Path group of order 5 (P_5)

7.24 WHEEL GRAPH

Definition 7.31. Let G be a cycle graph of order $(n - 1)$. The graph obtained by joining a single new vertex v to each vertex of G is called a wheel graph of order n.

A wheel graph of order n is denoted by W_n. The new vertex v added is called the "hub". The graphs shown in Fig. 7.58 are wheel graphs.

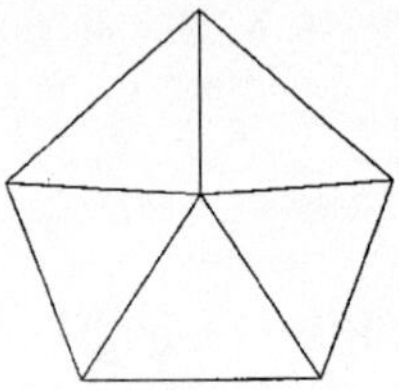

(a) W_6

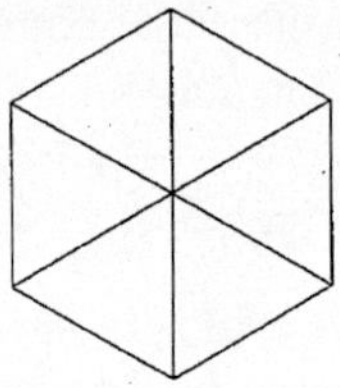

(b) W_7

Fig. 7.58 Wheel graphs of order 6 and 7

Theorem 7.9. If G is a graph with 6 points, then G or $\overline{G}$ contains a triangle.

Proof. Let $G = (V, E)$ be a graph with six points, and $v \in V$. then the vertex v is adjacent either in G of in $\overline{G}$, to the other five points of G. Let us assume that u_1, u_2 and u_3 are adjacent vertices of V in G. If any two of there vertices are adjacent; then the two adjacent vertices and v form a triangle. If no two of the points u_1, u_2 and u_3 are adjacent then they are adjacent in $\overline{G}$ and form a triangle in $\overline{G}$.

7.25 BIPARTITE GRAPH

Definition 7.32. A group G is called a bipartite graph if its vertex set V can be partitioned into disjoint subset A and B such that every edge in G, joins a vertex in A to a vertex in B.

The graphs shown in Fig. 7.59 are bipartite.

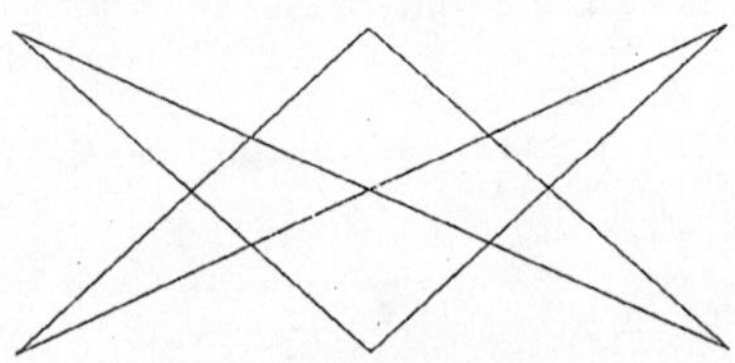

Fig. 7.59 Bipartite graph

7.25.1 Complete Bipartite Graph

Definition 7.33. A bipartite graph G in which every vertex of A is adjacent to every vertex in B is called a complete bipartite graph, where A and B are partitioned subsets of the vertex V of G.

If $|A| = m$ and $|B| = n$, then the complete bipartite graph is denoted by $k_{m,n}$ and has mn edges. The graphs in Fig. 7.60 are complete bipartite graphs.

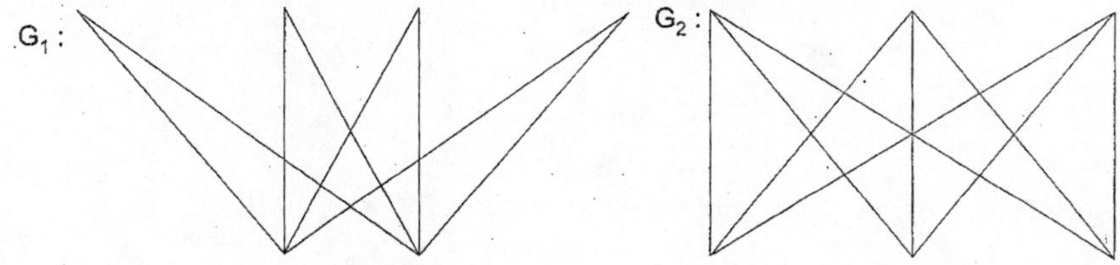

(a) $k_{4,2}$ (b) $k_{3,3}$

Fig. 7.60 Complete bipartite graphs

7.25.2 Star Graph

Definition 7.34. A complete bipartite graph $k_{1,n}$ is called a star graph (see Fig. 7.61).

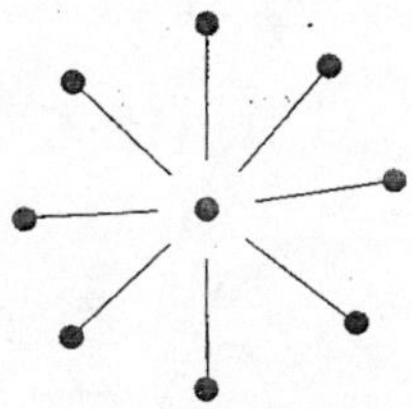

Fig. 7.61 Star graph $k_{1,8}$

7.26 SOLVED EXAMPLES

Example 7.26.1. Draw simple unlabeled graphs of 3 vertices.

Solution. The graphs shown in Fig. 7.62 are simple graphs with three vertices.

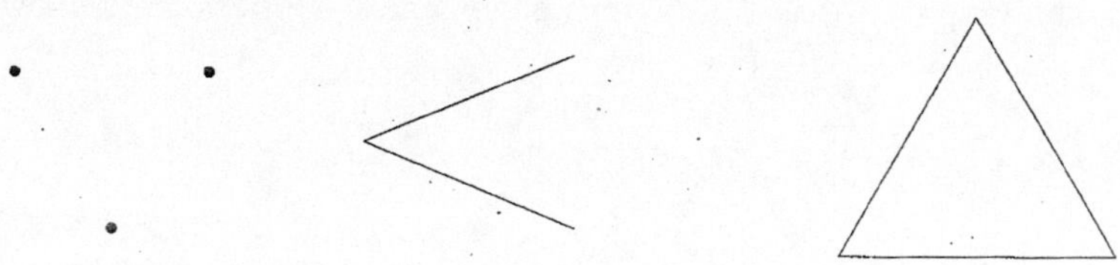

(a) (b) (c)

Fig. 7.62 Simple unlabeled graphs

Example 7.26.2. Find the number of connected graphs with four vertices and draw them.

Solution. There are five connected graphs with five vertices (see Fig. 7.63)

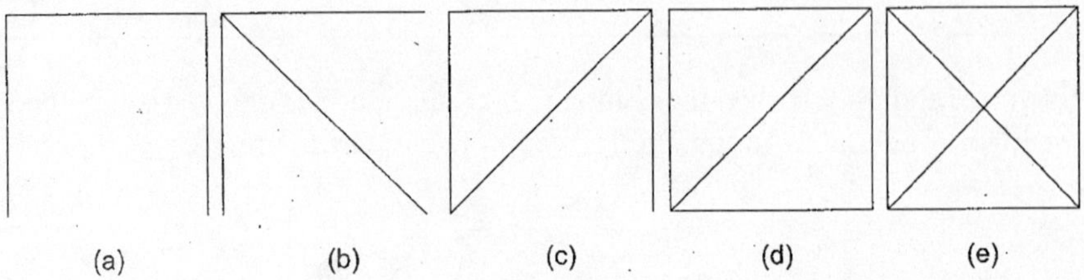

Fig. 7.63

Example 7.26.3. Draw the complement of the graph G shown in Fig. 7.64.

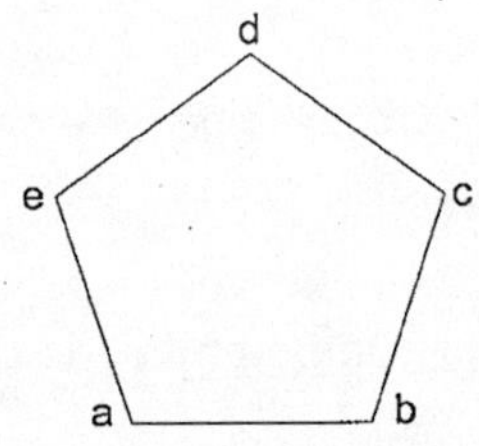

Fig. 7.64

Solution. The complement of G is shown in Fig. 7.65.

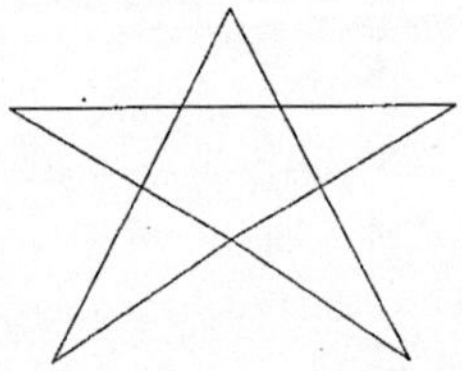

Fig. 7.65 Complement of G

Example 7.26.4. Draw $k_{2,5}$.

Solution. The graph $k_{2,5}$ has $2 \times 5 = 10$ edges and 7 vertices. It is shown in Fig. 7.66.

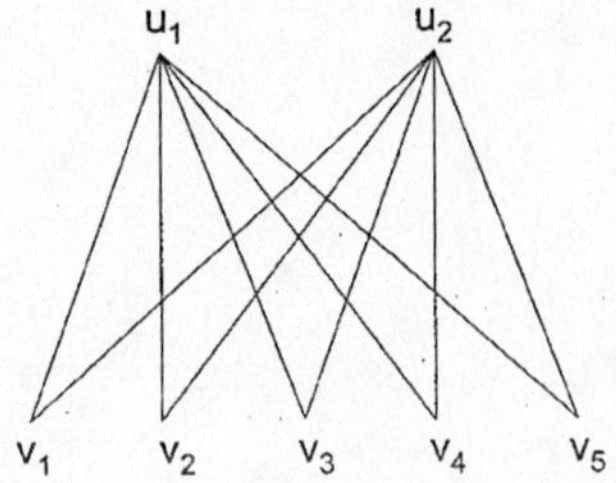

Fig. 7.66

Example 7.26.5. Show that the maximum number of edges in a complete bipartite graph is $n^2/4$.

Solution. Let G be a complete bipartite graph with n vertices. Vertex set of G can be partitioned into 2 blocks such that each block contains $n/2$ edges and each vertex of one block is adjacent to each vertex of another block. Hence the number of edges in G is $n^2/4$. Now to show that the maximum number of edges is $n^2/4$. We assume that one block has m vertices and another block has $(n-m)$ vertices. Then, the total number of edges is given by

$$f(m) = m\cdot(n-m) = nm - m^2$$

$$\Rightarrow \qquad f'(m) = n - 2m$$

for maxim we have $f'(m) = 0 \Rightarrow n - 2m = 0 \Rightarrow m = n/2$ hence, the maximum occurs when $m = n/2$.

i.e. when each block has $n/2$ vertices.

Hence, the maximum number of edges in $G = \frac{n}{2}\cdot\frac{n}{2} = \frac{n^2}{4}$.

7.27 FOREST

Definition 7.35. A graph is acyclic if it has no cycles. A forest is an acycle graph.

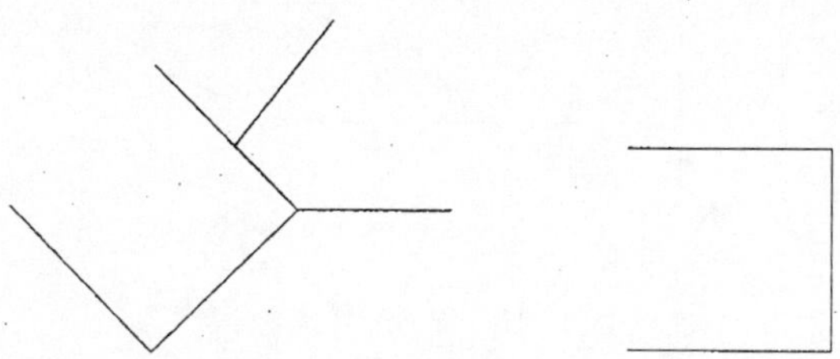

(a) (b)

Fig. 7.67 A Forest

7.28 CUT VERTEX

Definition 7.36. Let G be a connected graph. If v is a vertex of G such that $G - v$ is not a connected graph, then the vertex v is called a cut vertex.

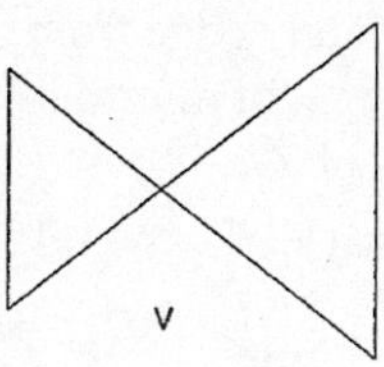

Fig. 7.68

If v is cut vertex of G. Then the removal of the vertex v increases the number of components in G. A cut vertex is also called a cut point.

Theorem 7.10. A vertex v in a connected graph G is a cut vertex if and only if there exists vertices u and w distinct from v such that every path containing u and w contains the vertex v.

Proof. Let v be a cut vertex in a connected graph G. Then $G - v$ is a disconnected graph and $G - v$ contains at least two components says A and B. Let u be a vertex of A and w be a vertex of B. There is no path in $G - v$ connecting u and w. Since G is connected there exists a path P from u to w in G.

If this path does not contain u, then the removal of v from G will not disconnect the vertices u and w which is a contradict to the fact that u and v lie in two different components of $G - v$.

Conversely. If every path from u to w, contains the vertex v_1 the removal of v from the graph G disconnects u and w. Hence u and w lie is different components of G which shows that $G - v$ is a disconnected graph. Thus v is a cut vertex of G.

7.29 CUT EDGE (BRIDGE)

Definition 7.37. Let G be a connected graph. If e is an edge of G. Such that $G - e$ is not connected, then the edge e is called a cut edge (or bridge).

In the graph of Fig. 7.69, the edge e is a cut edge.

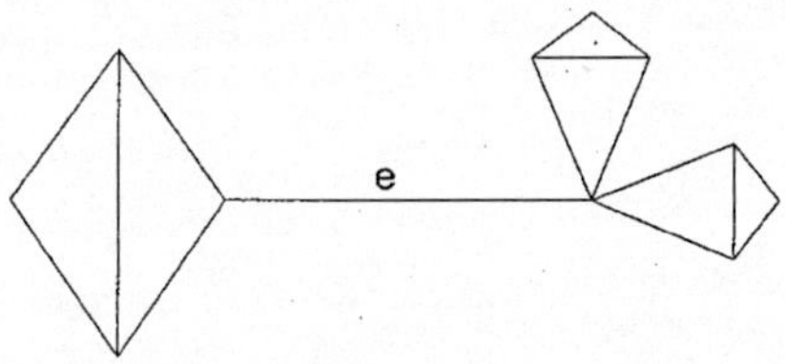

Fig. 7.69

7.30 CUT SET

Definition 7.38. Let G be a connected graph. A cut set in G is a set of edges whose removal from G leaves the graph G disconnected; provided no subset of these edges disconnects the graph.

In the graph shown in Fig. 7.70, the set of edge $\{a, c, d, f\}$ is a cut set.

A cut set in a graph always breaks the graph G into two parts. Every edge in a tree is a cut set, since the removal of any edge from a tree breaks the tree into parts. A cut set may also be defined as the set of minimum number of edges of a graph G, whose removal disconnects the graph G.

If A is a cut set in a graph $G = (V, E)$, then A is a subset of E and $E - A$ is a disconnected set of edges in G.

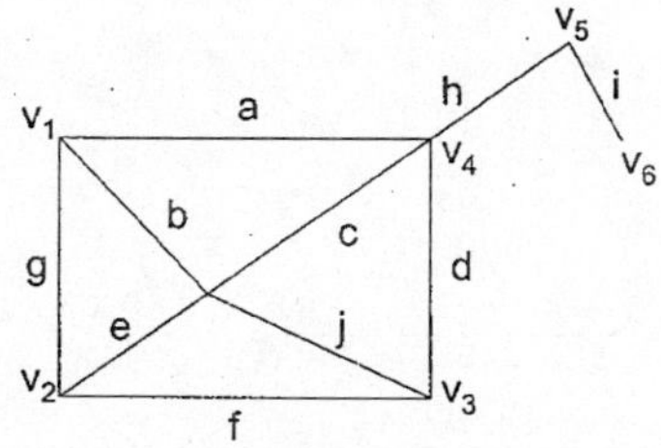

Fig. 7.70

7.31 SEPARABLE AND NON-SEPARABLE GRAPHS

Definition 7.39. A connected graph with at least one cut vertex is called a separable graph.

If G is a separable graphs, then G is connected and there exists a subgraph H of G such that H and $\overline{H}$ (complement of H in G) have only one vertex is common.

Definition 7.40. A graph G is called a non-separable graph, if it is connected, non-trivial and has no cut points. It is often called a **block**.

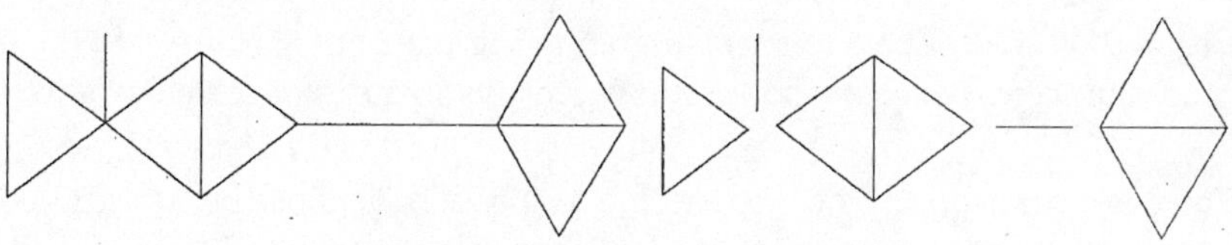

Fig. 7.71 A graph and its blocks

7.32 LABELED AND WEIGHTED GRAPHS

A graph in which each vertex is assigned a unique label is called a labeled graph. Labeled graphs are useful in counting the number of different graphs. The graph in Fig. 7.72 is a labeled graph.

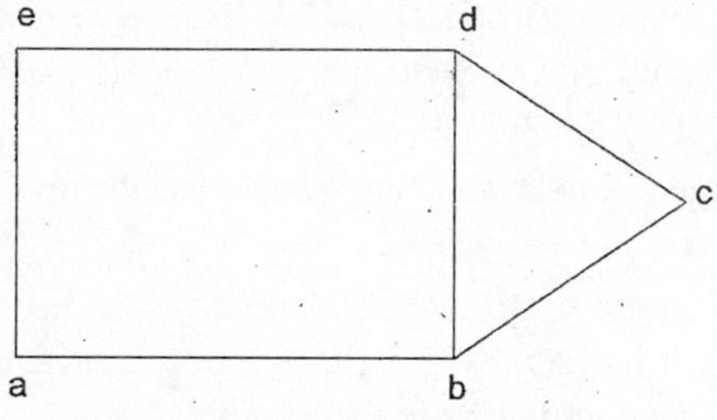

Fig. 7.72 A labeled graph

A graph G is called a weighted graph, if each edge e is assigned a non-negative number $w(e)$ called the weight of e. The graph shown in Fig. 7.73 is a weighted graph.

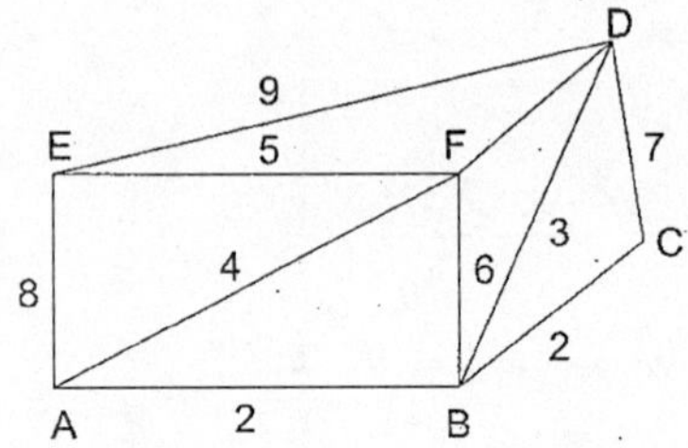

Fig. 7.73 A weighted graph

If G is a weighted graph and P is a path in G, then the weight of the path P is the sum of weights of the edges in the path.

7.33 CONNECTIVITY

7.33.1 Edge Connectivity

Definition 7.41. Let G be a connected graph. The edge connectivity of G is the minimum number of edges, whose removal results in a disconnected or a trivial graph.

The edge connectivity of a connected graph G is denoted by $\lambda(G)$. If G is disconnected graph then $\lambda(G) = 0$. If G is a connected graph and G has a bridge then $\lambda(G) = 1$. From the definition edge connectivity G the minimum cardinality of a cut set among all the cut sets of a connected graph.

7.33.2 Vertex Connectivity

Definition 7.42. Let G be a connected graph. The minimum number of vertices whose removal results in a disconnected or a trivial graph is called the vertex connectivity of G.

The vertex connectivity of G is denoted by $k(G)$. If G has a bridge then $k(G) = 1$. The vertex connectivity of a cycle graph C_n is two. If G is the complete graph (k_n) with n vertex then $k(G) = n - 1$.

Theorem 7.11. The edge connectivity of a connected graph G cannot exceed the minimum degree of G. i.e. $\lambda(G) \leq (G)$.

Proof. Let G be a connected graph and v be a vertex of minimum degree in G. Then the removal of edge incident with v disconnects the vertex v from the graph G. Thus the set of all edges incident with the vertex v form a cut set of G. But from the definition edge connectivity is the minimum number of edges whose removal result in a disconnected graph. Therefore edge connectivity cannot exceed the minimum degree of v i.e. $\lambda(G) \leq \delta(G)$.

Theorem 7.12. The vertex connectivity of a graph G is always less than or equal to the edge connectivity.

i.e. vertex connectivity of G cannot exceed the edge connectivity of G. ($k(G) \leq \lambda(G)$).

Proof. If G is a disconnected graph or a trivial graph then $k(G) = \lambda(G) = 0$.

If G is connected and has a bridge, then $\lambda(G) = 1$. In this case $k(G) = 1$, since either G has a cut point incident with e or G.

$\therefore \qquad k(G) \leq \lambda(G)$ when $\lambda(G) = 0$ or 1

Finally, let us suppose that $\lambda(G) \geq 2$. Then G has λ lines whose removal disconnects G. Clearly $\lambda - 1$ of these edges produces a graph with a bridge $e = \{u, v\}$. For each of these $\lambda - 1$ edges select an incident point which is different from u or v. The removal of u and v will result in either a disconnected or a trivial graph. Hence $k \leq \lambda$ in each case and this completes the proof of the theorem.

Thus, the vertex connectivity of a graphs G does not exceed the edge connectivity.

Theorem 7.13. For any graph G, $k(G) \leq \lambda(G) \leq \delta(G)$.

Proof. From theorem 7.12 $k(G) \leq \lambda(G)$ and from theorem 7.11 we have $\lambda(G) \leq \delta(G)$. Hence the theorem.

A graph G is n-connected if $k(G) \geq n$, and n-edge connected if $\lambda(G) \leq n$. Thus a non-trivial graph is 1-connected if and only if it is connected and 2-connected if and only if it is a block having more than one edge.

A maximal n-connected subgraph G is called an n-component of G. Two distinct n-components of a graph G have at most $n - 1$ points in common. The graph shown in Fig. 7.74 is a 3-component graph.

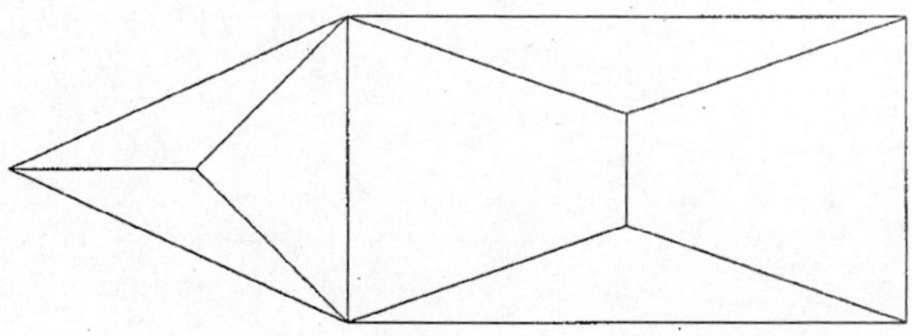

Fig. 7.74 A graph with 3-components

Example 1. If G is (p, q) graph, prove that if G is k-connected then $q \geq \dfrac{pk}{2}$

Solution. Let G be k-connected graph

Then $$q \leq \delta(G)$$

$\Rightarrow$ $$q \geq \frac{1}{2} \sum \deg(v_i) \leq \frac{1}{2}\, p\, \delta(G)$$

$\therefore$ $$q \geq \frac{pk}{2}$$

Example 2. Find the edge connectivity and the vertex connectivity of the graph G shown in Fig. 7.75.

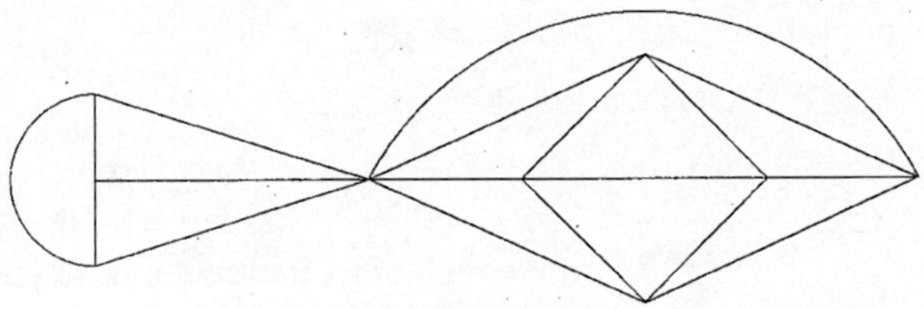

Fig. 7.75

Solution. The minimum number of edges whose removal disconnects the graph is 3, and minimum number of vertices whose removal disconnects the graph is one

$\therefore$ Edge connectivity $= \lambda(G) = 3$

and Vertex connectivity $= k(G) = 1$

Example 3. For the graph given below (Fig. 7.76) find all the fundamental cut sets.

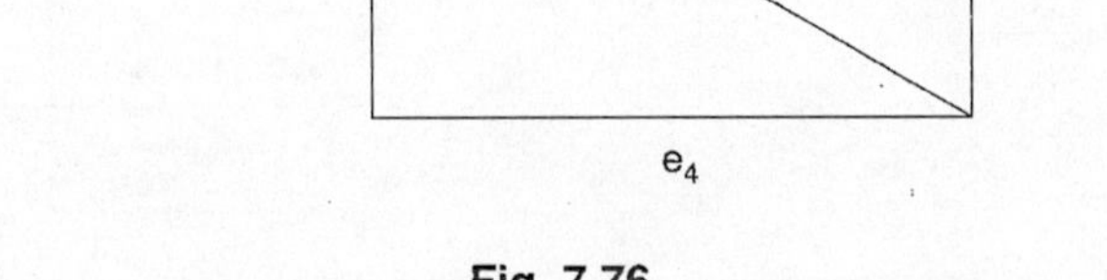

Fig. 7.76

The fundamental cut sets are

$$\{e_8\}, \{e_1, e_2\}, \{e_3, e_5\}, \{e_4, e_6\}, \{e_3, e_4, e_7\}$$

$$\{e_6, e_7, e_5\}, \{e_6, e_7, e_3\}, \{e_5, e_7, e_4\}$$

EXERCISE 7.2

1. Define:
   (a) Complement of a graph
   (b) Cycle graph
   (c) Path graph
   (d) Bipartite graph
   (e) Complete bipartite graph
   (f) Connected graph (MKU, MCA 2002)

   Give examples.
2. Define the terms:
   (a) Forest
   (b) Cut vertex
   (c) Cut edge
   (d) Cut set
   (e) Separable and non-separable graphs
3. Define: (1) Edge connectivity
   (2) Vertex connectivity
4. Construct a graph G with edge connectivity G, vertex connectivity 3 and vertex degree ≥ 5.
5. Prove that the vertex connectivity of a graph G can never exceed the edge connectivity of G.
6. Define connectivity, separability and cut vertex. (J.N.T.U. June 2003)
7. Prove that the edge connectivity of a graph G cannot exceed the minimum degree of G. (J.N.T.U. June 2003)

7.34 TREES AND SOME PROPERTIES OF TREES

The concept of a trees is important for those interested in applications of graph theory. The important applications of trees include searching, sorting, syntax checking and data base management. In this section we shall define a tree and study of its properties.

Definition 7.43. A tree is a connected graph without any circuits.

From the above definition it is clear that a tree is a connected acyclic graph. It has neither selfloops nor parallel edges and is denoted by the symbol T. Since trees are acycle. We adopt G convention similar to that used for Hasse diagrams. Trees may be directed or non-directed.

7.34.1 Directed Tree

Definition 7.44. A connected, acyclic directed graph is called a directed tree.

The graph in Fig. 7.77(a) is a non-directed tree and the graph shown in Fig. 7.77(b) is directed tree.

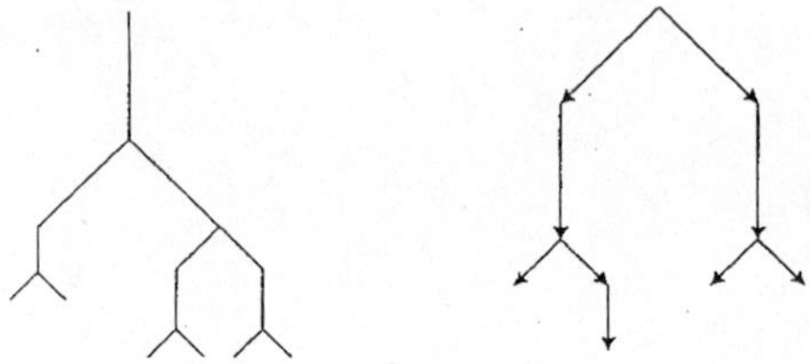

(a) (b)

Fig. 7.77 Trees

If T is a tree, then it has a unique simple non-directed path between each pair of vertices. A tree with only one vertex is called a trivial tree. If T is not a trivial tree then it is called a non-trivial tree. The vertex set (i.e. the set of nodes) of a tree is a finite set. In most cases the vertices of a tree are labeled.

Theorem 7.13 A simple non-directed graph G is a tree if and only if G is connected and has no cycles.

Proof. Let G be a tree. Then each pair of vertex of G are joined by a unique path. Therefore G is connected. Let u and v be two distinct vertices of G, such that G contains a cycle containing u and v. Then u and v are joined by at least two simple paths. One along one portion of the cycle and the other path completing the cycle. This contradicts over hypothesis that there is a simple unique path between u and v. Hence tree has no cycle.

Conversely, let G be a connected graph having no cycles. Let v_1 and v_2 be any pair of vertices of G and let there be two difference simple paths P_1 and P_2 from v_1 to v_2. Then we can find a cycle in G as follows.

Since the paths P_1 and P_2 are different there must be a vertex say u, which is on both P_1 and P_2 but its successor on P_1 is not on P_2. If it is the next point on P_1 which is also on P_2, the segment of P_1 and P_2 which are between P_1 and P_2 form a cycle in G leading to a contradiction. Hence there is at most one path between any two vertices of G, which shows that G is a tree.

Theorem 7.14. Any non-trivial tree has at least one vertex of degree one.

Proof. Let G be a non-trivial tree. Then G has no circuits. Let v_1 be any vertex of G. If $\deg(v_1) = 1$ the theorem is at once established. Let $\deg(v_1) \neq 1$. Move along any edge to a vertex v_2 incident with v_1. If $\deg(v_2) \neq 1$. Then continue to another vertex say v_3 along a different edge. Continuing the process we get a path $v_1 - v_2 - v_3 - \ldots$, in which none of the v_i's is repeated. Since the number of vertices in a tree is finite the path must end somewhere. The vertex at which the path ends is of degree one, since we can enter the vertex but cannot the leave the vertex.

Theorem 7.15. A tree T with vertices has exactly $(n - 1)$ edges.

Proof. The theorem will be proved by induction on the number of vertices of a tree.

If $n = 1$, then there are no edges in T. Hence the result is trivial.

If $n = 2$, then the number of edges connecting the vertices is one i.e. $1 = 2 - 1 = n - 1$.

Therefore the theorem holds for $n = 1$ and $n = 2$.

Assume that the theorem holds for all trees with fewer than n vertices. Consider a tree T with n vertices. Let v be a vertex of degree 1 in T and let T', denote the graph obtained by removing the vertex v and the edge e associated with it from T.

Consider $T' = T - v$.

T' has $n - 1$ vertices and fewer edges than T. If v_1 and v_2 are any two vertices in T', there is unique simple path in T' since there a unique path between v_1 and v_2 in T. Therefore T' is connected. The path between v_1 and v_2 is not affected by the removal of the vertex v and the edge e. There are no cycles in T', since there are no cycles in T. Thus T' is a tree and the number of edges in T' is $n - 2$. Therefore the number of edges in T is $n - 2 + 1 = n - 1$.

Theorem 7.16. If G is a non-trivial tree than G contains at least 2 vertices of degree 1.

Proof. Let T be a non-trivial tree with n-vertices. Then by the sum of degrees formula, we have

$$\sum_{i=1}^{n} \deg(v_i) = 2 \text{ (number of edges in } T)$$

$$= 2(n - 1) = 2n - 2 \qquad \text{...(1)}$$

Now, let us assume that there only one vertex say v_1, of degree 1 in T. Then

$$\deg(v_i) \geq \text{ for } i = 2, 3, \ldots, n$$

$$\therefore \quad \sum_{1}^{n} \deg(v_i) \equiv \deg(v_1) + \sum_{1=2}^{n} \deg(v_i) \geq 1 + 2\,(n - 1)$$

$$\Rightarrow \quad \sum_{1}^{n} \deg(v_i) \geq 1 + 2\,n - 2$$

$$\Rightarrow \quad 2n - 2 \geq 2n - 1 \qquad \text{(using (1))}$$

$$\Rightarrow \quad -2 \geq -1, \text{ a contradiction.}$$

Our assumption that T contains exactly one vertex of degree 1 leads to a contradiction. Therefore, T contains at least two vertices of degree 1.

Theorem 7.17. A graph G is a tree if and only if G has no cycles and $|E| = |V| - 1$.

Proof. Let G be a tree, then by theorem the number of edges in G is $|V| - 1$

i.e. $$|E| = |V| - 1$$

Conversely let G be a graph such that G has no cycle and $|E| = |V| - 1$. The graph G is a tree if we show that G is connected. To prove this, let us assume that $G_1, G_2, G_3, ..., G_k$ are k-components of G where $k \geq 1$.

Let $|V_i|$ denote the number of vertex in G_i. G is contains no cycles. Therefore each G_i is connected and each G contains no cycles in it.

Therefore each G_i is a tree.

Thus, the number of edges in each G_i is $|V_i| - 1$.

$\therefore$ Number of edges in G

$$\equiv |V_1| - 1 + |V_2| - 1 + ... + |V_k| - 1$$

$$= |V_1| + |V_2| + ... + |V_k| - k = |V| - k$$

by hypothesis number of edges in G is $|V| - 1$ $\quad (\because |V_1| + |V_2| + ... + |V_k| = V)$

$\therefore$ $$|V| - 1 = |V| - k$$

$\Rightarrow$ $$-1 = -K$$

$\Rightarrow$ $$k = 1$$

The number of components in G is one and G is connected.

$\therefore$ G is a tree.

7.35 DISTANCE

If u and v are two vertices of a connected graph G, there may be more than one path joining the vertices u and v. Various concepts can be defined based on the lengths of such paths between the vertices of G. The simplest is given below.

Definition 7.45. If G is a connected graph and u, v are any two vertices of G, then the length of the shortest path between u and v is called the **distance** between u and v. It is denoted by $d(u, v)$.

The distance function defined above has the following properties. If u, v and w are any three vertices of a connected graph G, then

(i) $d(u, v) \geq 0$ and $d(u, v) = 0$ if $u = v$

(ii) $d(u, v) = d(v, u)$

and (iii) $d(u, v) \leq d(u, w) + d(w, v)$

From the above properties, it is clear that the distance d in a graph is a metric.

Example. In the graph shown in Fig. 7.78

$$d(v_1, v_2) = 1, d(v_1, v_4) = 2$$

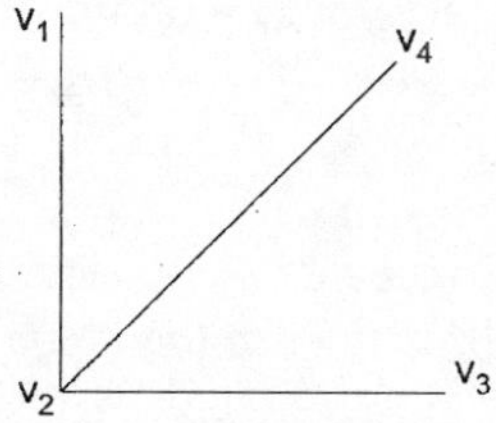

Fig. 7.78

Definition 7.46. (Eccentricity): Let G be a connected graph. For any vertex v in G, the eccentricity of v is denoted by $e(v)$ is

$$e(v) = \max \{d(u, v) : u, v \in V\}$$

from the definition, $e(v)$ is the length of the longest path in G starting from the vertex v.

Example. In the graph shown Fig. 7.79. $e(v_1) = 3$, $e(v_2) = 2$, $e(v_3) = 2$

$$e(v_4) = 3, e(v_5) = 3$$

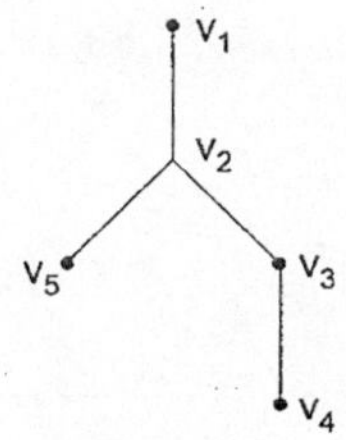

Fig. 7.79

Definition 7.47. The **diameter** of a connected graph G is defined as the maximum eccentricity among all vertices of the graph G. It is denoted by d.

$\therefore$ d = diameter if $G = \max \{e(v) : v \in V\}$

Example. Consider the tree T shown in Fig. 7.80.

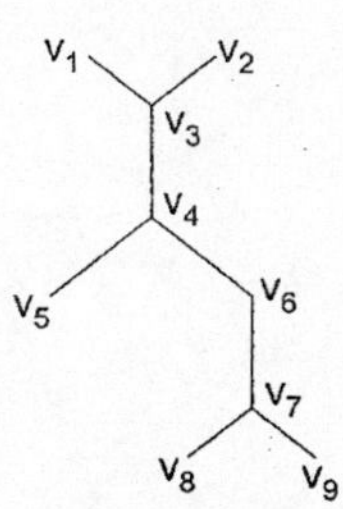

Fig. 7.80

We have $e(v_1) = 5,\ e(v_2) = 5,\ e(v_3) = 4,\ e(v_4) = 3,\ e(v_5)$

$e(v_6) = 3,\ e(v_7) = 4,\ e(v_8) = 5,\ e(v_9) = 5$

$$\max\{5, 5, 4, 3, 3, 4, 5, 5, 5, 8\} = 5$$

$\therefore$ The diameter of $T = 5$.

Definition 7.48. (radius): The radius of a connected graph G is defined as the maximum eccentricity among all vertices of the graph G. It is denoted by r.

$$\text{radius of } G = r = \min\{e(v) : v \in V\}$$

Note. The radius of a connected graph may not be half of its diameter.

Example. In the graph shown in Fig. 7.80. The radius of the graph in 3.

Definition 7.49. (Center of a graph) The centre of a connected graph G is defined as the set of vertices having minimum eccentricity among all vertices of the graph G. It is denoted by $C(G)$ (or by C) centre of $G = C(G) = \{v \in V : e(v) = r\}$

Example 1. Consider the graph shown in Fig. 7.81.

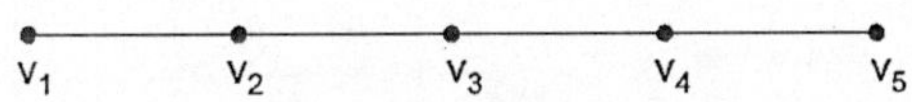

Fig. 7.81

We have, $e(v_1) = 4,\ e(v_2) = 3,\ e(v_3) = 2,\ e(v_4) = 3,\ e(v_5) = 4$

radius of $G = r = 2$

$\therefore$ Centre of $G = \{v_3\}$

Example 2. The centre of the graph in Fig. 7.82 is $C(G) = \{v_3, v_4)$

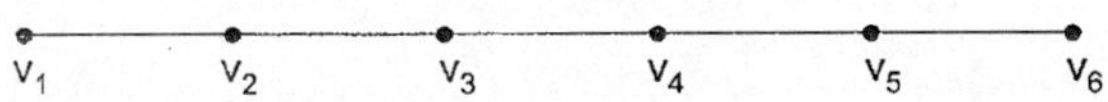

Fig. 7.82

Example 3. Find the centre of graph shown in Fig. 7.83.

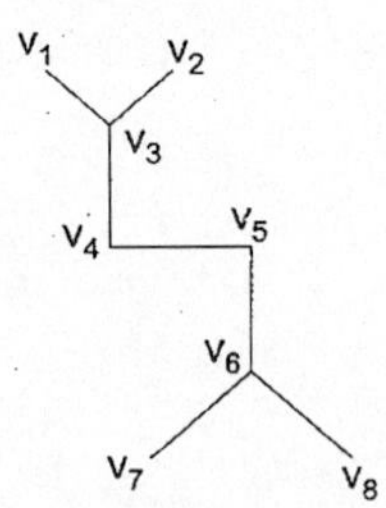

Fig. 7.83

Solution. We have $e(v_1) = 5$, $e(v_2) = 5$, $e(v_3) = 4$

$$e(v_4) = 3,\ e(v_5) = 3,\ e(v_6) = 4,\ e(v_7) = 5$$

$$e(v_8) = 5,\ \min |e(v_i) = v_i \in V| = 3$$

Occurs at v_4 and v_5.

$\therefore$ Centre of $G = \{v_4, v_5\}$.

Example 4. If r is the radius and d is the diameter of a connected graph G, then show that

$$r \le d \le 2r$$

Solution. From the definition of r and d, we have $r \le d$. ...(1)

Let u, v be the ends of a diametral path and w be the centre vertex, then

$$d(u, v) \le d(u, w) + d(w, v)$$

$\Rightarrow$ $$d \le r + v$$

i.e. $$d \le 2r \quad \text{...(2)}$$

From (1) and (2)

We have $$r \le d \le 2r$$

If G is a connected graph and $v_1, v_2, \ldots, v_n$ vertices of G, then

$$e(v_1),\ e(v_2), \ldots, e(v_n)$$

is called the eccentricity sequence of G.

Note.

(i) If $C(G) = V(G)$, then the graph G is called self-centred graph.

(ii) The distance between two adjacent vertices of a connected graph G is 1.

(iii) The maximum distance from each vertex of G, occurs at a pendant vertices of G.

(iv) If P is a path of even length, then P has only one vertex at the centre.

(v) If P is a path of odd length, then the centre of P contains two adjacent vertices.

7.35.1 Spanning Tree

Definition 7.50. Let G be a connected graph, and H be a subgraph of G. H is called a spanning tree of G. If

(i) H is a tree

and (ii) H contains all the vertices of G.

Example 1. In the graph shown in Fig. 7.84, H is a spanning tree of G.

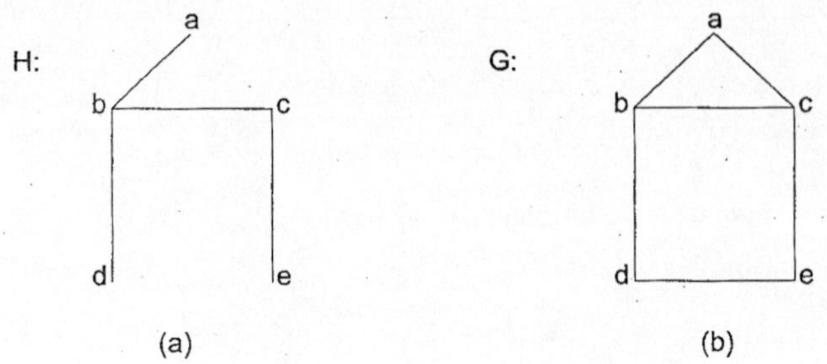

Fig. 7.84 Spanning tree

Example 2. Consider the graph shown in Fig. 7.85.

H_1 and H_2 are spanning trees of G.

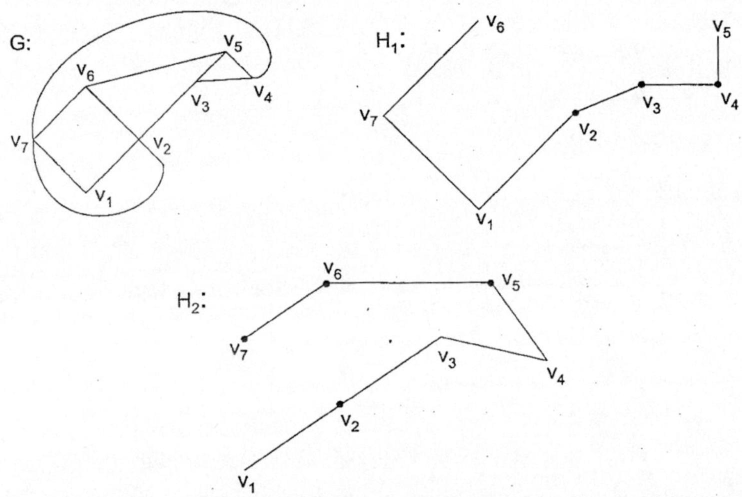

Fig. 7.85

7.35.2 Rank and Nullity

If G is a connected graph and T is a spanning tree of G, edges of G present in T are called the branches of G with respect to T, and edges which do not belong to T are called the chords of G with respect to T. If G has n vertices and e edges. Then the number of branches with respect to spanning tree T of G is $n - 1$, and the number of chords is $e - n + 1$.

The number of branches in a connected graph G is called the rank of G, and the number of chords is called the nullity of G. If G has k components then the rank of G is defined as the sum of ranks of the components. i.e.

$$\text{rank}(G) = \sum_{i=1}^{n} \text{rank}(G_i)$$

$$r = \sum_{i=1}^{k} (n_i - 1) = \sum_{i=1}^{k} n_i - \sum_{i=1}^{k}$$

$$r = n - k$$

where $G_i, i = 1, 2,... k.$

are components of G.

and nullity of $G = \sum_{i=1}^{k} \text{nullity}(G_i)$

$$\mu = \sum_{i=1}^{k} (e_i - n_i + 1)$$

$$\mu = \sum_{1}^{k} e_i - \sum_{1}^{k} n_i + \sum_{1}^{k} 1$$

$$\mu = e - n + k$$

where r is the rank of G and μ is the nullity of G.

For a tree; $r = n - 1$ and $\mu = e - n + 1$

Theorem 7.18. A non-directed graph G is connected if and only if G contains a spanning tree.

Proof. Let T be a spanning tree of G. Then there exists a path between any pair of vertices in G along the tree T. Therefore G is connected.

Conversely, let G be a connected graph and k be the number of cycles in G. If $k = 0$, then G has no cycles and G is connected. Therefore G is a tree when $k = 0$.

Let us suppose that all connected graphs with fewer than k cycles have a spanning tree. Let G be a connected graph with n cycles. Let e be an edge in one of the cycles. Then $G - e$ is a connected graph and $G - e$ contains all the vertices of G. Then G contains a spanning tree if $G - e$ is also a spanning tree.

7.36 BREADTH-FIRST SEARCH AND DEPTH-FIRST SEARCH

In this section, we define algorithms for finding a spanning tree of connected graph. These algorithms are known as breadth-first search and depth-first search algorithms.

7.36.1 Depth-first Search

Let G be a connected simple graph. In depth-first search we construct a spanning tree for G by successively adding the edges as follows.

Algorithm: (Depth-first search)

Input: A connected graph G, whose vertices are labeled as $v_1, v_2,..., v_n$

Output: A spanning tree T for G.

Step 1. Let v_1 be the root and set $L = v_1$ (where v_2 stands for the vertex last visited)

Step 2. For all vertices adjacent to v_2 choose the edge $\{v_l, v_k\}$, where k is the minimum index such that adding $\{L, v_k$ to the tree T does not create a cycle where v_k is an unvisited vertex and the edge.

$\{v_l, v_k\}$ is an unexamined edge.

If no such edge exists, go to step 3. Otherwise add edge $\{v_l, v_k\}$ to T and set $v_l = v_k$, repeat step 2 at the new value for V_l.

Step 3. Back traking and terminating: If u is the parent of v_l in T, Set $v_l = u$ and apply step 2 at the new value of v_l. If on the other hand v_l has no parent os T then the depth-first search terminates and T is a spanning tree for G.

The edges selected by a depth-first search of a graph G are called tree edges. All the other edges of G must connect a vertex to an ancestor or descendant. Of this vertex in the tree. Depth-first search is also called back tracking. In depth-first search, we choose an ordering of the vertices of the graph G and select one of the vertices of G as the root of the spanning tree.

7.36.2 Breadth-first Search (BFS)

In this type of search we visit all vertices of G sequentially on a given level before going onto the next level. The breadth-first search algorithm is given below.

Algorithm: Breadth-first search for a spanning tree.

Input: A connected graph G, with vertices labeled $v_1, v_2,..., v_n$.

Output: A spanning tree T for G.

Step 1. (Start) Let v_1 be the root of the spanning tree T form the set $V = \{v_1\}$.

Step 2. (Add new edges) Consider the vertices of V in order consistent with original labeling. Then for each vertex $u \in V$, add the edge $\{u, v_\alpha\}$ to T where α is the minimum index such that adding the edge $\{u_1, v_\alpha\}$ to T does not produce a cycle. If no edge can be added, then step 0; T is a spanning tree for G. After all the vertices of V are have been considered in order, go to step 3.

Step 3. (Update V): Replace V by all children v in T of the vertices u of V where the edges $\{u, v\}$ were added in step 2. Go back and repeat step 2 for the new set V.

Example 1. Consider the graph G, shown in Fig. 7.86(a) we apply depth-first search algorithm to the graph and construct a spanning tree.

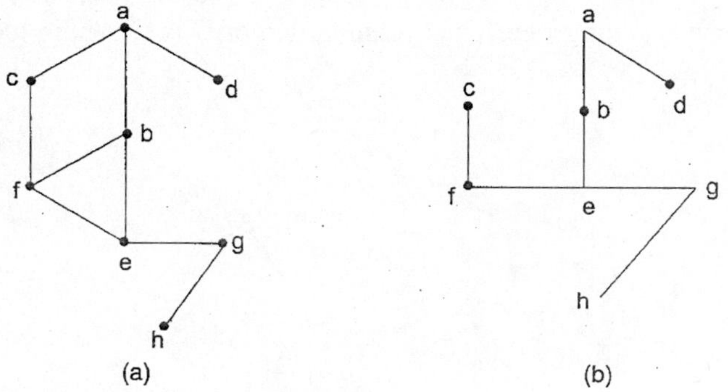

Fig. 7.86

We arbitrarily start with the vertex a. A path is constructed by successively adding the edges incident with vertices not already in the path, as long as possible. This produces a path $a - b - e - g - h$. Next back track to g. There is no path beginning at g containing vertices not already visited. So we back track to e and form a path $e - f - c$. We then back track to f (section of the edge $\{f, b)$ forms a cycle, hence the edge (f, b) is not selected we then back track to the vertex e and then to the vertex b. Form b we back track to a and form a path $a - c$ (note selecting the edge $\{a, c)$ for the path, forms a cycle, hence $\{a, c\}$ is not selected. The spanning tree produced is shown in Fig. 7.86(b).

Example 2. Consider the graph shown in Fig. 7.87, we construct a spanning tree for the graph shown in the Figure 7.87.

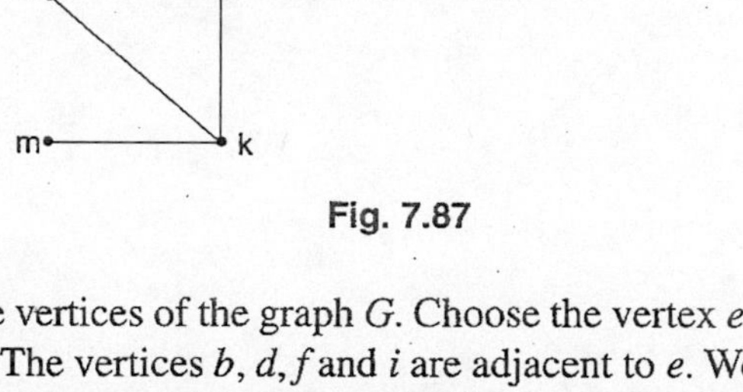

Fig. 7.87

Consider the vertices of the graph G. Choose the vertex e to be the root of the spanning tree T. The vertices b, d, f and i are adjacent to e. We add the edges b, d, f and i are added. These vertices are at level 1 in T. Consider the vertex b. The vertices a and c are adjacent to b. Hence the edges $\{b, a\}$ and $\{b, c\}$ are added. Next, we consider the vertex d and add the edge $\{d, h\}$. Similarly we add the edges $\{i, h\}$, $\{i, j\}$, are $\{i, k\}$. Next we add the edge $\{j, g\}$. The vertices a, c, h, j, g and k are at level 2.

Finally, we consider the edges from these vertices to the adjacent vertices not already in the tree. The edges are $\{g, l\}$ and $\{k, m\}$. We add these edges to construct the spanning tree. The spanning tree of G is shown in the figure given below.

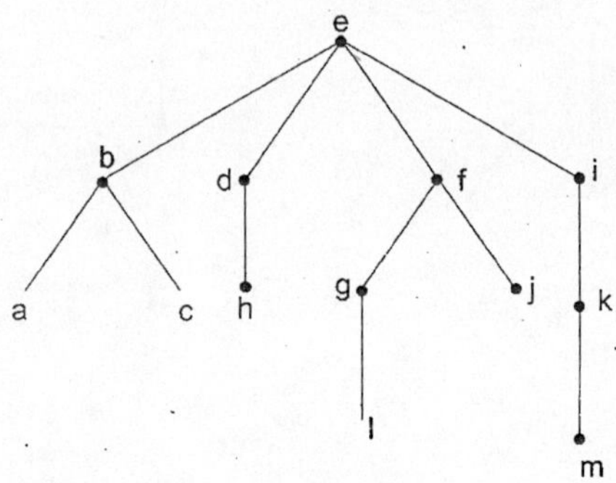

Fig. 7.88

7.37.1 Minimal Spanning Tree

Let G be a connected weighted graph. A minimal spanning tree of G is a spanning tree of G whose total weight is as small as possible.

There are various methods to find a minimal spanning tree in a connected weighted graph. In this section, we consider algorithms for generating minimal spanning trees in given graphs.

7.37.2 Algorithm

A connected weighted graph G, with n vertices.

Step 1. Arrange the edges of G in the order of decreasing weights.

Step 2. Produced sequentially, and delete each edge of G, that does not disconnect the graph G, until $n - 1$, edges remain.

Step 3. Exit.

Example. Consider the graph G given below. (see Fig. 7.89).

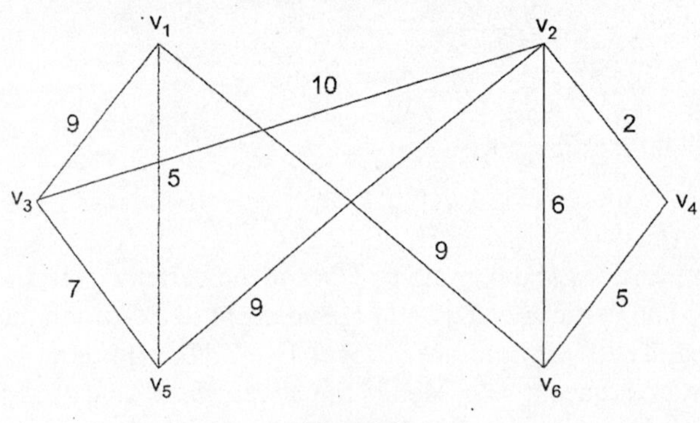

Fig. 7.89

Number of vertices in $G = 6$.

We apply the algorithm given above.

We order the edges by decreasing weights and delete edges of G until $n - 1 = 6 - 1 = 5$ edges remain.

Edges	$\{v_2, v_3\}$	$\{v_1, v_6\}$	$\{v_1, v_3\}$	$\{v_2, v_5\}$	$\{v_3, v_5\}$	$\{v_2, v_6\}$	$\{v_1, v_5\}$	$\{v_4, v_6\}$	$\{v_2, v_4\}$
Weight	10	9	9	9	7	6	5	5	2
Delete	yes	yes	yes	no	no	yes	no	no	no

The minimal spanning tree of G is shown in Fig. 7.90.

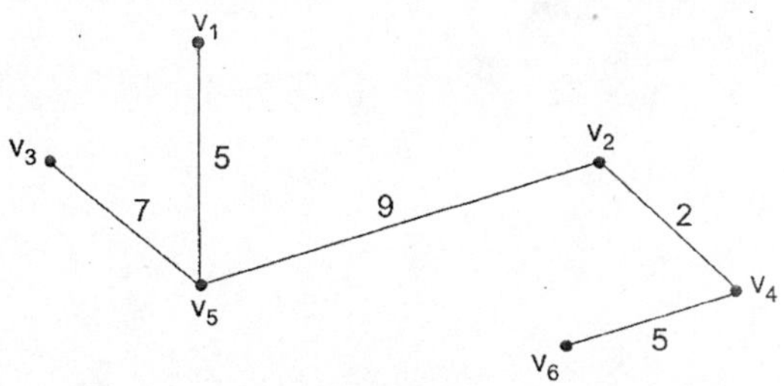

Fig. 7.90

The weight of the minimal spanning tree $= 9 + 7 + 5 + 5 + 2 = 28$

7.37.3 Kruskal's Algorithm

Input a connected weighted graph G with n vertices.

Step 1. Arrange the edges in order of increasing weights and select the edges with minimum weight.

Step 2. Proceed sequentially add each edge which does not result in a cycle until $n - 1$, edge selected.

Step 3. Exit.

Example. Consider the graph shown in Fig. 7.86.

Solution. We have $n = 6$

We order the edges by increasing weights $\{v_2, v_4\}$ is an edge in the graph with minimum weight. We select the edge $\{v_2, v_4\}$ and successively add edges $\{v_2, v_4\}$ without forming a cycle until $6 - 1 = 5$ edges are selected. This yields:

Edges	$\{v_2, v_4\}$	$\{v_1, v_5\}$	$\{v_4, v_6\}$	$\{v_2, v_6\}$	$\{v_3, v_5\}$	$\{v_1, v_3\}$	$\{v_1, v_6\}$	$\{v_2, v_5\}$	$v_2, v_3\}$
Weight	2	5	5	6	7	9	9	9	10
Add	yes	yes	yes	no	yes	no	yes	no	no

Edges in the minimum spanning tree are

$$\{v_2, v_4\}, \{v_1, v_5\}, \{v_4, v_6\}, \{v_3, v_5\}, \{v_1, v_6\}$$

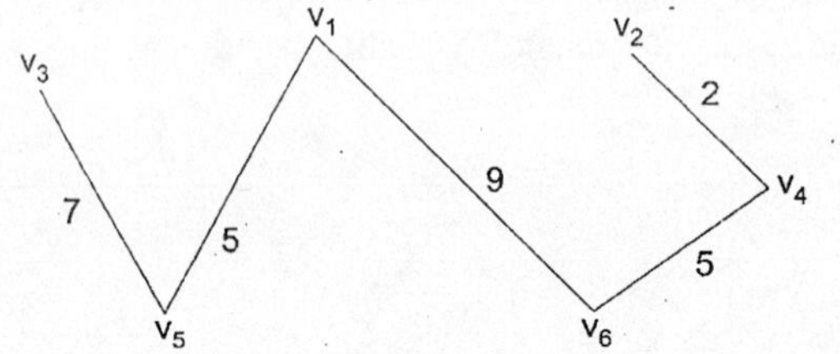

Fig. 7.91

The resulting minimal spanning tree is shown in Fig. 7.91. We apply the step of Kruskal's algorithm to the graph of 7.91 as follows.

$\{v_2, v_4\}$ is the edge with minimum weight, therefore we select the edge $\{v_2, v_4\}$.

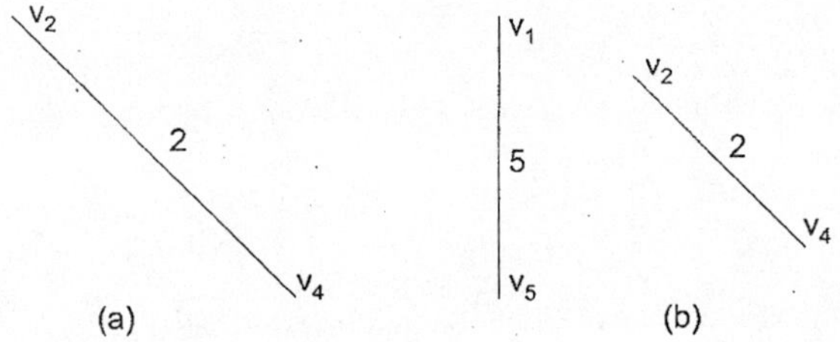

Fig. 7.92

The next edge with minimum weight is $\{v_1, v_5\}$.

Selection of $\{v_1, v_5\}$ does not result in a cycle. Therefore edge $\{v_1, v_5\}$ is selected.

The edge to be considered next is $\{v_4, v_6\}$.

Inclusion of the edge $\{v_4, v_6\}$ in the spanning tree does not result in a cycle. Therefore, we select the edge $\{v_4, v_6\}$.

Selection of the edge $\{v_2, v_6\}$ for the spanning tree results in a cycle. Therefore $\{v_2, v_6\}$ is not selected. In the next step we consider the edge $\{v_3, v_5\}$. Selecting of the edge $\{v_3, v_5\}$ does not result in a cycle. Hence the edge $\{v_3, v_5\}$ is selected.

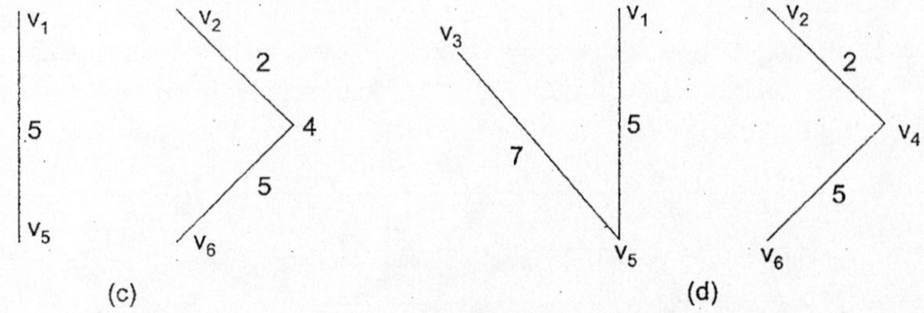

Fig. 7.92

Next we consider the edge $\{v_1, v_3\}$ from the list. Selection of $\{v_1, v_3\}$ results in a cycle. Therefore edge $\{v_1, v_3\}$ is not selected. Consider the edge $\{v_1, v_6\}$. Selecting of the edge $\{v_1, v_6\}$ does not result in a cycle. Hence $\{v_1, v_6\}$ is selected.

The number of edges selected is 5. Here we stop and obtain the spanning tree as shown in Fig. 7.92 (e).

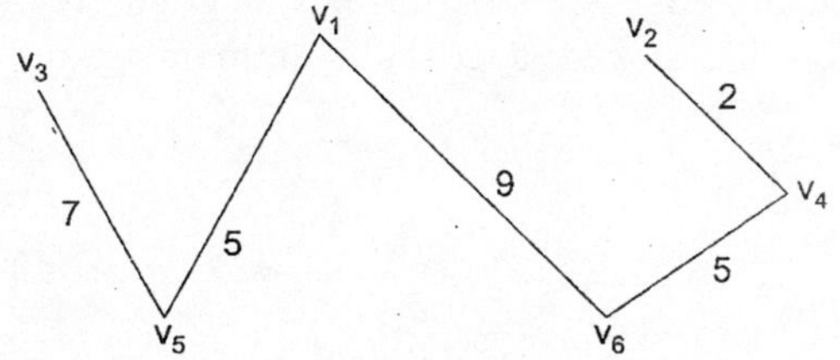

(e)

Fig. 7.92

The weight of the minimal spanning tree

$$= 2 + 5 + 5 + 7 + 9 = 27.$$

7.37.4 Prims Algorithm

Input: A connected weighted graph G with n vertices.

Step 1. Select an arbitrary vertex v_1 and an edge e_1 with minimum weight incident vertex v_1.

Step 2. Having selected the vertices $v_1, v_2, ..., v_i$ and $e_1, e_2, ..., e_{i-1}$. Select an edge e_i such that e_i connects a vertex of the set $\{v_1, v_2, ..., v_i\}$ and a vertex of $v - \{v_1, v_2, ..., v_i\}$ and of all such edges e_i has the minimum weight.

Step 3. Stop, it $n - 1$, edges are selected else go to step 2.

Example. Consider the graph shown in Fig. 7.93.

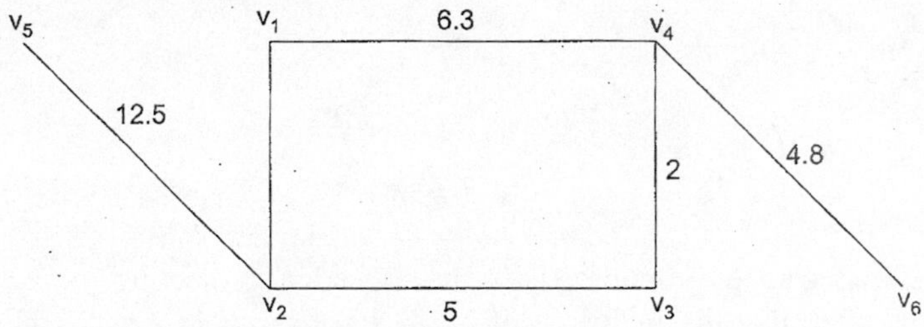

Fig. 7.93

Let $e_1 = (v_1, v_2)$, $e_2 = (v_2, v_3)$, $e_3 = (v_3, v_4)$

$e_4 = (v_4, v_1)$, $e_5 = (v_2, v_5)$ and $e_6 = (v_4, v_6)$

denote the edges of G.

We apply prims algorithm to the graph as follows:

The edge $e_3 \equiv (v_3, v_4)$ is an edge with minimum weight. Hence we start with the vertex v_3 and select the edge e_3 incident with v_3.

After selecting the edge e_3, we consider the edges connecting a vertex of $\{v_3, v_4\}$. We observe that e_6 is the edge with minimum weight.

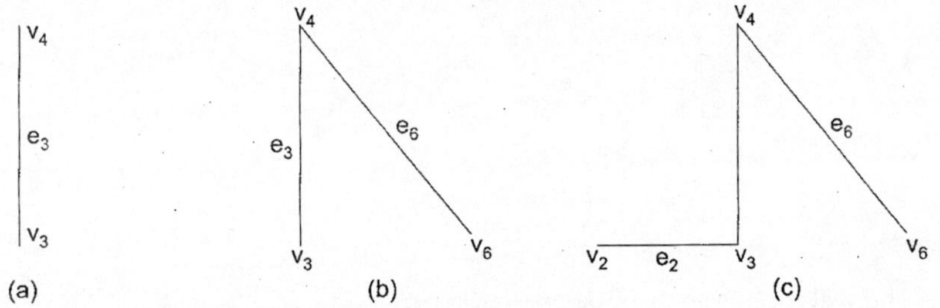

Fig. 7.94 (a, b, c)

Consider the edges connecting the vertices of the set $\{v_3, v_4, v_6\}$ with the vertices of $V - \{v_3, v_4, v_6\}$. The edge e_2 has the minimum weight. Hence we select the edge e_2 of the edges connecting the vertices $\{v_2, v_3, v_4, v_6\}$, with the vertex set $V\{v_2, v_3, v_4, v_6\}$, the edge e_4 has minimum weight. Therefore the edge e_4 is selected.

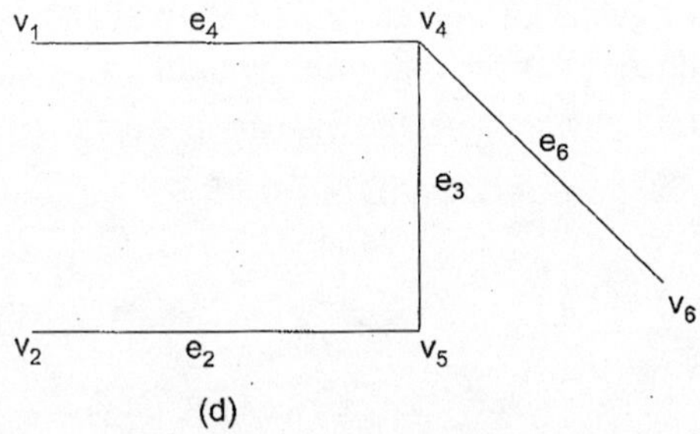

Fig. 7.94 (d)

e_1, e_5 are the edges remaining. e_5 is the only edge connecting $\{v_1, v_2, v_3, v_4, v_5, v_6\}$ and v_5, such that the inclusion of e_5 does not result in a cycle. Hence e_5 is selected. Since the number of edges selected is 5, we stop.

The minimal spanning tree obtained is shown in Fig. 7.94 (e).

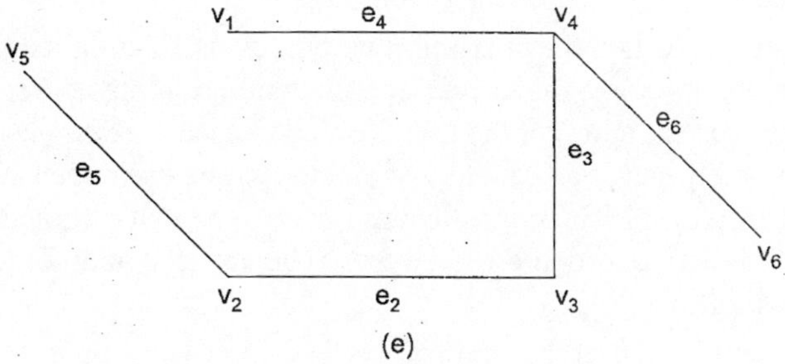

Fig. 7.94 (e)

Weight of the minimal spanning tree

$$\equiv 2 + 4.8 + 5 + 6.3 + 12.5 = 30.6$$

7.38 ROOTED TREE

A rooted tree is a tree with a designated vertex called the root of the tree.

Any tree may be made into a rooted tree by selecting one of the vertices as the root. A rooted tree is a directed tree if there is a root from which there is a directed path to each vertex of the tree. The graphs in Fig. 7.95 are rooted trees in which the root of each is at the top.

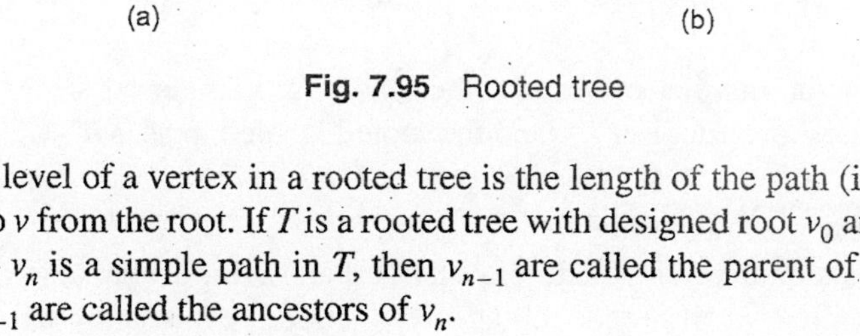

Fig. 7.95 Rooted tree

The level of a vertex in a rooted tree is the length of the path (i.e. number of edges) to v from the root. If T is a rooted tree with designed root v_0 and $v_0 - v_1 - v_2$ ... $v_{n-1} - v_n$ is a simple path in T, then v_{n-1} are called the parent of v_n and $v_0, v_1, v_2, \ldots, v_{n-1}$ are called the ancestors of v_n.

v_1 is a child of v_0,

v_2 is a child of $v_1, \ldots$

Let T be a rooted tree with designated vertex v_0. If u and v are two vertices (nodes) in T, then

(i) u is called a leaf of T, if it has no children i.e. leaves of T are vertices of T with degree 1.

(ii) v is descendant of u, if u is an ancestor of v.

(iii) v is an interval vertex of T, if v is not a leaf of T, and

(iv) The subgraph of T consisting of v and all of its descendants with v as the designated root is a such tree of T rooted at v.

If T is a rooted tree, then the maximum vertex level of T is called the depth of the tree. We usually adhere to the universal conversion of representing the root of the tree as the top vertex (apex) of the tree. Rooted trees are useful in enumerating all the logical possibilities of a sequence of events, where each event can occur in a finite number of ways. If the edges leaving each vertex of a rooted tree. T are labeled then T is called an ordered rooted tree. The vertices of an ordered rooted can be labeled as follows:

We assign 0 to the root of the tree.

We next assign 1, 2, 3,... to the vertices immediately following the root of T according as the edges where ordered. The remaining vertices can be ordered as follows:

If p is the label of a vertex v of T, then p_1, p_2, p_5 are assigned to the vertices immediately following v according as the edges were ordered. The tree in Fig. 7.96 is an ordered tree.

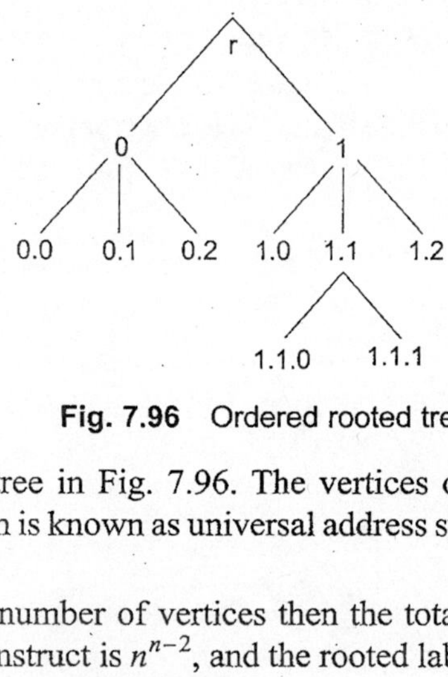

Fig. 7.96 Ordered rooted tree

r is the root of the tree in Fig. 7.96. The vertices of T are labeled with their addresses. The system is known as universal address system for an ordered rooted tree.

If n denotes the number of vertices then the total number labeled different trees, that one can construct is n^{n-2}, and the rooted labeled trees is n^{n-1}.

7.39 EXPRESSION TREES

Algebraic expression involving addition, subtraction, multiplication and division can be represented as ordered rooted trees called expression trees. The arithmetical expression $3 + 5 \times 9 - 7 \times 6^2$ can be respresented as a tree as shown in Fig. 7.97.

The variables in the algebraic expression appear as the other vertices. In the polish prefix representation, we place binary operational symbol before the argument and avoid parentheses. The expression $(a - b)/((c \times d) \times e)$ can be expressed as $/- ab + \times cde$.

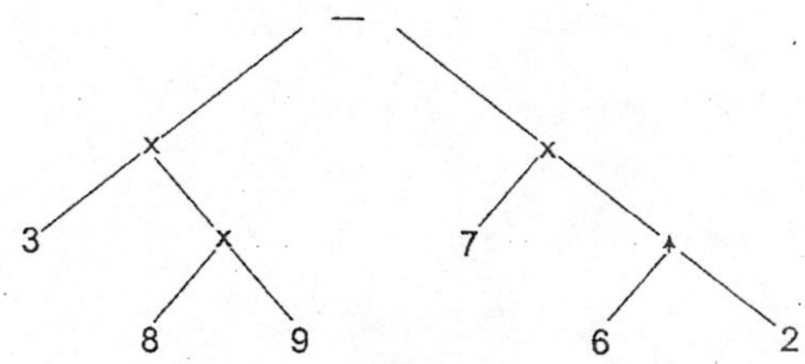

Fig. 7.97

Example. Write the following expression as a tree $[((a + b) \times c) \times (d + c) - (f - (g \times h))]$

Solution. The given expression can be expressed as a tree as shown in Fig.7.98

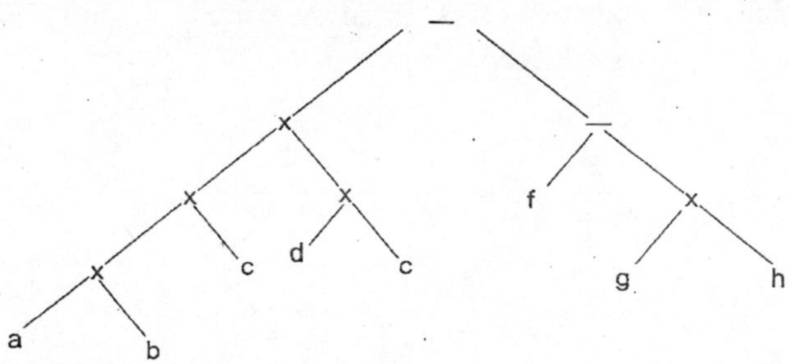

Fig. 7.98

7.40 BINARY TREES

So far we have discussed the tree and its properties. Now we shall study a special class of rooted trees known as binary trees, which play an important role in decision making. Binary trees are extensively used in the study of computer methods, binary identification problems and coaching theory.

Definition 7.51. A tree in which there is exactly one vertex of degree two and each of the remaining vertex of degree one or three is called a binary tree.

If T is a binary tree, the vertex of degree two, which is distinct from all the other vertices of T, serves as a root of T. Thus every binary tree is a rooted tree. The vertices of degree one in a binary tree are called external vertices and all the remaining vertices are called internal vertices. The number of internal vertices in a binary tree is one less than the number of pendant vertices.

The leaves of a binary tree are vertices of degree one. Usually the roots in graph theory are portrayed with the root at the top and the leaves at the bottom. The direction from the leaves to the root is taken as the up direction. The number of internal vertices in a binary tree is one less than the number of external (pendant) vertices.

A binary tree can also be defined as follows:

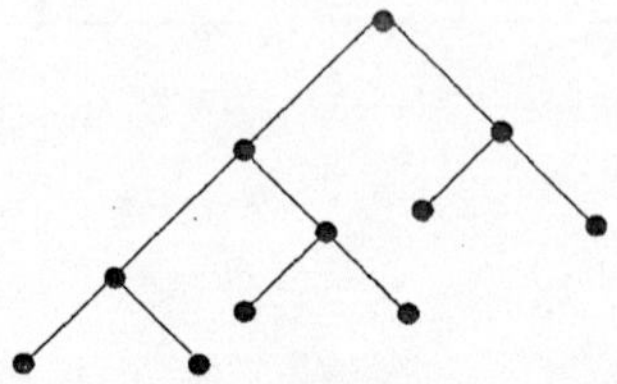

Fig. 7.99 Binary tree

Definition. 7.52. A binary tree is a directed tree $T = (V, E)$ together with an edge labeling $f : E \to \{0, 1\}$ such that every vertex of T has at most one edge incident from it labeled 0, and at most one edge incident from it labeled 1.

If T is a binary tree, than each edge (u, v) labeled 0, is called a left edge and u is called the parent of v and v is called a child of u. Each edge labeled with 1 is called a right edge in T. The vertex u is called the parent of v and u is called the right child of u.

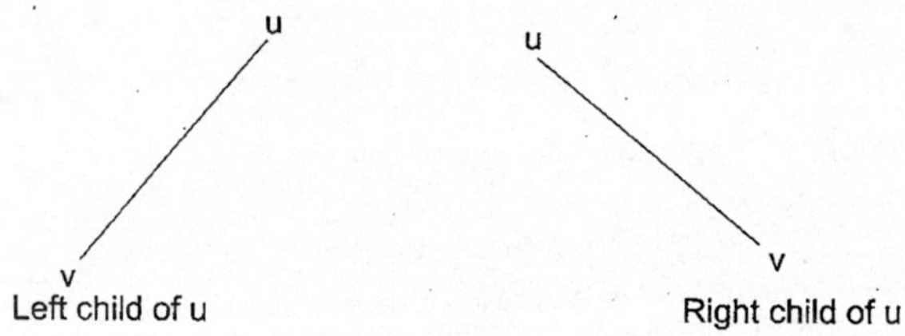

Fig. 7.100

If v_i is a vertex of a binary tree v_i is said to be at a level l_i if v_i is at a distance l_i from the root of the binary tree. Thus the root of a binary tree is at level 0.

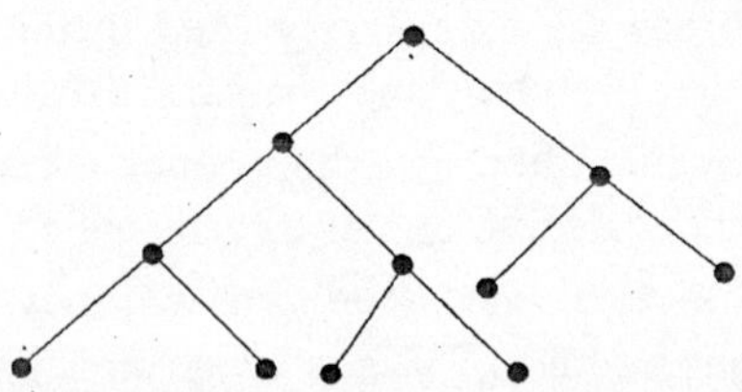

Fig. 7.101 A 11-vertex 3-level binary tree

The maximum level occurring in a binary tree is called the height of the binary tree. A binary tree with minimum height contains maximum of vertices at each level. The root of binary tree is at level 0 and there can be only one vertex at 0 level. The maximum number of vertices at level 1 is 2^1, at level 2 is 2^2, and so on. By induction we can prove that the maximum number of vertices possible at level k is a binary tree is 2^k. We now state the following theorem on the minimum possible height of a binary tree.

Theorem 7.19. The minimum height of a binary tree T on n vertices is $\lceil \log_2 (n+1) - 1 \rceil$, (where $\lceil m \rceil$ is the smallest integer $> n$) and maximum possible height is $\frac{n-1}{2}$.

Proof. The root of T is at level 0. We know that every vertex of T at level k can have 2^k successions. Therefore we have

$$2 \text{ vertices at level } 1$$
$$2^2 \text{ vertices at level } 2$$
$$\cdots$$
$$\cdots$$

Therefore, the maximum number of vertices in the binary tree of height l is

$$1 + 2 + 2^2 + \ldots + 2^l$$

but T has n vertices

$$\therefore \quad 1 + 2 + 2^2 + \ldots + 2^l > n$$

$$\Rightarrow \quad \frac{2^{l+1} - 1}{2 - 1} > n$$

$$2^{l+1} - 1 > n$$

$$\Rightarrow \quad 2^{l+1} > n + 1$$

$$\therefore \quad l > \log_2 (n+1) - 1$$

But l is an integer, therefore, the smallest possible value for 1 is $\lceil \log_2 (n+1) - 1 \rceil$.

Hence, the minimum possible height of a binary tree T is $\lceil \log_2 (n+1) - 1 \rceil$

Now, let l denote the maximum possible height of T.

We have the root of T at 0 level

$$2 \text{ vertices at level } 1,$$
$$2 \text{ vertices at level } 2, \ldots$$

when T is of height l, we have at least

$$1 + (2 + 2 + \ldots l \text{ times}) \text{ vertices in } T$$

i.e. $$l + 2l \text{ vertices in } T$$

$$\therefore l + 2.l \leq n$$

$$\Rightarrow 2l \leq n - 1$$

$$\Rightarrow l \leq \frac{n-1}{2}$$

but n is odd and $\frac{n-1}{2}$ is an integer.

$\therefore$ The maximum possible value of l is $\frac{n-1}{2}$

Thus we have $\min l = \lceil \log_2 (n+1) - 1 \rceil$

$$\max l = \frac{n-1}{2}$$

SOLVED EXAMPLES

Example 1. Suppose that a tree T has N_1 vertices of degree 1, 2 vertices of degree 2, 4 vertices of degree 3 and 3 vertices of degree 4, find N_1.

Solution. Number of vertices in $T = N_1 + 2 + 4 + 3 = N_1 + 9$

Number of edges in $T = N_1 + 9 - 1 = N_1 + 8$

$$\sum_{i=1}^{9} \deg(v_i) = N_1 \cdot 1 + 2.2 + 4.3 + 3.4$$

$$= N_1 + 28$$

$$\sum_{1}^{9} \deg(v_i) = 2. \quad \text{(Number of edges in } T)$$

$\Rightarrow$ $N_1 + 28 = 2\,(N_1 + 8)$

$\Rightarrow$ $N_1 + 28 = 2\,N_1 + 16$

$\Rightarrow$ $N_1 = 12$

Example 2. If l denotes the height of a binary tree with n vertices, find the minimum and maximum values of l; when $n = 11$.

Solution. $\min l = \lceil \log_2 (n+1) - 1 \rceil = \lceil \log_2 (11+1) - 1 \rceil$

$$= \lceil \log_2 12 - 1 \rceil$$

$$\max l = \frac{n-1}{2} = \frac{11-1}{2} = 5$$

Example 3. Show that the number of vertices in a binary tree is odd.

Solution. Let T be a binary tree with n vertices. Then T contains exactly one vertex of degree 2 and the remaining vertices of T are of degree one or three.

Therefore the number of odd degree vertices in T is $n-1$. But the number odd degree vertices in a grâph is even. Therefore $n-1$ is even. Hence n is odd.

Example 4. T is a binary tree on n vertices and p is the number of pendent vertices in T. Show that the number of vertices of degree 3 in T is $n-p-1$.

Solution. T has p vertices of degree 1, and one vertex in T is of two. Hence the number of remaining vertices is $n-p-1$.

Therefore $n-p-1$ vertices of degree 3 in T.

Example 5. T is a binary tree on n vertices. Show that the number of pendant vertices in T is $\frac{n+1}{2}$.

Solution. Let p denotes the number of pendant vertices in T. The number of edges in T is $n-1$.

$\therefore$ The degree sum in $T = 2(n-1)$.

$\Rightarrow \quad p.1 + (n-p-1).\,3 + 2 = 2(n-1)$

$\Rightarrow \quad p + 3n - 3p - 3 + 2 = 2n - 2$

$\Rightarrow \quad 2p = n + 1$

$\Rightarrow \quad p = \frac{n+1}{2}$

Example 6. Find the maximum possible height of a binary tree with 13 vertices and draw the graph of the tree.

Solution. We have $n = 13$

Maximum possible height of the binary tree $= \frac{n-1}{2} = \frac{13-1}{2} = 6$

The graph of the tree is shown in Fig. 7.102.

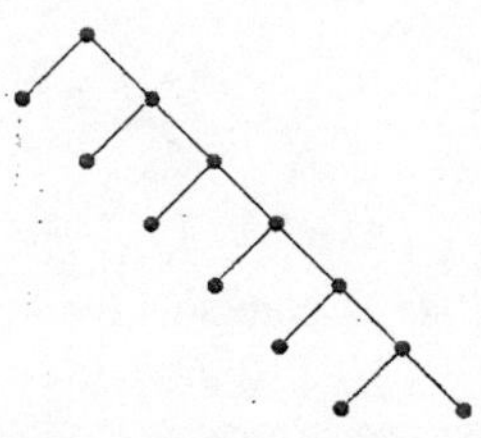

Fig. 7.102 Binary tree of maximum height with 13 vertices

Example 7. Find the minimum height of the tree with 9 vertices

Solution. We have $n = 9$

The minimum height of the binary tree

$$= \lceil \log_2 (n+1) - 1 \rceil = \lceil \log_2 (9+1) - 1 \rceil = 3$$

Example 8. Draw all trees with exactly 6 vertices.

Solution. Trees with exactly 6 vertices can be drawn as shown in Fig. 7.103.

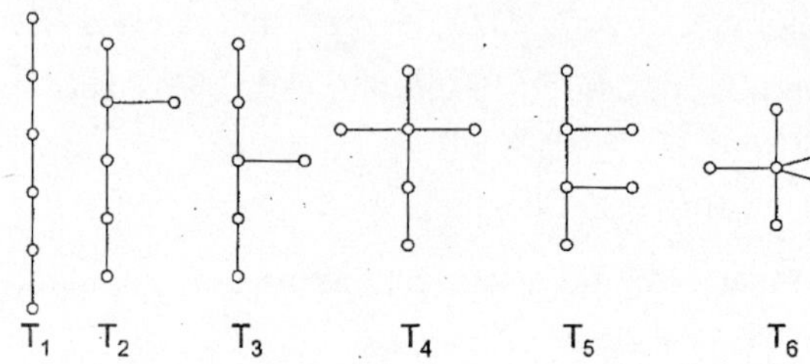

Fig. 7.103

T_1 is a tree whose diameter is 5.

The diameter of T_2 is 4, and the diameter of T_3 is also 4. T_4 and T_5 are trees with diameter 3.

T_6 has diameter 2.

Therefore, these are six such trees.

7.40.1 Complete Binary Tree

A binary tree for the level order indices of the vertices from a complete integral of 1, 2, ..., n of integers is called a complete binary tree.

If T is complete binary tree, then all its levels except possibly the last, will have the maximum number of possible vertices and all the vertices at the last level appear as far left as possible. The tree shown in Fig. 7.104 is a complete binary tree.

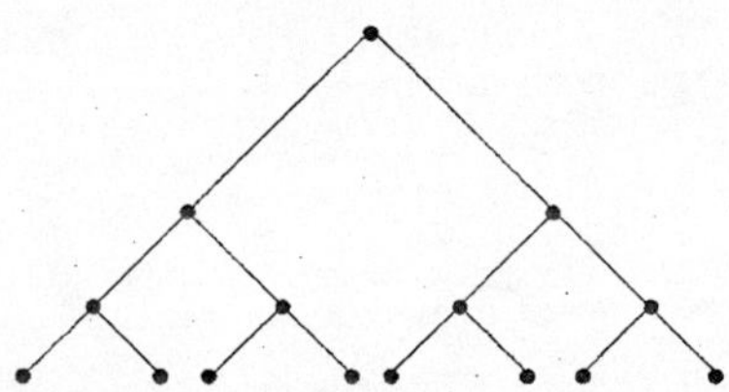

Fig. 7.104 A complete binary tree

If T is a complete binary tree with n vertices, then the vertices at any level 1 are given the label numbers ranging from 2^l to $2^{l+1} - 1$ or from 2^l to n if n is less than $2^{l+1} - 1$.

7.40.2 Height Balanced Binary Tree

A binary tree T in which the heights of left and right sub-trees of every vertex differ by at most one is called a height balanced binary tree.

Every complete binary tree is a height balanced binary tree. Height balanced trees are important in computer science and are more general than complete binary trees.

The height of a height-balanced tree is always O(log n) with respect to the number n of vertices in the tree.

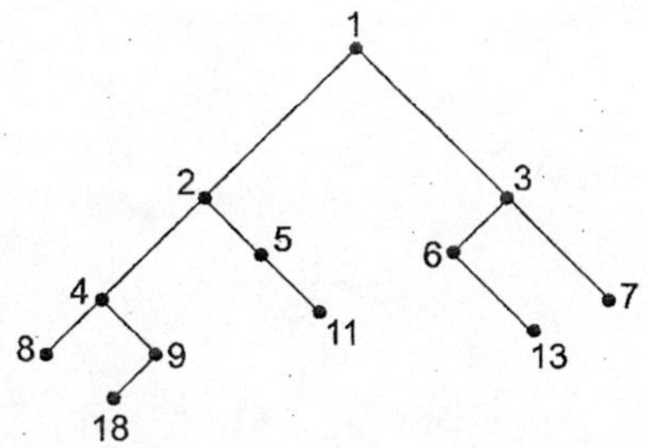

Fig. 7.105 Height balanced binary tree

Theorem 7.20. There are at least $\frac{1}{\sqrt{5}}\left[(1+\sqrt{5})/2\right]^{h+3} - 2$ vertices in any height balanced binary tree with height h.

Proof. Let $V(h)$ denotes the least number of vertices in a height balanced tree of height h. Then clearly $V(0) = 1$ and $V(1) = 2$.

For $h > 1$, there must be a root and two sub-trees where the height of one sub-tree must be $h - 1$ and the height of the other must be $h - 2$. The total number of vertices in such a tree is given by the recurrence relation

$$V(h) = V(h-1) + V(h-2) + 1 \qquad \text{...(1)}$$

The equation (1) is a seconds order recurrence relation. The auxiliary equation of (1) is

$$m^2 - m - 1 = 0 \qquad \text{...(2)}$$

Solving the equation (2) we get

$$m = (1+\sqrt{5})/2,\ (1-\sqrt{5})/2$$

The complementary function of (1) is

$$V(h) = a_1\left((1+\sqrt{5})/2\right)^h + a_2\left((1-\sqrt{5})/2\right)^h$$

The particular integral is obtained by setting $V(h) = p$ where p is a constant.

Substituting $V(h) = p$ in (1) we get

$$p = p + p + 1$$

$$\Rightarrow \qquad p = -1$$

The general solution of (1) is

$$V(h) = a_1 \left((1+\sqrt{5})/2\right)^h + a_2 \left(1-\sqrt{5}/2\right)^h - 1 \qquad \text{...(3)}$$

Since $V(0) = 1$, and $V(1) = 2$, the equation (3) given

$$a_1 = \left[\frac{5+2\sqrt{5}}{5}\right], \quad a_2 = \frac{5-2\sqrt{5}}{5}$$

∴ The solution can be written as

$$\therefore \qquad V(h) = \frac{1}{\sqrt{5}}\left((1+\sqrt{5})/2\right)^{h+3} - \frac{1}{\sqrt{5}}\left((1-\sqrt{5})/2\right)^{h+3} - 1 \qquad \text{...(4)}$$

The value of $(1+\sqrt{5})/2$ lies between 1.61803 and 1.61804

$$\therefore \qquad -1 < \left((1-\sqrt{5})/2\right)^{h+3} < 1$$

and in particular we have

$$-\frac{1}{\sqrt{5}}\left((1-\sqrt{5})/2\right)^{h+3} > -1$$

so that

$$V(h) > \frac{1}{\sqrt{5}}\left[(1+\sqrt{5})/2\right]^{h+3} - 2$$

Example 1. Draw a binary tree with level order indices

1, 2, 3, 4, 5, 6, 7, 8, 9, 10, 11, 15, 20, 30

Solution. Binary tree with given level order indices can be drawn as shown in Fig. 7.106.

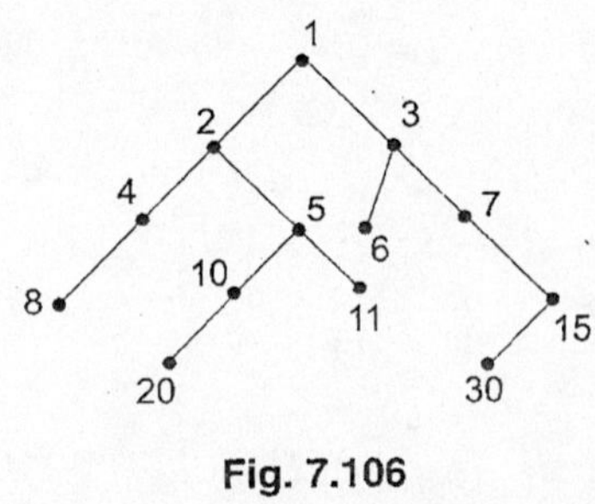

Fig. 7.106

Example 2. If $H(n)$ denotes the maximum possible height of a complete binary tree with n vertices then show that $H(n) = O(\log n)$.

Solution. We have $\quad 2^h \le n \le 2^{h+1} - 1 < 2^{h+1}$

$$\Rightarrow \qquad h \log_2 2 \le \log_2 n < (h+1) \log_2 2$$

$$h < \log_2 n < h + 1$$

or $$H(n) = O(\log_2 n)$$

Example 3. Draw the graph of a 13-vertex; 5-level binary tree.

Solution. The graph of a 13-vertex, 5-level binary tree is shown in Fig. 7.107.

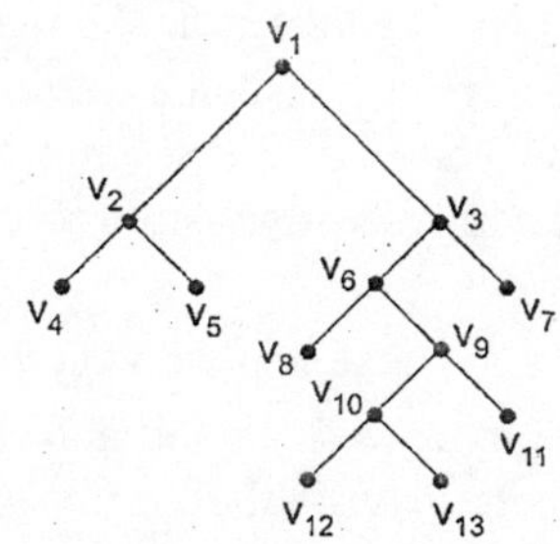

Fig. 7.107

7.41 TREE TRAVERSAL

In this section we discuss procedures for visiting each vertex of an ordered rooted tree. In many algorithms based on binary trees, it is necessary to visit each vertex (node) exactly once.

There are three basic algorithms for binary tree traversal namely: Preorder; in-order and post order.

We shall discuss the recursive algorithms for all three types of traversal.

7.41.1 Preorder Traversal

Definition 7.53. Let T be an order-rooted tree with root r.

If T consists only of r, then r is the preorder traversal of T. Otherwise, suppose $T_1, T_2, \ldots, T_n$ are the sub-trees at the root r from left to right in T. The preorder begins by visiting r. It continues traversing T_1, in preorder; then T_2 in preorder and so on, until T_n is traversal in preorder.

In preorder traversal of a tree with root, we first visit the root r, then visit left sub-tree T_1 in preorder, we next visit T_2 in preorder. The process is continued until T_n is visited in preorder. The number of steps involved is $n + 1$.

Example 1. Consider the tree shown in Fig. 7.108.

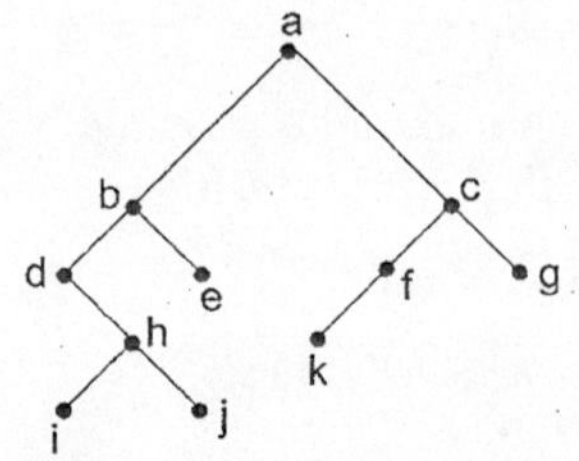

Fig. 7.108

The vertex a is the root of the tree. We begin our preorder traversal of the tree by visiting a. Upon completion of the visit, we visit vertex b. Then we visit the vertex d. The left child of d does not exist. Therefore we are directed to visit the right child of d. Therefore we visit the vertex n. We then visit the vertex i. The vertex i has no children. Therefore we visit the vertex j. The vertex j has no children. Therefore we visit the vertex e. We attempt to visit e's left and right children, which do not exist. Having already visited root a, and d's left child, we now visit c.

We then visit the vertex f. In the next step we visit the vertex k.

k has no children, hence we try to visit f's left and right children. Since f has no children, the algorithm directs to visit the vertex g. The preorder traversal terminates when we have attempted to visit g's left and right children.

∴ The preorder traversal of the given tree is

$$a, b, d, h, i, j, e, c, f, k, g$$

Example 2. The preorder traversal of the tree shown in Fig. 7.109 (a) is

$$v_1\ v_2\ v_4\ v_5\ v_6\ v_3\ v_7\ v_8\ v_{10}\ v_{12}\ v_{11}$$

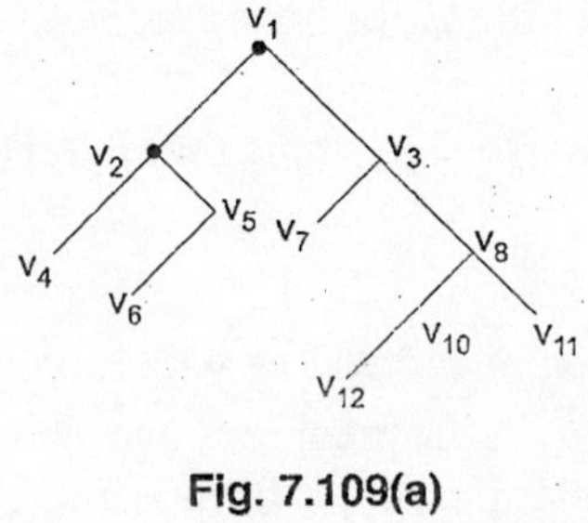

Fig. 7.109(a)

Example 3. Write the steps for the in order, pre-order and post-order traversal of trees. Also find the in-order, pre-order and post-order traversal for the tree shown in Fig. 7.109(*b*)

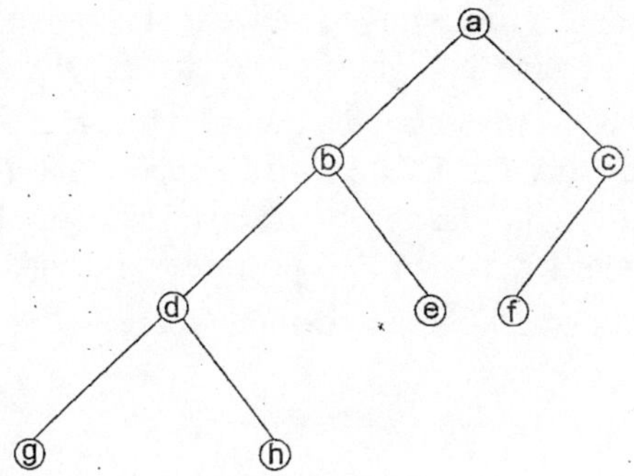

Fig. 7.109(b)

Solution. (i) In-order traversal:

Step 1. Traversal left tree

Step 2. Visit root

Step 3. Traverse right sub-tree

∴ In-order for the given tree is *g d h b e a f c*

(ii) Pre-order traversal:

Step 1. Visit root

Step 2. Traverse left sub-tree

Step 3. Traverse right sub-tree

∴ Pre-order for the given tree is *a b d g h e c f*

(iii) Post-order

Step 1. Traverse left sub-tree

Step 2. Traverse right sub-tree

Step 3. Visit root

∴ Post-order for the given tree is *g h d e b f c a*

7.41.2 Inorder Traversal

Using inorder traversal, we visit each vertex once and only once. If r denotes the root of the tree, we first perform inorder traversal on the r's left sub-tree and then visit r. In the final step we perform inorder traversal on r's right sub-tree.

Definition 7.54. Let T be an order rooted tree with root r. If T consist only of r; then r is inorder traversal of T. Otherwise, suppose that $T_1, T_2, \ldots, T_n$ are the subtrees at r from left to right. The inorder traversal begins by traversing T_1 in inorder then visiting r. The process continues by traversing T_2 in inorder, then T_3 in inorder, ..., and finally T_n in inorder.

The inorder traversal of the tree shown in Fig. 7.108 is

$$d, i, h, j, b, e, a, k, f, c, g$$

7.41.3 Postorder Traversal

In the postorder traversal of a tree with root r, we first traverse the left sub-tree of r in postorder and then traverse the right sub-tree of r. Finally we visit the root r.

Definition 7.55. Let T be an ordered rooted tree with root r. If T consists only of r, then r is the postorder traversal of T. Otherwise, suppose that $T_1, T_2, \dots, T_n$ are the sub-trees at r from left to right. The postorder traversal begins by traversing T_1 in postorder then T_2 in postorder,..., then T_n is postorder and ends by visiting r.

The postorder tree traversal of the tree given in Fig. 7.108 is

$$i, j, h, d, e, b, k, f, g, c, a$$

EXERCISE 7.3

1. Define the terms
   (i) Connected graph.
   (ii) Tree
   (iii) Rooted tree (MKU, MLA, May 2002)
2. Define
   (a) Cut vertex (b) Cut edge
   (c) Cut set, and give examples.
3. Show that every non-trivial tree has at least one vertex of degree 1.
4. Prove that a tree with n vertices has exactly $(n - 1)$ edges.
5. Define:
   (a) Spanning tree (b) Binary tree (c) Eccentricity
   (d) Leaf (e) Forest (f) Centre of a tree
   (g) Radius of a tree (h) Diameter of a tree (MKU, MCA, May 2002)
6. Show that the number of vertices of a binary tree is odd.
7. Define: (a) Minimal spanning tree (b) Complete binary tree.
8. Define the Height balanced binary tree and show that there are at least

   $$\frac{1}{\sqrt{5}}\left[\frac{1+\sqrt{5}}{2}\right]^{h+3} - 2.$$

   Vertices in a height balanced tree of height h.
9. State Kruskal's algorithm for finding a minimal spanning in a graph G. (O.U. MCA, 1998)
10. Define the terms:
    (i) Connected graph (ii) Spanning.
    And prove that every connected graph has one spanning tree. (O.U. MCA, 1991)

11. How many non-isomorphic trees are there with the number of vertices equal to

(a) 2 (b) 3 (c) 4 (d) 5

12. Show the sequential orders in which the vertices of the following trees are visited in a preorder, an inorder and a postorder traversal

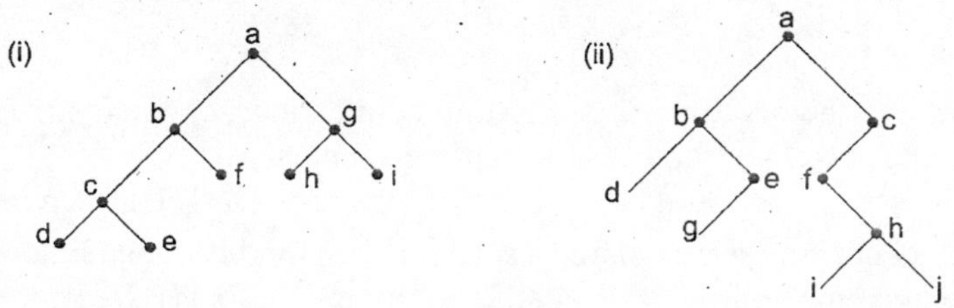

Fig. 7.110

13. Draw the binary tree whose level order indices are

$$\{1, 2, 4, 5, 8, 10, 11, 20\}$$

14. Draw a height balanced tree of height 4 with minimum number of vertices. (O.U. MCA, 1996)

15. Show that any connected graph with a vertices and $n - 1$, edges is a tree.

16. Let H be a sub-graph of a connected graph G. Show that H is a sub-graph of same spanning tree T if H contains no cycles.

17 Using Kruskal's algorithm, find minimal spanning tree of (JNTU, June 2002)

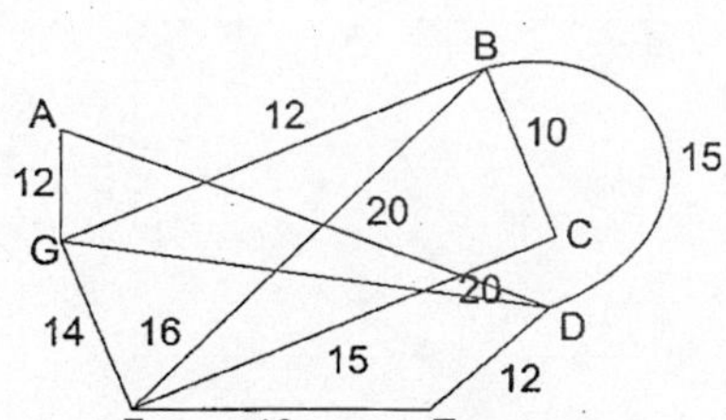

Fig. 7.111

18. What is directed tree? With the help of necessary illustrations explain the significance of directed trees. (JNTU, June 2002)

19. What is minimum spanning tree ? What are the different ways of creating spanning trees. (JNTU, June 2003)

20. Find the minimal spanning tree of the graph G, given below:

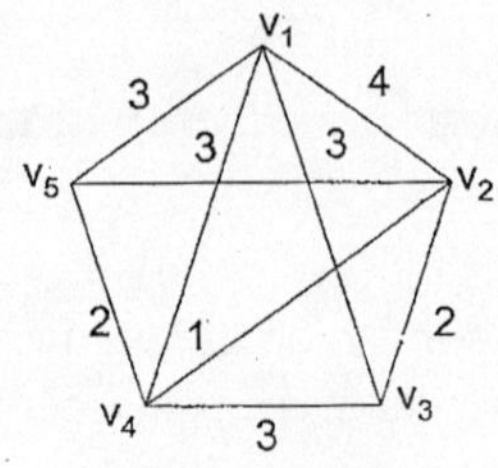

Fig. 7.112

21. What is a binary tree. Describe the applications of binary trees. Also describe necessary conditions to represent a binary tree.

(JNTU, December 2002)

22. Use a depth-first search to find a spanning tree for the graph G shown in the figure given below: (JNTU, December 2002)

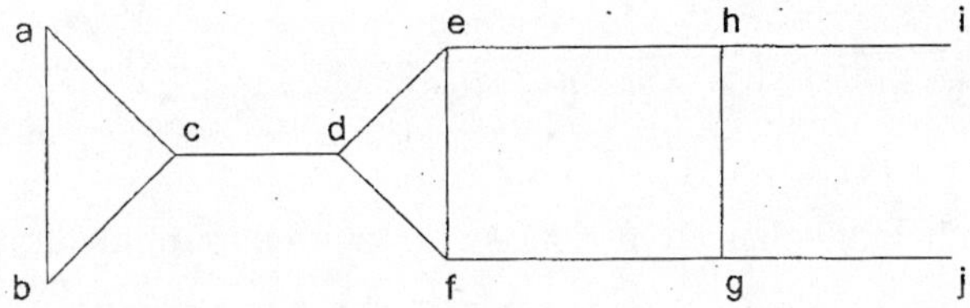

Fig. 7.113

23. Find all the fundamental cut-sets of the following graph with respect to the spanning tree $\{b, c, e, h, k\}$

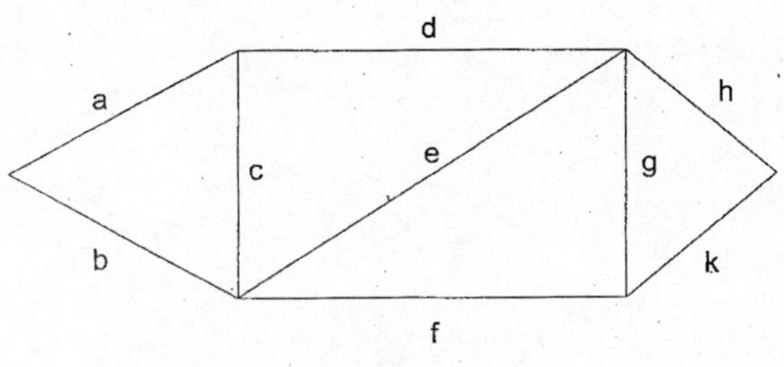

Fig. 7.114

24. Define: Fundamental cut-set. Give an example.

25. What is the prefix form for $[(x + y) \uparrow 2] + [(n - y/3)]$

26. What is value of the prefix expression

$$+ - \times 235 / \uparrow 234$$

27. Using Kruskali algorithm, find network of minimal cost for the cities in the Figure given below

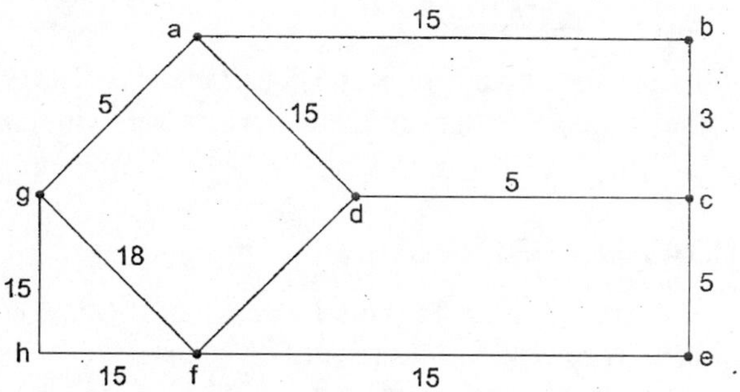

Ans.

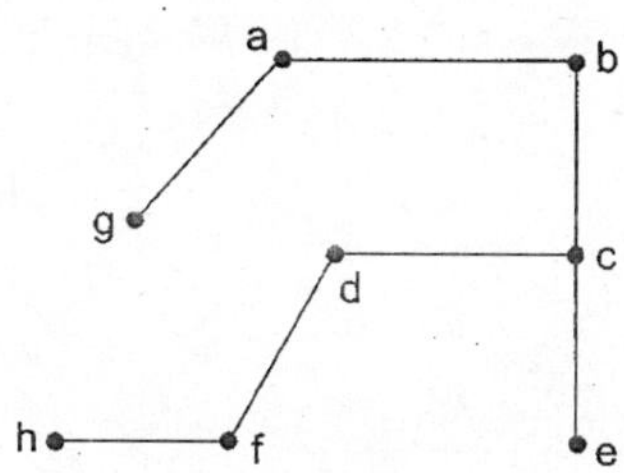

Minimal cost = 52

PLANAR GRAPHS

In this section we discuss about graphs, which can be drawn on a plane such that no two edges of the graph intersect. The points of intersection of edges in a graph are called crossovers. The edges in a graph *G*, which intersect are said to crossover each other. A graph *G* is said to be embedded in a surface *S*, when it can be drawn on *S*, such that no edges intersect.

Definition 7.56. A graph *G* is said to be planar if it can be drawn on a plane without crossovers.

A graph which cannot be drawn on a plane without crossovers between its edges is called a non-planar graph.

A graph *G* is said to be embedded in a surface *S* when it is drawn on the surface *S*, show that no two edges intersect. From the definition, given above, a graph *G* is planar if it can be embedded in the plane.

7.42 PLANE GRAPH

Definition 9.57. An embedding of a planar graph is called a plane graph.

A plane graph partitions the plane into several regions. These are called faces (also called windows or meshes). Each region is characterized the set of edges forming its boundary. Each plane graph determines a region of infinite area called the exterior region of G. If G is a connected graph, then the boundary of a region R is a closed path in which each edge counted twice. The boundary of a region R is a cycle if the boundary of R contains no cut edges of G.

7.43 DEGREE OF A FACE

Definition 7.58. Let G be a plane graph and f be a face of G. The degree of the face f is defined as the number of edges in the boundary of f, with cut edges counted twice.

7.44 CRITICAL PLANAR GRAPH

Definition 7.59. A graph G is said to be critical planar if G is non-planar, but any sub-graph obtained by removing a vertex of G is planar.

For $n \leq 3$, $k_{n,3}$ is critical planar. The graphs shown in Fig. 7.115 are planar.

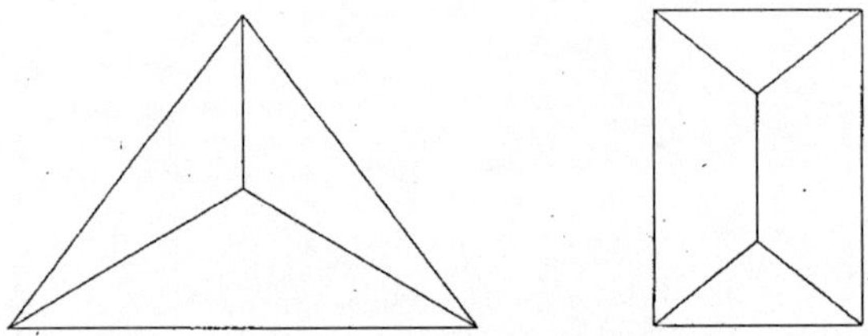

Fig. 7.115 Planar graphs

If G is a graph, then we denote the embedding of G on a surface S by $S(G)$. The vertices edges and faces of $S(G)$ constitute a map on the surface S. If S is the plane π, then the vertices, edges and faces of $S(G)$ constitute a planar map. It is denoted by $\pi(G)$. In a planar embedding of $\pi(G)$, all faces except one are bounded. The unbound face is called the exterior face. It is also called the outer face or infinite face. Region in a map which have at least one common edge are called adjacent regions.

Definition 7.60. (Polyhedral Graph): A plane connected graph in which each region has degree equal to or greater than 3 and each vertex has degree equal to or greater than 3 is called a polyhedral graph.

A planar graph may have different planar representation. The number of regions resulting from each embedding is always the same. If G is a planar graph, then the number of vertices, number of edges and the number of regions in G are interconnected. The number of regions in G can be computed by Euler's formula.

7.45 EULER'S FORMULA

Theorem 7.21. If G is a connected graph, then $|V| - |E| + |R| = 2$.

where $|V|$ denotes the number of vertices is G, $|E|$ denote the number of edges and $|R|$ denotes the number of region in G.

Proof. We prove the theorem by mathematical induction on number of regions determined by G. We first prove the theorem for a tree. i.e. for $k = 1$. A tree is determined by only one region. Therefore, we have

$$|E| = |V| - 1, \quad |R| = 1 \text{ for a tree}$$

$$\therefore \quad |V| - |E| + |R| = |V| - \{|V| - 1\} + 1 = 2$$

$\therefore$ The result holds for $k = 1$.

Let us assume that the theorem holds for $k \geq 1$. Let G be a connected plane graph determining $k + 1$, region. Deleting an edge common to the boundary of two regions in G we get a graph G'. If V', E', R' denote the number of vertices, the number of edges and the number of region respectively in G', then

$$|V'| - |E'| + |R'| = 2 \qquad \text{...(1)}$$

Also we have $\quad |V'| = |V|, \; |E'| = |E| - 1 \quad \text{and } |R'| = |R| - 1$

Therefore,
$$\begin{aligned} |V| - |E| + |R| &= |V'| - \{|E'| - 1\} + |R'| - 1 \\ &= |V'| - |E'| + 1 + |R'| - 1 \\ &= |V'| - |E'| + |R'| \\ &= 2 \qquad \text{(from (1))} \end{aligned}$$

By mathematical induction the result holds for all connected graphs.

Corollary. If G is a simple connected planar graph with $|E| > 1$, then

(i) $|E| \leq 3|V| - 6$

(ii) There exists a vertex v of G, such that $\deg(v) \leq 5$.

Proof. (i) G is a simple connected graph. Therefore each region of G is bounded by at least three edges and each edge belongs exactly to two regions.

$$\therefore \quad 2|E| \geq 3|R|$$

$$\Rightarrow \quad |R| \leq \frac{2}{3}|E|$$

$$\Rightarrow \quad |V| + |R| \leq |V| + \frac{2}{3}|E| \qquad \text{...(1)}$$

by Euler's formula, we have

$$|V| - |E| + |R| = 2$$

$$\Rightarrow \quad |V| + |R| = |E| + 2$$

Substituting in the L.H.S. of (1), we get

$$|E| + 2 \leq |V| + \frac{2}{3}|E|$$

$$\Rightarrow \qquad 3\,|E| + 6 \le 3\,|V|V + 2\,|E|$$

$$\Rightarrow \qquad |E| \le 3\,|V| - 6$$

(ii) Let each vertex of G have degree ≥ 6

$$\therefore \qquad \sum_{v \in v} \deg v = 2\,|E|$$

$$\Rightarrow \qquad 6\,|v| \le 2\,|E|$$

$$\Rightarrow \qquad |v| \le \frac{2}{6}\,|E|$$

$$\text{or} \qquad |v| \le \frac{1}{3}\,|E|$$

$$\text{but} \qquad |R| \le \frac{2}{3}\,|E|$$

$$\therefore \qquad |V| + |R| \le \frac{1}{3}\,|E| + \frac{2}{3}\,|E|$$

$$\Rightarrow \qquad |V| + |R| \le |E| \qquad \text{...(1)}$$

by Euler's formula,

$$|V| - |E| + |R| = 2$$

$$\Rightarrow \qquad |V| + |R| = |E| + 2 \qquad \text{...(2)}$$

from (1) and (2), we have

$$|E| + 2 \le |E|$$

$$\Rightarrow \qquad 2 \le 0$$

a contradiction.

Therefore, the graph G cannot have a vertex of degree ≥ 6.

Hence, each vertex of G is of degree ≤ 5.

Example 1. Count the number of vertices, the number of edges and the number of regions of each map and verify Euler's formula

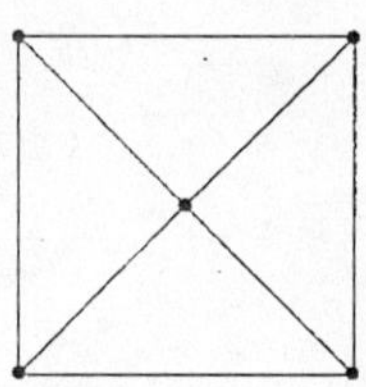

(a)

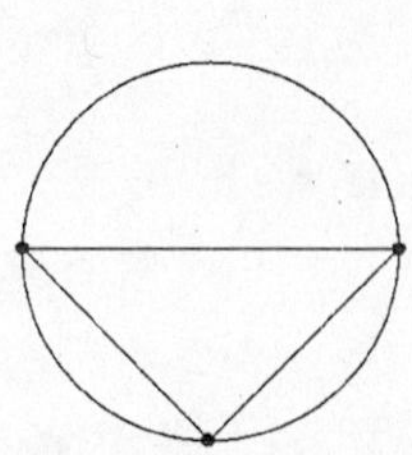

(b)

Solution. (a) Number of vertices = $|V| = 5$

Number of edges = $|E| = 8$

Number of regions = $|R| = 5$

$$|V| - |E| + |R| = 5 - 8 + 5 = 2$$

Eular's formula is verified

(*b*) We have $|V| = 3, |E| = 6, |R| = 5$

$\therefore \quad |V| - |E| + |R| = 3 - 6 + 5 = 2$

Eular's formula is verified

Example 2. Prove that k_5 is non-planar.

Solution. Number of vertices in $k_5 = |V| = 5$

Number of edges in $k_5 = |E| = 10$

If G is planar then $|E| \leq 3|V| - 6$

$\Rightarrow \quad 10 \leq 3.5 - 6$

$\Rightarrow \quad 10 \leq 9$

a contradiction.

$\therefore$ k_5 is non-planar.

Example 3. Show that the following graphs are planar

(i) Graph of order 5 and size 8

(ii) Graph of order 6 and size 12

Solution. (i) Let G denotes the graph of order 5 and size 8

Then we have $\quad |V| = 5, \quad |E| = 8$

and $\quad 3|V| - 6 = 3.5 - 6 = 9$

$\Rightarrow \quad |E| = 8 < 9 = 3|V| - 6$

The condition $|E| \leq 3|V| - 6$ is satisfied.

Therefore G is planar. The planar representation of G is

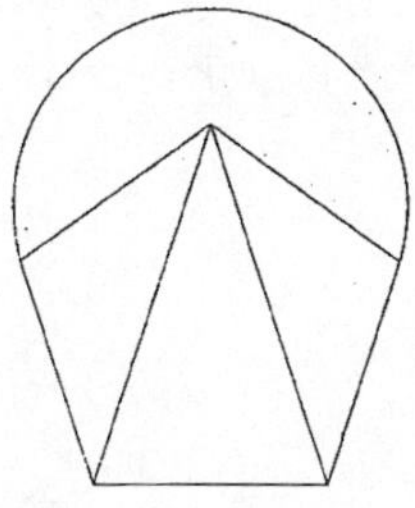

Fig. 7.116

(ii) Let G denote the graph of order 6 and size 12

$\therefore$ We have $\quad |V| = 6, \quad |E| = 12$

and $\quad 3|V| - 6 = 3.6 - 6 = 12$

i.e. $\quad |E| = 6 \leq 6 = 3|V| - 6$

$|E| \leq 3|V| - 6$

Therefore G is planar. The planar representation of G is

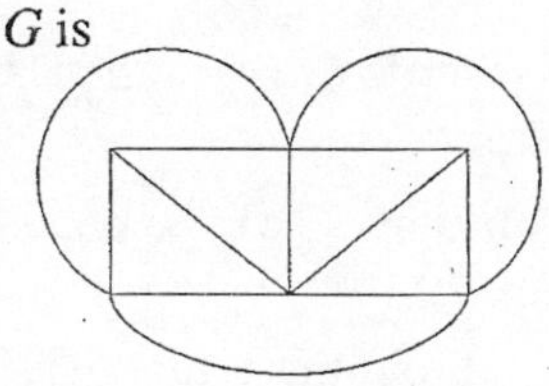

Fig. 7.117

Example 4. Show that $k_{3,3}$ satisfies the in equality

$$|E| \leq 3\,|V| - 6$$

But it is non-planar.

Solution. For the graph $k_{3,3}$ we have

$$|E| = 9, \quad |V| = 6$$

$\therefore$ $$3\,|V| - 6 = 3.6 - 6 = 12$$

$$|E| = 9 < 12$$

i.e. $$|E| \leq 3\,|V| - 6$$

$\therefore$ $k_{3,3}$ satisfies the inequality $|E| \leq 3\,|V| - 6$

If the graph is planar we must have

$$2\,|E| \geq 4\,|R|$$

Substituting for $|R|$, from Euler's formula we get

$$2\,|E| \geq 4\,\{|E| - |V| + 2\}$$

$\Rightarrow$ $$2.9 \geq 4\,[9 - 6 + 2]$$

$\Rightarrow$ $$18 \geq 20$$

a contradiction.

$\therefore$ $k_{3,3}$ is not planar.

Note: The graph k_5 is called Kuratowski's first graph and $k_{3,3}$ is called Kuratowski's second graph

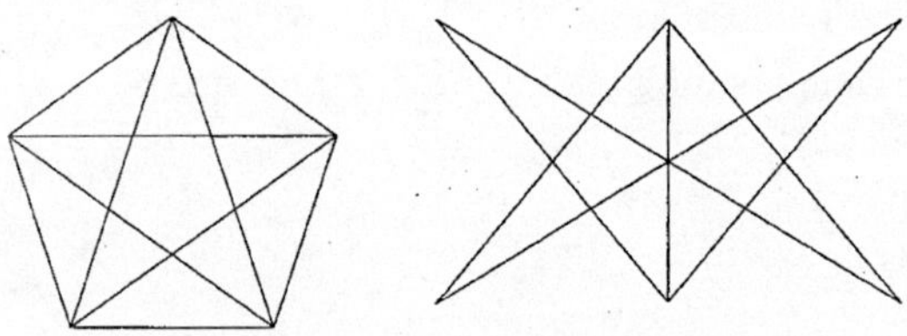

Fig. 7.118 Kuratowski's graphs

Example 5. Show that a complete graph of 4 vertices is planar.

Solution. K_4 can be drawn (without crossovers) as shown in Fig. 7.119.

From the figure 7.119, it is clear that k_4 can be drawn without crossovers.

Hence K_4 is a planar.

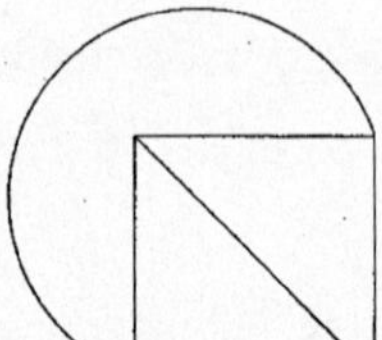

Fig. 7.119

Example 6. Show that $k_{2,2}$, $k_{2,3}$ and $k_{2,4}$ are planar.

Solution. The graphs of $k_{2,2}$, $k_{2,3}$ and $k_{2,4}$ are shown in Fig. 7.120.

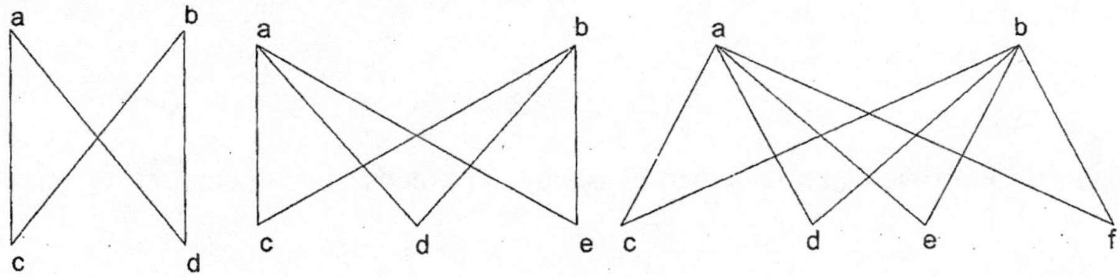

Fig. 7.120

The planar embeddings of $k_{2,2}$, $k_{2,3}$ and $k_{2,4}$ are shown in Fig. 7.21 in which the edges of the graphs are drawn without crossovers.

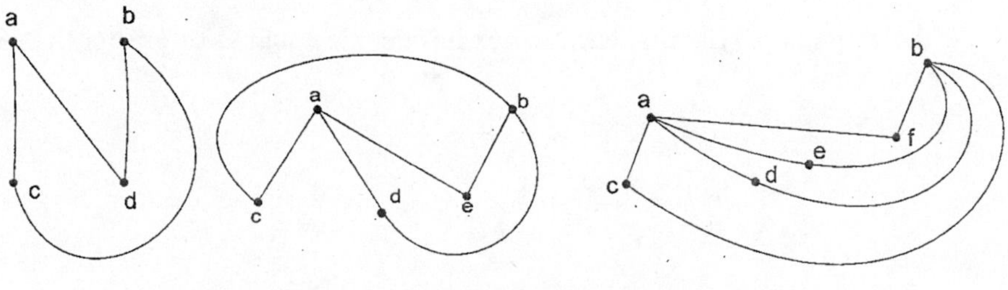

Fig. 7.121 Planar embeddings of $k_{2,2}$, $k_{2,3}$ and $k_{2,4}$

Example 7. Show that G shown in Fig. 1.122 is planar.

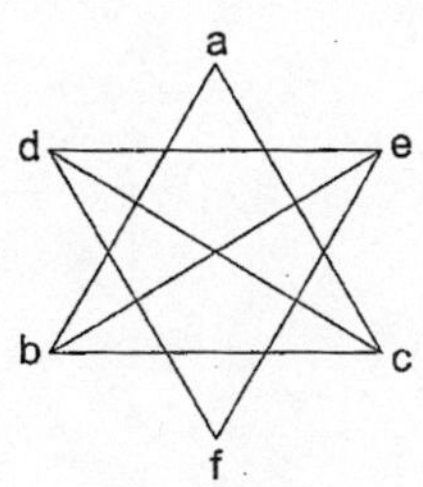

Fig. 7.122

Solution. The edges of G shown in Fig. 7.123 can be drawn without crossovers. The planer embedding of G is shown in Fig. 7.123.

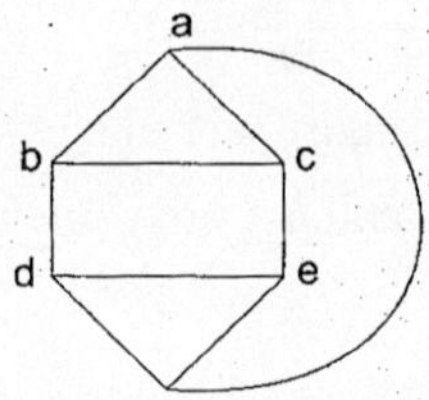

Fig. 7.123

Example 8. Prove that no polyhedral graph with exactly 30 edges and 11 regions exists.

Solution. We have $|E| = 30, \quad |R| = 1$

from Euler's formula

$$|V| - |E| + |R| = 2$$

$$\Rightarrow \quad |V| = |E| - |R| + 2$$

$$= 30 - 11 + 2 = 21$$

If the graph is polyhedral, each vertex has degree equal to or greater than 3 and we must have

$$3\,|V| \leq 2\,|E|$$

i.e. $$3.21 \leq 2 - 30$$

i.e. $$63 \leq 60$$

a contradiction.

∴ There cannot be a polyhedral graph with exactly 30 edges and 11 regions.

Example 9. Show that the graphs G_1, G_2, and G_3 shown in Fig. 7.124 are planar.

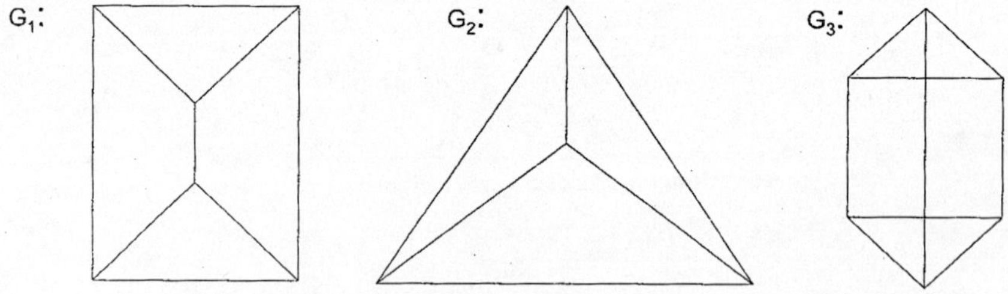

Fig. 7.124

Solution. The planar embeddings G_1, G_2, and G_3 are:

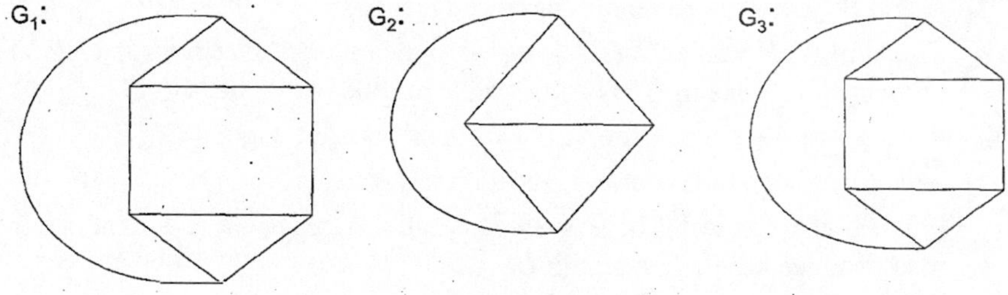

Fig. 7.125

7.46 HOMEOMORPHIC GRAPHS

Two graphs are said to be homeomorphic if one graph can be obtained from the other by the creation of edges in series or by the merger of edges in series.

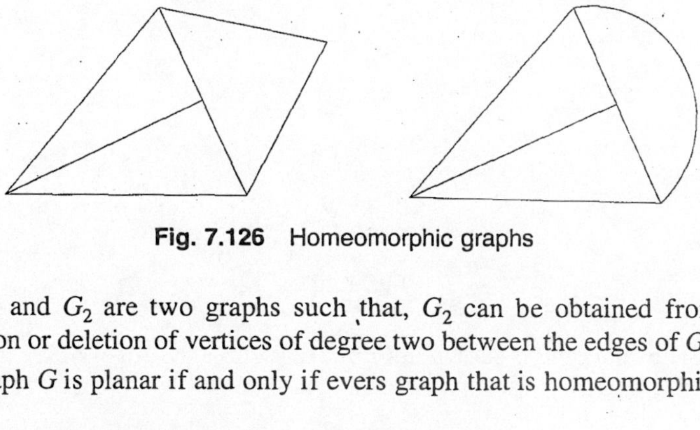

Fig. 7.126 Homeomorphic graphs

If G_1 and G_2 are two graphs such that, G_2 can be obtained from G, by intersection or deletion of vertices of degree two between the edges of G_1.

A graph G is planar if and only if evers graph that is homeomorphic to G is planar.

We now state the following theorem without proof.

Theorem (kuratowski's): A graph G is a planar if and only if it has no sub-graph homeomorphic to k_5 or $k_{3,3}$.

7.47 DUAL OF A GRAPH

7.47.1 Geometric Dual

Let G be a planar graph and F_1, F_2,..., F_k denote the faces or regions of G. Place vertices v_1, v_2,..., v_k one in each of the regions. If two regions F_i and F_j are adjacent draw an edge joining the vertices v_i and v_j that intersects the common edge between F_i and F_j exactly once. If the number of edge between F_i and F_j is more than one, then draw one edge between v_i and v_j for each of the common edge. If an edge e lies entirely in one region say F_i draw a self loop at the vertex v_i lying in F_i intersecting e exactly once. Denote the new graph obtained by G^*.

The graph G^* is called a dual of G. We observe the following in G and G^*.

(i) G and G^* have the same number of edges.

(ii) The number of vertices in G^* is equal to the number of regions in G and the number of regions in G^* is equal to the number of vertices in G.

(iii) An edge forming a self loop in G yields a pendant edge in G^*.

(iv) A pendant edge in G yields a self loop in G^*.

(v) Edges that are in series in G produce parallel edges in G^* and parallel edges in G produce edges in series in G^*.

(vi) G^* is planar

(vii) If G^* is a dual of G, then G is a dual of G^*.

(viii) The number of a graph G, need not be unique.

(ix) The nullity of G = The rank of G^*.
and, the nullity of G^* = the rank of G.

Example 1. Find the dual of the graph G shown in Fig. 7.127.

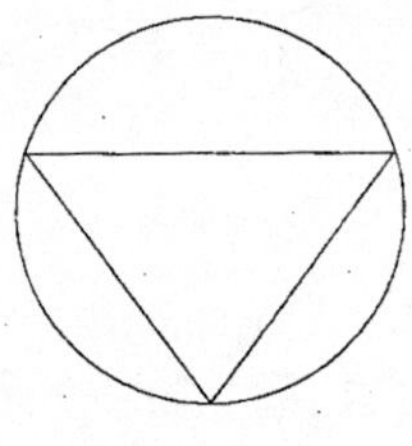

Fig. 7.127

Solution. The dual of G is shown in Fig. 7.128 where the edges of G^* are shown by dashed edges.

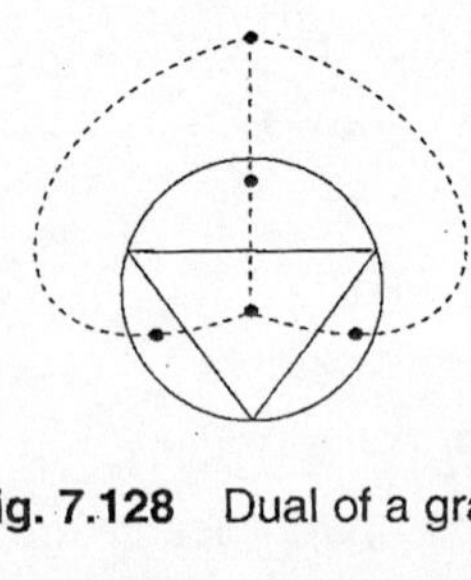

Fig. 7.128 Dual of a graph

A maximal planar graph is one to which no edge can be added without losing planarity.

(i) If G is a maximal plane graph with p vertices and q edges (lines), then every face is a triangle and $q = 3p - 6$. If G is a plane graph in which every face is a 4-cycle, then $q = 2p - 4$.

(ii) If G is a planar graph with the number of edges $p \geq 3$, then $q \leq 3p - 6$.

(iii) If G is a planar graph with the number of edges $p \geq 3$, and if G has no triangles, then $q \leq 3\,p - 4$.

Example 1. Show that k_5 is non-planar.

Solution. We have $p = 5$, (number of vertices in k_5) and

$$q = 10 \text{ (number of edges in } k_5)$$

$$3p - 6 = 3.5 - 6 = 15 - 6 = 9$$

If k_5 is planar we must have $q \leq 3\,p - 6$

but in k_5 we have $q = 10$

and $\qquad 3p - 6 = 9$

which gives $\qquad q = 10 > 9 = 3p - 6$

i.e. $\qquad q > 3p - 6$

a contradiction.

Hence k_5 is non-planar.

Example 2. Show that $k_{3,3}$ is non-planar.

Solution. $k_{3,3}$ has q edges and 6 vertices

i.e. $\qquad |v| = n = p = 6,\ q = |E| = 9$

we have $\qquad 2p - 4 = 2.6 - 4 = 8,\quad \text{and } q = 9$

$\therefore \qquad q = 9 \neq 8 = 2p - 4$

i.e. $\qquad q \neq 2p - 4$

$\therefore$ The condition $q = 2p - 4$ is not satisfied.

Hence, $k_{3,3}$ is non-planar.

Theorem 7.47.1. A graph is planar if and only if, it has a dual

Proof. The theorem is proved, if we show that no non-planar graph has a dual.

Let G be a non-planar graph. Then G contains a sub-graph homeomorphic to k_5 of $k_{3,3}$ ... we know that k_5 and $k_{3,3}$ have no duals. Moreover edges in series of a graph yield only parallel edges in its dual, therefore parallel edges in k_5 or $k_{3,3}$ does not support to construct a dual. Hence G has no dual.

7.47.2 Combinational Dual

Definition 7.61. Let G and G' be planar graphs. G and G^* are said to be combinational duals of each other if there is a one-one correspondence between the edges of G and G^* such that if G_1 is any sub-graph of G and G_1^* is the corresponding sub-graph G^*. Then,

Rank of $(G^* - G_1^*)$ = rank of G^* – nullity of G_1

The above definition is an abstract formulation of the concept of geometrical dual and was given by Whitney. He proved that a graph G is planar if and only if it has a combinational dual.

7.47.3 Self Dual Graphs

Let G be a planar graph, and G^* denote the dual of G. If G and G^* are isomorphic to each other then G is called a self dual graph.

Example 1. The tetrahedron is a self dual.

Example 2. The following graphs are self duals.

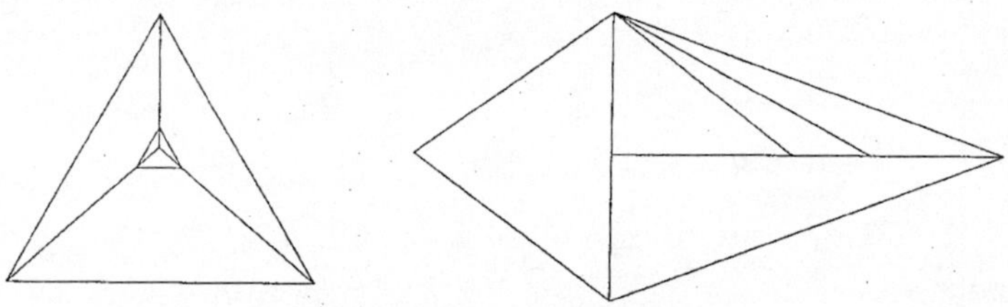

Fig. 7.129 Self duals

7.48 THICKNESS OF A GRAPH

Definition 7.62. The minimum number of planar sub-graphs whose union is the given in graph G, is called the thickness of G. It is denoted by $\theta(G)$.

The thickness of a planar graph is one.

7.49 CROSSING NUMBER

Definition 7.63. Let G be a graph. The minimum number of pairwise intersections of its edges when G is drawn in the plane is called the crossing number of G.

The crossing number of each of the Kuratowski's graphs is one. The crossing number of a planar graph is zero.

7.50 OUTER PLANAR GRAPH

A planar graph is outer planar if it can be embedded in the plane so that all its vertices lie on the boundary of a region of G.

The graph shown in Fig. 7.130 is outer planar.

We now state the following theorem without proof.

Theorem: Two planar graphs G and G' are duals of each other if there exists one-to-one correspondence between the edges of G and G' such that circuits in G correspond to cut-set in G'.

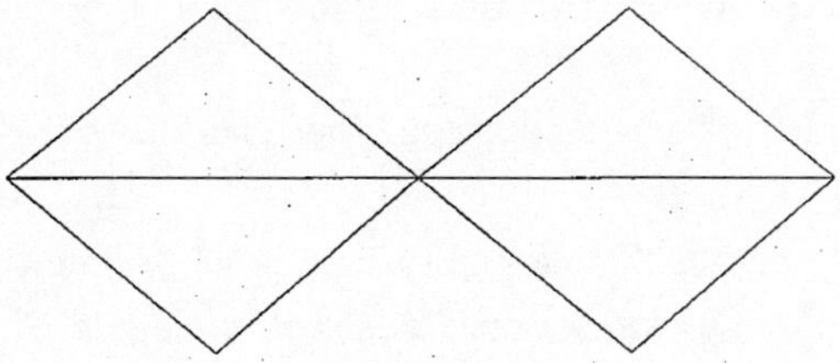

Fig. 7.130

Example. Find the dual of the graph shown in Fig. 7.130.

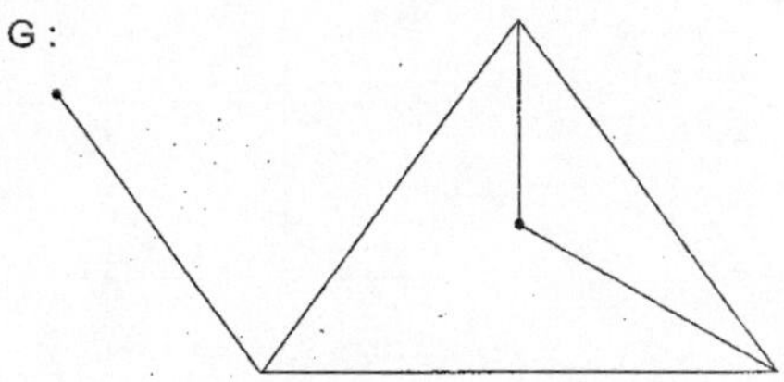

Fig. 7.131

Solution. The dual of the given graph G can drawn as shown in Fig. 7.132 where the edges of G^* are shown by dashed edges.

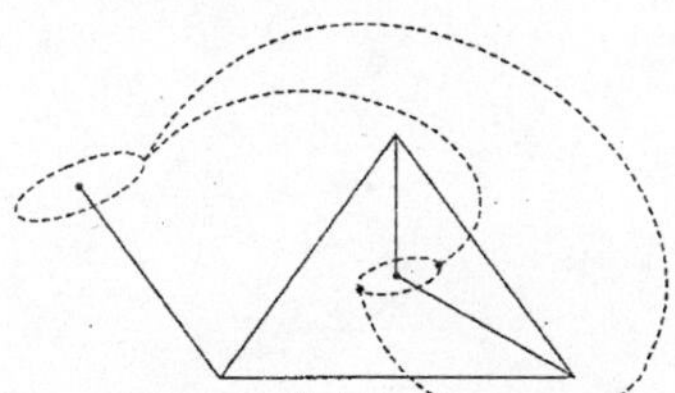

Fig. 7.132 Dual of G

EXERCISE 7.4

1. Draw a planar representation of the following graphs:

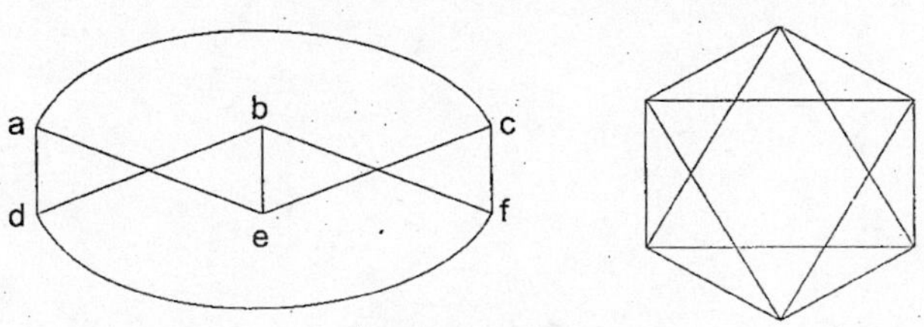

Fig. 7.133

2. Show that k_5 is non-planar.
3. Show that plane connected graph with less than 30 edges has a vertex of degree ≤ 4. (O.U. MCA, 1999)
4. Give an example of a self-dual graph other than k_4. (MKU, MCA, May 2002)
5. Show that $k_{3,3}$ satisfies the inequality $|E| \leq 3|V| - 6$ but it is non-planar.
6. Give an example of a plane connected graph such that $|E| = 3|V| - 6$.
7. Count the number of vertices, number of edges and the number of regions of each map and verify Euler's formula.

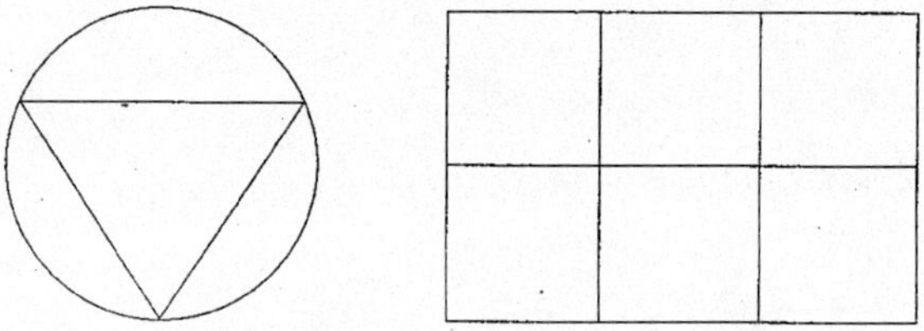

Fig. 7.134

8. Is the following graph planar

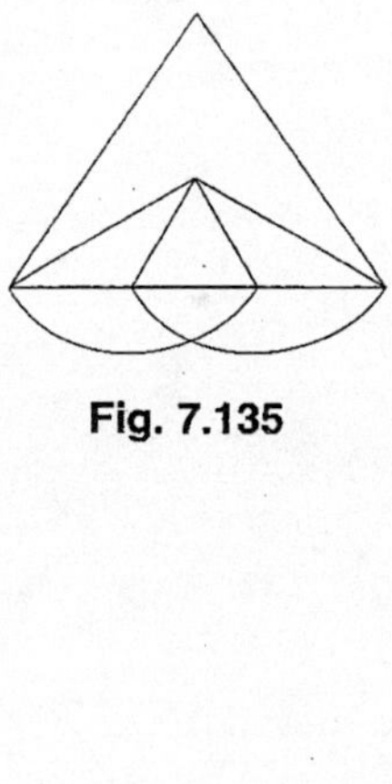

Fig. 7.135

CHAPTER 8

Application of Graph Theory

8.1 INTRODUCTION

In this chapter we study some applications of graph theory. We introduce the concepts of adjacency matrix and incidence matrix. We discuss the concepts of isomorphism, Eulerian paths, Eulerian circuits, Hamiltonian paths and cycles in graphs. We also study vertex colouring and edge colouring of graphs.

8.2 MATRIX REPRESENTATIONS OF GRAPHS

A graph can be represented by a matrix. There are several ways of representing graph by a matrix, namely

(i) Adjacency matrix (ii) Incidence matrix

(iii) Circuit matrix (iv) Cut set matrix and (v) Path matrix

8.2.1 Adjacency Matrix of a Non-directed Graph

Definition 8.1.

Let G be a graph with n vertices and no parallel edges. The adjacency matrix of G is a $n \times n$ symmetric matrix

$$X(G) = [x_{ij}]_{m \times n}$$

Where $x_{ij} = 1$, if v_i and v_j are adjacent in G (i.e. if $(v_i v_j) \in E(G)$)

$= 0$, if v_i and v_j are not adjacent in G

v_i and v_j are vertices of G.

The adjacency matrix $X(G)$ of a graph G with n vertices is a square binary matrix. It is a symmetric matrix, and the ith row sum (or i^{th} column sum) of $X(G)$ is the degree of the vertex v_i. The number of vertices in $X(G)$ is having self loop is equal to the number of non-zero diagonal entries. A graph G is connected it and only if there no labeling of the vertices of G such that its adjacency matrix $X(G)$ has the reduced form $X(G) = \begin{bmatrix} X_{11} & 0 \\ 0 & X_{12} \end{bmatrix}$.

Example 1. The adjacency matrix of a simple graph G shown in Fig. 8.1(a) and Fig. 8.1(b).

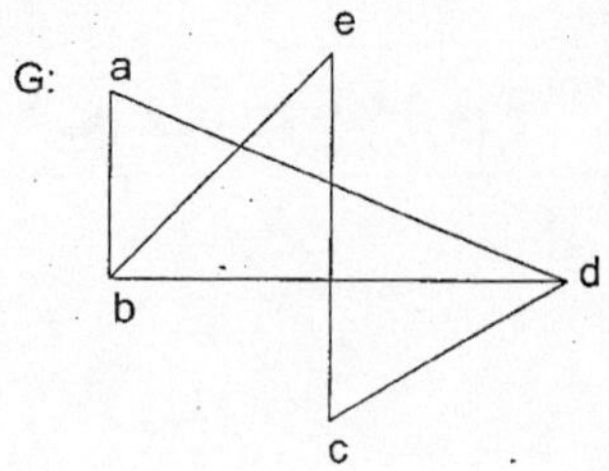

(a)

$$\text{Adjacency matrix } X(G) = \begin{array}{c} \\ a \\ b \\ c \\ d \\ e \end{array} \begin{bmatrix} a & b & c & d & e \\ 0 & 1 & 0 & 1 & 0 \\ 1 & 0 & 0 & 1 & 1 \\ 0 & 0 & 0 & 1 & 1 \\ 1 & 1 & 1 & 0 & 0 \\ 0 & 1 & 1 & 0 & 0 \end{bmatrix}$$

(b)

Fig. 8.1 Graph and its adjacency matrix

Example 2. Write the adjacency matrix of the graph given below

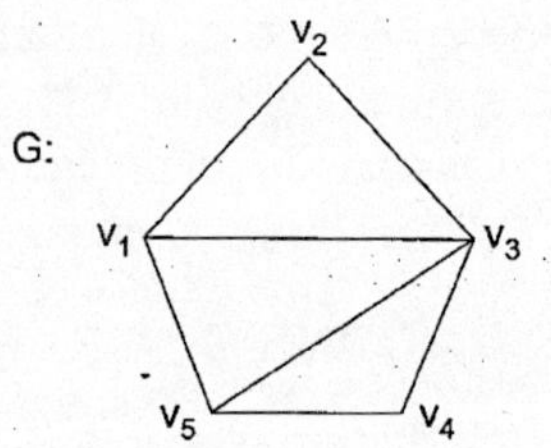

Fig. 8.2

Solution. The adjacency matrix of the given graph is

$$X(G) = \begin{bmatrix} 0 & 1 & 1 & 0 & 1 \\ 1 & 0 & 1 & 0 & 0 \\ 1 & 1 & 0 & 1 & 1 \\ 0 & 0 & 1 & 0 & 1 \\ 1 & 0 & 1 & 1 & 0 \end{bmatrix}$$

Example 3. Construct a graph whose adjacency matrix is given below.

$$X(G) = \begin{bmatrix} 0 & 1 & 1 & 0 & 0 \\ 1 & 0 & 1 & 0 & 1 \\ 1 & 1 & 0 & 1 & 0 \\ 0 & 0 & 1 & 0 & 1 \\ 0 & 1 & 0 & 1 & 0 \end{bmatrix}$$

Solution. Let v_1, v_2, v_3, v_4, and v_5 denote the vertices of G. The graph G, corresponding to $X(G)$ is shown in the Fig. 8.3.

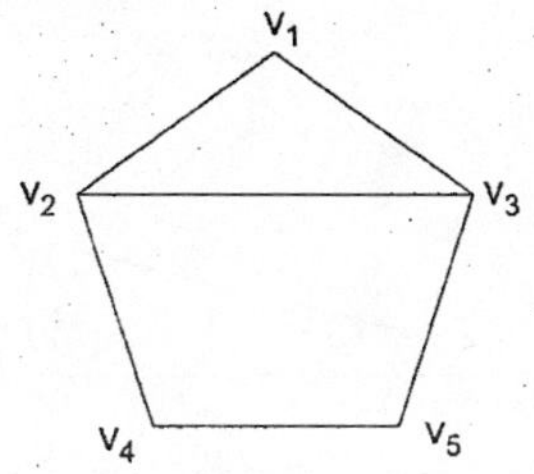

Fig. 8.3

8.2.2 Adjacency Matrix of a Digraph

Definition 8.2. Let G be a digraph with n vertices containing non-parallel edges. The adjacency matrix $X(G)$ of the digraph G is an $n \times n$ matrix defined by

$$X(G) = [a_{ij}]_{n \times n}$$

where $v_j = 0$, if there is an edge directed from v_i to $a_{ij} = 1$, otherwise

Example. Consider the digraph G, shown in Fig. 8.4.

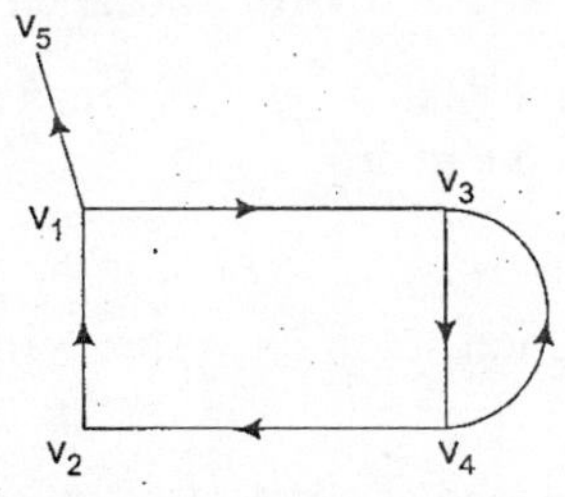

Fig. 8.4

The adjacency matrix of G is

$$X(G) = \begin{bmatrix} 0 & 0 & 1 & 0 & 1 \\ 1 & 0 & 0 & 0 & 0 \\ 0 & 0 & 0 & 1 & 0 \\ 0 & 1 & 1 & 0 & 0 \\ 0 & 0 & 0 & 0 & 0 \end{bmatrix}$$

The adjacency matrix of a graph is also called the connection matrix. It has different names in different disciplines. In the adjacency matrix of a graph, every non-zero element on diagonal represents a self loop at the corresponding matrix. We can find the adjacency matrices of multigraphs, but avoid parallel edges in the definition of $X(G)$.

Example. Find the adjacency matrix of the graph shown in Fig. 8.5

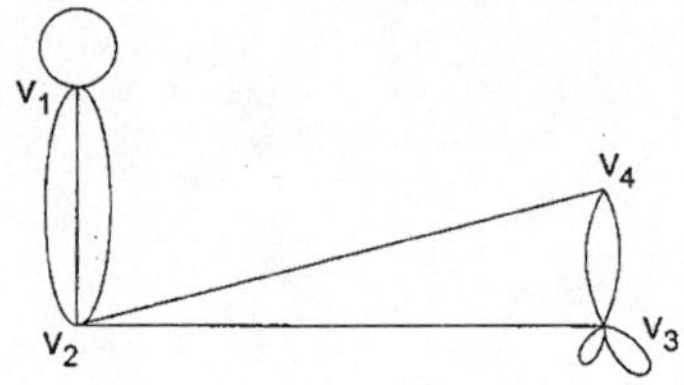

Fig. 8.5

Solution. The adjacency matrix $X(G)$ of the given multigraph is

$$X(G) = \begin{bmatrix} 1 & 3 & 0 & 0 \\ 3 & 0 & 1 & 1 \\ 0 & 1 & 2 & 2 \\ 0 & 1 & 2 & 0 \end{bmatrix}$$

8.2.3 Incidence Matrix of a Non-directed Graph (The matrix A)

Definition 8.3. Let G be a graph with n vertices and m edges. The incident matrix denoted by $A(G)$ is defined as the matrix

$$A(G) = [a_{ij}]_{n \times m}$$

where $a_{ij} = 1$, if an edge e_j is incident on ith vertex v_i in G, and
$= 0$, otherwise

$A(G)$ is an n by m matrix whose n rows correspond to the n vertices and m columns correspond to m edges. A graph and its incidence matrix are shown in Fig. 8.6.

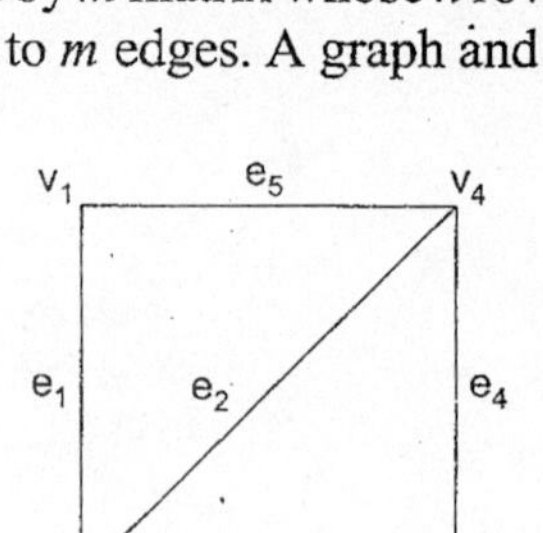

$$A(G) = \begin{array}{c} \\ v_1 \\ v_2 \\ v_3 \\ v_4 \end{array} \begin{bmatrix} e_1 & e_2 & e_3 & e_4 & e_5 \\ 1 & 0 & 0 & 0 & 1 \\ 1 & 1 & 1 & 0 & 0 \\ 0 & 0 & 1 & 1 & 0 \\ 0 & 1 & 0 & 1 & 1 \end{bmatrix}$$

Fig. 8.6

The incidence matrix contains only two elements 0 and 1. Each column in the incidence matrix of a graph has exactly two 1's appearing in that column. The sum of 1's in each row represents the degree of a vertex corresponding to the row. *A* row with all 0's in the incidence matrix represents an isolated vertex in *G*. Two identical columns in an incidence matrix correspond to parallel edges in the graph *G*.

Let *G* be a disconnected graph. If G_1, and G_2 are two components of *g*, then the incidence matrix *ACG*, of *G* can be written as

$$A = A(G) = \left[\begin{array}{c|c} A(G_1) & \\ \hline & A(G_2) \end{array}\right]$$

where $A(G_1)$ and $A(G_2)$ are the incidence matrices of components of G_1 and G_2. If $G_1, G_2, \ldots, G_m$ are *m* components of a disconnected matrices of these matrices. Then the block diagonal forum of $A(G)$ is

$$A(G) = \left[\begin{array}{c|c|c|c|c|c|c} A(G_1) & O & & O & & & \\ \hline & A(G_2) & & & & & \\ \hline & & & & & & \\ \hline & & & & & & \\ \hline & & & & & & \\ \hline & & & & & & A(G_m) \end{array}\right]$$

8.2.4. Nullity of a Graph

Let $G = (V, E)$ be a graph. The number of edger that are to be removed from *G*, so as to make *G* an acyclic graph is called the nullity of the graph *G*.

The nullity of a graph *G* is denoted μ

The nullity of *G* is given by

$\mu = |E| - |V| + 1$

if $|E| = e,\ |V| = n$, then

$\mu = e - n + 1$

8.2.5. Rank of a Graph

Let *G* a graphic the complement of nullity of a graph *G* is called the rank of *G*. The rank of graph *G* is equal to the number of chords in any spanning tree of the graph *G*.

The rank of *G* is denoted by γ.

γ is given by

$\gamma = |V| - 1$ (i.e. $\mu = n - 1$)

Therefore

rank of G + nullity of G

$= |G| = e$

= total number of edges in G

Theorem: The rank of an incidence matrix of a connected simple graph of order n is $n-1$.

Proof: Let G be a connected simple graph and $A(G)$ denote the incidence matrix of G. Each row in $A(G)$ may be regarded as a vector over $GF_{(2)}$ ($A(G)$ is the matrix defined over Galois field module 2 i.e. $GF_{(2)}$.

Let A_1 denote the vector in the first row,

A_2 denote the vector in the second row

$\vdots$

A_n denote the vector in the nth now, then

We have

$$A(G) = \begin{bmatrix} A_1 \\ A_2 \\ \vdots \\ A_n \end{bmatrix} \qquad \text{...(1)}$$

Since, each column of A has exactly two, is the sum of all these vectors is O. Therefore the vector $A_1, A_2, ..., A_n$ are not linearty independent and the rent of $A = r < n$

i.e. rank of $A < n-1$ ...(ii)

We now prove that rank of $A \geq n-1$.

Consider the sum of any m of these n vectors (i.e. $A_1, A_2, ..., A_n$)

Where $m \leq n-1$. If the graph G is connected, then $A(G)$ cannot be partitioned as in (1). i.e. for $m \leq n-1$, we cannot find a m by m submatrix of $A(G)$

Such that the module 2 sum these m row is equal to zero.

Since there only two elements 0 and 1 in the field, the additions of all vectors taken at a time for $m = 1, 2, ..., n-1$ row vectors. Thus we have shown that no linear combinations of m row vectors of a where m s $n-1$, can be equal to zero.

$\therefore$ rank of $A(G) > n-1$ ...(2)

from (1) and (2), we have

$$A(G) = n-1$$

Corollary: The reduced incidence matrix of a tree is non-singular.

Proof: Let G be graph is with a vertices, then G is disconnected. In which case the rank of the incidence matrix will be less then $n-1$. Therefore, the $(n-1)$ by $(n-1)$ reduced incidence matrix of G will not be non-singular. In other words, the reduced incidence matrix of a graph G is non-singular if and only if G is a tree.

8.2.6. Circuit Matrix

Let G be a graph. Let number of edges in G be e (i.e. $|E| = e$) and the number of different circuits in G be q. Then the circuit matrix $B = [b_{ij}]$ of G is a q by e, (0, 1)-matrix, defined as follows

$$b_{ij} = 1 \quad \text{if } i^{\text{th}} \text{ circuit includes } J^{\text{th}} \text{ edge,}$$
$$= 0, \quad \text{otherwise}$$

The circuit matrix B of G can also be written as $B(a)$

Example. Consider the graph G given below

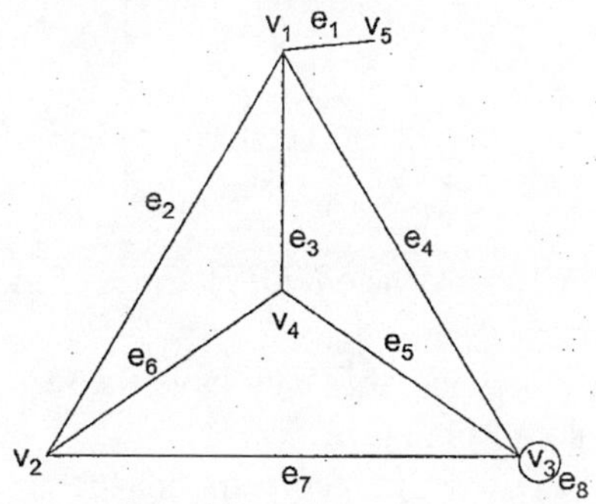

Fig. 8.7.

Solution: There is a circuit of length 1 (i.e. loop) in G, four circuits of length 3, and three circuits of length 4 is G

Let $C_1, C_2, C_3, C_4, C_5, C_6, C_7, C_8$ denote the circuits in G. then, we have

$C_1 = v_3 - v_3$ $\quad$ $C_2 = v_1 - v_2 - v_3 - v_1$

$C_3 = v_1 - v_2 - v_4 - v_1,$ $\quad$ $C_4 = v_2 - v_3 - v_4 - v_1,$

$C_5 = v_1 - v_4 - v_3 - v_1,$ $\quad$ $C_6 = v_1 - v_2 - v_3 - v_4\ v_1,$

$C_7 = v_1 - v_2 - v_4 - v_3 - v_1$ and $C_8 = v_1 - v_4 - v_2 - v_3 - v_1$

$\therefore$ The circuit matrix of the graph G is

$$B = B(G) = \begin{array}{c} \\ C_1 \\ C_2 \\ C_3 \\ C_4 \\ C_5 \\ C_6 \\ C_7 \\ C_8 \end{array} \begin{array}{c} \begin{matrix} e_1 & e_2 & e_3 & e_4 & e_5 & e_6 & e_7 & e_8 \end{matrix} \\ \begin{bmatrix} 0 & 0 & 0 & 0 & 0 & 0 & 0 & 1 \\ 0 & 1 & 0 & 1 & 0 & 0 & 1 & 0 \\ 0 & 1 & 1 & 0 & 0 & 1 & 0 & 0 \\ 0 & 0 & 0 & 0 & 1 & 1 & 1 & 0 \\ 0 & 0 & 1 & 1 & 1 & 0 & 0 & 0 \\ 0 & 1 & 1 & 0 & 1 & 0 & 1 & 0 \\ 0 & 1 & 0 & 1 & 1 & 1 & 0 & 0 \\ 0 & 0 & 1 & 1 & 0 & 1 & 1 & 0 \end{bmatrix} \end{array}$$

The following deservation can be made about B, (the circut matrix of graph G.)

1. Each row of $B = B(G)$ is a circuit vector
2. The number of 1's in a row of a circuit matrix B is equal to the number of edges in the corresponding circuit (i.e. length of the corresponding circuit)
3. If a column has only 0s (zeros) then it corresponds to a non-circuit edge
4. A circuit matrix of a graph G is capable of representing a self-loop
5. If G is a disconnected graph consisting of this components say G_1 and G_2 then $B(G)$ can be written as

$$B(G) = \begin{bmatrix} B(G_1) & 0 \\ 0 & B(G_2) \end{bmatrix}$$

   where, $B(G_1)$, and $B(G_2)$ are the circuit matrices of the components (blocks) G_1 and G_2 respectively.
6. If G_1 and G_2 are two graphs, this they have the same circuit matrix if and only if they are 2-isomorphic
7. Permutation of any two rows in B corresponds to the relabeling of circuits in G
8. Permutation of any two columns in B corresponds to the relabeling of edges in G

Theoram. Let G be a self-loop-free graph. Let B be the circuit matrix, and A be the incidence matrix (of G), whose columns are arranged using the same order of edges. Then

$$A\,B^T = B\,A^T = O. \qquad \text{(Mod 2)}$$

(i.e. every row of B is orthogond to every row of A)
where A^T and B^T are the transposed matrices of A and B respectively.

8.2.7 Fundamental Circuit Matrix

Let G be a connected graph and T be a spaning tree of G. The independent circuits in G are called fundamental circuits in G.

A submatrix of a circuit matrix B in which all rows correspond to a set of fundamental circuits is called a fundamental circuit matrix. It is denoted by B_f

Example: Consider the graph G given below (see Fig. 8.8)

If $T : v_1 - v_2 - v_3 - v_4$ (shown by bold lines) in the Figure 8.8 is a spanning tree of G. Then there are three fundamental circuits in G. If c_1, c_2 and c_3 denote the circuits then we have

$c_1 : v_1 - v_2 - v_3 - v_1$ (due to the chord e_5)
$c_2 : v_3 - v_2 - v_4 - v_3$ (due to the chord e_6)
$c_3 : v_3 - v_2 - v_4 - v_5 - v_3$ (due to the chord e_7)

$$E(T) = \{e_1, e_2, e_3, e_4\}$$

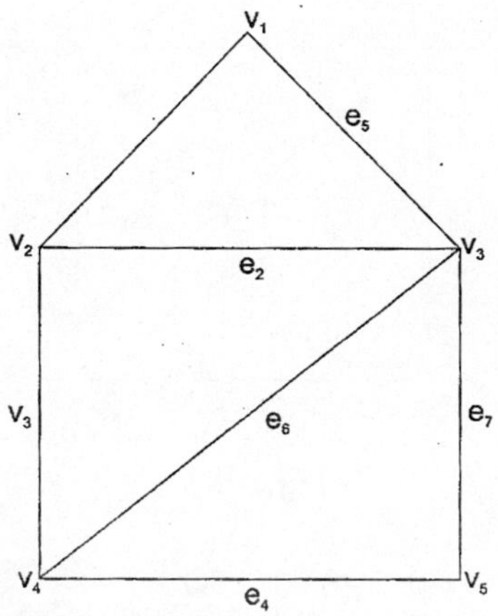

Fig. 8.8

∴ The fundamental circuit matrix corresponding to T is

$$B_t = \begin{matrix} & e_1 & e_2 & e_3 & e_4 & e_5 & e_6 & e_7 \\ c_1 & 1 & 1 & 0 & 0 & 1 & 0 & 0 \\ c_2 & 0 & 1 & 1 & 0 & 0 & 1 & 0 \\ c_3 & 0 & 1 & 1 & 1 & 0 & 0 & 1 \end{matrix}$$

Theorem: Let G be a connected graph with n (i.e. $|v|$) vertices and e (i.e. $|E|$) edges and B be a circuit matrix of G. Then the rank of B is $e - n + 1$

Proof: Let A be an incidence matrix of G

Then we have $A\,B^T = 0. = A^T B$ (mod 2)

By Sylvestis theorem

rank of A rank of $B < e$

i.e. rank of $< e -$ rank of A

$>$ rank of $< e - (n - 1)$ (∵ rank of $A = n - 1$)

$>$ rank of $B < e - n + 1$ ...(1)

But rank of $B > e - n + 1$ ...(2)

from (1) and (2), we have

rank of $B = e - n + 1$

8.2.8 Cut-Set Matrix

Let G be a graph. Let $C = [c_{ij}]$ be a matrix in which the rows correspond to the cut-sets and columns to the edges of G such that

$$c_{ij} = 1, \quad \text{if } i^{th} \text{ cut-set contains } j^{th} \text{ edge, and}$$
$$= 0, \quad \text{otherwise}$$

Then the matrix c is called a cut-set matrix of G.

A cut-set matrix of a graph G is also denoted by $C(G)$

Consider the graph G shown in Fig. 8.9

G:

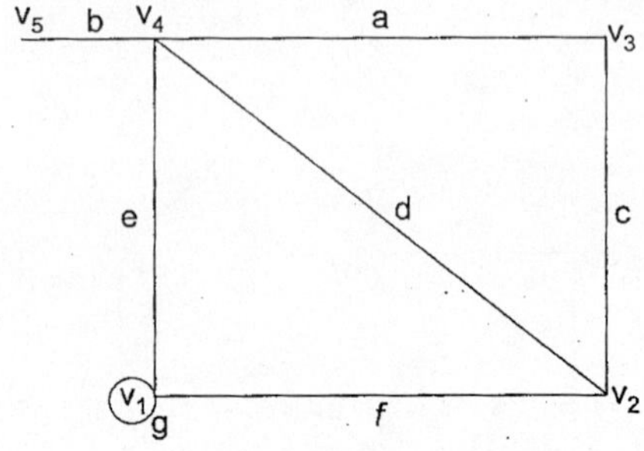

Fig. 8.9

$$A = \{a, c\}, \quad B = \{a, d, e\}, \quad C = \{b\}$$

$D = \{c, d, f\}$, and $E = \{e, f\}$ and time cut-set of G

$$\therefore\ C = C(G) = \begin{matrix} A \\ B \\ C \\ D \\ E \end{matrix} \begin{bmatrix} 1 & 0 & 1 & 0 & 0 & 0 & 0 \\ 1 & 0 & 0 & 1 & 1 & 0 & 0 \\ 0 & 1 & 0 & 0 & 0 & 0 & 0 \\ 0 & 0 & 1 & 1 & 0 & 1 & 0 \\ 0 & 0 & 0 & 0 & 1 & 1 & 0 \end{bmatrix}$$

The following observation may be made about a out-set matrix C i.e. $C(G)$:

1. Each row in a cut-set matrix $C(G)$ is a cut-set vector.
2. A column with all 0s corresponds to an edge forming a self in G.
3. Permutation of rows of C, corresponds to a r labeling of renaming of cut-sets in G.
4. Permutation of columns of C, corresponds to a relabeling of edges in G.
5. Parallel edges produce identical columns in the cut-set matrix.
6. If G is a non-separable graph, then $C(G)$ contains $A(G)$ where $A(G)$ is the incidence matrix of G.
7. If G is separable graph, then the incidence matrix of each block is contained in the cut-set matrix of G.
8. If $C(G)$ in the cut-set matrix and $A(G)$ is the incidence matrix of a graph G, then we have rank of $C(G) \geq$ rank of $A(G)$.
9. It G is a connected graph with n vertices then,
$$\text{rank of } C(G) \geq n - 1$$
10. The number of edges common to a cut-set and a circuit in a graph G is always even if the edges B (the circuit matrix of G) and the edges in C are arranged in the same order then we have,
$$BC^T = CB^T = 0 \pmod 2$$

Theorem: The rank of a cut-set matrix $C(G)$ is equal to the rank of the incidence matrix $A(G)$; which is equal to the rank of the graph G.

Proof: For a graph G we have

$BC^T = 0 \pmod 2$ ...(1)

$A(G)$ is a submatrix of $C(G)$, therefore we have rank of

$C(G) \geq$ rank of $A(G) \geq n-1$ ...(2)

on applying Sylvestis theroem to equation (1) we get

rank of B+ rank of $C \leq c$

$\Rightarrow$ rank of $C \leq e -$ Rank of B

$\Rightarrow$ rank of $C(G) \leq r$, the rank of G.

$\Rightarrow$ rank of $C(G) \leq n-1$ $\quad (\because r = n-1)$ ...(3)

from (2) and (3), we get

rank of $C(G) = n-1$, = the rank of G.

8.2.9 Fundamental Cut-Set Matrix

Let G be a connected graph with a vertices and edges. A fundamental cut-set matrix C_f is an $(n-1)$ by e submatrix of C such at the rows corresponding to the set of fundamental cut-sets with respect to some spanning tree.

Consider the graph G (shown in Figs. 8.10) of previous example:

Let a, b, d be the edges of a spanning tree T in G. i.e. $E(T) = \{a, b, d, c\}$

The fundamental cut-sets in G are

$$A = \{a, c\}, C = \{b\}, D = \{c, d, f\}, E = \{e, f\}$$

The fundamental cut-set matrix is:

$$C_f = \begin{array}{c} \\ A \\ C \\ D \\ E \end{array} \begin{array}{c} \begin{array}{ccccccc} a & b & c & d & e & f & g \end{array} \\ \begin{bmatrix} 1 & 0 & 1 & 0 & 0 & 0 & 0 \\ 0 & 1 & 0 & 0 & 0 & 0 & 0 \\ 0 & 0 & 1 & 1 & 0 & 1 & 0 \\ 0 & 0 & 0 & 0 & 1 & 1 & 0 \end{bmatrix} \end{array}$$

8.2.10 Reduced Incidence Matrix

Let $A(G)$ be the incidence matrix of a connected graph with n-vertices. The matrix obtained by removing anyone row from $A(G)$ is called a reduced incidence matrix. It is denoted by A_f.

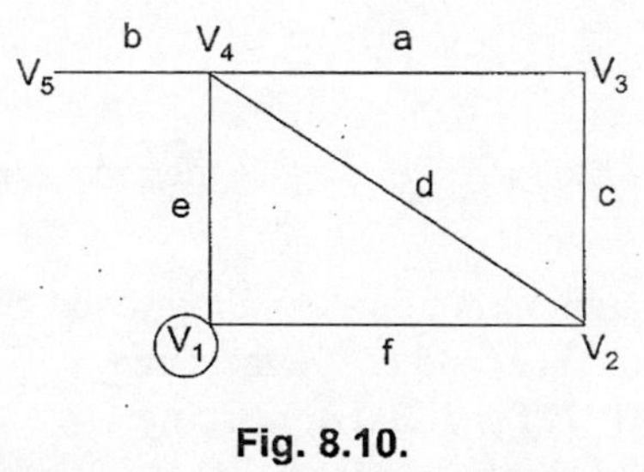

Fig. 8.10.

The vertex corresponding to the deleted row of A is called A_f reference vertex.

A_f is an $(n-1)$ by e submatrix of A.

The rows of A_f are linearly in dependent.

(A_f is non-singular matrix whenever A_f is a square matrix).

We now show that,

"The reduced incidence matrix of a graph is non-singular if and only if the graph G is a tree".

We know that a tree is a connected graph with n vertices and $(n-1)$ edges. The reduced matrix of a tree with n vertices is a sequare matrix of order $(n-1)$ and rank $(n-1)$. Therefore it is a non-singular matrix. Thus we have shown that the reduced incidence matrix of a tree is non-singular.

Now, let G be a graph with n vertices and $(n-1)$ edges G is not connected. If G is not a tree, then G is not connected, (i.e. disconnected), and the rank of the incidence matrix of G will be less than $n-1$. Therefore, the $(n-1)$ by $(n-1)$ reduced incidence matrix of G will not be non-singular. Thus (we have shown that) the reduced incidence matrix of a graph is non-singular if and only if the graph is a tree.

8.2.11 Relationships among A_f, B_f and C_f

Let B_f denotes the fundamental circuit matrix of a connected graph G. Let us rearrange the columns, such that all the $n+1$ thereby.

If n is the number of vertices in G and e is the number of edges in G, then B_f is an $(e-n+1)$ by matrix. Let us arrange the columns in B_f, such that all the $e-n+1$ chords correspond to the first $e-n+1$ columns. Also if we arrange the rows such that the first row corresponds to the fundamental circuit made by the chord is the first column, the second row to the fundamental circuit made by the second, so on. A matrix B_f thus arranged can be written as a partitioned matrix:

$$B_f = [I_\mu : B_f]$$

where I_μ is the identity matrix of order $\mu = e - n + 1$, and B_f is the remaining μ by $(n-1)$ submatrix. Corresponding to the branches of the spanning tree. Similarly we can partition the fundamental cut-set matrix C_f and write:

$$C_f = [C_c : I_{n-1}]$$

where the last $(n-1)$ columns forming the identity matrix correspond to the $(n-1)$ branches of the spanning tree and first $e-n+1$ columns forming the matrix C_i correspond to the chords:

Consider

$$B_f = [I_\mu : B_f] \qquad ...(1)$$

where f denotes the submatrix corresponding to the branches of a spanning tree T.

and $$C_f = [C_i : I_{n-1}] \qquad ...(2)$$

where C_i denotes the submatrix corresponding to the chords.

Let the spanning tree T in (1) and (2) be the same.

Also let order of eges in (1) and (2) be the same.

Consider A_f the reduced incidence matrix of size $(n-1)$ by e.

Let the columns (i.e. edges) be arranged in the same order as in B_f and C_f. Partition A_f into two submatrices:

$$A_f = [A_t : A_t] \qquad ...(3)$$

where A_t consists of the $(n-1)$ columns corresponding to the branches of the spanning tree and A_c is the remaining submatrix corresponding to $e-n+1$ chords:

∴ we have

$$A_f.B_f^T = 0 \text{ (nod 2)} \qquad (\because \text{ the columns in } A_f \text{ and } B_f \text{ are arranged in the same order})$$

i.e.

$$[A_c : A_t].\left[\frac{I_\mu}{B_t^T}\right] = 0$$

and

$$A_c + A_t B_t^T = 0 \qquad ...(4)$$

A_t is non-singular, therefore A_f^{-1} exists:

Premultiplying equation (G) by A_f^{-1} we get

$$A_t^{-1}[A_c + A_t B_f^T] = A_t^{-1} = 0$$

$$\Rightarrow A_t^{-1} A_c + A_t^{-1} A_t B_f^T = 0$$

$$\Rightarrow A_t^{-1} A_c + I B_f^T = 0$$

$$\Rightarrow A_t^{-1} A_c + - B_f^T = 0 \qquad ...(5)$$

$$\Rightarrow A_t^{-1} A_c + B_f^T = 0 \; (\because -1 = 1 \text{ in modular 2 Arithmetic}$$

$$\text{i.e. } -B_f^T = B_t^T \text{ (mod 2))} \qquad ...(6)$$

$$\therefore B_f^T = A_t^{-1} A_c$$

Similarly, we have,

$$C_f B_f = 0 \qquad \text{(mod 2)}$$

$(\because$ the columns of B_f and C_f are arranged in the same order)

$$\Rightarrow C_c : I$$

$$\Rightarrow [C_c : I_{n-1}]\left[\frac{I_\mu}{B_t^T}\right] = 0$$

$$\Rightarrow C_c : -B_t^T$$

$$\Rightarrow C_c : B_t^T \qquad ...(7)$$

$$\Rightarrow C_c : A_t^{-1} A_c \qquad \text{(from (61)) (8)}$$

Thus we have proved that,

$$B_t^T = A_t^{-1} A_c$$

$$B_t^T = C_c$$

and

$$B_t^T = A_t^{-1} A_c$$

$$B_t^T = C_c = A_t^{-1} A_c$$

where B_t^T denotes the transposed matrix of B_t

From the above relations the following conclusions can be drawn

1. If A or A_f of a graph G in given them we can readily construct the matrices B_f and C_f, starting from an arbitrary spanning tree and its subgraphs A_t and in A_f.
2. If either B_f or C_f is given of graph G is given, then we can construct the other.

   B_f determines a graph upto 2-isomorphism therefore C_f also determine graph upto 2-isomorphism.
3. If either B_f or C_f of G is given, then A_f in general cannot be determined completely.

8.2.12 Path matrix

Let G be a connected graph and let it V be a pair of vertices in G. Then a path matrix $P(u, v) = [pij]$ is defined as follows

$$p_{ij} = 1 \qquad \text{if } j^{\text{th}} \text{ edge lies in } i^{\text{th}} \text{ path otherwise}$$
$$= 0$$

The rows in a path matrix $P(u, v)$ correspond to different paths between the vertices u and v, and the columns in $P(u, v)$ correspond to the edges in G.

Example: Consider the graph G is shown in Fig. 8.11.

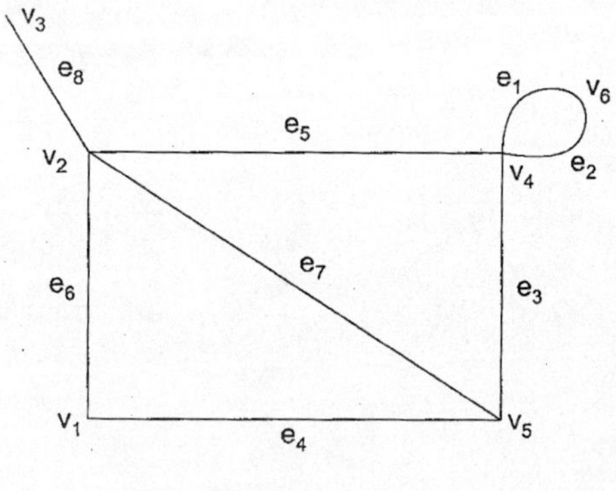

Fig. 8.11

Consider the vertices V_3 and V_4.

There were three different paths between the vertices V_3 and V_4.

The paths are $\qquad P_1 = \{e_8, e_5\}$

$\qquad P_2 = \{e_8, e_3, e_3\}$

$\qquad P_3 = \{e_8, e_6, e_4, e_3\}$

The path matrix $P_3(v_3, v_4)$ is given below:

$$P(v_3, v_4) = \begin{array}{c} \rho_1 \\ \rho_2 \\ \rho_3 \end{array} \overset{\begin{array}{cccccccc} \rho_1 & \rho_2 & \rho_3 & \rho_4 & \rho_5 & \rho_6 & \rho_7 & \rho_8 \end{array}}{\begin{bmatrix} 0 & 0 & 0 & 0 & 1 & 0 & 0 & 1 \\ 0 & 0 & 1 & 0 & 0 & 0 & 1 & 1 \\ 0 & 0 & 1 & 1 & 0 & 1 & 0 & 1 \end{bmatrix}}$$

If $P(u, v)$ is a path matrix if a graph G, then we observe the following:

1. There is no row with all 0s (zeros)
2. The path matrix of a tree is a row matrix.
3. A column of all 0s correspond to an edge that does not lie in any path u and v.
4. A column of all is correspond to an edge that lies in every path between the vertices u and v.
5. The ring sum of any two rows gives a union of dispoint circuits of G.
6. Path matrices are defined for connected graphs. A path matrix connot be defined for these vertices lying in different components.
7. The number of is in a row of a path matrix is equal to the length of the path corresponding to the row.

Theorem 8.1. Let G be a labeled graph with adjacency matrix $A(G)$. Then the i, j entry of $A^n(G)$ in the number of walks of length n from the vertex v_i to the vertex v_j.

Corollary 8.1.1. If G is a connected graph, the distance between the vertices v_i and v_j for $i \neq j$ is the least positive integer n for which the i, j entry of $A^n(G)$ is non-zero.

Example. Find the number distance walks of length three from the vertex v_1 to v_2 in the graph shown in Fig. 8.12.

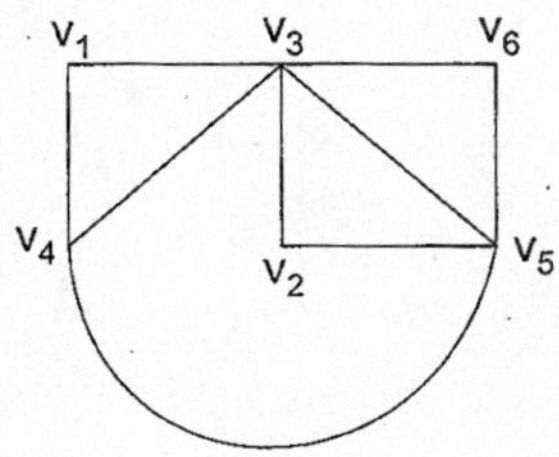

Fig. 8.12

Solution. Let $X(G)$ denote the adjacency matrix of G

Then
$$X(G) = \begin{array}{c} v_1 \\ v_2 \\ v_3 \\ v_4 \\ v_5 \\ v_6 \end{array} \overset{\begin{array}{cccccc} v_1 & v_2 & v_3 & v_4 & v_5 & v_6 \end{array}}{\begin{bmatrix} 0 & 0 & 1 & 1 & 0 & 0 \\ 0 & 0 & 1 & 0 & 1 & 0 \\ 1 & 1 & 0 & 1 & 1 & 1 \\ 1 & 0 & 1 & 0 & 1 & 0 \\ 0 & 1 & 1 & 1 & 0 & 1 \\ 0 & 0 & 1 & 0 & 1 & 0 \end{bmatrix}}$$

We have
$$X^2(G) = \begin{bmatrix} 2 & 1 & 1 & 1 & 2 & 1 \\ 1 & 2 & 1 & 2 & 1 & 2 \\ 1 & 1 & 5 & 2 & 3 & 1 \\ 1 & 2 & 2 & 3 & 1 & 2 \\ 2 & 1 & 3 & 1 & 4 & 1 \\ 1 & 2 & 1 & 2 & 1 & 2 \end{bmatrix}$$

and
$$X^3(G) = \begin{array}{c} \\ v_1 \\ v_2 \\ v_3 \\ v_4 \\ v_5 \\ \\ \end{array} \begin{array}{c} \begin{array}{cccccc} v_1 & v_2 & v_3 & v_4 & v_5 & v_6 \end{array} \\ \begin{bmatrix} 2 & 3 & 7 & 5 & 4 & 3 \\ 3 & 2 & 8 & 3 & 7 & 2 \\ 7 & 8 & 11 & 9 & 9 & 8 \\ 7 & 3 & 9 & 4 & 9 & 3 \\ 4 & 9 & 9 & 9 & 6 & 7 \\ 3 & 2 & 8 & 3 & 7 & 2 \end{bmatrix} \end{array}$$

$\therefore$ The number of walks of length 3 from v_1 to v_2 = 1, 2, entry of $X^3(G)$
$= 3$

Example 2. Show that there is no connected graph corresponding to the incidence matrix

$$A(G) = \begin{bmatrix} 1 & 0 & 1 & 0 \\ 0 & 1 & 0 & 0 \\ 1 & 0 & 1 & 0 \\ 0 & 1 & 0 & 0 \end{bmatrix}$$

Solution. Consider $A(G) = \begin{bmatrix} 1 & 0 & 1 & 0 \\ 0 & 1 & 0 & 0 \\ 1 & 0 & 1 & 0 \\ 0 & 1 & 1 & 0 \end{bmatrix}$

The last column of $A(G)$; does contains no 1's. Therefore, the graph corresponding to $A(G)$ does not exist.

EXERCISE 8.1

1. Write down the adjacency matrices of the following graphs

(a)

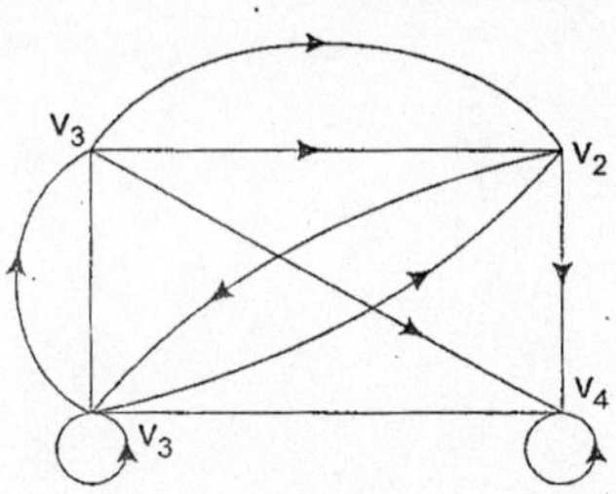

(b)

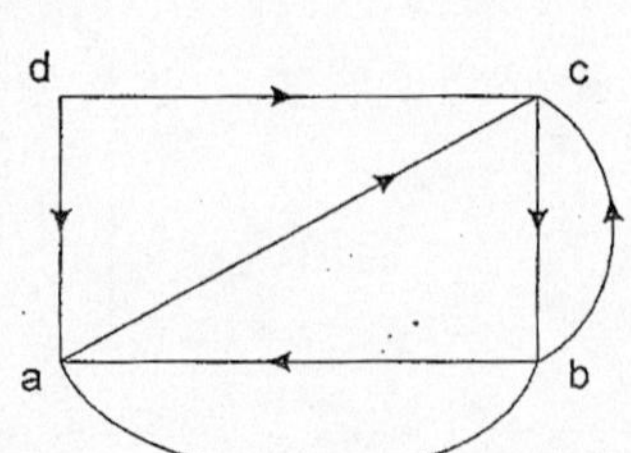

(c)

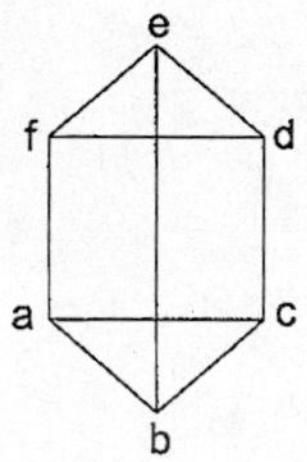

(d)

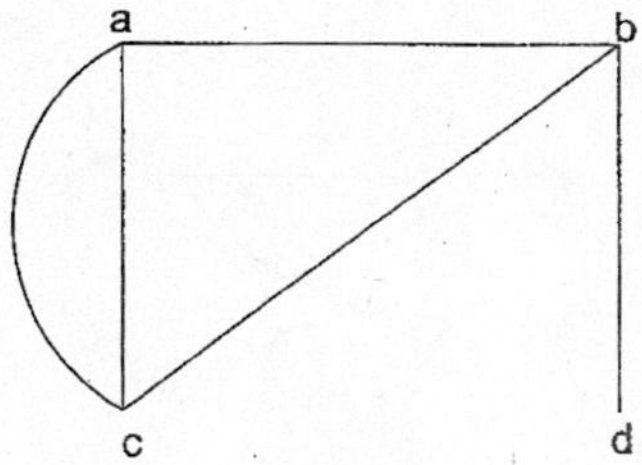

(e)

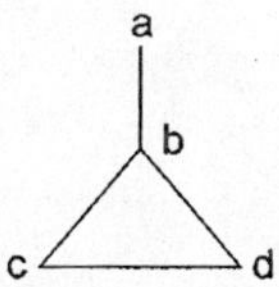

(f)

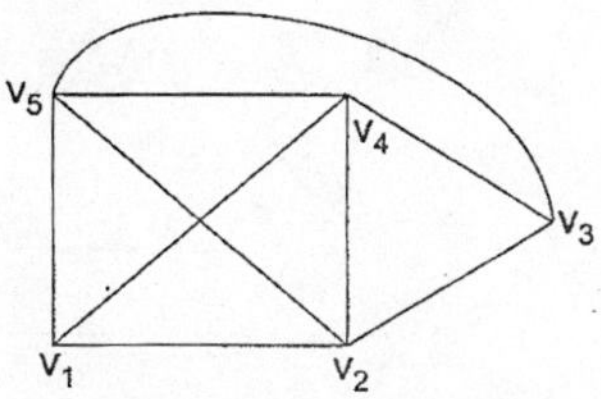

(g)

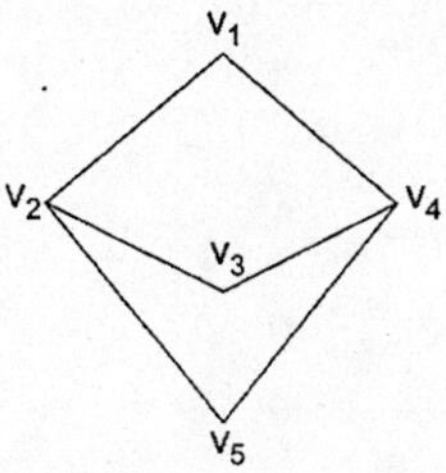

(h)

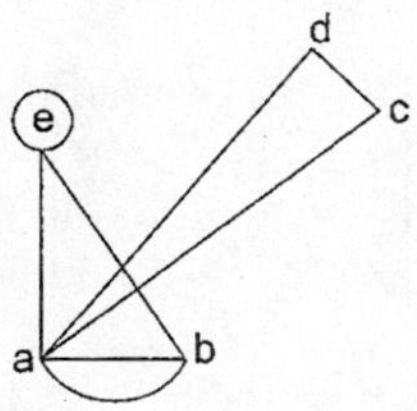

(i)

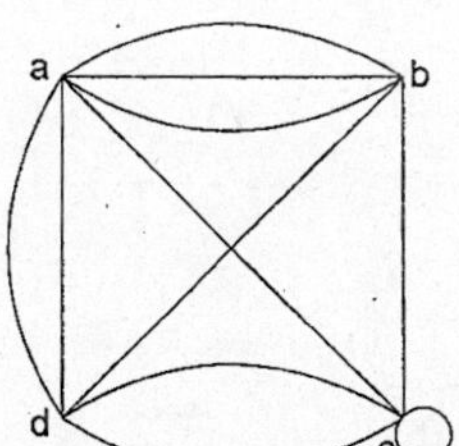

(j)

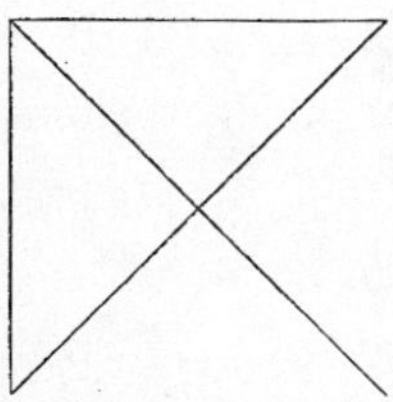

2. Write the incidence matrices for the following graphs:

(a)

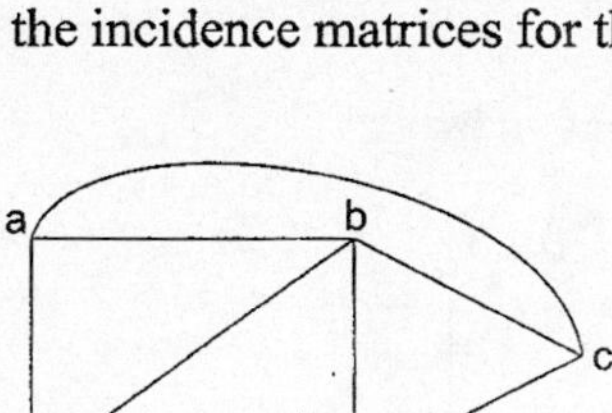

(b)

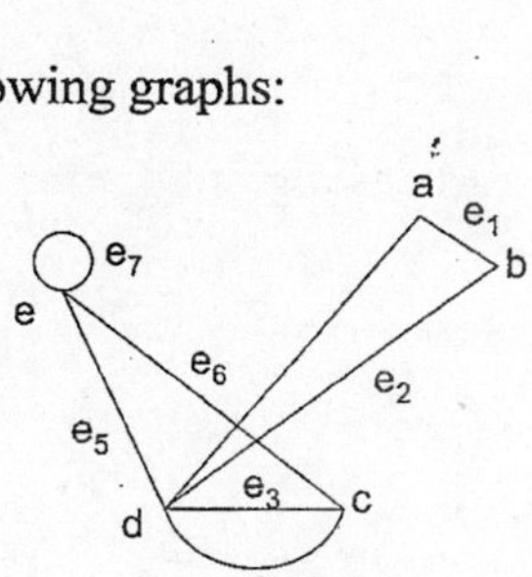

(c) 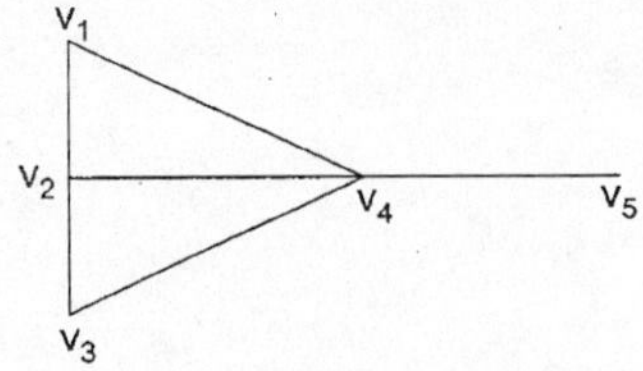

(d) 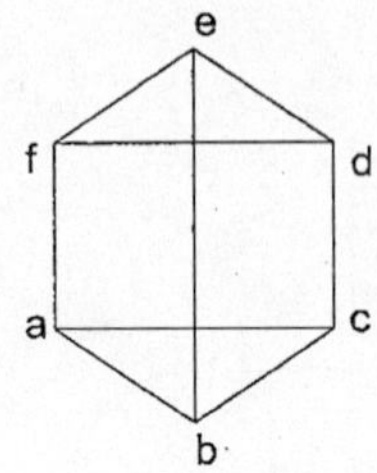

(e) 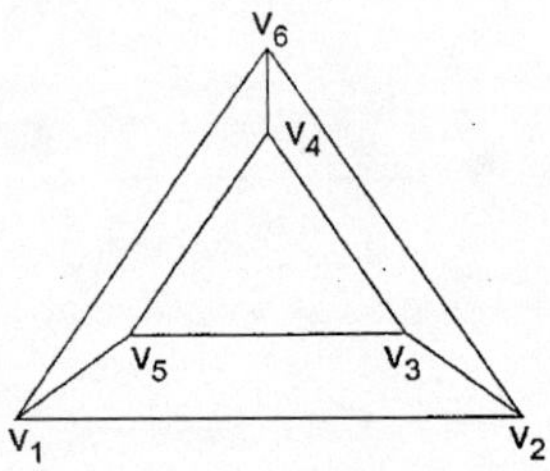

(f) 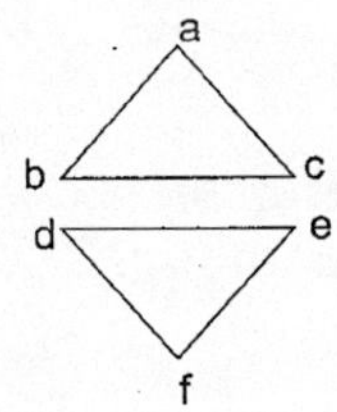

3. Draw the graph, corresponding to each adjacency matrix:

(a) $\begin{bmatrix} 0 & 1 & 1 & 1 & 0 \\ 1 & 0 & 0 & 1 & 0 \\ 1 & 0 & 0 & 1 & 1 \\ 1 & 1 & 1 & 0 & 1 \\ 0 & 0 & 1 & 1 & 0 \end{bmatrix}$ (b) $\begin{bmatrix} 1 & 0 & 0 & 1 \\ 0 & 0 & 2 & 1 \\ 0 & 2 & 0 & 0 \\ 1 & 1 & 0 & 1 \end{bmatrix}$

(c) $\begin{bmatrix} 0 & 1 & 1 & 0 & 0 & 0 & 0 \\ 1 & 0 & 1 & 1 & 0 & 0 & 0 \\ 1 & 1 & 0 & 0 & 0 & 0 & 0 \\ 0 & 1 & 0 & 0 & 1 & 1 & 0 \\ 0 & 0 & 0 & 1 & 0 & 1 & 1 \\ 0 & 0 & 0 & 1 & 1 & 1 & 1 \\ 0 & 0 & 0 & 0 & 1 & 1 & 0 \end{bmatrix}$ (d) $\begin{bmatrix} 0 & 1 & 0 & 1 & 0 & 0 \\ 1 & 0 & 1 & 0 & 1 & 0 \\ 0 & 1 & 0 & 0 & 1 & 0 \\ 1 & 0 & 0 & 0 & 1 & 0 \\ 0 & 1 & 1 & 1 & 0 & 1 \\ 0 & 0 & 0 & 0 & 1 & 1 \end{bmatrix}$

4. Construct a digraph for the adjacency matrix:

$$\begin{bmatrix} 0 & 1 & 0 & 0 & 0 \\ 0 & 0 & 1 & 0 & 0 \\ 0 & 0 & 0 & 1 & 1 \\ 1 & 1 & 0 & 0 & 0 \\ 0 & 0 & 0 & 0 & 0 \end{bmatrix}$$

5. Show that there exists no connected graph G, corresponding to the incidence matrix:

$$\begin{bmatrix} 1 & 0 & 1 \\ 0 & 1 & 0 \\ 1 & 0 & 1 \\ 0 & 1 & 0 \end{bmatrix}$$

6. Show that there exists a connected graph G, corresponding to the incidence matrix:

$$\begin{bmatrix} 1 & 0 & 1 & 0 \\ 1 & 0 & 0 & 1 \\ 0 & 1 & 1 & 0 \\ 0 & 0 & 0 & 1 \\ 0 & 1 & 0 & 0 \end{bmatrix}$$

7. Draw the graph corresponding to the adjacency matrix given below:

$$\begin{bmatrix} 0 & 0 & 0 & 0 & 1 \\ 0 & 0 & 1 & 1 & 1 \\ 0 & 1 & 0 & 1 & 0 \\ 0 & 1 & 1 & 0 & 1 \\ 1 & 1 & 0 & 1 & 0 \end{bmatrix}$$

Ans.

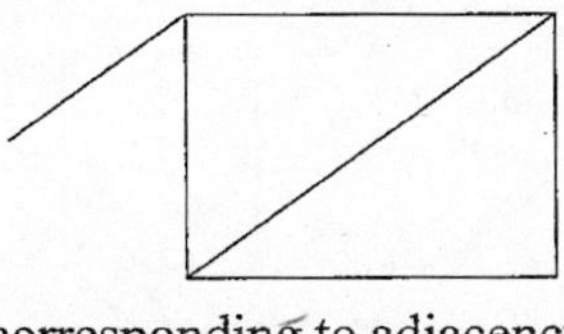

8. Draw the corresponding to adjacency matrix given below:

$$\begin{bmatrix} 0 & 1 & 1 & 0 & 0 \\ 1 & 0 & 1 & 0 & 0 \\ 1 & 1 & 0 & 0 & 0 \\ 0 & 0 & 0 & 0 & 1 \\ 0 & 0 & 0 & 1 & 0 \end{bmatrix}$$

Ans.

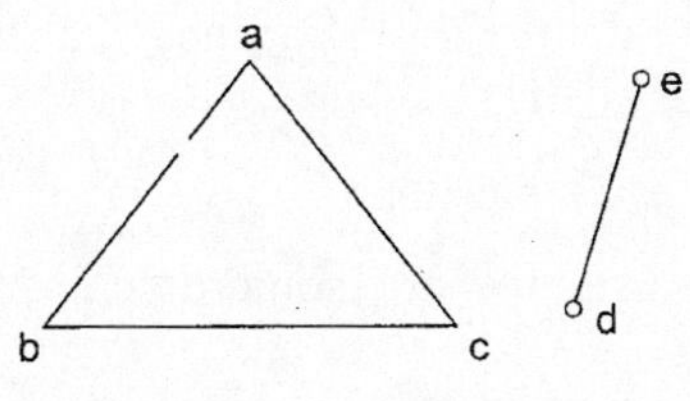

8.3 ISOMORPHISM

Definition 8.4. Two graph G is G' are said to be isomorphic if there is a function $f: V(G) \to V(G')$ from the vertices of G to the vertices of G' such that

(i) f is one-one (ii) f is onto

and (iii) for each pair of vertices u, and v of G, $\{u, v\} \in E(G)$ if and only if $\{f(u), f(v)\} \in E(G')$

(i.e. f – preserves adjacency)

If $f: G \to G'$ is an isomorphism, then G and G' are said to be isomorphic, and if two graphs G and G' are isomorphic then there may be several isomorphisms from G to G'.

If G and G' are isomorphic, then we write $G \cong G'$.

If two graphs G and G' are isomorphic and $f: G \to G'$ is an isomorphism, then

(i) $|V(G) = |V(G')|$

(ii) $|E(G)| = |E(G')|$

(iii) The degree sequences of G and G' are the same

and (iv) If $v_0 - v_1 - v_2 - \ldots - v_{k-1} - v_k$ is a cycle in G, then $f(v_0) - f(v_1) - f(v_2) - \ldots - f(v_{k-1}) - f(v_k)$ is a cycle in G'.

Two graphs G and G' may have the same number of vertices and the same degree sequence but they may not be isomorphic. The graphs G and G' shown in Fig. 8.13 have the same degree sequence and same number of vertices but they are not isomorphic.

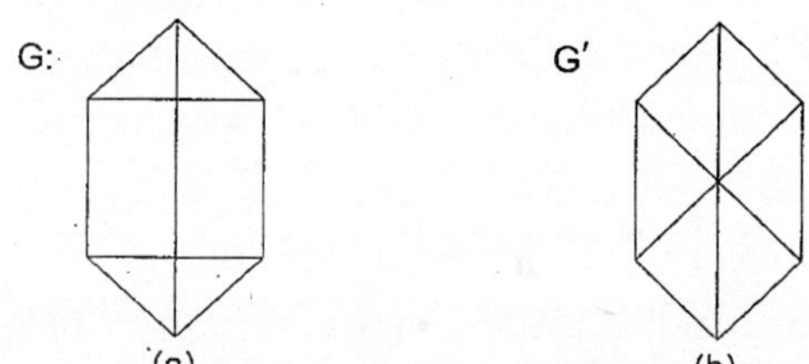

Fig. 8.13 Nonisomorphic graphs with same degree sequence

Usually, we employ the adjacency matrix to check whether or not the given graphs G and G' are isomorphic.

Suppose $f: G \to G'$ is a one-one onto function. Let $v_0 - v_1 - v_2 - \ldots - v_n$ be the vertex ordering of G, and $f(v_0) - f(v_1) - \ldots - f(v_n)$ be the corresponding vertex ordering of G'. Let $A(G)$ denote the adjacency matrix for the vertex ordering $v_0 - v_1 - \ldots - v_n$ of G and $A(G')$ denote the adjacency matrix for the vertex ordering $f(v_0) - f(v_1) - \ldots - f(v_n)$ of G'. If

$$A(G) = A(G')$$

Then we conclude the graphs G and G' are isomorphic and if $A(G) \neq A(G')$, then f is not an isomorphism.

Theorem 8.2. Let $G=(V,E)$ and $G'=(V',E')$ be any two graphs and $f: G\to G'$ be an isomorphism. If $v\in V$ then $\deg(v)=\deg f(v)$

Proof. G and G' are isomorphic therefore two points u' and v' of G are adjacent in the graph G if and only if $f(u')$ and $f(v')$ are adjacent in G'.

Therefore the number of vertices which are adjacent to $v\in V$ is equal to the number of vertices in G' which are adjacent to $f(v)$. Hence $\deg(v)=\deg f(v)$

Definition 8.5. (Automorphism): An isomorphism of a graph G onto itself is called an automorphism of G.

SOLVED EXAMPLES

Example 1. Show that the graphs G and G' shown in Fig. 8.14 are isomorphic.

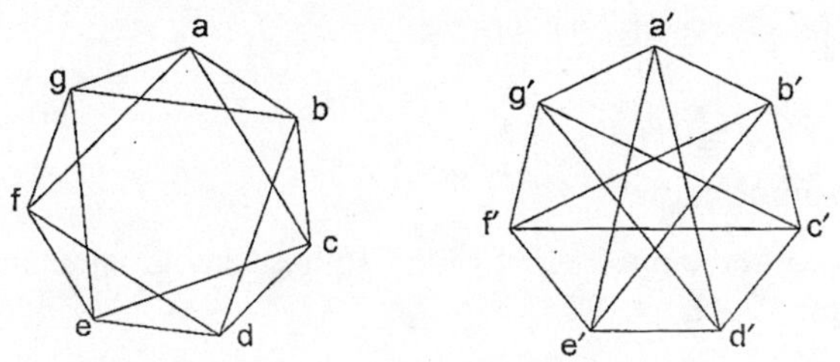

Fig. 8.14

Solution. Number of vertices in $G=|V(G)|=7$

Number of vertices in $G'=|V(G')|=7$

$$\therefore \qquad |V(G)|=|V(G')|$$

$$E(G)=14, E(G')=14 \Rightarrow |E(G)|=|E(G')|$$

Each vertex in G is of degree 4 and each vertex of G' is also of degree of 4.

Define a mapping $\phi : G\to G'$, such that

$$\phi(a)=a', \phi(b)=c', \phi(c)=b', \phi(d)=f', \phi(e)=c', \phi(f)=g' \text{ and } \phi(g)=d'$$

clearly the mapping ϕ in one-one and onto.

$\therefore$ ϕ preserves adjacency

Hence $\phi : G\to G'$ is an isomorphism and the graphs G, G' are isomorphic.

Example 2. Show that the graphs G and G' (in Fig. 8.15) are isomorphic.

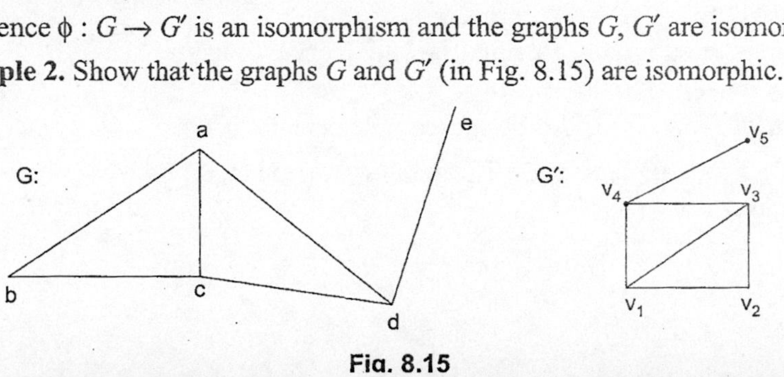

Fig. 8.15

Solution. We have number of vertices in G = number of vertices in G' = 6 number of edges in G = number of edges in G

i.e. $$|V(G)| = |V(G')| \text{ and } |E(G)| = |E(G')|$$

both the graphs have the same degree sequence. Consider the map $f: G \to G'$ defined as follows $f(a) = v_1$, $f(b) = v_2$, $f(c) = v_3$, $f(d) = v_4$, $f(e) = v_5$

The adjacency matrix of G for the ordering a, b, c, d, e and the adjacency matrix of G' for the ordering

$$a \to v_1, b \to v_2, c \to v_3, d \to v_4, e \to v_5 \text{ is the matrix}$$

$$\begin{bmatrix} 0 & 1 & 1 & 1 & 0 \\ 1 & 0 & 1 & 0 & 0 \\ 1 & 1 & 0 & 1 & 0 \\ 1 & 0 & 1 & 0 & 1 \\ 0 & 0 & 0 & 1 & 0 \end{bmatrix}$$

i.e. $$X(G) = X(G')$$

$\therefore$ G and G' are isomorphic.

Example 3. Show that the graph G in Fig. 8.16 has a subgraph isomorphic to $k_{3,3}$ Identity the subgraph

G:

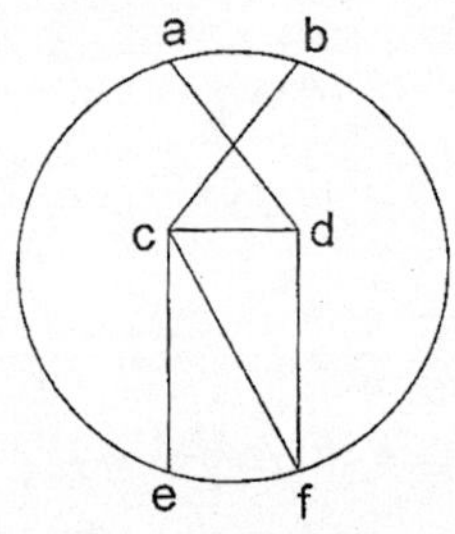

Fig. 8.16

Solution. Deleting the edge $\{e, f\}$ from the graph G, we get a subgraph of G whose vertex set can be partitioned into the sets.

$A = \{a, c, f|$, $B = \{b, d, e|$ and the subgraph obtained can drawn as follows:

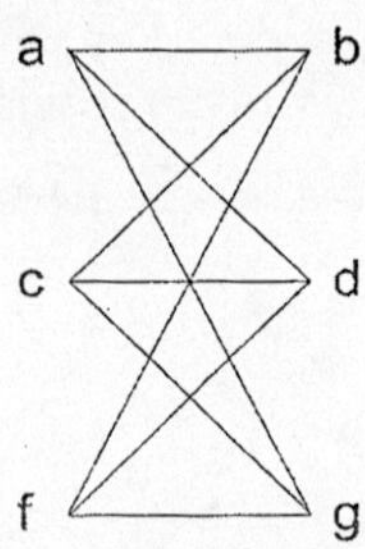

Fig. 8.17

Clearly the subgraph obtained is $k_{3,3}$. Hence G has a subgraph, which is isomorphic to $k_{3,3}$.

Definition. A graph is said to be self-complementary, if G is isomorphic to its complement $\overline{G}$. The graph shown in Fig. 8.17 is self-complementary.

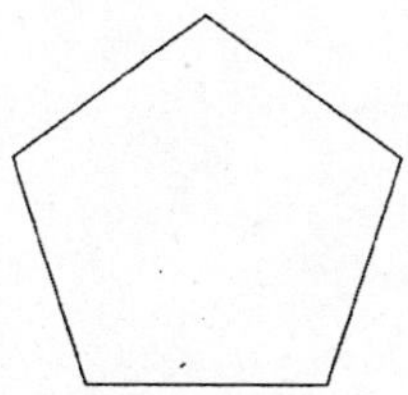

Fig. 8.18

Two simple graphs are isomorphic if and only if their complements are isomorphic.

SOLVED EXAMPLES

Example 1. Show that a graph G is self-complementary if it has $4n$ or $4n + 1$ points (vertices) where n is non-negative integer.

Solution. Let $G = (V, E)$ be a self-complementary graph with m points

Since G is self-complementary, G is isomorphic to $\overline{G}$

$$\therefore \quad |E(G)| = |E(\overline{G})|$$

$$\text{Also} \quad |E(G)| + |E(\overline{G})| = \frac{m(m-1)}{2}$$

$$\Rightarrow \quad 2|E(G)| = \frac{m(m-1)}{2}$$

$$\Rightarrow \quad |E(G)| = \frac{m(m-1)}{4}$$

$\frac{m(m-1)}{4}$ is an integer and one of m or $(m - 1)$ is odd.

$\therefore$ m or $(m - 1)$ is a multiple of 4.

Hence m is of the form $4n$ or $4n + 1$.

Note. From the above example it is clear that a graph G with n vertices is isomorphic to its complement if n or $(n - 1)$ is a multiple of 4 and the number of edges in $G = \frac{n(n-1)}{4}$.

Example 2. Can a graph with seven vertices be isomorphic to its complement.

Solution. We have $n = 7$

$$\therefore \quad n - 1 = 7 - 1 = 6$$

Neither 7 nor 6 is a multiple of 4. Therefore a graph with 7 vertices cannot be isomorphic to its complement.

Example 3. Show that the graphs shown in Fig. 8.19 are isomorphic.

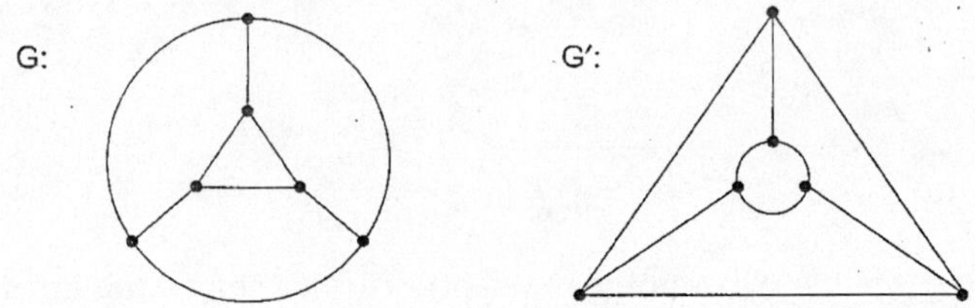

Fig. 8.19

Solution. Both the graphs G and G' have same number of vertices, same number of edges and same degree sequence

i.e. $\quad |V(G)| = |V(G')|, \quad |E(G)| = |E(G')|$

and degree sequence of $G = (3, 3, 3, 3, 3, 3)$ = degree sequence of G'.

But the graphs are not labeled. We label the vertices of the graphs as shown in Fig. 8.20.

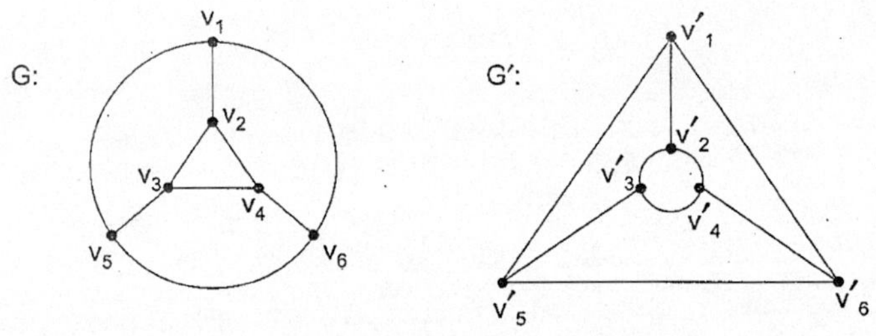

Fig. 8.20

We defined a mapping $\phi : G \to G'$ by $\phi(v_i) = v_i'$

The mapping ϕ – preserves adjacency

$\therefore$ G and G' are isomorphic.

Example 4. Show that the graphs G_1 and G_2 are not isomorphic (see Fig. 8.21).

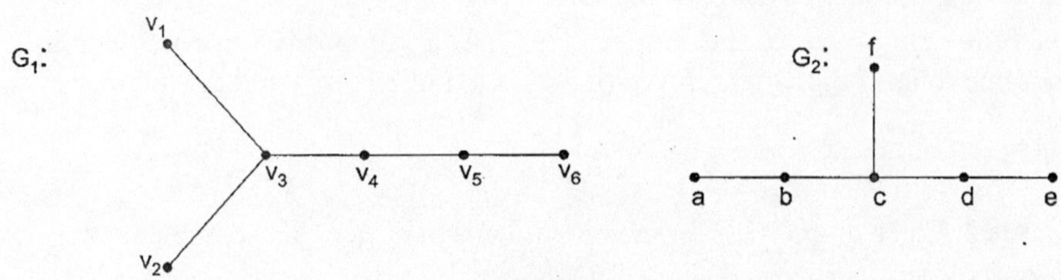

Fig. 8.21

Solution. We have

Number of vertices in G_1 = Number of vertices in G_2

Number of edges in G_1 = number of edges in G_2

In the graph G_1, the vertex v_3 is of degree 3. There are two pendant vertices adjacent to v_3 in G_1.

In the graph G_2, the degree of the vertex C is 3. There is only pendant vertex adjacent to C in G_2. Hence adjacency is not preserved in the graphs G_1 and G_2.

$\therefore$ G_1 and G_2 are not isomorphic

EXERCISE 8.2

1. Define: (a) Isomorphic graphs (b) Self-complementary graphs.
2. Show that the graphs given below are isomorphic.

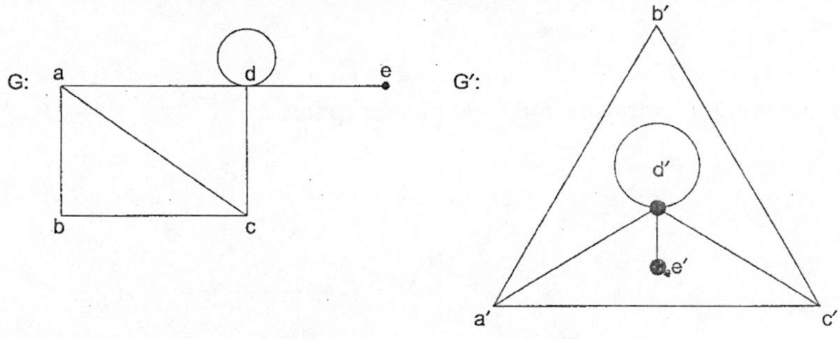

3. Show that the following graphs are isomorphic

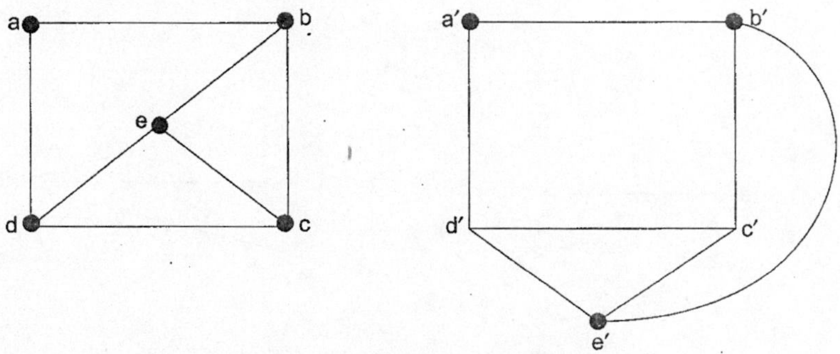

4. Draw the complementary of the graphs given below:

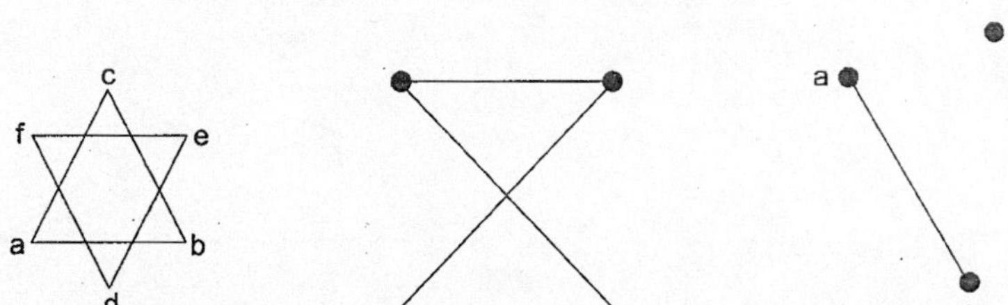

5. Show that the graphs G and G' are isomorphic:

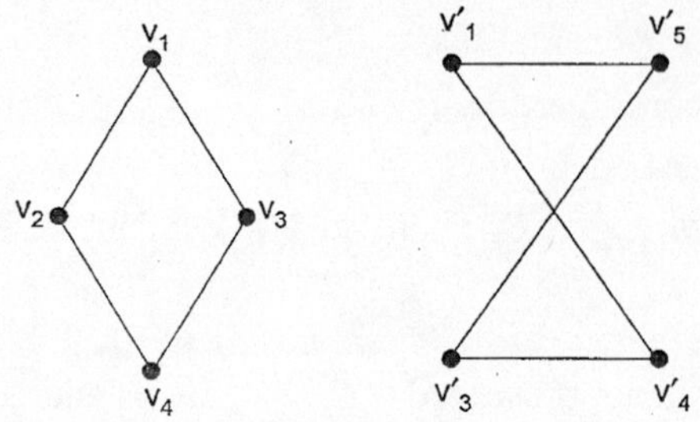

6. Show that the graphs G and G' have the same degree sequence but are not isomorphic:

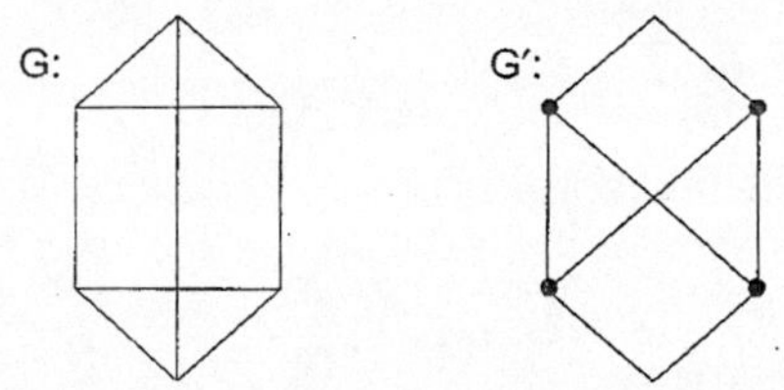

7. Show that the graphs G and G' are isomorphic:

(i)

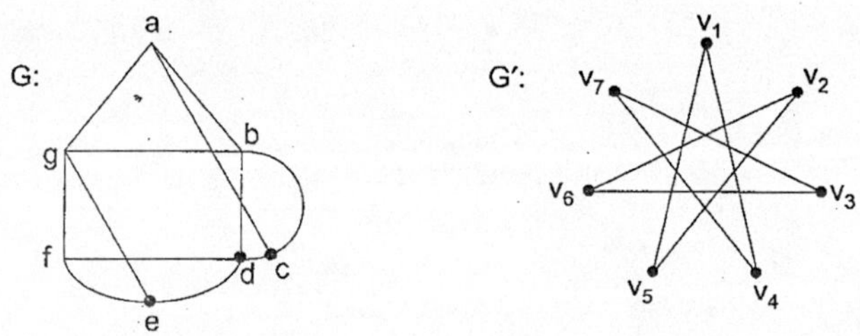

(ii)

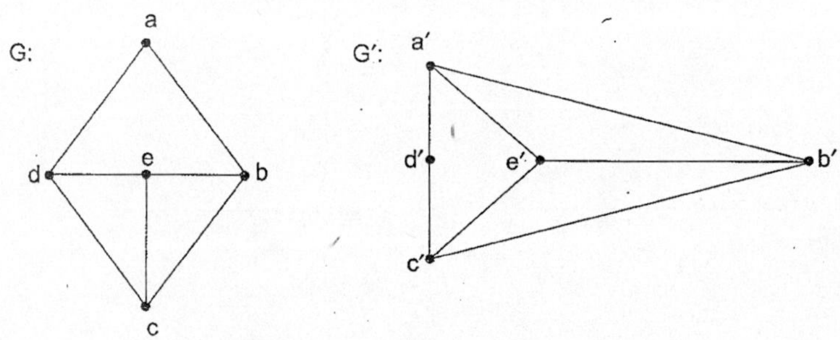

(iii)

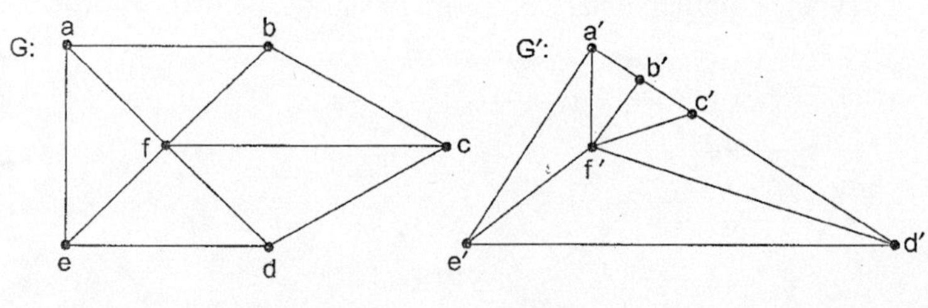

(iv)

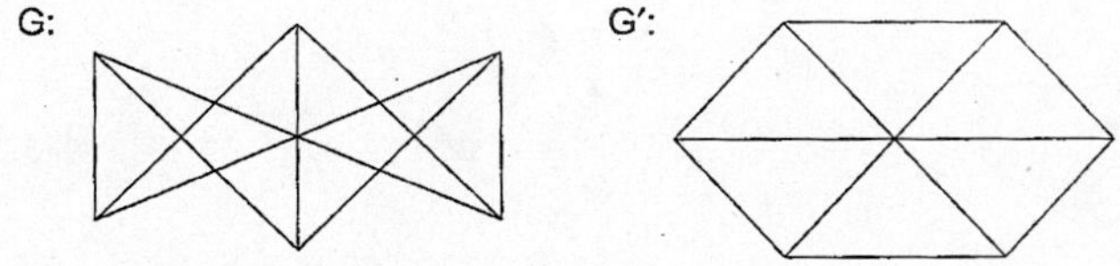

8. Which of the following pairs of graphs are isomorphic. Justify your answer:

(i)

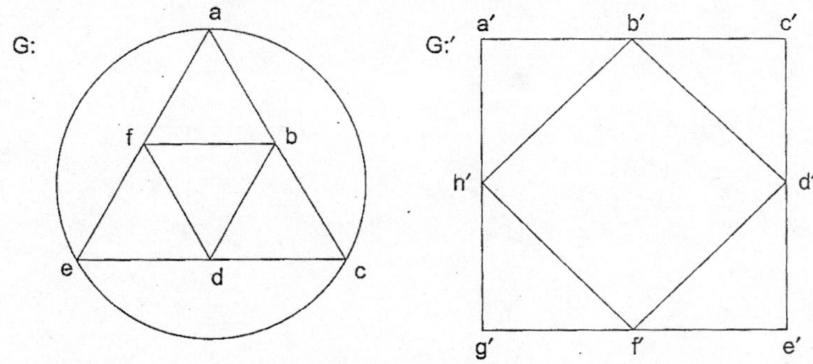

(ii)

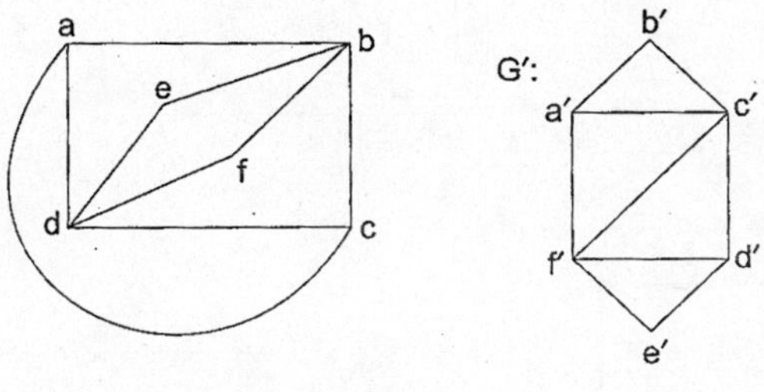

(iii)

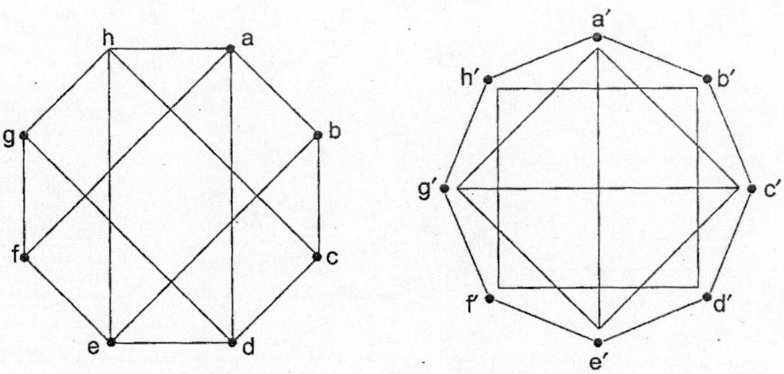

(iv)

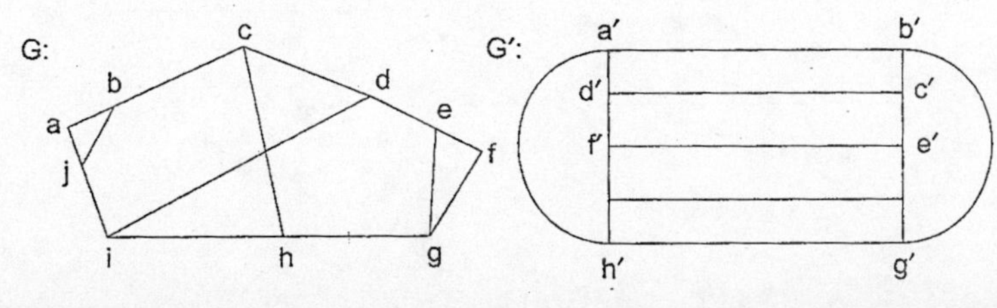

(v)

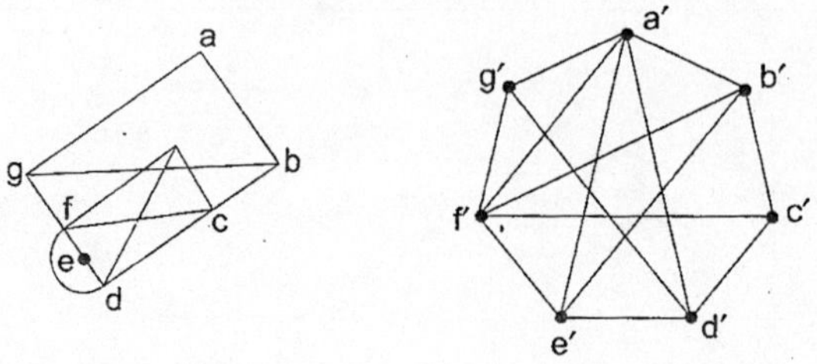

(vi)

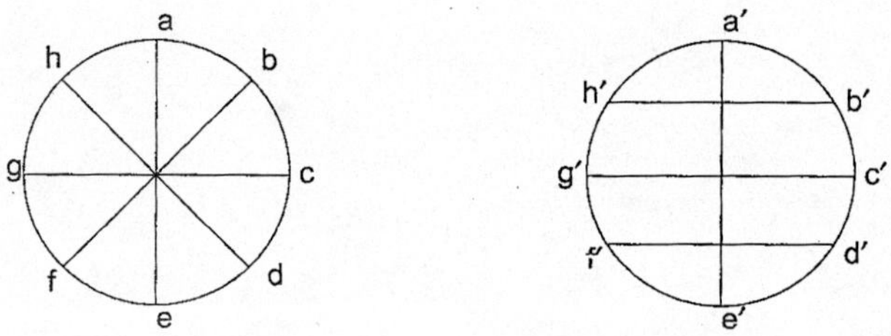

(vii)

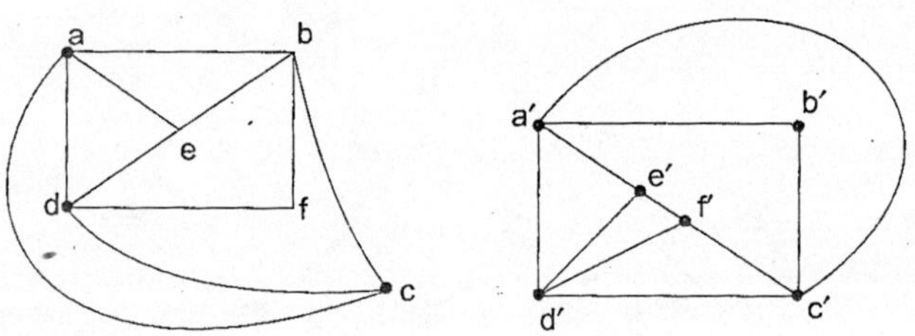

(viii)

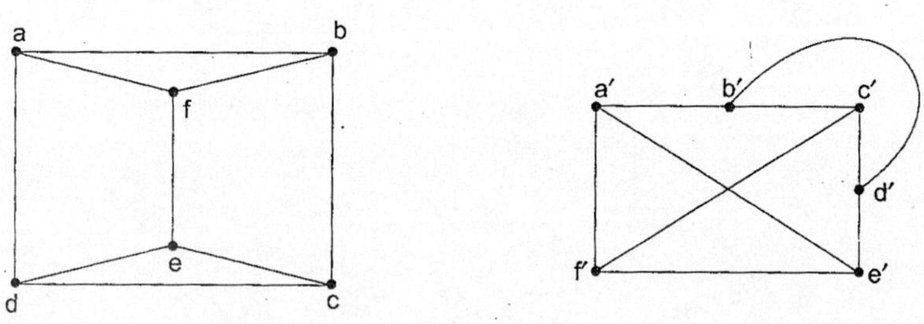

(ix)

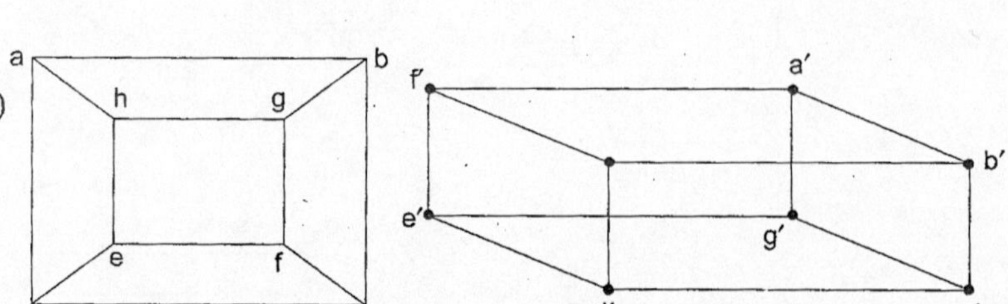

9. Give an example in which two non-directed graphs with 4 vertices and 6 edges are not isomorphic.

10. Show that C_5 is the only cycle graph and 5 vertices isomorphic to its complement.

11. G_1 and G_2 are two isomorphic graphs and G_1 is a connected graph. Show that G_2 is also connected.

12. Show that the three graphs G_1, G_2 and G_3 given below are isomorphic:

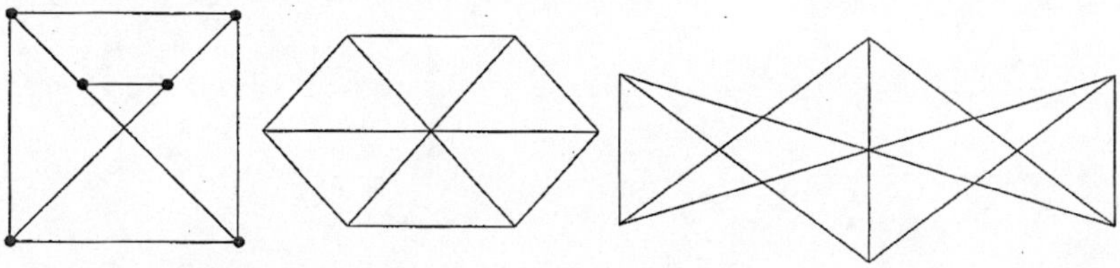

13. Let G be a simple graph all of whose vertices have degree 3 and $|E| = 2|V| - 3$. What can be said about G.

14. Draw all isomorphic trees with 6 vertices.

15. Give two non-isomorphic graphs of order 6, size 6 and with degree sequence (2, 2, 2, 2, 2, 2).

ANSWERS

8. (i) Not isomorphic since $|V(G)| \neq |V(G')|$

(ii) Not isomorphic, Adjacency is not preserved

(iii) Not isomorphic, degree sequences are not equal

(iv) Not isomorphic, degree sequences are not equal; G has cycles of length 3. G' has no cycles of length 3.

(v) Isomorphic

(vi) Not isomorphic, adjacency is not preserved

(vii) Not isomorphic, adjacency is not preserved in the graphs.

(viii) and (ix) are isomorphic

9.

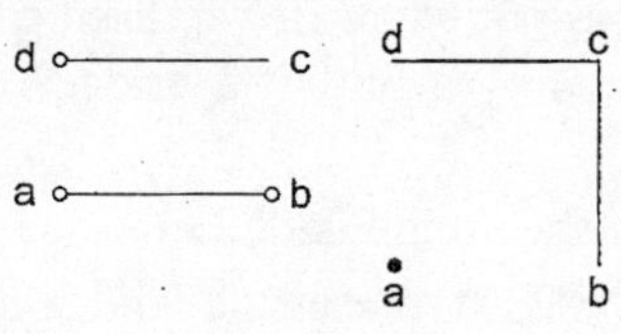

13. G is isomorphic to $k_{3,3}$ and the graph

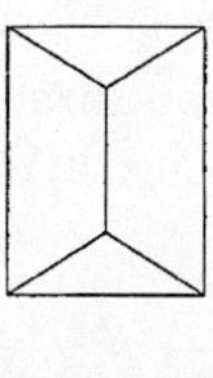

14.

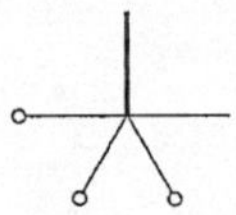

(a)

(b)

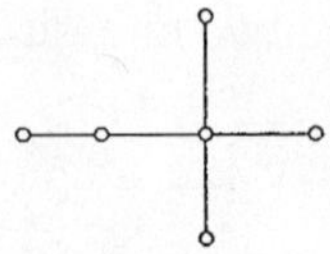

(c)

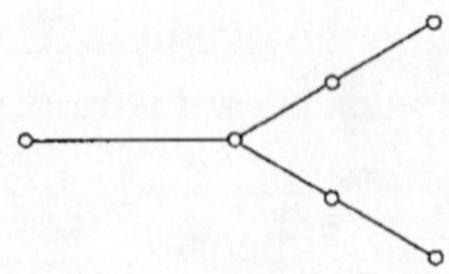

(d)

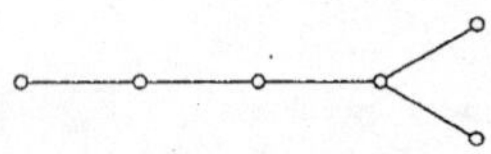

(e)

(f)

15.

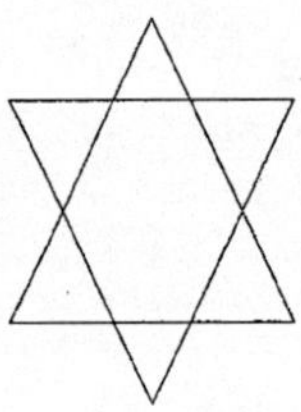

(a)

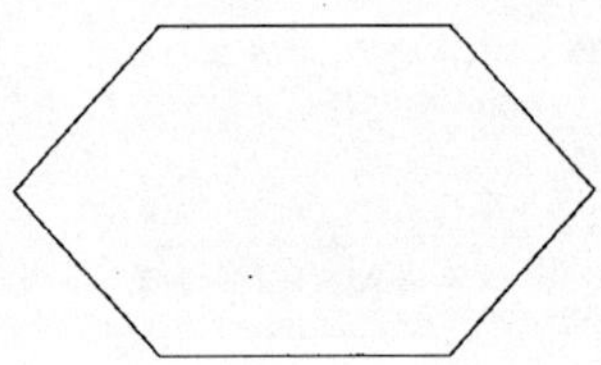

(b)

8.4 EULER'S GRAPHS

In this section we shall study graphs known as Euler's graphs. Euler is called the father of graph theory. In 1736 he solved a long-standing problem called the Konigsberg bridge problem. There were two islands linked to each other and to the banks of Pregel river by seven bridges as shown in Fig. 8.22. The problem was to begin at any of the four land areas, walk across each bridge exactly once and return to the starting point. Many attempts were made to solve this famous problem. Euler proved that a solution to this problem does not exist. He represented each land area by a point and joined two such points by a line, if a bridge existed connected the land areas.

The graph of Konigsberg bridge problem is shown in Fig. 8.23.

Each point in the graph (A, B, C, D in Fig. 8.23) represents a land area, and each line in the graph represents a bridge.

Euler generalized this problem and developed criterion for a given graph to be traversable satisfying the given conditions.

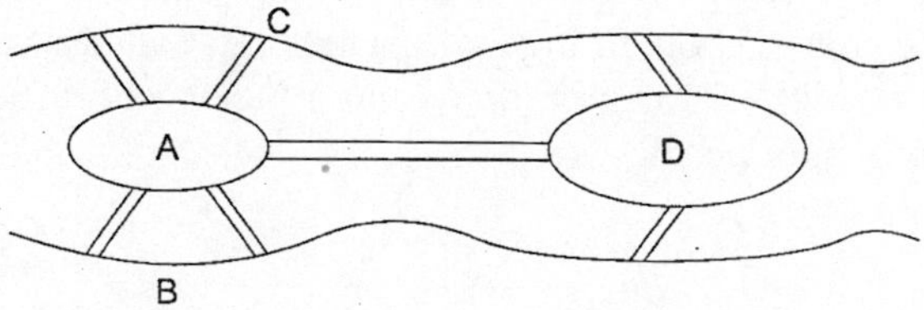

Fig. 8.22 Konigsberg bridge problem

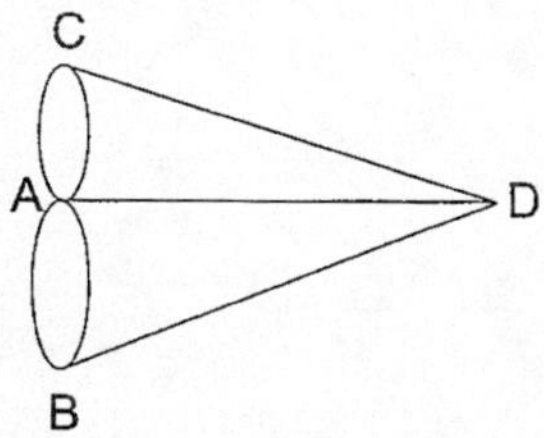

Fig. 8.23 The graph of Konigsberg bridge problem

8.5 TRAVERSABLE GRAPH

Definition 8.6. A graph G is traversable if it has a path.

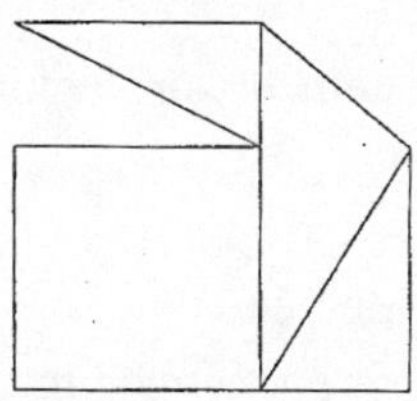

Fig. 8.24 A traversable graph

8.6 EULERIAN PATH

Definition 8.7. Let G be a multigraph. A path in G that includes each edge of G exactly once and intersects each vertex of G at least once is called an **Euler path** in G.

An **Euler circuit** in G is an Euler path in G whose end points are identical.

8.7 EULERIAN GRAPH

Definition 8.8. A graph is said to Eulerian if it has an Eulerian circuit.

From the above definitions, it is clear that a trail in a multigraph G is Eulerian if it contains all the edges of the graph G and if a graph G contains a closed Eulerian trail then it is called an Eulerian graph. In the graph shown in Fig. 8.25. The sequence $v_1 - v_2 - v_3 - v_4 - v_2 - v_5 - v_1$ is a closed Eulerian trail. It contains all the vertices and edges of G. If G is a connected Euler graph; then starting from any vertex say v_0 of G, entire graph G can be drawn without lifting the pencil (or pen), ending at v and vice versa. If any graph G can be drawn starting from a point and ending at the same point without lifting the pen, then G is called a traceable graph. The multigraphs in Fig. 8.25 can be drawn on a paper with a pen (or pencil) without lifting the pen (or pencil).

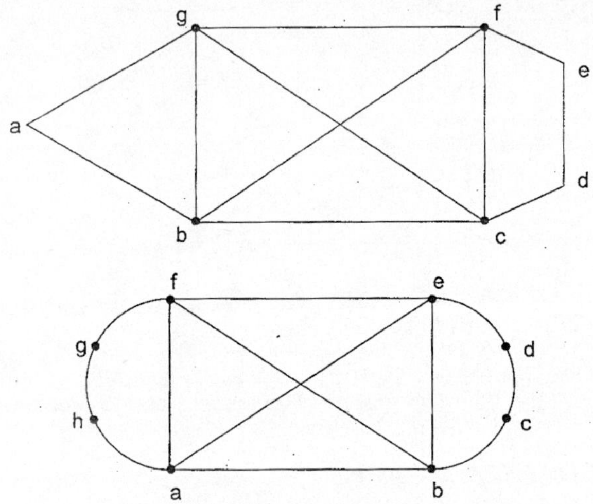

Fig. 8.25 Eulerian graphs

Theorem 8.3. The following statements are equivalent for a connected multigraph G.

(1) G is Eulerian

(2) Every point of G has even degree

(3) The set of edges of G can be partitioned into cycles

Proof. (1) $\Rightarrow$ (2): Let P be an Eulerian path in G. Each occurrence of given a vertex in P contributes two to the degree of that vertex and since each edge of G appears exactly once in P, every vertex must have even degree

(2) $\Rightarrow$ 3:G is connected and non-trivial, therefore degree of every vertex in G is at least two and G contains a cycle say z_1. The removal of the edges of z, results in a spanning sub-graph G_1, in which every vertex has an even degree. If G_1 has no edges (3) already holds; otherwise a representation of the argument applied to G_1 results in a graph G_2 in which again all points are even. By continuing this process we get a totally disconnected graph G_n, which shows that the edges of G can be partitioned into cycles.

(3) $\Rightarrow$ (1): Let z_1 be one of the cycles of the portion of G. If consists of only this cycle, then G is obviously Eulerian otherwise these is another cycle z_2 with a vertex v in common with z_1. The walk beginning at v and consisting of the cycles

z_1 and z_2 in succession is a closed path containing the edges of these two cycles. By continuing this process, we can construct a closed path containing all the edges of G. Hence G is Eulerian. For example consider the graph G shown in Fig. 8.21. G has an Eulerial path (trail).

Every vertex of G has even degree, and the set of edges can be partitioned into cycles.

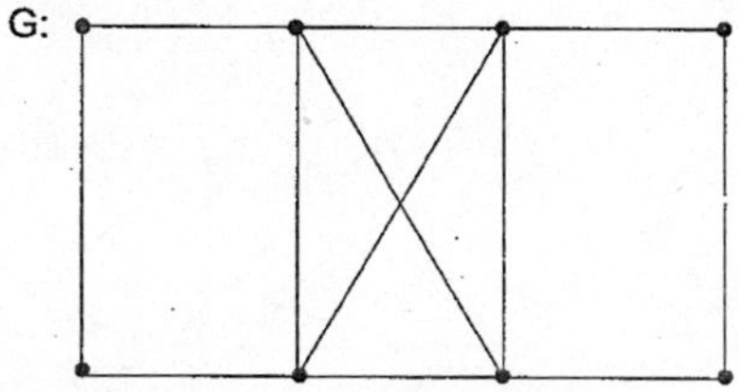

Fig. 8.21 An Eulerian graph

Theorem 8.4. If a graph G has more than

(i) Two vertices of odd degree, then there can be no Euler path in G

(ii) If G is a connected graph and has exactly two vertices of odd degree, then there is an Euler path in G. Any Euler path in graph G must begin at vertex of odd degree and end the other.

Proof. (i) Let G be a graph having more than two vertices of odd degree, and let v_1, v_2, v_3 be the vertices of odd degree in G. If there is an Euler path in G, then it must leave (arrive) each of the vertices. Let one of the vertices say v_1 be the beginning of the Euler path in G and another vertex say v_2 be the end of the path. But this leaves the vertex v_3 at one end of an untravelled edge. Thus there can be no Euler path.

(ii) Let u and v be two vertices of odd degree in G. By adding the edge $\{u, v\}$ to G, we can produce a connected graph say G', all of whose vertices are of even degree. Since G' is connected and every vertex of G' we can find an Euler circuit e in G'. Deleting the edge $\{u, v\}$ from c, we get an Euler path that begins at 4(or v) and ends at v(or u).

If G is a graph in which there are 0 or two vertices of odd degree then G has an Euler path and the graph G is traversable. If all the vertices are of even degree, then G has an Euler circuit and G is traversable. The graphs shown in Fig. 8.22 are Eulerian.

An Euler's circuits in a graph is a closed walk containing every edge exactly once. If may pass through a given vertex any number of times. Suppose, there exists a vertex v such that an Euler's circuit is always obtained when one walks from the vertex v following the rule, that one selects any edge that has not traversed before, then the graph is said to be arbitrarily traceable from the vertex v. For in Fig. 8.23 is arbitrarily traceable from the vertex v but not from any other vertex in the graph.

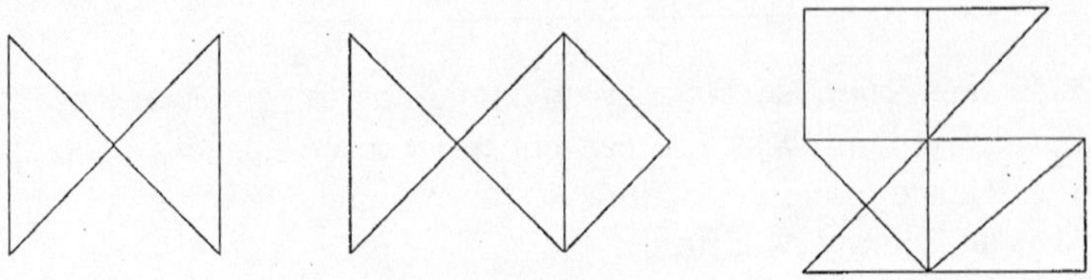

Fig. 8.27

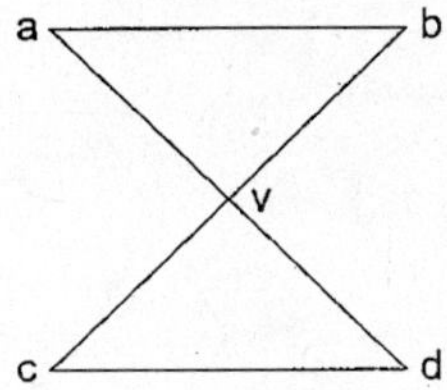

Fig. 8.28

EXAMPLES

Example 1. Show that the graph G (Fig. 8.29) is Eulerian and find an Eulerian circuit in G.

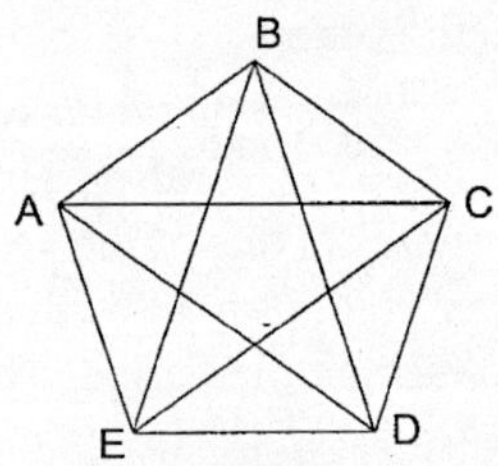

Fig. 8.29

Solution. Each vertex is of even degree in G. Therefore G is Eulearian.

$A\,B\,C\,D\,E\,A\,C\,E\,B\,D\,A$ is an Eulerian circuit in G.

Example 2. Show that the graph G shown in Fig. 8.30 is not Eulerian.

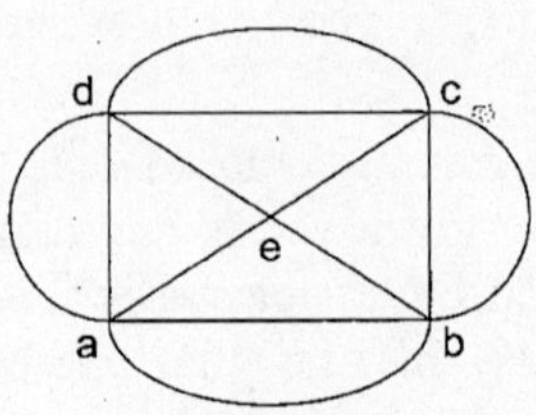

Fig. 8.30

Solution. There are four vertices of degree 5 in the graph.

We have $\deg(a) = \deg(b) = \deg(c) = \deg(d) = 5$

i.e. there are more than two vertices of odd degree in G, therefore G is not Eulerian.

EXERCISE 8.3

1. Show that the following graphs are Eulerian

(a)

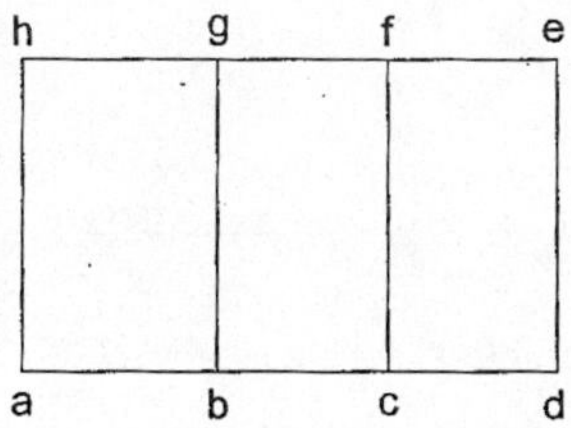

(b)

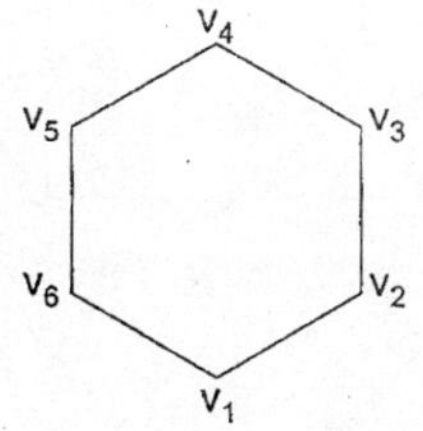

(c)

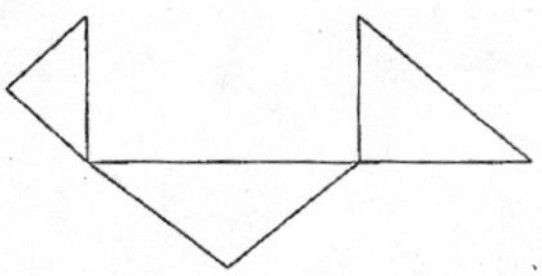

(d)

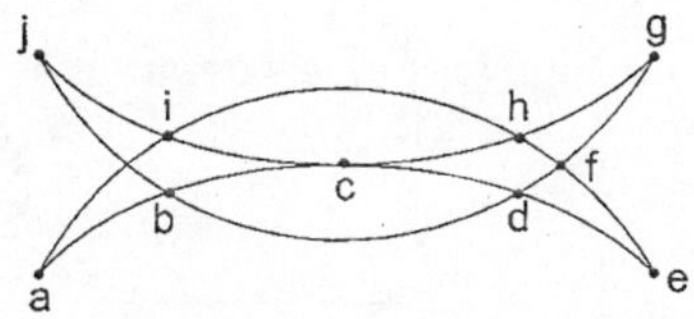

(e)

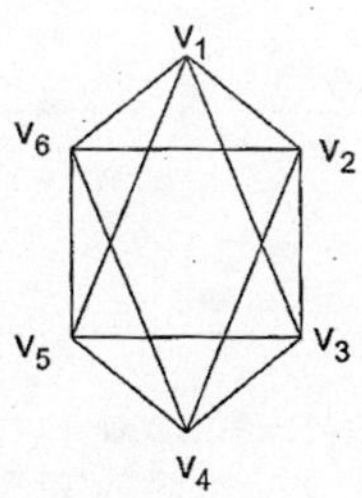

2. Find an Euler's path in the graph:

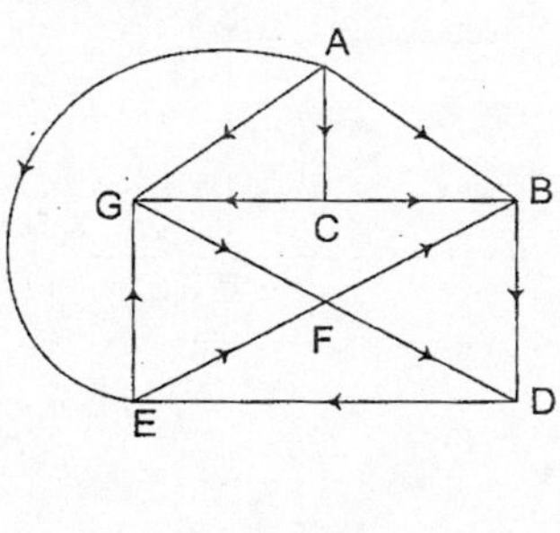

3. Show that the following graphs are not Eulerian:

(i)

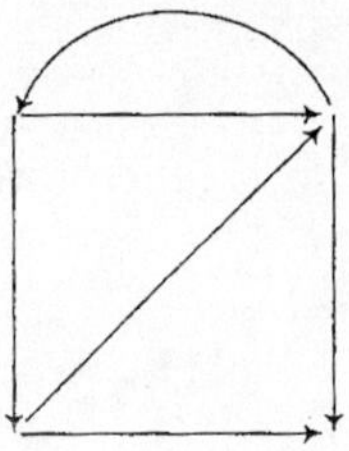

(ii)

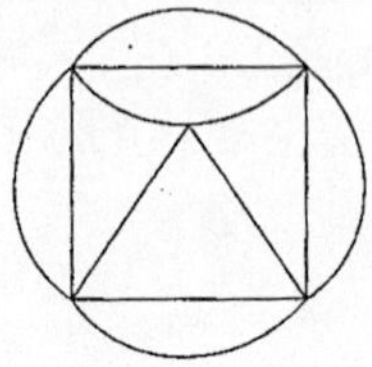

(iii)

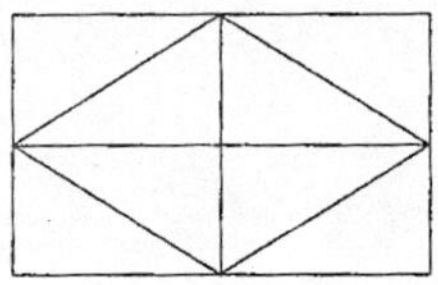

(iv)

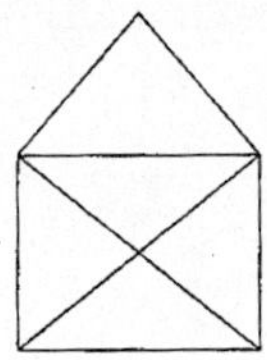

(v)

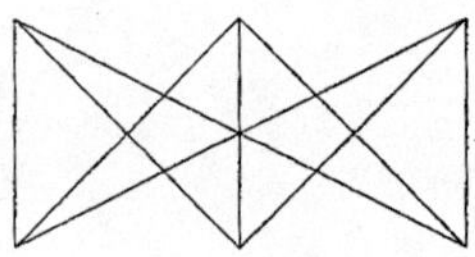

4. Find an Euler's circuit in the graph given below:

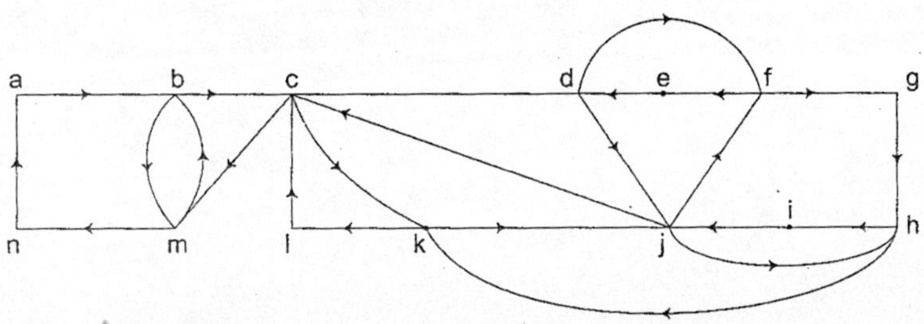

Ans. *n a b c d e f g h i j f e d j h k l c k j c m b m n*

5. Prove that a graph with no vertices of odd degree and no vertices of 0 degree must contain a Euler's cycle. (MKU Sep. 1999)

6. Find an Euler path in the graph given below:

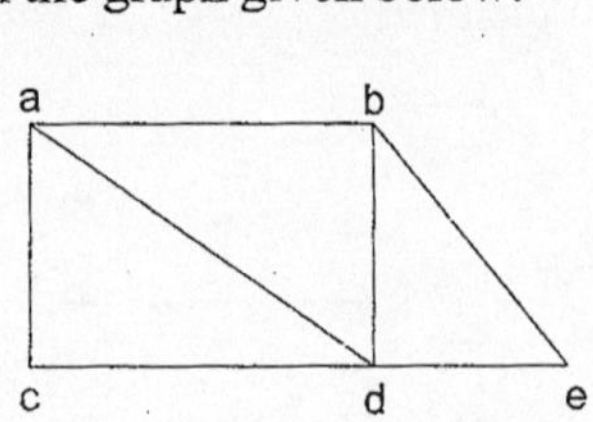

Ans. *a c d e b d a b*

7. Find an Euler circuit in the graph G.

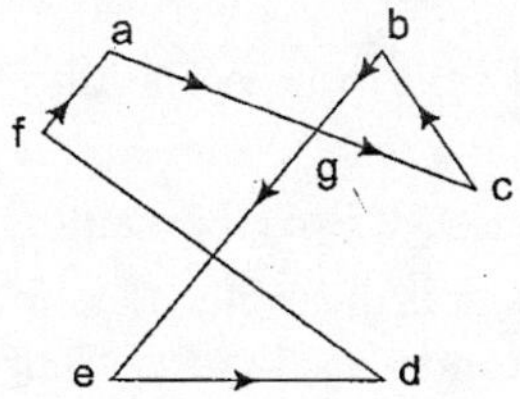

Ans. $a\ g\ c\ b\ g\ e\ d\ f\ a$

8. Find an Euler path in the graph given below:

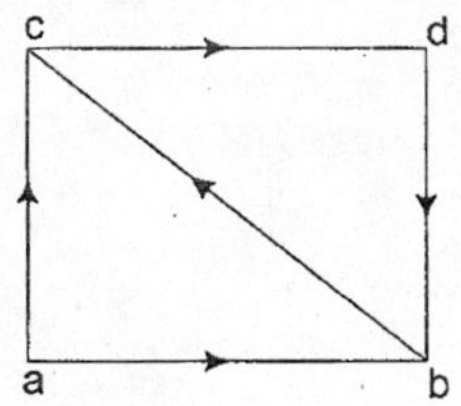

Ans. $c\ a\ b\ c\ d\ b$

9. Find an Euler path in the graph:

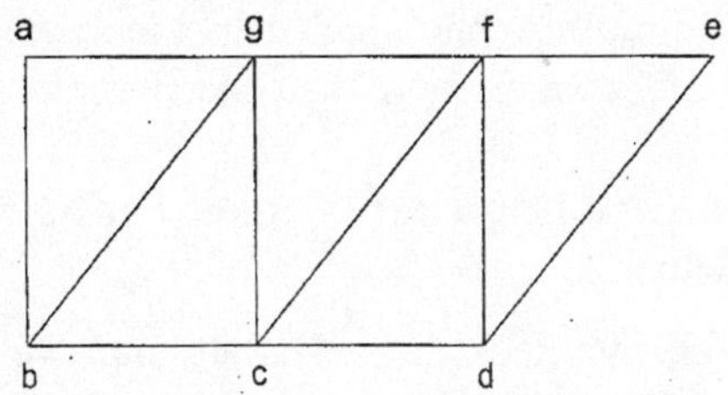

Ans. $b\ a\ g\ f\ e\ d\ c\ g\ b\ c\ f\ d$ is an Euler path in the graph.

8.8 HAMILTON GRAPHS

In section 8.4 we have seen that an Eulerian path in a connected graph traverses every edge of the graph. In this section; we discuss about the graphs called the Hamiltonian graphs which contain a closed walk; that traverses every vertex of the graph exactly once, except the starting vertex, at which the walk terminates, while studying non-commutative algebra Sir Willian Rowan Hamilton has invented a game called 'The Icosian game'. The same in its simplest forms asks to find a cycle containing all the vertices of the graph of Dodecahedron. Hamilton graphs have many applications. We begin over study with definitions.

Definition 8.9. A path is a graph *G* is called a Hamiltonian path if it contains every vertex of *G*.

Definition 8.10. A cycle in *G* is said to a Hamiltonian cycle if it contains every vertex of *G*.

Definition 8.11. A graph *G* is said to be a Hamiltonian graph; if it contains a Hamiltonian cycle.

i.e. *G* is Hamiltonian; if there exists a cycle containing every vertex of *G*.

From the above definitions, it is clear that a Hamiltonian path in a graph *G* is always a subgraph of a Hamiltonian cycle in *G*. Hamiltonian path in *G* may be obtained by deleting an edge from a Hamiltonian cycle in *G*. Every graph which has a Hamiltonian cycle (circuit) contain a Hamiltonian path, but a graph containing a Hamiltonian path may not have a Hamiltonian cycle and there are many graphs with Hamiltonian paths that have no Hamiltonian cycles. The graphs shown in Fig. 8.31 are Hamiltonian.

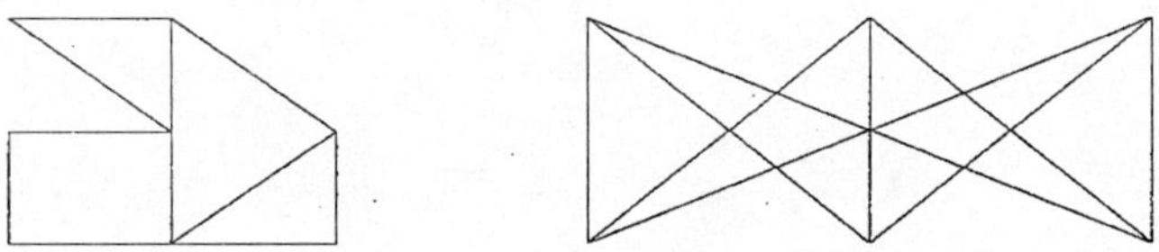

Fig. 8.31 Hamiltonian graphs

A Hamiltonian cycle in a graph of *n* vertices consists of exactly *n* edges. Usually we consider only simple graphs which do not include self loops of parallel edges. A given graph *G* may contain more than one Hamiltonial cycle.

8.8.1 Rules for Constructing Hamiltonian Paths, and Cycles in a Graph G

Rule 1. If *G* is a graph with *n* vertices, then a Hamiltonian cycle in *G* must contain exactly *n* edges and a Hamiltonian path must contain $n - 1$ edges.

Rule 2. There cannot be more than three or more edges incident with one vertex in Hamiltonian cycle in *G*. Every vertex *u* in a Hamiltonian cycle of *G* must contain exactly 2 edges incident on *v*. If *v* is a vertex in *G* then a Hamiltonian path in *G* must contain at least one edges incident on *v* and at most 2 edges incident on *v*.

Rule 3. A Hamiltonian path or cycle constructed in *G* must contain all the vertices of graph *G*.

Rule 4. Let *v* be a vertex of *G*. Once a Hamiltonian cycle we are constructing has passed through *v*, then all the unused edges incident on *v* can be deleted.

Example. Consider the graph G shown in Fig. 8.32.

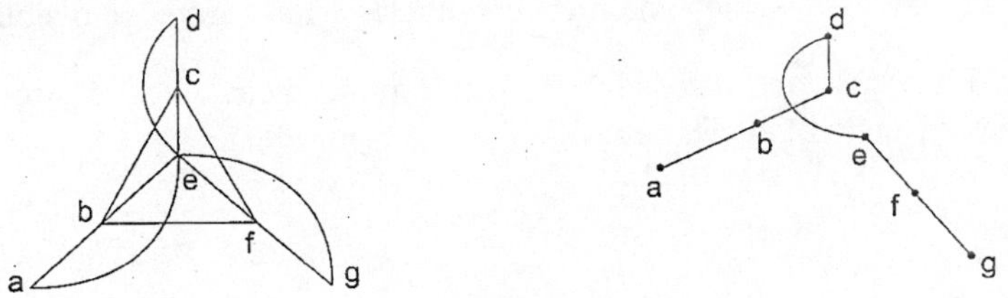

Fig. 8.32

$\{a, b\}$, $\{b, c\}$, $\{c, d\}$, $\{d, e\}$, $\{e, f\}$, $\{f, g\}$ is a Hamiltonian path in G. The graph has no Hamiltonian circuit in G.

8.8.2 Number of Hamiltonian Cycle in a Graph

A given graph G may contain more than one Hamiltonian circuit. In general the determination of number of Hamiltonian cycles (i.e. circuits) in a graph is an unsolved problem. However the number of edge disjoint a Hamiltonian circuits in a complete graph with n vertices where n is odd (n and $n \geq 3$) is given by the following theorem.

Theorem 8.5. In a complete graph with n vertices, there are $(n-1)/2$ edge disjoint Hamiltonian cycles, if n is an odd number, ≥ 3.

Proof. Let G be a complete graph with n vertices where $n \geq 3$, and n is odd. Then G has $n(n-1)/2$ edges. A Hamiltonian cycle in G must contain n edges, Therefore the number of edge disjoint Hamiltonian circuits in G cannot exceed $\frac{n-1}{2}$. i.e. there are $\frac{n-1}{2}$ edge disjoint Hamiltonian circuits in G, when n is odd. Consider the subgraph of G, shown in Fig. 8.33. The subgraph is a Hamiltonian cycle in G.

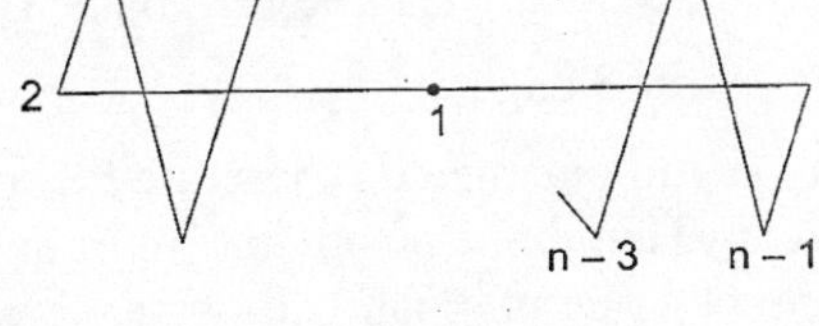

Fig. 8.33

Keeping the vertices fixed on a circle rotate the polygonal clockwise by $\frac{360}{n-1}, 2.\frac{360}{n-1}, 3.\frac{360}{n-1}, \ldots, \frac{n-3}{2} \cdot \frac{360}{n-1}$ degrees observe that each rotation produces a Hamiltonian circuit, that has no edge in common with any of the previous ones.

Thus we have $\frac{n-3}{2}$, new Hamiltonian cycles which are edge disjoint from the one in Fig. 8.33; and also edge disjoint among themselves. Therefore the number of edge disjoint cycles in G is

$$\frac{n-3}{2}+1=\frac{n-1}{2}$$

Hence proved.

Example 1. Show that the graph G in Fig. 8.34 is not Hamiltonian.

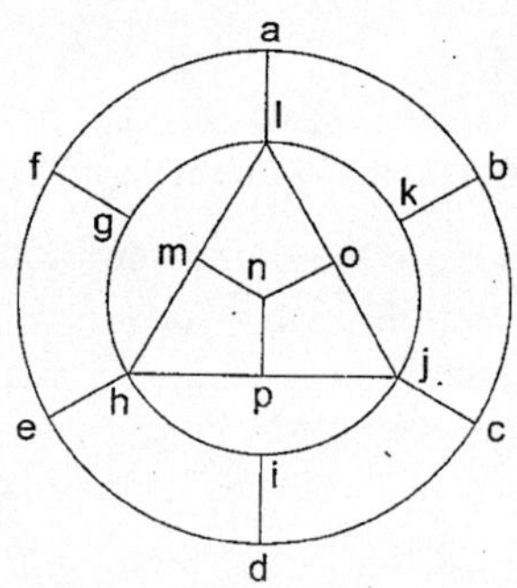

Fig. 8.34

Solution. G has 27 edges and 16 vertices, therefore any Hamiltonian cycle in G must have 16 edges and Hamiltonian path in G must contain exactly 15 edges. In the graph G we have

$$\deg(l)=\deg(h)=\deg(j)=5$$

i.e. there are three vertices of degree 5. Therefore at least three edges on l cannot be included in any Hamiltonian path. The same is true for the vertices h and J. There are 13 vertices of degree 3 in G. Now consider the vertices b, d, f and n. Such of these vertices of is of degree 3. At least one of the three edges on each of these vertices cannot be included in any Hamiltonian path. Thus at least

$$3+3+3+4=13 \text{ edges}$$

cannot be included in any Hamiltonian path. Therefore the number of remaining edges is $27-13=14$. But any Hamiltonian path in G must have 15 edges. Hence; it is not possible to construct a Hamiltonian path with the remaining 14 edges. Thus, G has no Hamiltonian path and G is not Hamiltonian.

Example 2. Show that the graph shown in Fig. 8.35 has no Hamiltonian cycle. But the graph has a Hamiltonian path.

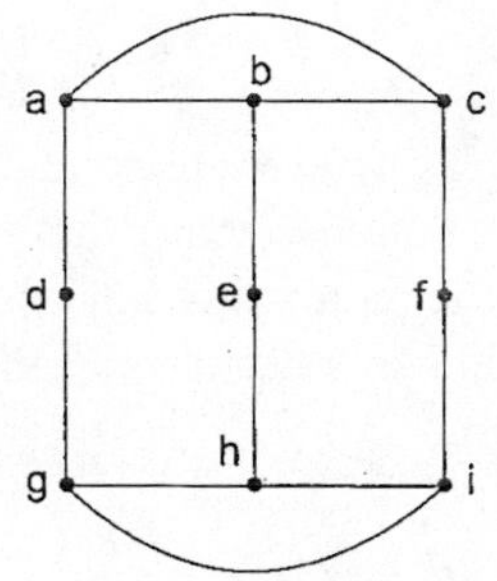

Fig. 8.35

Solution. G has 9 vertices, therefore any Hamiltonian cycle in must have 9 edges. There are 3 vertices of degree 2 in G. Therefore, all the edges incident on the vertices d_1 e and f must be included in any Hamiltonian cycle i.e. edges $\{a, d\}$, $\{d, g\}$, $\{b, e\}$, $\{c, f\}$, $\{f, i\}$ which are incident on d_1 e, and f must be incident in the Hamiltonian cycle. Now consider the vertex b: when we include the edge $\{b, e\}$ in the cycle, there are two edges, namely, $\{a, b\}$ and $\{b, c\}$ of which only one edge is to be consider for the Hamiltonian cycle in G. Let us include the edge $\{a, b\}$ in Hamiltonian cycle and delete the edge $\{b, c\}$. We must include the edge $\{a, c\}$ in the Hamiltonian cycle. Thus the degree of a would be 3. Hence, no Hamiltonian cycle exists in G. However, There exists a Hamiltonian path in G. The edges $\{a, b\}$, $\{d, g\}$, $\{g, h\}$, $\{e, b\}$, $\{b, c\}$, $\{c, f\}$, $\{f, i\}$ when included will give us a Hamiltonian path in G i.e. $a-d-g-h-e-b-c-f-i$, is a Hamiltonian path in G. It is shown in Fig. 8.36.

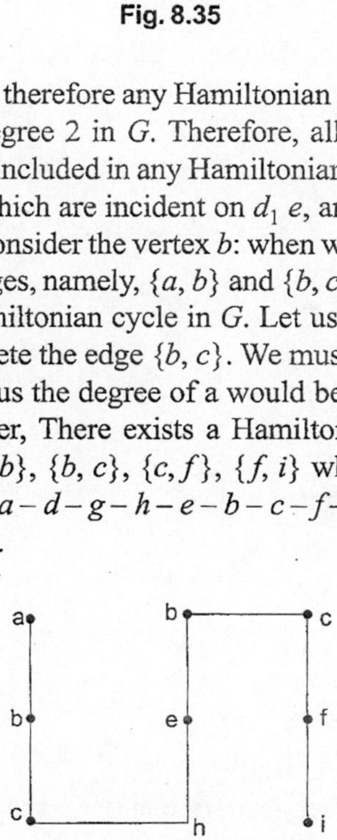

Fig. 8.36

Definition. 8.12 (Diagonal). Let G be a plane Hamiltonian graph and c be a fixed Hamiltonial cycle in G. A diagonal with respect to c is an edge of G that does not lie on G.

Grinberg's Theorem

Theorem 8.6. Let G be a simple plane graph with n vertices and C be a Hamiltonian cycle in G. If r' denotes the number of regions of G in the interior of C whose boundary contains exactly i edges and r_i denotes the number of regions of degree i, in the exterior of C, then

$$\sum_{i=3}^{n}(i-z)(r_i-r_i')=0$$

Proof. G is a simple planar graph, therefore none of the edges of G intersect. We first consider the interior of C. Suppose that exactly d, diagonals occur in the interior of C. A diagonal splits the region through which it passes into two parts. Each time a diagonal is inserted in the interior of C, the number of regions in C is increased by 1. Hence

$$\sum_{i=3}^{n} r_i = d+1 \Rightarrow d = \sum_{i=3}^{n} r_i - 1 \qquad ...(1)$$

If N denotes the sum of the degrees of all regions interior to C, then

$$N = \sum_{i=3}^{n} i\, r_i$$

N counts each diagonal twice and each edge of C exactly once, therefore

$$N = \sum_{i=3}^{n} i\, r_i = 2d+n$$

Substituting for d we get

$$\sum_{i=3}^{n} i r_i = 2\left[\sum_{i=3}^{n} r_i - 1\right] + n$$

$$= 2\sum_{i=3}^{n} ri - 2 + n$$

$$\therefore \qquad \sum_{i=3}^{n}(i-2)\,r_i = n-2 \qquad ...(2)$$

Similarly, by considering the regions external to C, we get,

$$\sum_{i=3}^{n}(i-2)\,r_1' = n-2 \qquad ...(3)$$

Substituting (3) from (2), we get

$$\sum_{i=3}^{n}(i-2)(r_i-r_1')=0$$

Example 1. Shown that there are no Hamilton planar graphs with regions of degree 5, 8, 9 and 11 with exactly one region of degree 9.

Solution. From Grinberg's theorem, we have

$$(5-3)(r_5 - r_5') = 0, \quad (8-2)(r_8 - r_8') = 0$$

$$(9-2)(r_9 - r_9') = 0, \quad (11-2)(r_{11} - r_{11}') = 0$$

Adding we get

$$2(r_5 - r_5') + 6(r_8 - r_8') + 7(r_9 - r_9') + 9(r_{11} - r_{11}') = 0$$

There is exactly one region of degree 9, therefore we get

$$3(r_5 - r_5') + 6(r_8 - r_8') + 7(\pm 1) + 9(r_{11} - r_{11}') \equiv 0$$

$$\Rightarrow \quad 3(r_5 - r_5') + 6(r_8 - r_8') + 9(r_{11} - r_{11}') = \pm 7$$

$$\Rightarrow \quad 3[(r_5 - r_5') + 2(r_8 - r_8') + 3(r_{11} - r_{11}')] = \pm 7$$

$\Rightarrow$ 3 is a divisor of ± 7

a contradiction.

Hence, there cannot be Hamilton planar graphs with regions 5, 8, 9 and 11, with exactly one region of degree 9.

Example 2. Show that the graph G in Fig. 8.37 is not Hamiltonian.

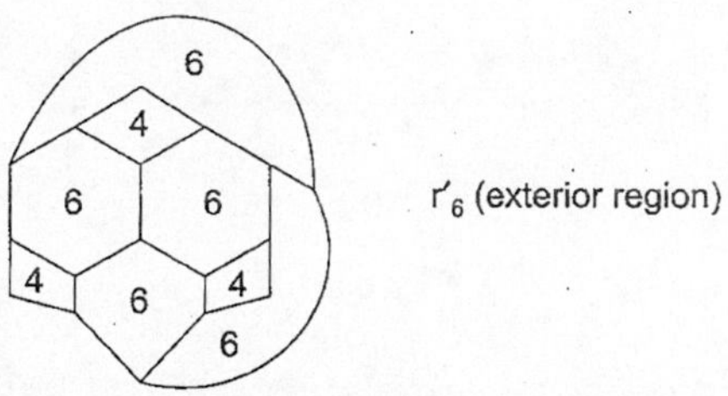

Fig. 8.37

Solution. Number of regions of degree 4 in $G = 3$

$$\therefore \quad r_4 + r_4' = 3$$

Number of regions of degree 6 is 6

$$\therefore \quad r_6 + r_6' = 6$$

applying Grinberg's theorem, we get

$$(4-2)(r_4 - r_4') + (6-2)(r_6 - r_6') = 0$$

$$\Rightarrow \quad 2(r_4 - r_4') + 4(r_6 - r_6') = 0$$

$$\Rightarrow \quad r_4 - r_4' + 2(r_6 - r_6') = 0$$

$$\Rightarrow \quad r_4 - r_4' = -2(r_6 - r_6')$$

∴ The difference between r_4 and r_4' must be a multiple of 2 (i.e. even).

But $r_4 + r_4' = 3$.

The possibilities are 0, 3 and 1, 2. The difference between 0 and 3 is not even. Similar by the difference between 1 and 2 is also not even. Therefore, there cannot be a Hamiltonian cycle in the graph.

Example 3. Find three distinct Hamiltonian circuits in the weighted graph given below. Also find their weights

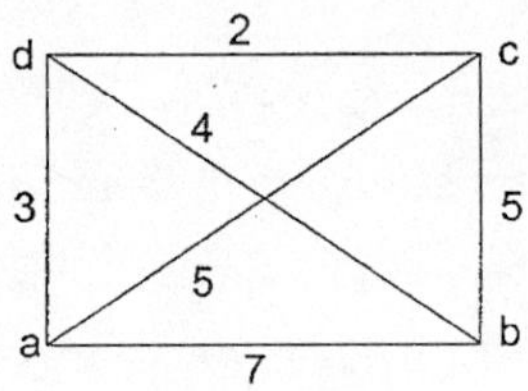

Fig. 8.38

Solution. The three distinct Hamiltonian circuits are:

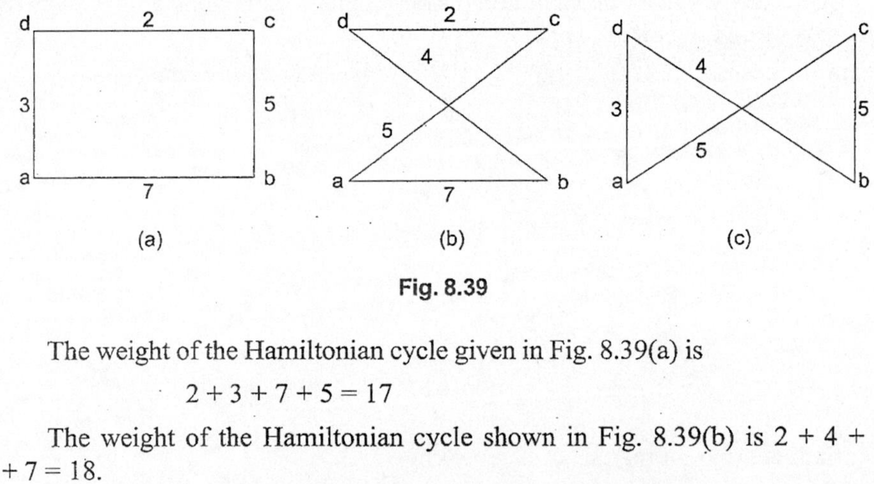

Fig. 8.39

The weight of the Hamiltonian cycle given in Fig. 8.39(a) is

$$2 + 3 + 7 + 5 = 17$$

The weight of the Hamiltonian cycle shown in Fig. 8.39(b) is 2 + 4 + 5 + 7 = 18.

The weight of the Hamiltonian cycle shown in Fig. 8.39(c) a is 3 + 4 + 5 + 5 = 17.

Theorem 8.7. (Dirace's Theorem). If G is a simple graph with n vertices, $(n \geq 3)$, such that the degree of every vertex in G is at least $n/2$, then G has a Hamiltonian circuit.

Theorem 8.8. (Ore's Theorem). If G is a simple graph with n vertices $(n \geq 3)$, such that $\deg(u) + \deg(v) \geq n$ for every pair of non-adjacent vertices u and v in G. Then G has a Hamiltonian circuit.

Hamiltonian paths and circuits can be a used in many practical problems. For example, consider the famous traveling salesman problem (*TSP*) in which the problem asks for the shortest route a traveling salesman should take to visit a set of cities. This problem reduces to finding a Hamiltonian circuit in a complete graph such that the total weight of its edges (i.e. distances between cities) as small as possible.

EXERCISE 8.4

1. Show that the following graphs are Hamiltonian.

(a)

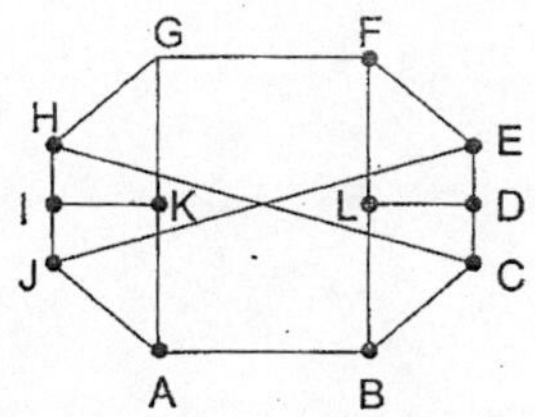

(b)

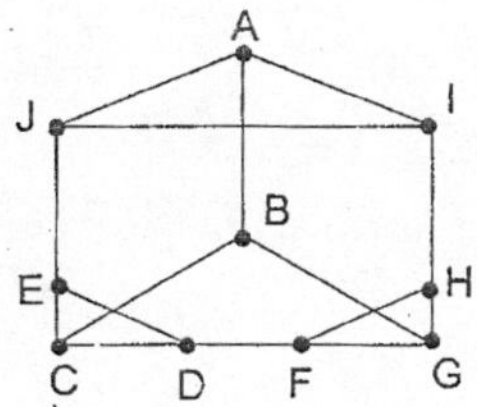

(c)

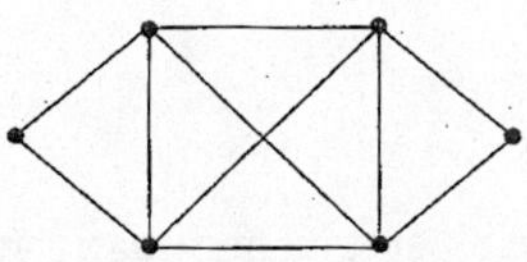

2. Show that the graph G is Eulerian but not Hamiltonian.

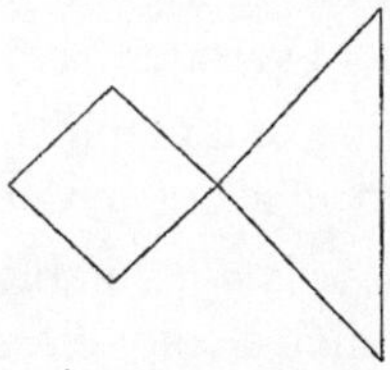

3. Show that Peterson's graph is non-Hamiltonian.
4. In the graph given below, find a Hamiltonian cycle or show that it does not (O.U. MCA, 1998)

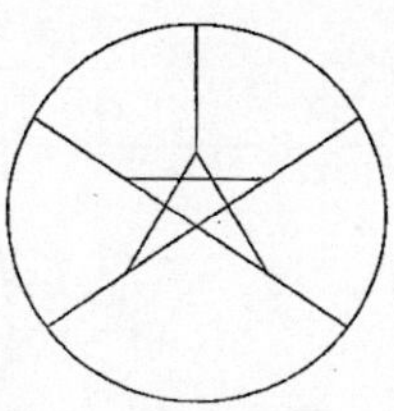

5. Are the graphs given below in figures (i) and (ii) Hamiltonian. If yes, find the Hamiltonian cycles. (O.U. MCA, 1991)

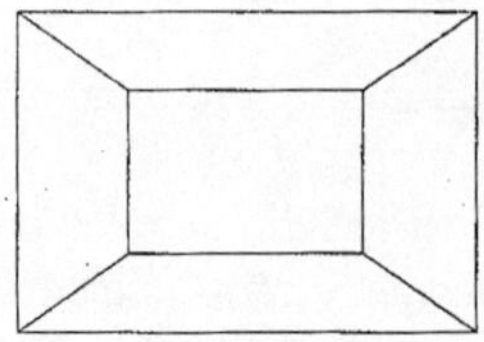

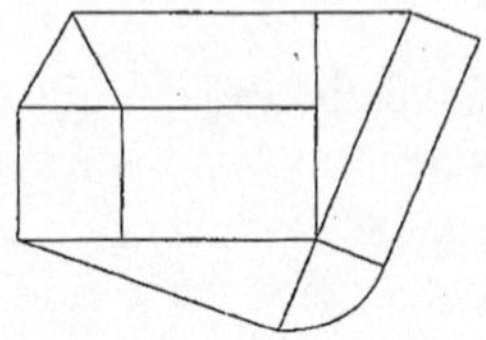

6. Show there are no planar Hamiltonian graphs of degree 4, 5 and 8 with only one regions of degree 4.
7. Define: (i) Hamiltonian path (ii) Hamiltonian cycle in graphs and give examples.
8. Show that the graph given below has neither a Hamiltonian cycle or a Hamiltonian path.

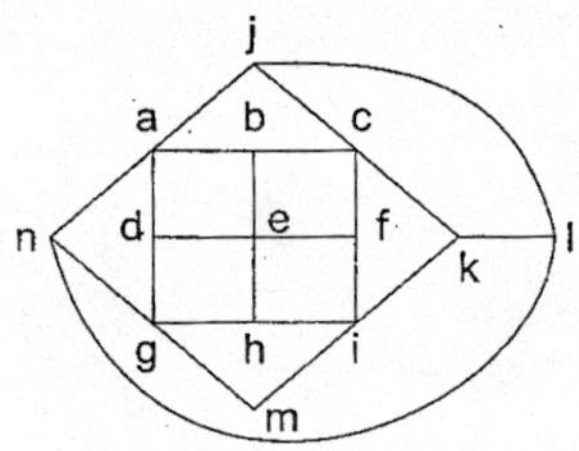

8.9 GRAPH COLORING

In this section, we describe the coloring of a graph and its chromatic number. We also discuss questions related to coloring of vertices, edges and regions.

Definition 8.12. A vertex coloring a graph G is an assignment of colors to its vertices, so that on two adjacent vertices have the same color.

Definition. A k-vertex coloring of a graph G is an assignment of k-colors to the vertices of G such that no two adjacent vertices receive the same color.

Every graph with n vertices is n-colorable. The set of all points with any one color is independent and is called a color class. If a graph is n-colorable, then the vertex set $V(G)$ can be partitioned with n color classes. The coloring of a graph is also called **proper coloring.** The graph shown in Fig. 8.40 is a properly colored graph.

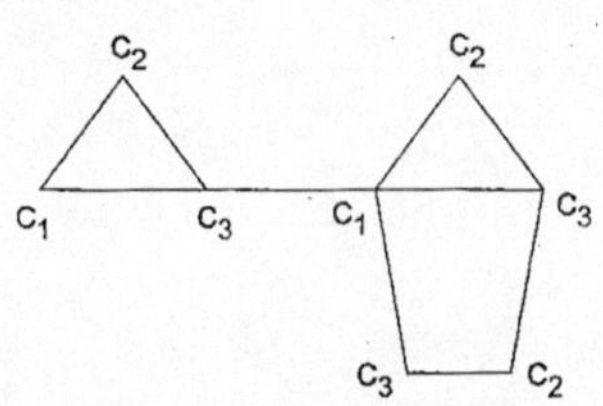

Fig. 8.40

Colors are usually represented by positive integers. Thus n-coloring of a graph is a function

$$f: U(G) \rightarrow \{1, 2, ..., n\}$$

such that $f(v_1) \neq f(v_2)$ whenever $\{v_1, v_2\} \in E(G)$

The vertex coloring of two different graphs are shown in the graphs of Fig. 8.41.

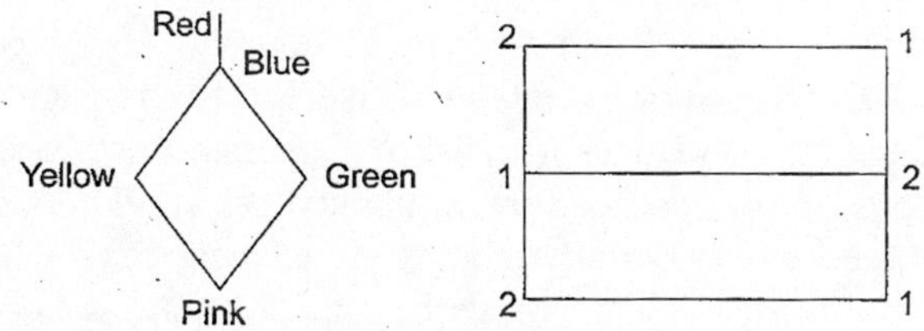

Fig. 8.41

8.10 CHROMATIC NUMBER

Definition 8.13. The chromatic number of a graph G, is the minimum number of colors required to color the vertices of G, such that no two adjacent vertices receive the same color.

The chromatic number of a graph G is denoted by $X(G)$. We observe the following:

1. The chromatic number of an isolated vertex is one
2. The chromatic number of a graph having at least one edge is at least two.
3. The chromatic number of a path P_n, $(n \geq 2)$ is two.
4. The chromatic number of a wheel graph is 3, if it has an odd number of vertices and if it has an even number of vertices.
5. If G is a graph consisting of simply a circuit with $n \geq 3$, then G is 2 chromatic if n is even and 3 chromatic if n is odd.

We can also define a chromatic number of graph as follows.

Definition 8.14. The chromatic number of a graph G is the least positive integer k, such that G has k-coloring.

Theorem 8.6. There exists a k-coloring of a graph G if and only if $V(G)$ can be partitioned into k subsets $V_1, V_2, ..., V_k$ such that no two veertices in V_i; $i = 1, 2, ..., n$ are adjacent.

Proof. Let $f: V(G) \rightarrow (1, 2, ..., k)$ be a coloring of G and let

$$V_i = \{v \in G : f(v) = i, 1 \leq i \leq k\}.$$

Therefore V_i denotes the set of vertices colored i. Then $V = V_1 \cup V_2 \cup ... \cup V_k$ forms a partition of V, such that no two vertices in V_i, $(1 \leq i \leq k)$ are adjacent.

Conversely, let $V = V_1 \cup V_2 \cup ... \cup V_k$ be a partition of $V(G)$, such that no two vertices in V_i are adjacent. Then the function $f: V(G) \rightarrow (1, 2, ..., n)$ defined by

$$f(v) = i, \text{ if } v \in V_i \text{ is a } k\text{-coloring of } G.$$

Theorem 8.9. If H is a subgraph of G, then $X(H) \leq X(G)$

Proof. Let $X(G) = k$ and f denote a k-coloring of G. Then, f restricted to $V(H)$ is a k-coloring of H.

Hence, $X(H) \leq k \Rightarrow X(H) \leq (G)$.

Corollary. If $G \geq k_n$, then $X(G) \geq n$, i.e. a graph G containing complete graph of k-vertices is at least k-chromatic.

Theorem 8.10. A graph G is 2-chromatic if and only if G is bi-paxtite.

Therorem 8.11. Every tree with 2 or more vertices is 2-chromatic.

Proof. Let T denote a given tree and v be any vertex in T. Assign color 1 to the vertex v. Assign all the vertices adjacent to v with color 2 and assign color 1 to all the vertices adjacent to the vertices of color 2. Continue this process till every vertex in T is colored. Now, we find that all the vertices at odd distances from v have color 2 and the vertices which are at even distances from v have color 1. Since T is connected, and has at least two vertices it contains at least one edge, T is not 1-colorable. Hence T can be properly colored with 2 colors. Therefore chromatic number of T is 2.

Theorem 8.11(a). If $\Delta(G)$ denote the max $\{\deg(v_i) : v_i \in V\}$ for any graph G, then $X(G) \leq \Delta(G) + 1$.

8.11 EDGE COLORING

Definition 8.15. Assignment of colors to the edges of a graph G, so that no two adjacent edges receive the same color is called an edge coloring of G.

k-edge coloring of a graph G is an assignment k-colors to the edges of G such that no two edges of G receive the same color.

Definition 8.16. A graph G is said to be k-edge colorable, if there exists k-edge coloring of G.

Definition 8.17. (Edge chromatic number). The minimum number k, such that a graph G has k-edge coloring is said to be the edge-chromatic number of G. The edge chromatic number of a graph G is denoted by $X'(G)$.

Thus $X'(G)$ denotes the minimum number of colors required to color the edges of the graph G, such that no two adjacent edges receive the same color.

Theorem 8.12. A graph G containing at least one edge is 2-chromatic if and only if it contains no odd cycles.

Proof. Let G be a connected graph with cycles of only even lengths. Consider a spanning tree T of G. We know that T can be properly colored with two colors. Now add the edges of G which are not in T, one by one. Since G has no cycles of odd length, the end vertices of every edge being replaced are differently colored in T. Thus G is properly colored with two colors. Hence T is 2-chromatic.

Conversely, let G be 2-colored graph with at least one edge. If G has a cycle of odd length, we would need at least three colors just for that cycle. But G is 2-chromatic, hence G contains no odd cycles.

Example 1. Find the chromatic number of the graph shown in Fig. 8.42.

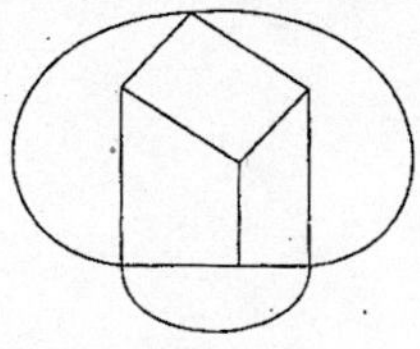

Fig. 8.42

Solution. The proper coloring of the graph of Fig. 8.42 is shown in Fig. 8.43, in which we have used the colors c_1, c_2, c_3 and c_4 shown in circles.

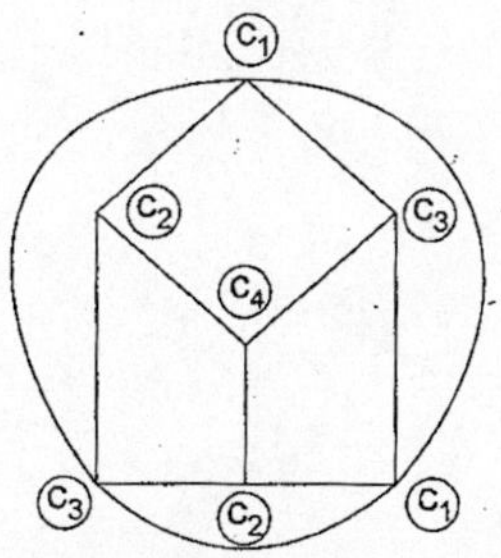

Fig. 8.43

∴ The chromatic number if the graph is 4.

Example 2. Find the chromatic number of the graph, given below.

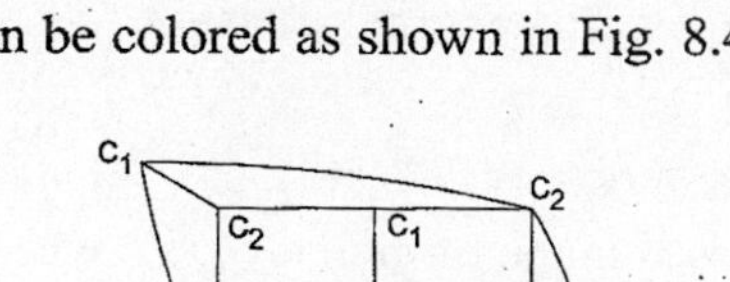

Fig. 8.44

Solution. The graph can be colored as shown in Fig. 8.45.

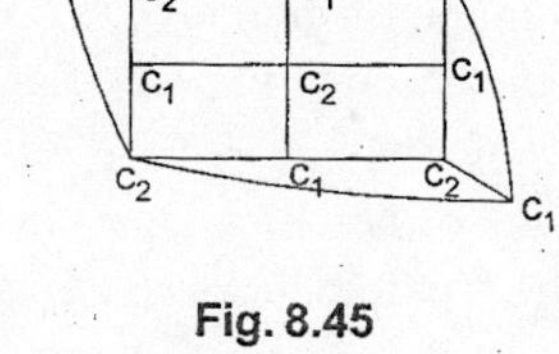

Fig. 8.45

c_1 and c_2 are the colors used for the proper coloring of G.

∴ The chromatic number of $G = 2$.

Example 3. Find the chromatic number of the graph given below.

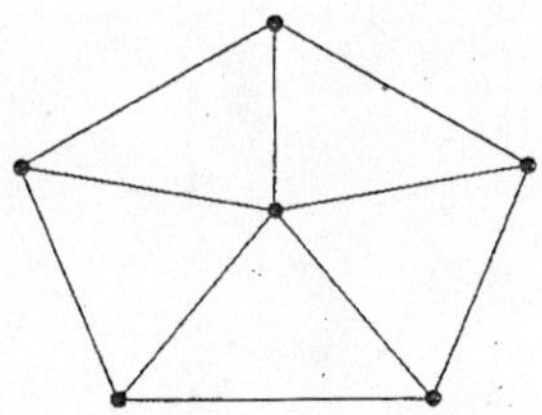

Fig. 8.46

Solution. The given graph is a wheel graph with 6 vertices.

Since the number of vertices is even, the chromatic number $X(G)$ is 4. The graph can be colored as shown in Fig. 8.47.

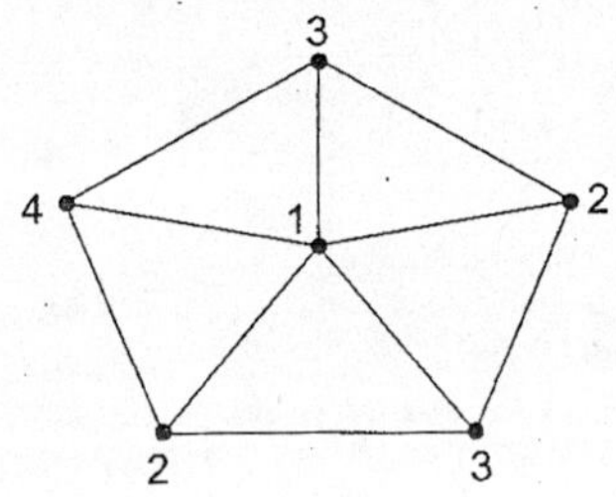

Fig. 8.47

where 1, 2, 3 and 4 are used denote the colors.

8.12 k-CRITICAL GRAPH

A graph G is said to be k-critical if $x(G) = k$, and $X(G - v) < x(G)$ for each vertex v of G.

Every k-chromatic graph has a k-critical subgraph.

Theorem 8.13. Let G be a k-critical graph. Then

(i) G is connected

(ii) The degree of each vertex of G is at least $k - 1$, that is $\delta(G) \geq k$.

(iii) G cannot be expressed in the form of $G_1 \cup G_2$ where G_1 and G_2 are graphs which intersect in complete graph. In particular, G contains no cut vertices.

Proof. (i) If the graph G is not connected, let C be any component of G with $x(G) = k$ and let v be any vertex of G which is not in the component C. But then $X(G - v) \geq k$, contradicting the fact that the graph G is k-critical.

(ii) G is k-critical, therefore $X(G-v) \leq k-1$ for each vertex v of G. If $\deg(v) < k-1$, then the neighbors of v will be colored with at most $k-2$ colors. But then it follows that any $(k-1)$ – coloring of $G-v$, can be extended to a $(k-1)$ – coloring of G by coloring the vertex by the color different from the colors on the neighbors of v, which contradict the fact that $X(G)$ $2k$.

(iii) Let $G = G_1 \cup G_2$ and $G_1 \cap G_2 = k_2$, then

$X(G_1) = k-1$, $X(G_2) \leq K-1$,since G_1 and G_2 are subgraph of G and G is k-critical. We can relabel the colors and color the vertices of $G_1 \cap G_2$, the same way in both the graphs. These two coloring can be combined to give a, $(k-1)$ coloring of the graph G, contradicting the fact that $X(G) = k$. By considering the case $r = 1$, it follows immediately that the graph G contains no cut vertex.

8.13 FIVE COLOR THEOREM

We shall now show that every planar graph can be properly colored with five colors.

Theorem 8.14. Every planar graph is 5-colorable.

Proof. We shall prove the theorem by induction on the number n of vertices. Since any graph with 1, 2, 3, 4, or 5 vertices can be properly colored with five colors, the theorem obviously holds if $n \leq 5$. Suppose $n > 5$; and the theorem holds for graph with less than n vertices. Since G is a planar graph with $n > 5$, G has vertex such that $\deg(v) \leq 5$. By induction subgraph $G' = G - v$ is 5-colorable. Assume one such coloring and let c_1, c_2, c_3, c_4 and c_5 denote the five colors used. If the vertices adjacent to the vertex v, useless than five colors. Then we simply assign v with one of the remaining colors and obtained a 5-coloring of the graph G. We are now left with the case in which $\deg(v) = 5$, i.e., the case in which v is adjacent to five vertices which are assigned different colors. Let us assume that the vertices moving counter clockwise about are $v_1, v_2, \ldots, v_5$. Let these vertices be assigned the colors $c_1, c_2, \ldots, c_5$ respectively (see Fig. 8.48).

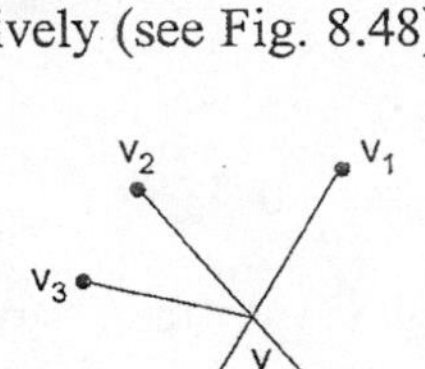

Fig. 8.48

Let H_1 be the subgraph of G, generated by the vertices assigned the colors c_1 and c_3. Clearly H_1 includes the vertices v_1 and v_3. If v_1 and v_3 belong to different components of H, then we can interchange the colors c_1 and c_3 in the component containing v_1 without destroying the coloring of $G-v$.

Then assign the color c_3 to the vertices v_1 and v_3. Now we get proper coloring of G by coloring v with the color c_1.

If v_1 and v_2 belong to the same component of H_1, then there exists a $v_1 - v_3$ path in G all of whose vertices are colored c_1 or c_3. This path together with the path $v_1 - v - v_3$ produces a cycle in G, which necessarily encloses the vertex v_2 or both the vertices v_4 and v_5. In any case, there exists no path between the vertices v_2 and v_4 all of whose vertices are colored c_2 or c_4. Hence if we let H_2 denote the subgraph $G - v$ induced by the vertices. Colored c_2 or c_4, then v_2 and v_4 belong to different components of H_2. Thus we interchange colors of the vertices in the component of H_2 containing V_2, a five coloring of $G - v$ is produced in which no vertex adjacent with v is colored c_2. We may then assign the color c_2 to the vertex v and obtain a 5-coloring of G.

8.14 FOUR COLOR CONJECTURE

So far we have discussed proper coloring of vertices and proper coloring of edges of a graph. Now we briefly consider the proper coloring of regions in planar graphs such that no two adjacent regions receive the same color.

The four color problem; states that "every plane map however complex, can be colored with four colors" fascinated mathematicians. This problems was solved by Appel and Hanken in 1976. However, this problem is infact equivalent to the statement of a conjecture.

Four color conjecture. Every planar graph is 4-colorable

Example. The graph k_4 is a planar graph and k_4 is 4-colorable.

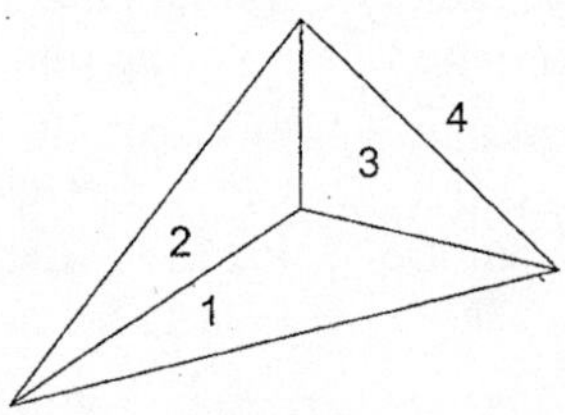

Fig. 8.49

Theorem. Every planar graph with fewer than 4 triangles is 3-colorable.

Corollary. Every planar graph without triangles is 3-colorable. All plane maps with up to 39 regions are 4-colorable. A plane map containing large number of regions is 5-colorable.

8.15 WELCH – POWELL ALGORITHM

We give an algorithm by Welch – Powell for a coloring of a graph G.

Algorithm. (Welch – Powell)

Input. A graph G.

Step 1. Order the vertices of G according to decreasing degrees.

Step 2. Assign the first color c_1 to the first vertex and then in sequential order assign c_1 to each vertex, which is not adjacent to a previous vertex, which was assigned the color c_1.

Step 3. Repeat the step 2. With a second color c_2 and the subsequence of non-colored vertices.

Step 4. Repeat step 3. With a third color c_3, then a fourth color c_4 and so on until all the vertices of G are properly colored.

Step 5. Exit

Note. Welch – Powell algorithm does not always yield a minimum coloring of a graph G.

Example. Use Welch – Powell algorithm to color the graph G and find the chromatic number of the graph.

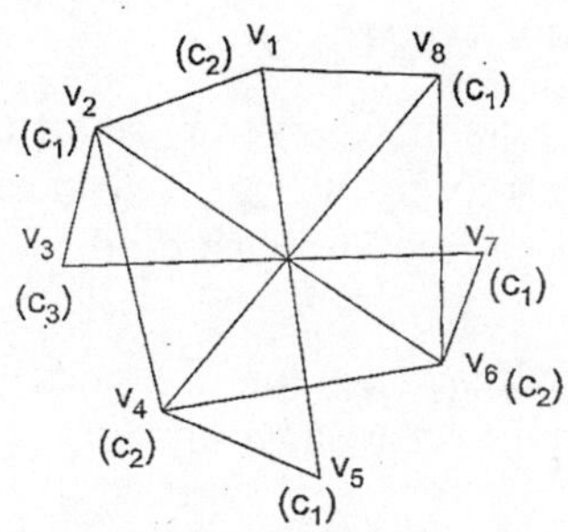

Fig. 8.50

Solution. We have $\deg(v_1) = d(v_1) = 4$

$$d(v_2) = 4,\ d(v_3) = 3,\ d(v_4) = 4,\ d(v_5) = 3,\ d(v_6) = 4$$

$$d(v_7) = 2 \text{ and } d(v_8) = 5$$

We first order the vertices to decreasing degree sequence to obtain the following sequence

$$v_8, v_1, v_4, v_6, v_2, v_3, v_5, v_7.$$

Let c_1 denotes the first color, c_2 denotes the second color,... which are used to paint the vertices of G.

We first use the color c_1 to paint the vertices v_8, v_2, v_5, and v_7 (since v_1, v_4, v_6 are adjacent to v_8 and each of the vertices and v_3 is connected to v_2, therefore we can assign the color c_1 to v_1, v_4, v_6, and v_2).

We use the color c_2 to paint the vertices v_1 and v_4. The remaining vertices are colored by using the color c_3. (see Fig. 8.51).

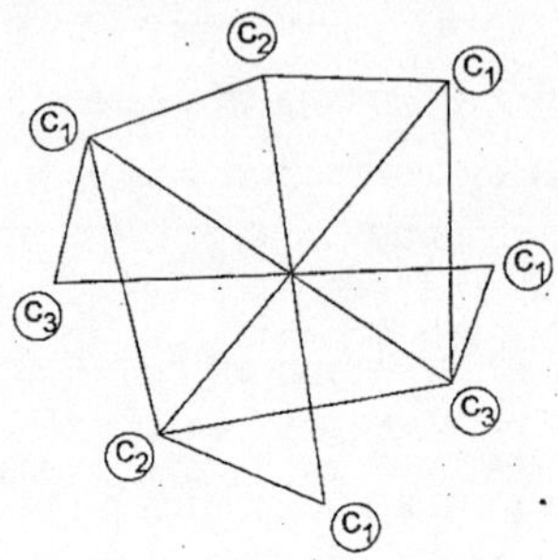

Fig. 8.51

∴ Number colors used for the coloring vertices of $G = 3$

∴ $X(G) = 3$

8.16 INDEPENDENT SETS, CHROMATIC PARTITION AND CHROMATIC POLYNOMIALS

Let G be a properly colored graph. The vertex set of G can be partitioned into different subsets. The proper coloring of G induces a partition of the vertices of G, such that no vertices of the same subset are adjacent. Such sets are called independent sets.

Definition 8.18. A set of vertices in a graph G is said to be an independent set of vertices if no two vertices in the set are adjacent.

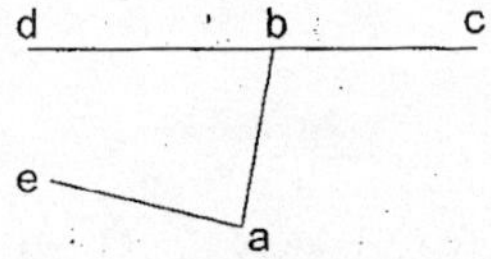

Fig. 8.52

In the graph of 8.52, the set $\{a, c, d\}$ and $\{b, e\}$ are independent sets.

Consider the graph shown in Fig. 8.53.

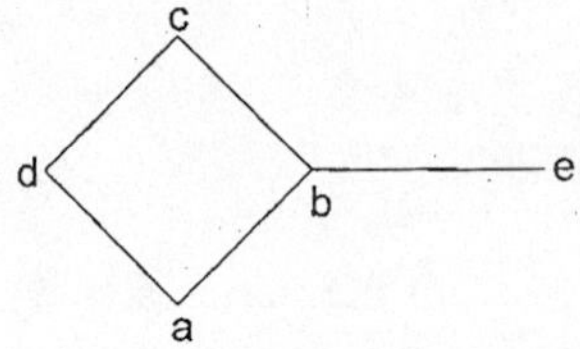

Fig. 8.53

In the graph $\{a, c, e\}$ is an independent set.

Note. A single vertex in any graph constitutes an independent set and every subset of an independent is independent.

Definition 8.19. Let G be any graph. A subset V' of the vertex set $V(G)$ is called a maximally independent set of G if and only if

(i) V' is an independent set of G and

and (ii) If V'' is any other independent set of G, then V^1 is not a proper subset of G.

Example. In the graph shown in Fig. 8.54 the sets $\{v_2, v_3, v_4\}$ and $\{v_2\}$ are maximally independent sets.

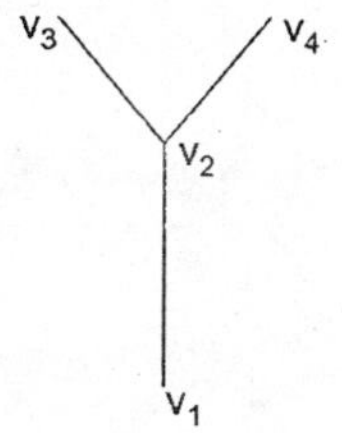

Fig. 8.54

8.17 CHROMATIC POLYNOMIALS

Birchoff and Lewis have introduced chromatic polynomials of graphs. Let G be a labeled graphs. A coloring of G from λ colors is a coloring of graph G which uses λ or fewer colors. If at least one of the vertices of G is assigned; different colors, then two colorings of G from λ colors are said to be different. A graph G with n vertices can be colored in many different ways using a sufficiently large number of colors. We use chromatic polynomials to find the number of ways of different proper coloring of a graph on n vertices with a maximum of λ colors.

Definition 8.20. The number of different colorings of a graph G on n vertices that can be obtained using λ colors or fewer colors can be expressed as a polynomial of G.

The chromatic polynomial of G is denoted by $P_n(\lambda)$. Let G be a graph on n vertices. If C_i denotes two different ways of properly coloring of G, using exactly i distinct colors, then these i colors can be chosen out of λ colors in $\binom{\lambda}{i}$ distinct ways. Therefore, there are $c_i.\binom{\lambda}{i}$ distinct ways of properly coloring the graph G using exactly i colors out of λ colors. Since i can be any positive integer from 1 to n, the chromatic polynomial is

$$\sum_{i=1}^{n}\binom{\lambda}{i}c_i$$

where each c_i has to be evaluated individually for the given graph. If G is a graph with at least one edge, then G requires at least two colors for proper coloring of G. Therefore $c_k = c_1 = 0$. If G has n vertices, then it can be properly colored using n colors in $n!$ ways. Therefore $c_n = n!$ If G is a complete graph (k_n) on n vertices, each vertex has $n - 1$, vertices adjacent to it. the vertices of k_n can colored be properly colored by exactly n colors. Since these n colors can be permuted in $n!$ ways, for k_n we have $c_n = n!$ and $c_i = 0$ for $i = 1, 2,..., n - 1$.

$$\text{Therefore } P_n(l) = \sum_{i=1}^{n}\binom{\lambda}{i}c_i$$

$$= \frac{\lambda(\lambda-1)(\lambda-2)\dots(\lambda-n+1)}{n!}\,.\,n!$$

$$= \lambda(\lambda - \lambda)(\lambda - 2) \dots (\lambda - n + 1)$$

If $G = k_4$ (i.e. complete graph on 4-vertices)

We have $c_1 = c_2 = c_3 = 0$, and 4!

∴ The chromatic polynomial of G is

$$P_4(\lambda) = \frac{\lambda(\lambda-1)(\lambda-2)(\lambda-3)}{4!}\,.\,4!$$

$$= \lambda(\lambda - 1)(\lambda - 2)(\lambda - 3)$$

SOLVED EXAMPLES

Example 1. Show that the chromatic polynomial of a graph G is

$$\lambda(\lambda - 1)(\lambda - 2) \dots (\lambda - n + 1)$$

if and only if G is a complete graph on n vertices.

Solution. With λ colors, the first vertex of G can be colored in λ ways. a second vertex can be colored in $(\lambda - 1)$ ways, the third vertex can be colored $(\lambda - 2)$ ways, if and only if the third vertex is adjacent to first two vertices ... and the nth vertex can be colored $\lambda - n + 1$ ways if and only if every vertex is adjacent to every other, i.e. if and only if G is complete.

∴ The chromatic polynomial of $G - k_n$ is

$$P_n(\lambda) = \lambda(\lambda - 1)(\lambda - 2) \dots (\lambda - \mathrm{n} + 1)$$

Example 2. Find the chromatic polynomial of the graph shown in Fig. 8.55.

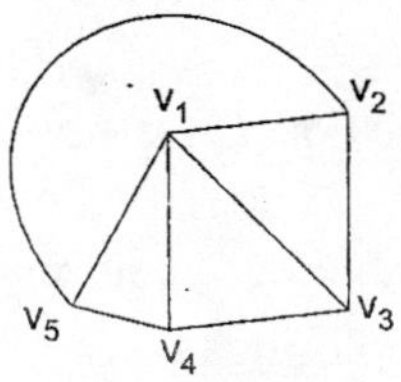

Fig. 8.55

Solution. G has a triangle, therefore G requires minimum 3 colors for proper coloring

$$\therefore \quad C_1 = 0, \quad C_2 = 0$$

When we have three colors, three vertices of a triangle can be properly colored in 3! ways. If v_1, v_2, v_3 are considered, then they can be properly colored in 3! ways. The vertex v_4 can be assigned the color of v_2 and v_5 the color of v_3. Thus we have $c_3 = 3! = 6$.

When we have 4 colors when take any four vertices say v_1, v_2, v_3 and v_4 of G and properly color then with 4 colors in 4! ways. The vertex v_5 can be assigned the color of either v_2 or v_3. Thus

$$C_4 = 4! \text{ and we have } C_5 = 5!$$

Hence, the chromatic polynomial of G is

$$P_5(\lambda) = 3!\,\frac{\lambda(\lambda-1)(\lambda-2)}{3!} + 2.4!,\ \frac{\lambda(\lambda-1)(\lambda-2)(\lambda-3)}{4!}$$

$$+ 5!\,\frac{\lambda(\lambda-1)(\lambda-2)(\lambda-3)(\lambda-4)}{5!}$$

$$= \lambda(\lambda-1)(\lambda-2) + 2\lambda(\lambda-1)(\lambda-2)(\lambda-3) + \lambda(\lambda-1)(\lambda-2)(\lambda-3)(\lambda-4)$$

$$= \lambda(\lambda-1)(\lambda-2)[1 + 2(\lambda-3)(\lambda-3)(\lambda-4)]$$

$$= \lambda(\lambda-1)(\lambda-2)(\lambda^2 - 5\lambda + 7)$$

Theorem 8.15. Let u and v be two non-adjacent vertices of a graph G and G' be a graph obtained by adding an edge between u and v (i.e. $G' = G + uv$).

Let $G'' = G\,.\,uv$ (i.e. G'' be the simple graph obtained from G by fusing the vertices u and v together and replacing set of parallel edges with single edge). Then

$$P_n(\lambda) \text{ of } G = P_n(\lambda) \text{ of } G' + P_{n-1}(\lambda) \text{ of } G'$$

Proof. The λ coloring of G is of two types:

(1) λ coloring of G assigning different colors to the vertices u and v.

(2) λ coloring the same color to u and v.

$\therefore$ A proper λ coloring of the first type is equivalent to a proper coloring of $G + uv$

and a proper λ coloring of the type (2) is equivalent to a proper λ coloring of $G\ u.v.$

Hence $P_n(\lambda)$ of G = the number of λ colorings of G

= The number of proper λ colorings of $G + uv$ + The number if proper λ colorings of $G . uv$

$= P_n(\lambda)$ of $G' + P_{n-1}$ (l) of G''

Hence proved.

The above theorem is often used in evaluating the chromatic polynomial of a graph. For example consider the graph shown in Fig. 8.56, illustrates how the chromatic polynomials of a graph G is expressed as a sum of chromatic polynomial of 5 complete graphs

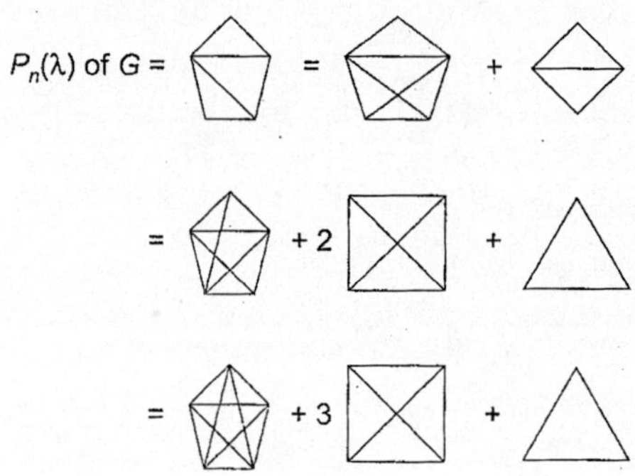

Fig. 8.56

$$= P_n(\lambda) \text{ of } k_5 = 3P_n(\lambda) \text{ of } k_4 + P_n(\lambda) \text{ of } k_3$$

$$P_5(\lambda) \text{ of } G = P_5(\lambda) \text{ of } k_5 + 3\ P_4(\lambda) \text{ of } k_4 + P_3(\lambda) \text{ of } k_3$$

Example. Find the chromatic polynomial of the graph given below.

G:

Fig. 8.57

Solution. The chromatic polynomial of the graph shown in Fig 8.57 can be obtained as the sum of chromatic polynomials of complete graphs as shown below.

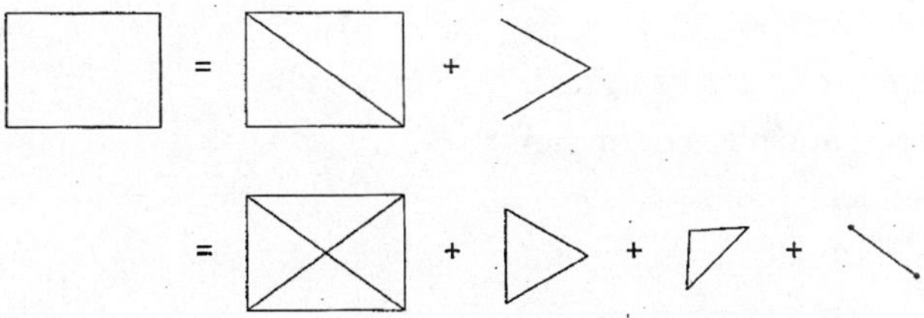

$$\therefore \quad P_n(\lambda) \text{ of } G = (\lambda) + P_{k_4}(\lambda) + P_{k_3}(\lambda) + P_{k_3}(\lambda) + P_{k_2}(\lambda)$$

$$= P_{k_4}(\lambda) + 2P_{k_3}(\lambda) + P_{k_2}(\lambda)$$

$$= \lambda(\lambda-1)(\lambda-2)(\lambda-3) + 2\lambda(\lambda-1)(\lambda-2) + \lambda(\lambda-1)$$

$$= (\lambda)(\lambda-1)\left[(\lambda-2)(\lambda-3)+2(\lambda-2)+1\right]$$

$$= (\lambda^2-\lambda)(\lambda^2-5\lambda+6+2\lambda-4+1)$$

$$= (\lambda^2-\lambda)(\lambda^2-3\lambda+3)$$

$$= \lambda^4-4\lambda^3+6\lambda^2-3\lambda$$

Theorem 8.16. The chromatic polynomial of a tree with n vertices is $\lambda(\lambda-1)^{n-1}$

Proof. We prove the theorem by induction on n, the number of vertices. When $n = 1$. We have $T = k_1$. and the chromatic polynomial of $T = k_1$ is $\lambda = \lambda(\lambda-1)^{1-1}$. Hence the theorem holds for $n = 1$. Assume that the theorem is true for all tree on $n-1$ vertices and T is a tree with n vertices. Let v be a pendant vertex of T and u be a unique vertex adjacent to v.

Let $T' = T - v$. Then T' is a tree with $n-1$ vertices therefore the chromatic polynomial of T' is $\lambda(\lambda-1)^{n-2}$. But each proper λ coloring of T' can be extended to a proper λ coloring of T by assigning to v any one of the $\lambda - 1$ colors not assigned to the vertex u in T'. Thus each proper coloring of T yields $\lambda - 1$.
$\lambda - 1$ proper λ colorings of T. Hence the chromatic polynomial of T is $\lambda(\lambda-1)^{n-2}$ $(\lambda-1)^{n+1}$.

Hence proved.

Example. Find the number of ways in which a tree T on 5 vertices can be properly colored with at most 4 colors.

Solution. We have $n = 5$ and $\lambda - 4$.

$\therefore$ The number of ways in which the tree T can be properly colored $= \lambda(\lambda-1)^{n-1}$

$$= 4(4-1)^{5-1}$$

$$= 4 \times 3^4 = 324$$

EXERCISE 8.5

1. Define the terms
   (i) Vertex coloring of a graph
   (ii) Chromatic number of graph
   (ii) k-critical graph
2. Find the chromatic number of the following graphs:

(i) 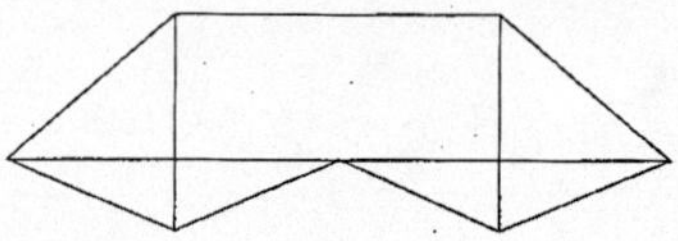

(ii) 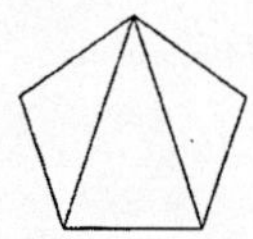

(iii) 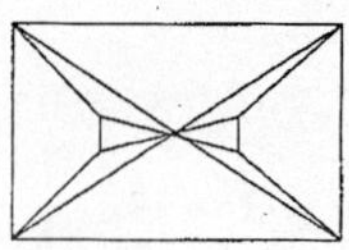

(iv) 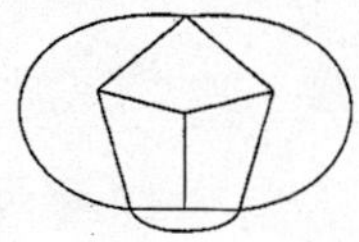

(v) 

(vi) 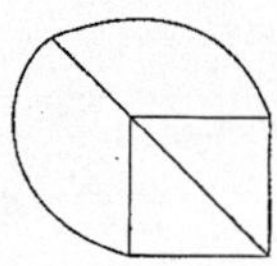

3. Write down the possible independent sets of the following graphs.

(i) 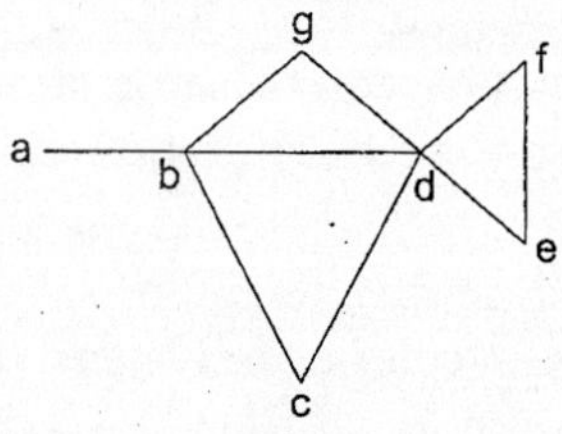

(ii) 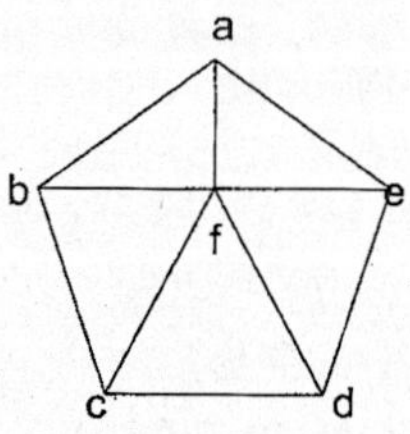

(iii) 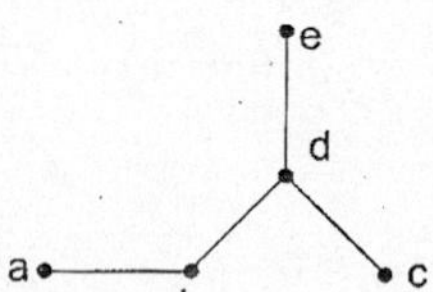

4. Show that the vertices of a planar graph with less than 30 edges is 4-colorable.
5. Show that a simple connected planar graph with 17 edges and 10 vertices cannot be 2-colorable.
6. Show that a graph dichromatic if and only if it has odd circuits.

7. Show that the regions of a planar graph can be 2-colored if each has degree which is even.
8. Find the chromatic polynomials of the following graphs.

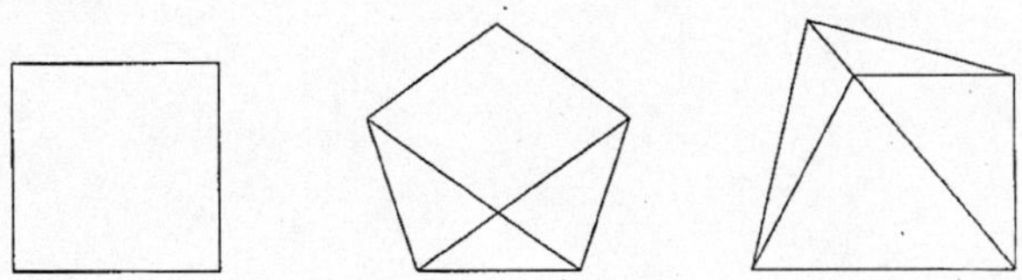

(a) (b) (c)

Bibliography

Barker. S.F. The elements of logic, 5th edition. MC Graw-Hill; Newyork, 1989

Bondy. J.A. and U.S.R. Murty, Graph theory with applications, Elsevier Newyork, 1977

Copi, I.M. and C. Cohen, introduction to logic, 10th edition Prentice-Hall, Calif 1986

Deo. N. Graph Theory and application to Engineering and Computer Science, Pretice-Hall, Englewood Cliffs, N.J. 1974

Edgar. W.J. The elements of logic, SRA Chicago, 1989

Fraleigh, John. B. A First course in Abstract Algebra 3rd edition, Addison-Wesley, Reading Hass, 1982

Gould. R. Graph theory, Benjamin, 1988 Reading, Mass 1969

Havary. F. Graph theory, Addison-Wesley J

John Truss. Discrete Mathematics, Computer Scientists, Pearson-publishing, 2001

Lipson. John. O. Elements of Algebra and Algebraic Computing, Addison-Wesley, 1981

Liu. C.L. Elements of Discrete mathematics, 2nd edition MC Graw-Hill, 1985

Stephen A.Wiitata, Discrete mathematics- A unified approach 1st edition, MC Graw-Hill, 1987

Tucker A. Applied Combinatorics; 3rd edition, Wiley, Newyork, 1995

West. O. Introduction to graph theory 2nd edition, Prentice-Hall upper Saddle River, N.J. 2000

Wilson. R.J. Introduction to graph theory, 4th edition, Addison-Wesley, Mass. 1996